Fields Institute Communications

VOLUME 74

W0037242

The Fields Institute for Research in Mathematical Sciences

The Fields Institute is a centre for research in the mathematical sciences, located in Toronto, Canada. The Institutes mission is to advance global mathematical activity in the areas of research, education and innovation. The Fields Institute is supported by the Ontario Ministry of Training, Colleges and Universities, the Natural Sciences and Engineering Research Council of Canada, and seven Principal Sponsoring Universities in Ontario (Carleton, McMaster, Ottawa, Queen's, Toronto, Waterloo, Western and York), as well as by a growing list of Affiliate Universities in Canada, the U.S. and Europe, and several commercial and industrial partners.

More information about this series at http://www.springer.com/series/10503

René Aïd • Michael Ludkovski • Ronnie Sircar
Editors

Commodities, Energy and Environmental Finance

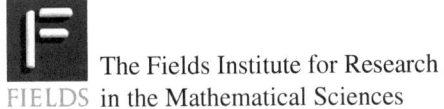

The Fields Institute for Research
in the Mathematical Sciences

Springer

Editors
René Aïd
EDF R&D, Clamart Cedex
Paris, France

Ronnie Sircar
ORFE Department
Princeton University
Princeton, NJ, USA

Michael Ludkovski
Department of Statistics
  and Applied Probability
University of California Santa Barbara
Santa Barbara, CA, USA

ISSN 1069-5265   ISSN 2194-1564   (electronic)
Fields Institute Communications
ISBN 978-1-4939-4987-8   ISBN 978-1-4939-2733-3   (eBook)
DOI 10.1007/978-1-4939-2733-3

Mathematics Subject Classification (2010): 60-XX, 62-XX, 91-XX, 93-XX

Cover illustration: Drawing of J.C. Fields by Keith Yeomans

Printed on acid-free paper

Springer Science+Business Media LLC New York is part of Springer Science+Business Media (www.springer.com)

Preface

This volume is an outgrowth of the Focus Program on Commodities, Energy, and Environmental Finance hosted by the Fields Institute during August 2013. The Focus Program addressed the interaction of markets and the environment, including analysis of the functioning of electricity markets, modeling of energy resources (such as wind speeds or exploration of oil fields), and study of the influence of traders on commodity prices.

The busy month had a variety of activities, including three summer school mini-courses, two research workshops, and a lively seminar series. The first mini-course by Glen Swindle (Scoville Risk Partners) gave an overview of practical issues of energy risk management, highlighting the complexity of calibrating models, dealing with spotty market data, worrying about unknown correlations, and pinpointing seasonal patterns. This material has since been published in the just-printed monograph on "Valuation and Risk Management in Energy Markets" (Cambridge University Press).

In parallel with Glen's mini-course, René Carmona (Princeton) gave a comprehensive overview of the emerging issue of financialization of commodities, whereby commodity price swings are linked to the ebb and flow of funds in the related contracts. These notes form the backbone of the first chapter of this volume. In the second half of his mini-course, René then addressed the latest developments in mean field games (MFG), especially as related to the (as yet unresolved) links between MFG and controlled McKean-Vlasov dynamics. The third mini-course of the Program was given by Fred E. Benth (Oslo) who concentrated on the modeling of electricity prices. Because electricity demand is largely driven by weather, stochastic modeling of related quantities, such as wind or solar intensity, is an emerging research area. Fred is a co-author on Chap. 5 of the volume.

About 30 participants attended each mini-course; throughout there was a lot of audience participation with many lectures ending with an extended Q&A discussion. All the presentation slides of the mini-courses have been archived and can be accessed by the public at the Program website http://www.fields.utoronto.ca/programs/scientific/13-14/envirofinance/.

The Focus Program also included two research workshops. The first workshop was devoted to modeling and risk-management in commodity markets and is reflected in Parts I and II of this volume. The second research workshop corresponds to Part IV and concentrated on new directions in stochastic analysis inspired by theories of mean field games and stochastic equilibria in energy production. Beyond the research workshops, further presentations were given under the auspices of the Program Visitor Seminars. There was also industry partnership, with several participants from energy and utilities companies, and sponsorships from Zerofootprint and Electricité de France (EDF).

Overall, over 40 different presentations were delivered during the Program. Nearly all the presentations are posted on the Program website and all talks were live-recorded and available for streaming via Fields*Live* (http://www.fields.utoronto. ca/live) providing a lasting record of the proceedings. Several of the volume chapters are an outgrowth of these talks, though, of course, not all of them could be included. For example, the volume only very briefly touches upon mean field games that were discussed among others by Minyi Huang (Carleton), Francois Delarue (Nice), and Daniel Lacker (Princeton) during the Focus Program.

Group photo during one of the Research Workshops, August 2013
Photography by Cedric Miao, Fields Institute

This volume is based on the presentations and discussions during the Focus Program. The contributions present certain aspects of a disparate and multi-faceted research area. The multi-disciplinary developments currently taking place in the subject mean that mathematicians, probabilists, statisticians, industrial and operations engineers, economists, and finance practitioners are all working simultaneously (and frequently together) on the same problems. This volume reflects this breadth, with authors including mathematicians, economists, statisticians, and market practitioners. The cross-disciplinary perspectives are unified by tools of mathematical finance and economics, especially stochastic methods comprising stochastic differential equations, dynamic game theory, stochastic control, and no-arbitrage theory.

A particular emphasis of the Focus Program and this volume was on making connections with the Mathematics of Planet Earth thematic year 2013. The volume highlights the growing involvement of the mathematics community in vital themes of sustainable development, effective risk management of weather events, and the role of finance in the production and consumption of energy. Other open mathematical challenges include analysis of best government policies to lean on markets in mitigating climate change, encouraging technological transition to renewable energy sources, and statistical investigations into weather data crucial for renewable power generation. Some more discussion on these issues can be found via the Mathematics of Planet Earth blog entries

http://mpe2013.org/2013/05/07/fields-institute-focus-program-on-commodities-energy-and-environmental-finance/

http://mpe2013.org/2013/05/27/mathematics-shines-some-light-on-the-growing-markets-for-solar-renewable-certificates/

http://mpe2013.org/2013/10/17/fields-institute-focus-program-on-commodities-energy-and-environmental-finance-2/

The volume is organized into four parts containing a total of 15 chapters. Part I on commodities and financial markets offers perspectives connecting mainstream financial mathematics with the world of commodities. The opening chapter by R. Carmona surveys the emerging topic of financialization, whereby commodity prices seem to be affected by the finance-based traders, altering market dynamics, and introducing new couplings between asset levels. The author presents a comprehensive review of the economic, statistical, and mathematical literatures that weigh in on this issue, and also provides a summary of the different commodity indices that have been a key driver in expanding the universe of commodity trading. The chapter by K. Guo and T. Leung examines the performance of commodity Exchange Traded Funds (ETFs), which have been a major investment vehicle in commodities and presumably driving the ensuing financialization. The authors analyze the empirical tracking performance of leveraged ETFs vis-a-vis their benchmark index and discuss the issues of volatility drag and realized effective fees. The third chapter by D. Lautier, J. Ling, and F. Raynaud illuminates the links between various commodity classes from the angle of systemic risk. The authors map out the network graph connecting different commodity futures (oil, agriculture, metals, etc.) and analyze its time series properties, especially through the financial crisis of Fall 2008. The last chapter of this section by H. Tuenter presents a new perspective on the problem of valuing spread options. After reviewing existing methods (including an elegant new proof of Margrabe's classic formula), a new closed-form approximation for spread options is obtained and shown to be highly accurate.

Part II covers electricity and related markets, in particular weather and emissions. The chapter by A. Veraart, F.E. Benth, and O. Barndorff-Nielsen presents a model based on ambit fields to capture the main stylized characteristics of energy futures prices, such as non-Gaussianity, stochastic volatility, and Samuelson effect. The model is set in a multi-dimensional framework that allows to take into account the dependencies between energy prices. As an application, the authors present a pricing formula for spread options that generalizes Margrabe's formula in the case of ambit fields. The work by N. Oudjane and A. Nguyen Huu addresses the problem of hedging a derivative upon a non-yet quoted asset. This situation is encountered in the case of derivatives on electicity futures, wherein the most desirable contracts are quoted only a short time before their maturity. The market being incomplete, a partial hedging of a contingent claim is proposed based on an adaptation of the stochastic target approach, including a detailed numerical example. In the chapter by O. Féron and E. Daboussi, the authors provide an overview of the calibration to data of different electricity price models, a point which is often ignored in the literature. Moreover, the authors offer the readers a comparison between reduced-form and structural models.

Part III focuses on real options, i.e., decision making associated with commodity management, such as starting new projects or expanding/curtailing production. S. Jaimungal and Y. Lawryshyn propose a new extension of this framework that allows the incorporation of subjective beliefs of the decision maker into the model. These beliefs are expressed in terms of a nontraded stochastic factor and are then optimally hedged through trading in a correlated liquid market index using indifference pricing. In the ninth chapter, M. Davison and C. Maxwell analyze valuation of real options under regulatory uncertainty. Motivated by the frequently changing government subsidies for renewable energy production, the authors examine the impact of unpredictable regulatory shocks on profitability and operational strategies. The chapter by E. Brigatti, F. Macias, M. Souza, and J. Zubelli proposes an application of the HMC numerical algorithm due to Potters et al. to the pricing of real options. The concluding chapter by R. Aïd and I. Ben Tahar offers an application of the theory to an issue in renewable energy. The authors propose a simple model to assess the effectiveness of subsidies to foster transition to electric mobility.

Part IV showcases new game-theoretic approaches to commodity markets, in particular addressing the questions of oligopolistic behavior and long-run equilibria among commodity producers. The survey by M. Ludkovski and R. Sircar offers a tour through game models of energy production, centering on the long-standing concern of exhaustible resources and "peak oil" within a competitive dynamic market. The chapter by M. Bossy, N. Maïzi, and O. Pourtallier considers game effects that link electricity and CO_2 emission markets, offering insights for the decisions of power generators that participate in bid auctions simultaneously for both markets. In the fourteenth chapter, M. Ludkovski and X. Yang describe competition between exhaustible and renewable producers under changing market conditions. Their analysis of the interaction between stochastic demand and resource exploration brings a new perspective to the cyclical exploration and production (E&P) investments by energy producers. The final chapter, by A. Dasarathy and R. Sircar, examines the impact on energy markets of variable costs in production from differing sources, within a game-theoretic oligopolistic framework. An example is oil production becoming more expensive, as we switch to shale oil, and renewables becoming cheaper, due to technological advances.

Commodities and environmental finance continues its rapid growth and is sure to bring forth many more developments in the near future.

Paris, France	René Aïd
Santa Barbara, CA, USA	Michael Ludkovski
Princeton, NJ, USA	Ronnie Sircar
January 2015	

Contents

Part I   Commodities and Financial Markets

Financialization of the Commodities Markets: A Non-technical Introduction 3
René Carmona

Understanding the Tracking Errors of Commodity Leveraged ETFs 39
Kevin Guo and Tim Leung

Integration of Commodity Derivative Markets: Has It Gone Too Far? 65
Delphine Lautier, Julien Ling, and Franck Raynaud

Margrabe Revisited 91
Hans J.H. Tuenter

Part II   Electricity and Related Markets

Cross-Commodity Modelling by Multivariate Ambit Fields 109
Ole E. Barndorff-Nielsen, Fred Espen Benth,
and Almut E.D. Veraart

Hedging Expected Losses on Derivatives in Electricity Futures Markets 149
Adrien Nguyen Huu and Nadia Oudjane

Calibration of Electricity Price Models 183
Olivier Féron and Elias Daboussi

Part III   Real Options

Incorporating Managerial Information into Real Option Valuation 213
Sebastian Jaimungal and Yuri Lawryshyn

Real Options with Regulatory Policy Uncertainty 239
Christian Maxwell and Matt Davison

A Hedged Monte Carlo Approach to Real Option Pricing 275
Edgardo Brigatti, Felipe Macías, Max O. Souza,
and Jorge P. Zubelli

Transition to Electric Mobility: An Optimal Price Subsidy Rule 301
René Aïd and Imen Ben Tahar

Part IV   Dynamic Games in Commodity Markets

Game Theoretic Models for Energy Production 317
Michael Ludkovski and Ronnie Sircar

**Game Theory Analysis for Carbon Auction Market Through
Electricity Market Coupling** 335
Mireille Bossy, Nadia Maïzi, and Odile Pourtallier

**Dynamic Cournot Models for Production of Exhaustible
Commodities Under Stochastic Demand** 371
Michael Ludkovski and Xuwei Yang

Variable Costs in Dynamic Cournot Energy Markets 397
Anirudh Dasarathy and Ronnie Sircar

Part I
Commodities and Financial Markets

Financialization of the Commodities Markets: A Non-technical Introduction

René Carmona

Abstract The goal of the first part of this chapter is threefold: (a) to introduce the term structure of forward/futures commodity prices, the contango/backwardation duality and the notion of rolling yield as it pertains to trading through commodity indexes; (b) to use principal component analysis and the computation of equity and commodity "betas" to provide empirical evidence of the dramatic changes which occurred in the mid-2000s; (c) finally, to review the major arguments which have been put forth in the debate over the *financialization of these markets*. While conspicuously absent from some of the English language dictionaries, the word *financialization* has been widely used to describe the increasing role of institutional investors in the commodity markets. Using econometric data analyses for the purpose of illustration, we concentrate on futures price data from the post-2004 period during which the commodity markets experienced a significant influx of new financial investors. As far as we know, mathematical models attempting to reproduce or illustrate (let alone explain) the empirical observations at the core of the debate are few and far between. As a result, our approach remains mostly descriptive of the data which have been used to back up the claims of the various sides of the argument. The originality of our contribution, if any, is the discussion of a new generation of *roll yield maximizing* commodity indexes, the empirical analysis of the term structure of open interest, and the possible connections between the two.

1 Introduction

The main goal of this chapter is to document the dramatic changes in commodity prices during the post-2004 period, when commodity markets experienced a large influx of new money, especially from institutional investors. For the sake of completeness, we review some of the idiosyncrasies of these markets as well as the main data analysis techniques used to study the term structure of forward prices,

R. Carmona (✉)
Department of ORFE, Bendheim Center for Finance, Princeton University,
Princeton, NJ 08544, USA
e-mail: rcarmona@princeton.edu

© Springer Science+Business Media New York 2015
R. Aïd et al. (eds.), *Commodities, Energy and Environmental Finance*,
Fields Institute Communications 74, DOI 10.1007/978-1-4939-2733-3_1

our objective being to focus on the changes which occurred over the last decade. We rely on economic studies to explain why, according to the restricted form of the financialization hypothesis, they produced changes in correlations, and rises in open interest and trading volume. Whether this increase in open interest and volume is due to index investing or herding behavior is still unclear. We demonstrate the increase in trading volume and open interest throughout the period, and we analyze the term structure of commodity open interest. We use Principal Component Analysis to demonstrate the shift of open interest down the curve. This increase in open interest along longer maturities coincides with the appearance of a new generation of commodity indexes optimizing the roll yield. While the compositions of the portfolio covered by these indexes is pretty much the same as the compositions of the traditional indexes, the spectrum of contract maturities they comprise is different because of the special nature of the rolling algorithms. While we cannot prove causality between the appearance of these new indexes and the sliding down the curve of the open interest, we highlight their simultaneity as food for thought.

First, we start by defining the meaning we shall give to the term *financialization* which according to the *New Oxford American Dictionary* means *the process by which financial institutions, markets, etc., increase in size and influence*. In this chapter, we talk about the financialization of commodities to mean the increased role of financial markets in the operation of the commodities markets. For the purpose of this chapter, we restrict the scope of this definition and use the terminology *financialization hypothesis* to mean that the sharp increase in volatility and the price hikes observed in the commodity markets between 2004 and 2008 are due to the overwhelming influence of large institutional investors using indexes to gain exposure to commodities, and not to an imbalance in supply and demand for physical commodities due to the growth in emerging economies such as China, India and Brazil.

While there is no clear rhyme or reason for the timing of the emergence of this financialization, it is widely accepted to be associated for the most part, with the appearance of a new class of large investors who chose to take positions on commodities as a group, in order to capture profits considered to be unattainable from investments in more traditional assets. Treating commodities at the same level as stocks, bonds, real estate, etc. they promoted commodities to the rank of a *new asset class*.

This spectacular increase in investment in the commodity markets by investors whose primary business or financial interests were not directly dependent upon changes in the prices of the physical commodities was treated as pure speculation, and has been the source of heated discussions among economists, policy makers as well as in the media. Case in point, the 2008 bubble in the prices of a wide range of commodities as shown in Fig. 1 with the plot of the evolution of a global commodity index representative of the spot prices of a large group of commodities. Details on the construction of the index plotted in Fig. 1 will be given in Sect. 3. As we are about to explain, this bubble has caught the attention of policy makers and focus their investigations on the roles of the various groups of financial investors in the commodity markets.

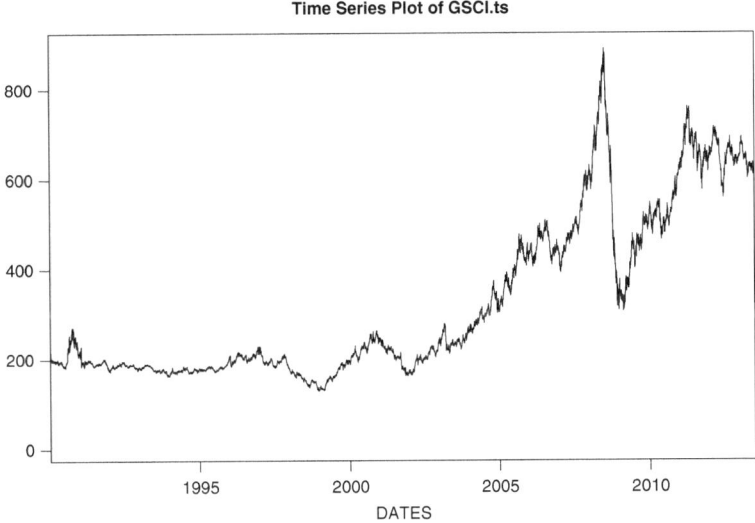

Fig. 1 Time series plot of the GSCI daily spot index

The emergence of specialized indexes and the growth in popularity of long-only index-fund investing are some of the remarkable differences between commodities and other asset classes. According to Barclays' internal reports, in 2006–2007, index fund investment increased from 90 billion to 200 billion USD. Simultaneously, commodity prices increased 71 % as measured by the Commodity Research Bureau. At the peak of the price bubble in 2008, commodity fund investors, including ETFs and hedge funds like *Soros Fund Management*, controlled a record 4.51 billion bushels of corn, wheat and soybeans through the futures markets of Chicago Board of Trade, equal to half the amount held in U.S. silos on March 1, 2008. In his testimony before the U.S. Senate Commerce Committee, George Soros stated that commodity investment, as a new venue for institutional investors, had become 'the elephant in the room' and as a result, investment in these assets might exaggerate price rises. After the price collapse which occurred between June 2008 and early 2009, many pundits referred to this boom and bust as a *bubble* as futures prices far exceeded fundamental values. The large scale speculative buying by index funds was held as culprit. A number of studies on financial markets have suggested that herd formation among large institutional investors may have destabilized market prices and created excess volatility (see for example Dennis and Strickland [10], and Gabaix et al. [12]). From these studies, one can argue that herd behavior in the commodity markets, as driven by financial investors moving funds in and out of commodities, was a contributing factor behind the booms and busts observed in a wide range of commodities.

On the other end, some economists (including Nobel Prize winner P. Krugman [22], Irwin and Sanders [16], Hamilton [13] and Kilian [19]) remained skeptic about the bubble theory. They argue that commodity price cycles are driven by fundamental factors like supply and demand, and that the temporary imbalances observed in 2008 are due to the spectacular growth in emerging economies. Adding support to this view, Buyuksahin and Harris [4] examine the trading positions of various types of traders in the crude oil market, and find little or no evidence that financial investors' position changes caused price changes in the oil market.

This did not stop commodity index investing from being under attack. Increased participation in futures markets by non-traditional investors was deemed *disruptive* and blamed for the 2007–2008 "Food Crisis" that is at the origin of the famous *"Casino of Hunger: How Wall Street Speculators Fueled the Global Food Crisis"* [11]. See also [3]. A report from the U.S. Senate Permanent Subcommittee on Investigation *". . . finds that there is significant and persuasive evidence to conclude that these commodity index traders, in the aggregate, were one of the major causes of unwarranted increases in the price of wheat futures contracts relative to the price of wheat in the cash market. . . ..."* . To add insult to injury, a group of 48 agriculture ministers meeting in Berlin said that they were *". . . concerned that excessive price volatility and speculation on international agricultural markets might constitute a threat to food security. . . .",* according to a joint statement handed out to reporters on January 22, 2011.

Broadly speaking, the *financialization of commodities* should refer to the increased leverage and the exponential growth of financially settled contracts dwarfing their physically settled counterparts. More recently, this term has also been used to refer to the significant impact of index trading on commodity prices, and even more narrowly speaking, to the increased correlations between the commodities included in the same index, and also between equity returns and commodity index returns. This last fact is illustrated in Fig. 2 which shows the time evolution, as given by a Kalman filter, of the time-dependent "beta" of the least squares linear regression of the Goldman Sachs Commodity Index Total Return against the returns of the S&P 500 index. Instantaneous "betas" are typically computed using Kalman linear filters as estimates of the slope of a local linear regression whose domain varies with time. See for example section 7.5.2 entitled *Linear Models with Time Varying Coefficients* of the textbook [6] for details. The standard commodity indexes are reviewed in Sect. 3, and a new generation of roll yield optimizing indexes is introduced in Sect. 5.

It is an empirical fact that return correlations are no longer what they used to be, and it is now commonly accepted that correlation in price changes for commodities included in the same index tightened before 2007. Tang and Xiong [32] argue that commodity index trading is responsible for this correlation *tightening*. See also [9], the works of Irwin and Sanders [16, 29], and especially [28, 30] for the impact of index trading on the agricultural commodities. This restrictive form of the financialization hypothesis is discussed in Sect. 4.

Note that it is likely that this correlation tightening is a scale dependent phenomenon. Indeed, it seems that high frequency traders do not see (and hence ignore) these correlation increases.

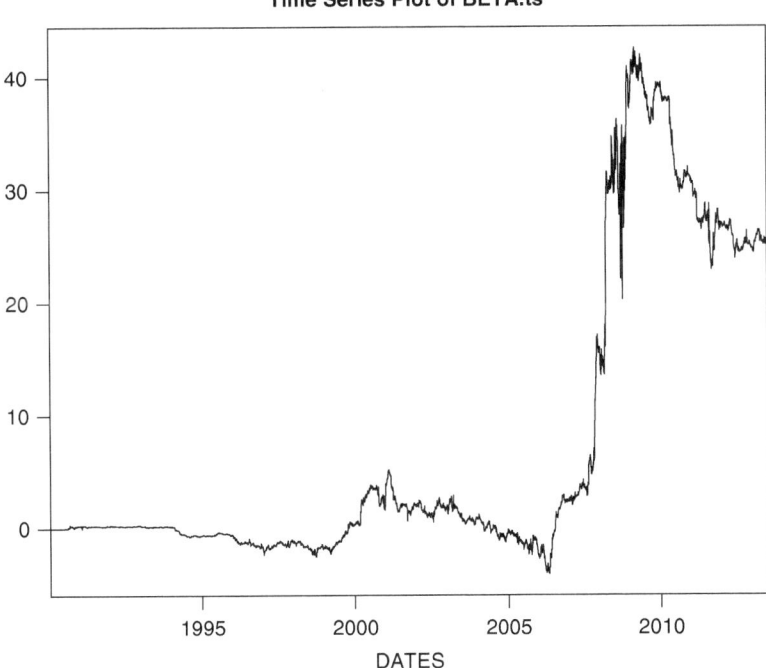

Fig. 2 Instantaneous dependence (β) of the daily GSCI-TR returns upon the corresponding S&P 500 returns

Commodity contract valuation is best understood by equilibrium arguments based on supply and demand for the physical commodities. In [7], we advocate structural models for the pricing of commodities and commodity derivatives. However, one of the main contention of the financialization of commodities is that the pricing models based on matching supply and demand are impaired by the overwhelming sizes of trades by institutional investors which increase price volatility and drive prices away from the levels predicted by fundamental supply and demand relationships. As a result, commodity price dynamics no longer merely reflect changes in fundamentals.

These conflicting views are yet to be reconciled, and investor behavior in the commodity markets needs to be further investigated, especially for the role it plays (if any) in the excessive price movements observed so frequently. The dramatic increase of commodity trading volume (often referred to as the financialization of commodity trading) occurred essentially at the same time as demand for physical commodities from emerging economies increased rapidly. The simultaneity of these two contributing factors make it extremely difficult to parse out their relative contributions to the increased volatility of the markets, and disentangle their respective price impacts.

We close this introduction with a short summary of the contents of the chapter. Section 2 offers a crash course focusing on the idiosyncrasies of the commodity markets, while Sect. 3 takes an historical perspective to introduce the traditional commodity indexes. The influx of institutional investors in the commodity markets and the changes they are responsible for are documented in Sect. 4 where a short review of the publications on the financialization debate is provided. Section 5 introduces the new generation of commodity indexes designed in most part for the purpose of maximizing the roll yield, and Sect. 6 documents the changes in the term structure of open interest as food for thought as a possible impact of the growing popularity of these new indexes.

2   Generalities on the Commodity Markets

In order to set the stage for our discussion of the *financialization hypothesis*, we review some of the basic idiosyncrasies of the commodity markets, focusing on those relevant to the debate. As already emphasized in the introduction, the role of institutional investors is paramount to the discussion of financialization of the commodity markets. The large number of commodities, the large number of venues on which these commodities are traded, added to the great variety of contract maturities, physical commodity grades and delivery locations offer a wide range of opportunities for hedgers and speculators. As a result, liquidly traded contracts represent a rather small part of the commodity world. However, they are most likely to be included in the commodity indexes, and traded for purely speculative purposes. Consequently, they will be our favorite targets when we look for illustrations of some of the claims used in the financialization debate.

2.1   The Markets and the Trades

Because of the physical nature of the interest underlying the contracts, commodity prices are determined by equilibrium arguments which involve matching supply and demand for the physical commodity itself. On the supply side, estimating and predicting inventories and quantifying the costs of storage and delivery are important factors which need to be taken into account. This is not always easy in the context of standard valuation methods which are mostly based on traditional finance theory (think for example of NPV which attempts to compute the present value of the flow of future dividends).

Whether they were called spot markets (when they involved the immediate delivery of the physical commodity), or forward markets (when delivery is scheduled at a later date), commodity markets started as physical markets. Trading volume exploded with the appearance of financially settled contracts. While forward contracts are settled Over the Counter (OTC), and as such, carry the risk that the counterparty may default and not meet the terms of the contract, most of the financially settled contracts are exchange-traded futures for which the exchange acts as clearing house controlling default risk by a system of margin calls, and attracting speculators to provide liquidity to the markets. While trading in physically and financially settled contracts were traditionally the two ways an investor could gain exposure to commodities, the creation of indexes and the increasing popularity of index tracking Exchange Traded Funds (ETFs) have offered a new way to gain exposure to commodities.

In the early 2000s, investing in commodities was promoted as a fool-proof portfolio diversification tool. After all, these financial interests were *believed to be uncorrelated or negatively correlated with stocks*. Case in point, the prospectus of the S&P GSCI (see the section on commodity indexes for details on the definition and the properties of this index) claims "*... and provides diversification with low correlations to other classes*".

The exponential growth of this new form of investment in commodities which took place over the last decade may have been a self-defeating prophecy as recent econometric studies have shown that this form of index trading has created new correlations between commodities and stocks, and between the commodities included in the same index [32]. Pushing the argument even further, one could posit that the influence of investors has overturned Keynes' *theory of normal backwardation*[1] (see for example [17], or [18] for a more modern account, and [8] for a discussion focused on agricultural commodities), causing a recent predominance of forward curves in contango, thus further weakening the attractiveness of investing in these markets. We explain the duality contango/backwardation in Sect. 2.4 below.

One of the many convenient features of commodity trading is the specialization of the exchanges, leading to simple correspondences between commodities and locations where they are traded. In other words, a given commodity is traded on *one or a small number* of specialized exchanges. This is in sharp contrast with the equity markets for which a given stock can be traded on many platforms, leading to subtle optimization problems as the choice of a particular venue for a trade can affect the profits or losses on the trade.

The following table gives a few examples of some of these exchanges in the US and in Europe.

---

[1]In Keynesian economics, the expected future spot price of a commodity should be higher than the forward price. Indeed, according to this theory, the producers of commodities are eager to sell, and willing to sell at a loss if necessary. As a result, the price of a forward or futures contract is below the expected spot price at contract maturity, and the resulting futures or forward curve is downward sloping (i.e. *inverted*), since contracts for further dates trade at lower prices. In practice, the term backwardation is often used to refer to situations when the current spot price exceeds the price of the future.

Exchange	Location	Contracts
CME Group		
CME	Chicago	Agriculture, weather
Chicago Board of Trade (CBOT)	Chicago	Agriculturals
COMEX	Chicago	Metals
NYMEX	New York	Energy, metals
Intercontinental Exchange (ICE)		
ICE	Atlanta, US	Energy, emissions, agricultural
NYSE.Liffe	London	Agricultural
NYSE.Euronext	Europe & US	Agricultural, energy
Kansas City Board of Trade (KCBT)	Kansas City	Agricultural
Climex (CLIMEX)	Amsterdam	Emissions
European Climate Exch. (ECX)	Europe	Emissions
London Metal Exch. (LME)	London	Industrial metals, plastics

2.2 Trading Commodities

Traditionally, the investment portfolios of large institutional investors (e.g. pension funds and endowment funds) included only stocks, bonds, and cash. The primary advantage of including commodities is that commodity returns are expected to be relatively uncorrelated with the returns of traditional asset classes. The absence of correlation is attributable in part to inflation. In fact, holding commodity futures is often considered to be an inflation hedge. Indeed, during periods of rising inflation, traditional asset categories like stocks and bonds perform poorly. Commodities, on the other hand, generally perform well during these periods. Indeed, increased demand for goods and services, typical in periods of rising inflation, usually implies increased demand for the commodities used in the production of those goods and services.

There are several ways in which traditional investors used to gain exposure to commodities.

1. The *old-fashioned way* to invest in commodities is to actually *purchase* the physical commodity itself. However most investors are not ready or equipped to deal with issues of transportation, delivery, storage and perishability. This form of involvement in commodities was created for, and is essentially limited to, the hedgers who mitigate the financial risks associated with uncertainties in the production and delivery of commodities relevant to their businesses.
2. Another way to gain exposure to commodities is to invest in stocks in commodity intensive businesses: for example buying shares of Exxon or Shell as a way to invest in oil. Many exchange traded funds (ETFs) are tracking portfolios of stocks of companies with well defined commonalities. The portfolios of a large number of these ETFs comprise only energy companies, and as such, they call

themselves commodity ETFs. They promote themselves as investment vehicles to gain exposure to commodities despite the fact that they are technically equity ETFs. However, this type of investment offers at best an indirect exposure as shares of natural resource companies are not perfectly correlated with commodity prices.

3. A more direct form of exposure to commodities is through straight investment in commodity futures and options. The exchanges offer transparency and integrity through clearing, and relatively small initial investments are needed to take large positions through leverage. However, this convenience comes at a serious price, as discovered by many *rookies* who ended up choking, unable to face the margin calls triggered by adverse moves of the values of the interests underlying the futures contracts. Also, purely speculative investments of this type may need to be structured with a careful *rolling forward* of the contracts approaching maturity in order to avoid having to take physical delivery of the commodity: trading wheat futures can be done from the comfort of an office set up in a basement, but taking physical delivery of one lot (i.e. 5,000 bushels) of wheat requires a large backyard! Consequences of some of the simplest rolling strategies are discussed in Sect. 2.4 below.

We first discuss the idiosyncrasies of commodity prices, and postpone to a later section the presentation of the more recent (and most relevant to the focus of this chapter) form of exposure to commodities based on index investing and/or tracking.

2.3 Data Used for Illustration Purposes

We use a specific set of commodities for the purpose of illustration. We chose *Crude Oil* because of its overwhelming impact on the global economy, and *Copper* as an example of *metal*. Copper is widely accepted by economists as a representative commodity because historically, it has been a consistent predictor of the health of the global economy, presumably because it is an important input in a huge number of industrial processes. Figure 3 gives a time series plot of the values of the nearest maturity Copper futures contract (as close as we can get from a spot price!).

We use Light Sweet Crude Oil futures price data from NYMEX (part of the CME group) provided by *Data Stream*. These prices serve as a key international benchmark. The contract sizes are for 1,000 barrels and the prices are quoted in US dollars and cents per barrel. Prices are quoted for monthly contracts with times to maturity up to 6 years. Trading in the nearest maturity contract ceases on the third business day prior to the 25th calendar day of the month preceding the delivery month. Delivery is *free-on-board* (FOB) at any pipeline or storage facility in Cushing, Oklahoma.

When discussing Copper, we use forward data, also from CME COMEX, and also provided by *Data Stream*. The contract sizes are for 25,000 pounds, and the

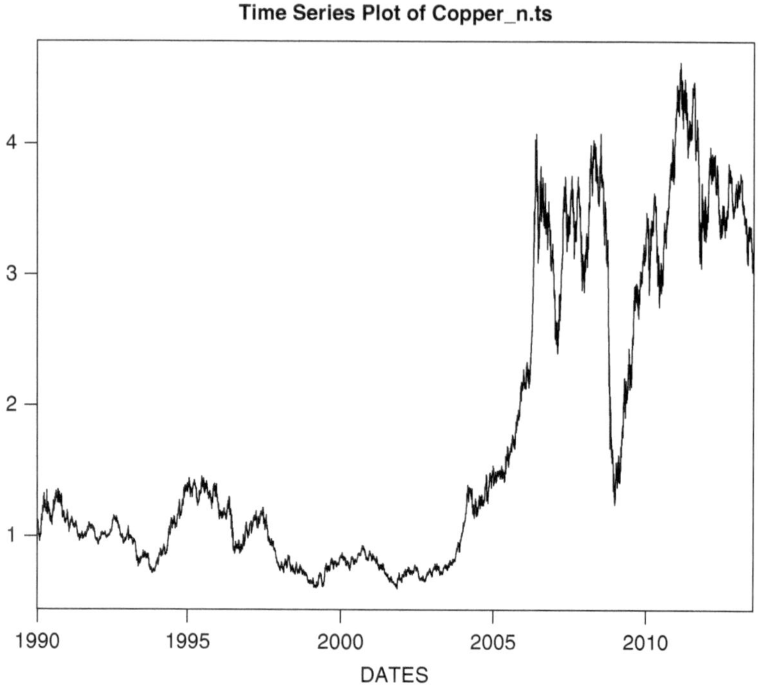

Time Series Plot of Copper_n.ts

Fig. 3 Time series plot of the daily price of the nearest copper futures contract between January 2, 1990 and September 9, 2013. Source: *Data Stream*

prices are quoted in US cent per pound. While more forward contract prices are listed, we shall only use the nearest 23 maturity months. Trading in a contract with a given delivery month (maturity) ends on the third last business day of the delivery month. Note that these contracts are also traded on the London Metal Exchange (LME) and the Shangai Futures Exchange.

During the period 1998 through 2007, the trading volume in exchange-traded commodity futures and futures options experienced a five-fold increase. As an example, Fig. 4 gives the time series plot of WTI Crude Oil daily open interest. This plot represents on each day, the total number of contracts, irrespective of their maturities, held by investors. Corresponding plots (see for example Fig. 10 for the case of Copper) for other commodities would show the same dramatic increase, attesting the significant influx of money in commodities.

However, most institutional investors do not have the sophisticated trading operations necessary to manage a complex portfolio of futures contracts: commodity index funds and OTC commodity return swaps appeared as attractive solutions. Both forms of investment are transparent and passive, so no need to monitor the market to identify underpriced commodities or timing profit opportunities.

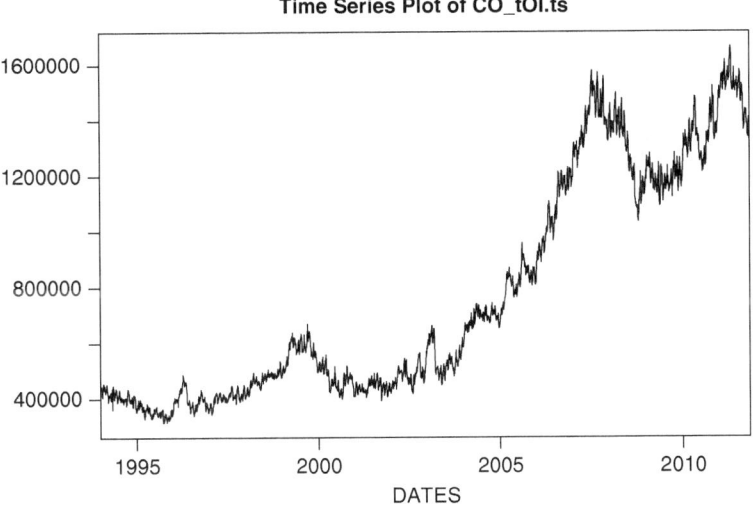

Fig. 4 Time series plot of the daily (total) open interest in WTI crude oil between January 3, 1994 and November 22, 2011. Source: *Data Stream*

2.4   *Contango, Backwardation and the Roll Yield*

We now introduce more of the jargon of the term structure of forward and futures prices in the form of a definition for easier reference.

Definition 1. We say that the market is in backwardation, or that the forward curve is backwarded, when futures prices are lower than the expected future values of the spot price.

Because the futures prices must converge toward the expected spot price when approaching maturity of the contract, futures prices are rising to get in line with the expected spot price. Typically backwardation occurs when the left most part of the curve is downward sloping.

Definition 2. We say that the market is in contango, or that the forward curve is in contango, when futures prices are higher than the expected future values of the spot price.

Because the futures prices must converge toward the expected spot price when approaching maturity of the contract, futures prices are falling to get in line with the expected spot price. Typically, contango occurs when the left most part of the curve is upward sloping.

We close this subsection with a formal definition of the roll yield, and a simple example showing that this yield is positive (resp. negative) when the forward curve is in backwardation (resp. contango).

Definition 3. The roll yield is the return (profit or loss) captured by a market participant liquidating a long position in a contract approaching maturity, and taking the same position in the new nearest maturity contract.

The typical example to keep in mind is the profit (in which case the roll yield is positive) gained in backwardation, by merely maintaining a long position in the nearest contract. Indeed, in the case of a backwarded forward curve we have $p_1 > p_2$ if we denote by p_1 and p_2 the prices of the forward/futures contracts with the shortest maturities $T_1 < T_2$ after the current time t. Consequently, maintaining a long position in the nearest contract is done by closing the current position (i.e. selling at the unit price p_1 the contract with maturity T_1 as t approaches T_1), and opening the same long position in the nearest maturity contract (i.e. buying the same amount of contracts with maturity T_2 at the price p_2), locking a profit, just for rolling the position to the new nearest maturity! So taking a long position in a backwarded market guarantees a positive roll yield, and hence a profit, just for rolling the same position from one maturity to the next when contracts approach maturity.

Similarly, maintaining (rolling) a long position in a contango market leads to losses, and hence a negative roll yield as a result. A transition from a backwarded market to a market in contango is one of the common fears of passive commodity traders.

Systematic studies of the nature of the roll yield can be found in the academic literature. As an example, the interested reader may want to look at [27].

2.5 The Term Structure of Forward Prices

On any given day t, the term structure of forward prices is given by the prices of the futures contracts for a given set $T_1, T_2, \cdots, T_n$ of maturity tenors. While t changes from one day to the next, the tenors $T_1, T_2, \cdots, T_n$ remain the same as long as $t < T_1$. While the actual values of the maturity dates T_i are crucial to understand seasonal commodities such as natural gas or most of the grains, they can be a hindrance for many statistical data analysis techniques which require some form of stationarity in time of the data. On any given trading day, say t, if price quotes $p_1, p_2, \cdots, p_n$ are available for maturity dates $T_1, T_2, \cdots, T_n$, the points

$$(T_1, p_1), (T_2, p_2), \cdots, (T_n, p_n)$$

in the plane offer a discrete sampling of an hypothetical forward curve $T \hookrightarrow f(t, T)$ which could be defined for $T > t$. One of the problems is that when time passes by and t becomes, $t + 1, t + 2, \ldots$, the maturity dates $T_1, T_2, \ldots$ do not change, and eventually t gets too close to T_1 and the contract with maturity T_1 ceases to be traded, and the nearest contract available for trading becomes T_2. To avoid this sudden change in the input data, it is often convenient to re-parameterize the term structure of forward prices by the time to maturity $\tau = T - t$ instead of the time of maturity T.

2.5.1 Data Pre-processing

Switching from the parameterization by time-of-maturity to time-to-maturity requires extrapolation/smoothing and resampling of the forward prices. Below, we describe the steps we took to produce the numerical illustrations given in this chapter. Other procedures have been proposed to solve this issue. The problem is especially delicate in illiquid markets with a small number of quoted forward prices, and in highly volatile markets like the electricity markets. For example, the reader is referred to Chapter 7 of [2] for a detailed discussion of the latter.

On any given trading day, say t, we replace the maturity times $T_1, T_2, \ldots$ after t by the corresponding times to maturity $\tau_1 = T_1 - t, \tau_2 = T_2 - t, \ldots, \tau_n = T_n - t$, and we plot the price quotes $p_1, p_2, \cdots, p_n$ against these values of τ. In other words, we consider the points

$$(\tau_1, p_1), (\tau_2, p_2), \cdots, (\tau_n, p_n)$$

as discrete sample observations of a hypothetical forward curve $\tau \hookrightarrow \tilde{f}(t, \tau)$ which could be defined for $\tau > 0$. The main advantage of this re-parameterization of the curve is that its domain of definition does not change with t, and it is thus easier to have meaningful comparisons between forward curves on different days. On each day t, this hypothetical forward curve $\tilde{f}(t, \cdot)$ is often called a continuous maturity forward curve. It can be estimated by regression. In all the examples considered in this chapter, we used a standard cubic spline regression to produce continuous maturity curves. Modelling the term structure of forward prices by parametric families of classical functions is very convenient. This approach was successfully implemented for the analysis of the term structure of interest rates, and central banks, regulators and fixed income desks of major banks have developed their own proprietary methods to do so. But from a practical point of view, handling functions of a continuous variable is not always easy, and it is natural to work with discrete subsamples

$$\tilde{p}_1 = \tilde{f}(t, \tilde{\tau}_1), \tilde{p}_2 = \tilde{f}(t, \tilde{\tau}_2), \cdots\cdots, \tilde{p}_1 = \tilde{f}(t, \tilde{\tau}_m),$$

for a fixed set $\tilde{\tau}_1, \tilde{\tau}_2, \ldots$ which will not change from day to day. The choice of these fixed values of the time to maturity often starts with values like $1mo, 2mo, \ldots$, but these values do not have to be regularly spaced, and they do not have to be in the same number as the number n of original observations. The discrete forward curve so obtained

$$(\tilde{\tau}_1, \tilde{p}_1), (\tilde{\tau}_2, \tilde{p}_2), \cdots, (\tilde{\tau}_m, \tilde{p}_m)$$

is called a constant maturity forward curve. Note that except for some exceptional cases, the prices $\tilde{p}_i$ are the results of data analysis, and they are not observed quotes from the market. So any conclusion drawn from the analysis of these modified prices is subject to artifacts created by the way we massaged the data, and should possibly be taken *with a grain of salt*!

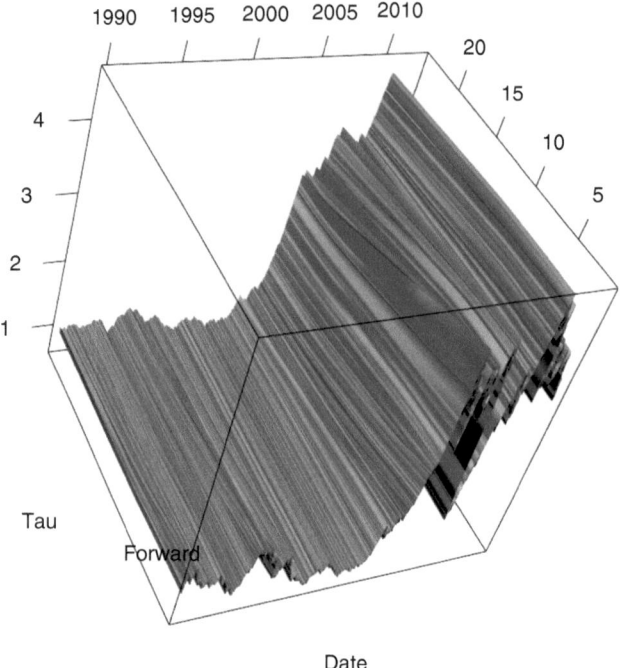

Fig. 5 Surface plot of the daily forward curves for copper between January 3, 1990 and July 7, 2013

Figure 5 gives the plot of the daily forward curves for Copper between January 3, 1990 and July 7, 2013. The trading days t appear on the axis labelled "Date" while the resampled time to maturity appears on the axis labelled "Tau". We express τ in months and in the particular case of Copper, we resampled the continuous maturity forward curve for the values $\tau = 1, \tau = 2, \cdots, \tau = 24$ months.

Principal Component Analysis (PCA) is the most basic data analysis technique to identify the effective dimension of multidimensional objects. It was successfully used by Litterman and Scheinkman in [24] to identity the main factors in the time evolution of the term structure of interest rates. Since then, it has been used systematically each time a financial engineer faces a forward curve of any kind. We performed PCA on the daily changes in the Copper constant maturity forward curves over two different periods, period P_1 ranging from January 3, 2000 to December 31, 2004, and period P_2 ranging from January 3, 2010 to July 7, 2013 (Fig. 6).

The first four loadings of each of the PCAs are reproduced in Fig. 7. While the shapes of the first loadings are strikingly similar (the first and main one representing a parallel shift, the second one corresponding to a tilt of the curve, while the third one provides convexity or concavity to the curve), the proportions of the variance explained by the factors which are given in Fig. 6 deserve some explanation. Despite the fact that the scales of the vertical axes partially mask the differences between the

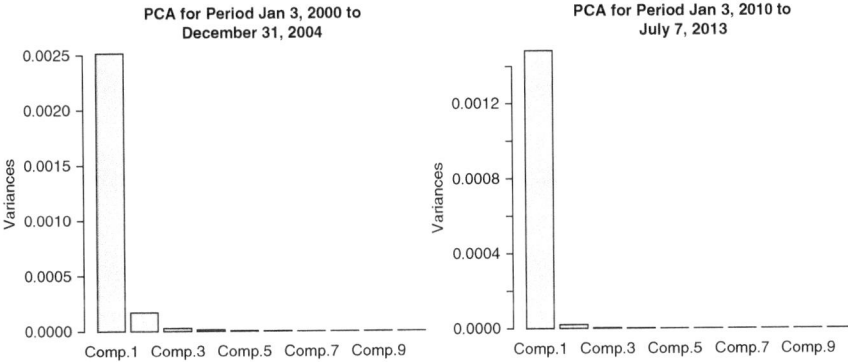

Fig. 6 Proportions of the variance explained by the loadings of the PCA of the copper forward curves for the period P_1 ranging from January 3, 2000 to December 31, 2004 (*left*) and the period P_2 ranging from January 3, 2010 to July 7, 2013 (*right*)

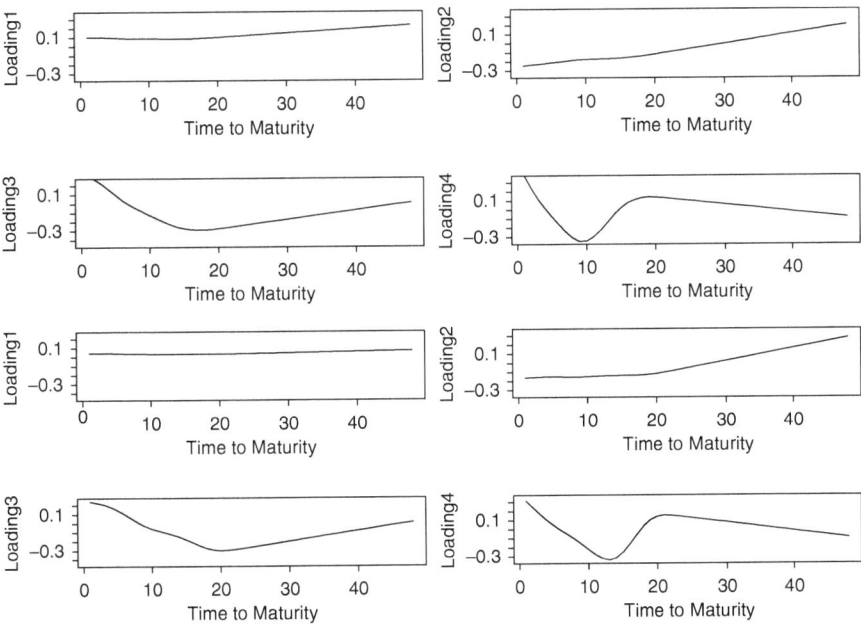

Fig. 7 Loadings of the PCA of the copper forward curves for the period P_1 ranging from January 3, 2000 to December 31, 2004 (*left*) and the period P_2 ranging from January 3, 2010 to July 7, 2013 (*right*)

two periods, it appears clearly that the term structure of forward prices is *stiffer* in the second period. By this we mean that a smaller number of factors explains the same proportion of the fluctuations in the time evolution of the forward curves. This phenomenon is widespread throughout the commodity markets and seems to have appeared in the mid-2000s. More on that later on.

2.6 Market Participants

The original *raison d'être* of the commodity markets was to facilitate price discovery and allow the transfer of price risk from producers and consumers to agents willing to assume that risk. Over the last decade, the growing financialization of these markets has dramatically changed this idealized picture, and the activity of these markets became increasingly murky and time and again more difficult to compartmentalize.

In its weekly Commitment of Traders (COT) reports, the CFTC provides information on the various categories of market participants which are active in the commodity markets. Originally, these participants could be organized in two major groups: hedgers trading in futures contracts to reduce an existing risk exposure in their commercial business (which is the reason why they are also called commercials), and speculators or non-commercials. However, through financialization, an increasing number of commodity index swap dealers who hedge to offset financial positions were categorized as commercials. To remedy this problem, starting in 2007, the CFTC added a supplementary Commodity Index Traders (CIT) report, and more recently, weekly Disaggregated Commitment of Traders (DCOT) reports. The five categories of market participants identified by the DCOT reports are given in Table 1. The reader interested into more details is referred to [35].

The swap dealer category is not limited to passive investors tracking commodity indexes. It includes swap dealers who do not have commodity index-related positions. On the other hand, money managers trade on short-term horizons and adopt active investment strategies.

Table 1 CFTC classification of commodity markets participants from its "Disaggregated Commitment of Traders Reports". See [35] for details

Trader categories	Description
Producers, merchants, processors, users (PMPU)	Entities that predominantly engage in the physical commodity markets and use the futures markets to manage or hedge risks associated with those activities
Swap dealers	Entities that deal primarily in swaps for a commodity and use the futures markets to manage or hedge the risks associated with those swap transactions. The bulk of these traders' clients are index investors who invest in commodity indexes
Money managers	Entities that manage and conduct organized futures trading on behalf of their clients. This category includes registered commodity trading advisers (CTAs), registered commodity pool advisers (CPOs), and unregistered funds identified by the CFTC. Hedge funds and large ETFs are part of this category
Other reporting traders	Every other reportable trader that is not included in one of the other three categories
Non-reporting traders	Smaller traders who are not obliged to report their positions

2.7   Exchange Traded Products and Index Investing

Exchange traded products (ETPs) include exchange traded funds (ETFs), exchange traded vehicles (ETVs), exchange traded notes (ETNs) and exchange traded certificates (ETCs). Many energy or commodity ETFs are tracking proprietary benchmark indexes measuring the aggregate performance of stocks in the energy or commodity sector. For example, the Vanguard Energy ETF (VDE) is a typical passively managed portfolio aiming at a full replication (whenever possible) of a portfolio of stocks of companies involved in the exploration and production of energy products such as oil, natural gas and coal. Most ETPs replicate the return on a single commodity, or a group of commodities. ETPs issue shares that are traded like a stock on a securities exchange. So the shares of ETPs are traded on equity markets. Some of them are easily accessible by small-scale investors, while others offer large single coupons, and are therefore more attractive to institutional investors such as pension funds. Apart from ETFs for precious metals, such funds have traditionally used futures contracts as collateral. But an important recent development is that some ETPs, such as those in copper and aluminum, are now backed by physical commodities. Futures-backed ETPs expose investors to counterparty risk, as transactions involving buying and selling of ETPs do not go through a clearing house on commodity exchanges. The rising importance of physically-backed ETPs indicates that risk aversion and growing concern with counterparty risk have made it more acceptable for financial investors to bear the storage cost of the physical commodities as they can be used as collateral. The currently very low interest rates, which reduce the cost of credit used to finance storage costs, has most likely also contributed to the increased importance of physically-backed ETPs. Returns on such products are determined by spot price movements, while the returns on futures-backed ETFs are largely influenced by the roll yield, and thus share the characteristics of traditional index investments.

2.7.1   ETVs

Exchange Traded Vehicles (ETVs) provide investors exposure to commodity futures contracts without actually trading futures or ever taking physical delivery of the underlying commodity. Most often, they track a single commodity, as opposed to an index computed on a portfolio of commodities. They are traded as equities on equity markets. They can be short-only as well as long-only

2.7.2   ETNs

Exchange Traded Notes (ETNs) are debt securities issued by banks. Up until they mature, their returns are based on the performance of an underlying index. They combine features of bonds and ETFs. ETNs' values are affected by the credit worthiness of the issuer. As a result, their values depend not only upon the value of the underlying portfolio of commodity contracts, but also on the credit rating of the issuer.

2.7.3 Commodity Mutual Funds vs ETFs

The reason for broad commodities mutual funds' popularity, say professional investors, is largely due to the fact that similar commodities ETFs hold futures contracts. This leaves ETFs more prone to so-called contango effects, as well as vulnerable to tax hits and front-running. As already mentioned earlier, precious metals ETFs, however, avoid these problems by directly owning their underlying commodities.

There are two types of commodity ETFs. Those which track an index computed from the performance of a portfolio of stocks of companies whose business is commodity intensive, and those which track the performance of a commodity index. We are mostly concerned with the latter. They usually hold futures contracts because the definitions of the indexes they track are based on the performances of specific contracts. But this can lead to problems, as the ETFs have fallen victim to contango when a fund loses money every time it rolls over from a near-month contract to a further-dated contract. See the example of UNG discussed below.

Some mutual funds, case in point PIMCO Commodity Real Return Fund PCRDX, have tried to avoid these pitfalls. Their strategy is to gain exposure to commodities through debt instruments such as swaps and pre-paid forward notes, rather than futures, in order to avoid the hit of a normal backwardation/congango transition.

2.7.4 Index Investing

The final way to gain exposure to commodity which we discuss in this chapter is investing directly in *Commodity Indexes* or in ETPs tracking these commodity indexes. For liquidity reasons, most ETPs simply invest in contracts with the shortest possible times to maturity. When the contracts they are holding approach maturity, in order to avoid delivery or settlement issues, they automatically *roll* their holdings by closing the positions in the contracts approaching maturity, and taking the same exact positions in the contracts available for trading with the shortest possible maturities. See the discussion of the example following the definition of roll yield in Sect. 2.4 and of the roll algorithm in Sect. 3.5 below. This form of *passive investment* (after all there is no need for a *Commodity Trading Advisor* (CTA) for that), has become very popular as a way to diversify an investment portfolio with an exposure to commodities without having to deal with the gory details of all the convoluted idiosyncrasies of the relevant markets. Nevertheless, an understanding of forward curve dynamics and the effect of frequent (typically monthly) rolls is still vital, as a recent investor in a natural gas ETF would undoubtedly agree: between June

2008 and March 2012 this ETF (called UNG) lost a shocking 96 % of its value, with roughly half attributable to the spot price drop and half to the steep contango witnessed throughout this period.

However, one the main original contribution of this chapter, if any, is to review and investigate the impact of a new generation of commodity indexes with a different roll mechanism, optimizing the roll yield. See Sect. 5 below for details.

2.8 Active Versus Passive Investing

Investing in a portfolio tracking the composition of an index like those discussed in Sect. 3 below, is often called *indexing*. It is a form of *passive investing* because managing such a portfolio does not require active involvement, except for setting up the portfolio and periodic re-balancings. The expected performance of indexing is no different from the performance of the benchmark index. This is in contrast with active investing whose objective is to outperform the market or a benchmark index. Depending upon their styles, active managers rely on fundamental analysis, technical analysis or macroeconomic analysis to identify inefficiencies and anomalies in the markets which they then try to exploit. In a recent report [33], the United Nations Conference on Trade and Development (UNCTAD) argued that between July 2009 and February 2011, the importance of index traders diminished at the expense of active investment strategies. Based on Bloomberg and CFTC data, it published Pearson correlation coefficients between prices in specific commodities (e.g. oil, cocoa, maze, sugar and wheat) and positions in these commodities by index investors on one hand, and money managers on the other hand. These numbers show a close correlation between commodity prices and the positions of financial investors that pursue an active trading strategy. See also the shorter and more aggressive policy brief [34] mostly focused on WTI Crude oil prices.

3 Commodity Indexes

Indexes can be traded through the use of index swaps which involve the exchange of a fixed payment for the value of the index at a pre-determined date. In most cases, this type of passive investment relies on ETPs, such as ETFs, backed by portfolios of futures contracts more often than individual futures contracts. The commodity-related assets under this form of management was at a historic high in March 2011, when it reached about $410 billion which is approximately the double of its pre-crisis level of 2007. While index investment accounted for 65–85 % of the total between 2005 and 2007 prior to the financial crisis, its relative importance fell to 45 % since 2008. This decline occurred despite a roughly 50 % increase in the value of index investments between 2009 and the end of 2010.

Table 2 Original commodity indexes

	CRB/CCI	GSCI	Rogers RICI	DJ-AIG
Started	1957/1986	1992	1996	1999
Exchange traded	Yes	Yes	No	No
Number of components	17	22	35	20
Energy (%)	18	50	44	31
Metals (gold)	24 6	12 2	21 3	29 9
Grains	18	18	21	21
Food/fiber	30	10	11	10
Livestock	12	11	3	9

3.1 Index Terminology

We now give specific definitions for some of the terms we already used when we commented on some of the figures at the beginning of the chapter.

- A *Spot Index* is based on the prices of the contracts included in the index;
- An *Excess Return Index* incorporates the returns of the corresponding spot Index as well as the discount or premium obtained by *rolling* hypothetical positions in contracts approaching their delivery dates;
- A *Total Return Index* incorporates the returns of the corresponding excess return index as well as the interests earned on fully collateralized contract positions on the commodities included in the index.

As for the original indexes introduced in Table 2, we briefly review the main features of CCI and RICI in Sects. 3.2 and 3.3 respectively, and we postpone the discussion of the major indexes GSCI and DJ-AIG to Sect. 3.4.

3.2 The Continuous Commodity Index (CCI)

The Continuous Commodity Index has been around since 1986 as a means to track the overall performance of the commodity markets and to offer investors a way to trade a diversified group of commodities under one contract. CCI is a broad grouping of 17 different commodity futures. It is one of many reincarnations of the original CRB Index that was developed in 1957. It is equally weighted. Each member commodity represents 5.88 % of the index. Over the years, some commodities have been dropped and replaced by new ones to give a better representation of the overall performance of commodities.

For the sake of completeness, we list by groups the commodities currently included in the Continuous Commodity Index CCI (Table 3):

Table 3 CCI composition

Energies: 17.64 %	Crude oil	Heating oil	Natural gas		
Grains: 17.64 %	Corn	Soybeans	Wheat		
Softs: 29.40 %	Coffee	Cocoa	Cotton	Orange juice	Sugar
Livestock: 11.76 %	Lean hogs	Live cattle			
Metals: 23.52 %	Copper	Gold	Platinum	Silver	

Notice that this partition of the index universe into commodity groups does not coincide with the partition given in Table 5 of the universes of the S&P-GSCI and DJ-UBSCI indexes into sectors. This is unfortunate, but typical of the historical lack of standardization of commodity indexes which change over time.

3.3 The Rogers International Commodity Index (RICI)

This total return index was designed by James B. Rogers, Jr. in the mid 1990s. It comprises futures contracts on 36 physical commodities ranging from agricultural to energy and metals products, quoted in four different currencies, listed on 12 exchanges in five countries. Its goal is to capture the price of raw materials throughout the world, and consumption patterns in developed as well as developing economies.

Over the past decade, two commodity indexes have emerged as industry *behemoths*: the Standard and Poor's-Goldman Sachs Commodity Index (S&P-GSCI), and the Dow Jones-UBS Commodity Index (DJ-UBSCI). They are marketed as tradable and for this reason, they are based on liquid commodity contracts traded on highly active futures markets.

3.4 The Two Major Commodity Indexes

In this section, we present the two major commodity indexes: the S&P-Goldman Sachs commodity index (SP-GSCI) and the Dow Jones-UBS commodity index (DJ-UBSCI). While they have been historically fierce competitors, The McGraw-Hill Companies owning the S&P indices and the CME Group, major shareholder in Dow Jones Indexes merged in the summer of 2012 to form the giant index provider S&P Dow Jones Indices.[2]

---

[2]On July 1st, 2014, 1 year after submission of the original version of this chapter, and 1 month before its revision, Bloomberg took over the calculation, distribution, governance and licensing of this index. In the process, it was renamed Bloomberg Commodity Index (BCOM). It is now part of the Bloomberg commodity index family.

3.4.1   The Dow Jones-UBS Commodity Index

Introduced on July 14, 1998, as the Dow Jones-AIG Commodity Index, this index
is rebalanced annually, the weights being based on production and liquidity as long
as, after each rebalancing, no commodity group constitutes more than 33 % of the
index, and no single commodity constitutes more than 15 % of the index. It was
acquired in May 2009 by the Swiss bank UBS AG.

3.4.2   The S&P-Goldman Sachs Commodity Index

Goldman Sachs published the GSCI starting 1991. It was acquired by S&P
Indices in February 2007 when it became the SP-GSCI. The weights used in the
computation of the index value are based on world production of the physical
commodities. When a futures contract included in the index approaches maturity,
a smoothed rolling procedure is implemented to replace the soon to mature contract
with the next to nearest maturity contract. As most commodity indexes, it comes in
three flavors: Excess Return, Total Return, and Spot. A time series plot of the spot
index was given in Fig. 1.

3.4.3   Comparison

Table 4 provides a detailed comparison, as of August 2013, of the compositions
of the two major commodity indexes. Table 5 provides a summary comparing the
weights given by the two indexes to the various commodity sectors.

3.5   The Roll Algorithm

While the composition of equity indexes can be relatively stable, commodity
indexes have to deal with the issue of maturing contracts. Even if the relative
proportions between the physical commodities entering an index can remain stable
over time, futures contracts approaching maturity need to be replaced by similar
contracts with longer maturities in order to avoid to have to take delivery of the
physical commodities. Each index prospectus describes the algorithm used to *roll*
the contracts approaching maturity into longer lived contracts. For the most part,
the indexes considered in the first part of this chapter use a simple roll strategy,
replacing contracts nearing delivery by contracts with the next maturity dates. There
are some exceptions, due mostly to liquidity considerations. These exceptions are
spelled out in documents publicly available, but the index boards reserve the right
to alter the rolling procedures on a case by case basis when exceptional market
conditions render the rolling algorithm unpractical.

Table 4 Side by side comparison of the two major commodity indexes

Sector	Commodity	Exchange	Ticker	S&P-GSCI weights (%)	DJ-UBSCI weights
Energy	Crude oil (Brent)	ICE-UK	LCO	22.34	
	Crude oil (WTI)	NYM / ICE	CL	24.71	11.16 %
	Unleaded gas	ICE-UK	QS		3.76
	Gasoil	ICE-UK	LGO	8.56	
	Heating oil	NYM	HO	6.17	3.88 %
	Natural gas	NYM / ICE	NG	2.0	12.41 %
	Oil (RBOB)	NYM	RB	5.90	2.58 %
Industrial metals	Aluminum	LME	MAL	2.13	4.58 %
	Copper	LME	MCU	3.28	6.78 %
	Lead	LME	MPS	0.40	
	Nickel	LME	MNI	0.58	1.91 %
	Zinc	LME	MZN	0.51	2.52 %
Precious metals	Gold	CMX	GC	3.00	9.73 %
	Silver	CMX	SI	0.49	3.23 %
Agriculture	Cocoa	ICE-US	CC	0.23	
	Coffee	ICE-US	KC	0.82	2.00 %
	Corn	CBT	C	4.66	5.26 %
	Cotton #2	ICE-US	CT	1.07	2.06 %
	Wheat (Chicago)	CBT	W	3.22	3.17 %
	Wheat (Kansas)	KBT	KW	0.88	1.22
	Soybean oil	CBT	BO		2.53 %
	Soybean meal	CBT	S		2.86 %
	Soybeans	CBT	S	2.62	5.70 %
	Sugar#11	ICE-US	SB	1.85	3.57 %
Livestock	Feeder cattle	CME	FC	0.52	
	Lean hogs	CME	LH	1.58	2.05 %
	Live cattle	CME	LC	2.62	3.32 %

Table 5 Sector by sector comparison of the two major commodity indexes

Sector	S&P-GSCI (%)	DJ-UBSCI (%)
Energy	69.71	37.47
Industrial metals	6.90	15.79
Precious metals	3.50	12.96
Agriculture	15.17	28.42
Livestock	4.73	5.36

As mentioned several times already, when the forward curve is in backwardation, replacing a maturing contract by the nearest maturity contract results in a net gain which is called the *roll yield*. However, when the curve is in contango, rolling contract is done at a cost. This simple fact needs to be kept in mind when one think about investment in commodity futures.

4 Review of the First Wave of Works on the Financialization Hypothesis

In [31], Singleton uses data from the 2008 boom—bust in oil prices to argue that flows from institutional investors have contributed significantly to the volatility of commodity prices.

In a decisive study [32], Tang and Xiong refute the idea that growing demand from emerging economies was the only driver of the commodity price burst in 2006–2008, and that commodity prices were influenced by financial factors and financial investor behavior. They use correlation coefficients computed in a trailing sliding window to argue that the co-movements between oil and other commodities rose dramatically following the inflow of institutional investors starting from 2004. Comparing with non-index commodities, they also demonstrate that this *correlation increase effect* is especially pronounced among commodities included in the same indexes. They show that the co-movements of the prices of different commodities increased after 2003–2004, and argue that this coincides with the beginning of significant position-taking by commodity index investors. A further evidence of that claim is the fact that for the commodities included in the major indexes this increase was significantly greater than for those not included.

We first illustrated the dramatic increase in return correlations between equities (as represented by the S&P 500 index) and commodities (as represented by the GSCI Spot index) in Fig. 2. There we can clearly see the increase in the instantaneous "beta" over the period 2006–2009. We further stress this claim by reproducing in Fig. 8 the time evolution of the instantaneous "betas" of Copper against the S&P 500. As expected this plot is noisier since we lost the averaging effect of the commodity index, but it is still providing a strong evidence for the tightening of the correlations between commodities and equities over that period.

Based on a thorough analysis of a proprietary dataset from the CFTC [5], Buyuksahin and Robe argue that the recent increase in the correlation between equity indices and commodities is due to the presence of hedge funds active in both equity and commodity markets.

In a recent study [15], Henderson, Pearson and Wang show that large investments in Commodity Linked Notes (CLNs) are the sources of hedges which cause significant price changes in the underlying futures markets.

However, not all the evidence point in the same direction. Surveys by Irwin and Sanders [16], and Fattouh, Kilian, and Mahadeva [21] argue against the claim that increased speculation in oil futures markets was an important factor in oil prices evolution. Furthermore, Kilian and Murphy [20] argue that the 2003–2008 oil price surge was due to global demand shocks rather than speculation. See also [26] and the technical report from the European Central Bank [25] for more balanced conclusions.

Following Kyle and Xiong [23], one can argue that portfolio rebalancing of commodity index funds can lead to correlated trades in related markets and thus create spillover effects across different commodities. In a recent econometric study

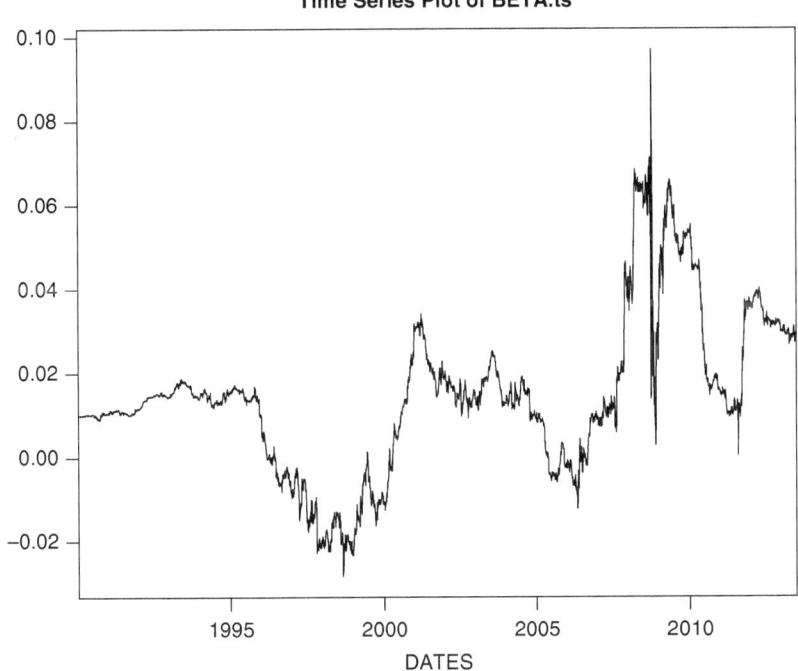

Fig. 8 Instantaneous dependence (β) of copper daily returns upon S&P 500 returns

of agricultural commodities, Hamilton and Wu [14] found no evidence that the positions of traders identified by the CFTC as *index traders* can help predict returns on the front month futures contracts.

While there is still lack of agreement on whether institutional investors affect commodity futures prices, it is well-established that institutional investors trades do affect stock prices. In the case of equity markets, several studies have analyzed the so-called *asset class effect* according to which correlations between assets belonging to the same index are higher than those between index and non-index assets. The co-movements associated to these unusually high correlations are attributed to the presence of institutional investors. This type of analysis was extended in [1] with an attempt to incorporate some of the idiosyncrasies of the commodity markets.

5   A New Generations of Indexes

Returns from investing in commodity futures contracts come typically from three different sources: spot, collateral and roll returns. New generations of indexes have chosen to optimize the roll return which was traditionally left to the backwardation/contango transitions. Among the most successful of this new breed of indexes, the Deutsche Bank Optimum Yield Commodity Index rolls according to a formula rather than simply rolling month to month. The formula seeks to achieve the best roll return possible given the shape of the forward curve at the time of the roll. Instead of rolling a contract nearing maturity into the nearest available maturity contract, the roll algorithm chooses the maturity with the best implied annual roll yield, as long as some liquidity constraints are satisfied.

We shall speculate on the possible impacts of the tracking of such indexes in our further look at consequences of this form of financialization.

5.1 Deutsche Bank PowerShare Optimum Yield Commodity Index

This Index comprises futures contracts on 14 heavily-traded physical commodities. with a distribution target of 55 % energy, 10 % precious metals, 12.5 % base metals, and 22.5 % agriculture. The weights are computed according to a combination of production and market liquidity. It is rebalanced annually in November. We give below a snapshot of its composition. The main originality of this index is the process used to implement the roll. As a general rule, commodity futures-based indexes replace contracts before they expire, and automatically buy into the next available maturity month. As explained earlier, this process is called "rolling" futures contracts forward. Instead of following this common practice, PowerShares DB Commodity Optimum Yield Index (and the ETFs tracking these indexes) use a procedure which is called Optimum Yield Roll process. As described in public prospectuses, it consists in choosing the maturity month among the next 13 maturity months available for trading at the time of the roll, which offers the best possible roll yield. As a result, the maturities of the futures contracts used in the computation of the PowerShares DB Commodity Index are not limited to the nearest month (Table 6). Accordingly, the portfolios of the corresponding Tracking Funds includes contracts with maturities *further down the curve*. While the details of the roll algorithm remain somehow mysterious due to the liquidity factor coming into the choice of the maturities to roll into, this roll strategy has been credited for out-performing the major indexes, both the SP GSCI and the Dow Jones-UBS commodity indexes, in the period 2006–2009 (Fig. 9).

Table 6 Composition of the
DB iShare index as of
09-Aug-2013 12:00 AM

Component	Contract date	Weight (%)
Aluminium	16-Oct-2013/OCT3	4.11
Brent crude	14-Mar-2014/APR4	13.62
Copper - Grade A	19-Mar-2014/MAR4	4.16
Corn	13-Dec-2013/DEC3	4.24
Gold	28-Apr-2014/APR4	6.39
Heating oil	31-Mar-2014/APR4	13.26
Light crude	20-Jun-2014/JUL4	14.64
Natural gas	26-Sep-2013/OCT3	4.88
RBOB gasoline	31-Oct-2013/NOV3	14.36
RBOB gasoline	29-Nov-2013/DEC3	0.14
Silver	27-Dec-2013/DEC3	1.33
Soybeans	14-Nov-2013/NOV3	5.26
Sugar #11	30-Sep-2013/OCT3	5.21
Wheat	14-Jul-2014/JUL4	4.10
Zinc	18-Dec-2013/DEC3	4.28

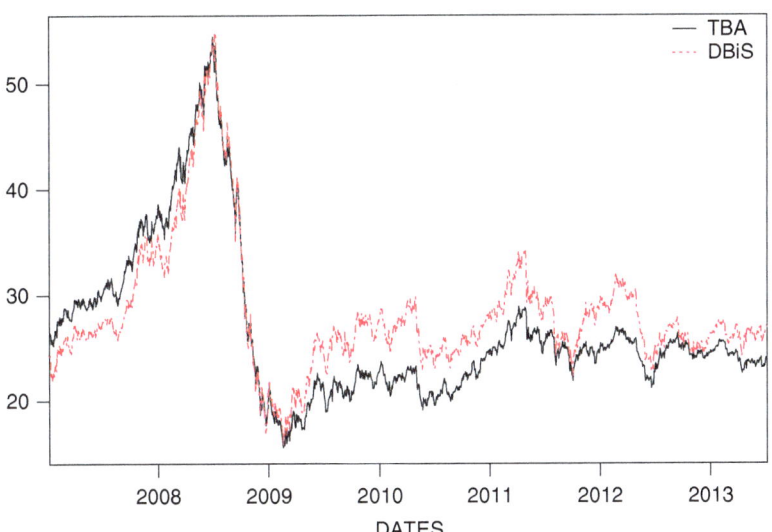

Fig. 9 Time series plot of the daily GSCI total return index (*black*) and Deutsche bank iShare DBiS (*red*)

The commodities included in the index are traded on the following futures exchanges:

- **NYMEX**: Light Sweet Crude Oil (WTI), Heating Oil, RBOB Gasoline and Natural Gas;
- **ICE Futures Europe**: Brent Crude;

- **Commodity Exchange NY**: Gold and Silver;
- **London Metal Exchange**: Aluminum, Zinc and Copper Grade A;
- **Chicago Board of Trade**: Corn, Wheat and Soybeans;
- **ICE Futures U.S.**: Sugar.

5.2 Dow Jones-UBS Roll Select Commodity Index

Deutsche Bank is by far not the only financial institution to have tried to capitalize on the attractiveness of the roll yield optimization. Indeed, a version of the Dow Jones-UBS Commodity Index was designed with the same goal in mind. Its goal is to mitigate the effects of contango on index performance. For each commodity included in the index, the roll algorithm chooses the futures contract (within the next nine maturity month available), which exhibits the most backwardation or least contango.

5.3 The UBS Bloomberg Constant Maturities Commodity Index

Partly motivated by the losses incurred by the traditional indexes in the recent contango period, the UBS Bloomberg Constant Maturity Commodity Index (CMCI) uses constant maturity contracts to provide diversification across maturity dates.

While the distribution of the relative weights across the sectors is not much different from the major commodity indexes, the goal of the index is to overload the diversification across the 28 commodities included in the index by a diversification across five constant maturities $\tau = 3mo$, $\tau = 6mo$, $\tau = 1Y$, $\tau = 2Y$ and $\tau = 3Y$, with weights varying with the commodities. The details are given in Tables 7 and 8.

5.4 Still More Commodity Index Rolling Down the Curve

The Credit Suisse Commodity Benchmark (CSCB) index is also a long-only index of commodities weighted by world production and liquidity. It is rebalanced

Table 7 Sector distribution of the UBS Bloomberg commodity indexes

Sector	UNS-Bloomberg CMCI weight (%)
Energy	36.3
Industrial metals	25.5
Precious metals	36.1
Agriculture	28.1
Livestock	4.0

Table 8 Composition of the UBS-Bloomberg constant maturities commodity index (Target weights H1-2013)

Sector	Commodity	Total weight (%)	Relative constant maturities weights				
			3mo (%)	6mo (%)	1Y (%)	2Y (%)	3Y (%)
Energy	Crude oil (Brent)	7.72	49.20	19.84	15.17	9.32	6.48
	WTI crude oil (NYMEX)	8.83	45.66	18.74	16.81	11.48	7.32
	WTI Crude oil (ICE)	3.45	44.98	20.86	16.21	10.82	7.13
	Heating oil	3.46	57.36	26.45	16.19		
	Gasoil	4.35	54.21	26.67	19.12		
	RBOB gasoline	4.21	69.37	30.63			
	Natural gas	4.37	48.57	22.39	15.34	7.87	5.83
Industrial metals	LME aluminum	6.71	34.84	21.85	19.50	14.09	9.72
	LME copper	9.18	30.65	21.01	22.85	15.94	9.55
	High grade copper	3.24	73.31	26.69			
	LME zinc	2.19	46.23	28.99	24.78		
	LME nickel	2.27	52.37	25.24	22.39		
	LME lead	1.29	50.98	27.75	21.28		
Precious metals	Gold	4.96	62.41	17.65	10.88	9.06	
	Silver	1.29	61.72	17.06	11.75	9.48	
Agriculture	SRW wheat	2.33	50.84	30.39	18.77		
	KCBOT HRW wheat	1.20	59.56	40.44			
	Corn	6.06	48.33	31.81	19.86		
	Soybeans	5.37	53.30	29.63	17.06		
	Soybean meal	1.33	63.73	36.27			
	Soybean oil	1.63	64.27	35.73			
	Sugar #11	4.62	41.77	35.90	22.33		
	Sugar #5	2.23	62.57	37.43			
	Cocoa	0.69	58.49	41.51			
	Coffee 'C'	1.32	57.96	28.41	13.63		
	Cotton	1.64	56.74	43.26			
Livestock	Live cattle	2.31	63.24	36.76			
	Lean hogs	1.75	62.50	37.50			

monthly, and contracts approaching maturity (starting 15 days prior to actual maturity) are rolled into equally weighted averages of the three contracts with maturities up to 3 months further out the term structure of forward prices.

Barclays has also a suite of exchange traded products tracking commodity indexes based on portfolios of futures contracts updated with optimized rolling algorithms. Among its many ETPs, the Barclays Capital Commodity Index Pure Beta TR and Barclays Capital Commodity Index TR are ETNs (iPath Pure Beta ETNs) tracking commodity indexes created by Barclay implementing a rolling algorithm involving varying expiration dates, typically choosing *at roll time*, the contract with the highest positive implied roll yield when the curve is backwarded or the lowest negative return when the forward curve is in contango.

We claim that the presence of these funds pushed the open interest down the curve, phenomenon which we now demonstrate in Sect. 6 devoted to a discussion of the impact of the financialization of commodities on open interest data.

6 Commodity Open Interest

So far, our discussion has been mostly concentrated on prices. We switch gear and turn our attention to two important variables whose values and changes can shed informative light on the future evolutions of prices. The first of these variables is volume. On any given day, and for each contract maturity, volume quantifies the trading activity in this particular contract. It provides a measure of the amount of contracts that have changed hands, the amounts of new positions open or closed for this specific maturity date. While a good indicator of the volatility of the market, it may not be as representative of economic fundamentals as it is of trader sentiments and behaviors. We choose to study open interest instead. On any given day, and for each contract maturity, open interest is the total number of outstanding contracts with that specific maturity that are held by market participants on that day. These numbers are often aggregated over the set of all maturities available for trading and a total open interest figure is given as the total number of outstanding contracts held by market participants on that day. We used this aggregate open interest for Crude Oil earlier in the chapter (recall Fig. 4) to illustrate the influx of investments over the period 2004–2009. We give the corresponding plot for Copper in Fig. 10 below.

6.1 The Term Structure of Open Interest

The purpose of this section is to demonstrate the changes in the term structure of open interest which occurred in the mid 2000s. Our contention is that open interest slid down the curve as investment in longer maturity contracts increased. We illustrate these claims with a close look at the two commodities we followed throughout the chapter: WTI crude oil and copper. While crude oil may have a seasonal component, it is not strong enough to overwhelm the features we are looking for. The same analysis would have been more difficult with natural gas.

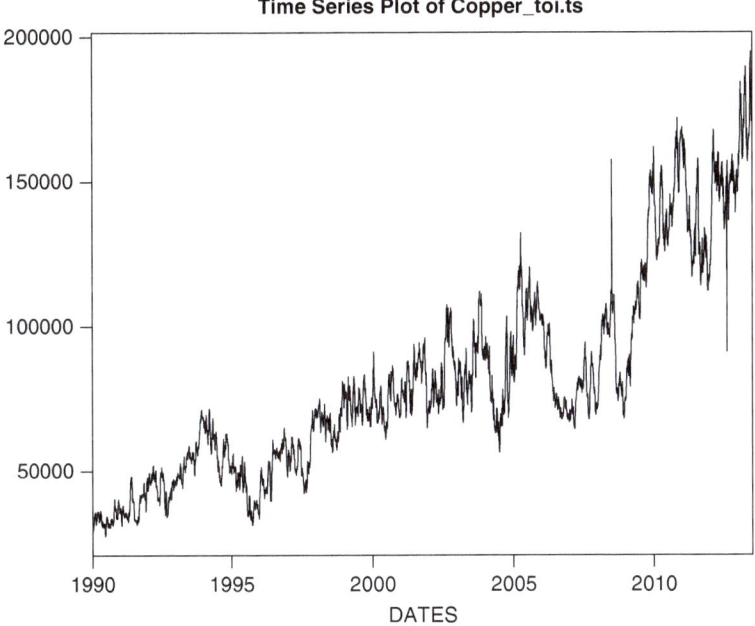

Fig. 10 Time series plot of the daily open interest in copper futures contracts between January 2, 1990 and September 9, 2013. Source: *Data Stream*

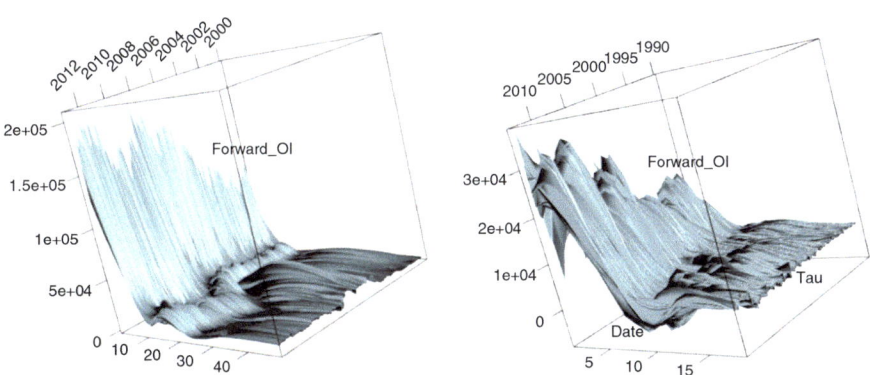

Fig. 11 Surface plot of the daily term structure of open interest for WTI crude oil (*left*) and copper (*right*) futures contracts between January 3, 1990 and July 7, 2013

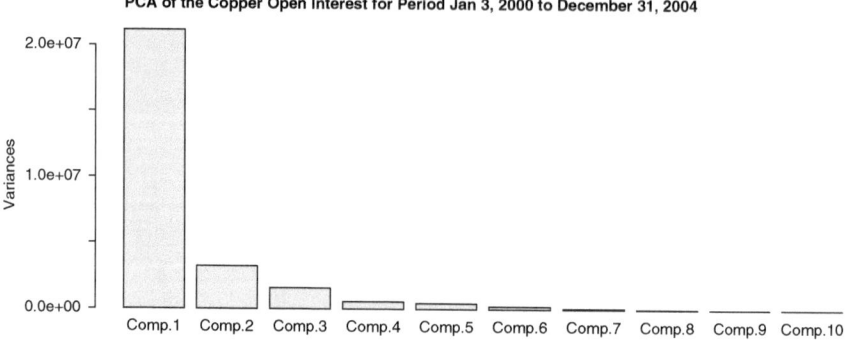

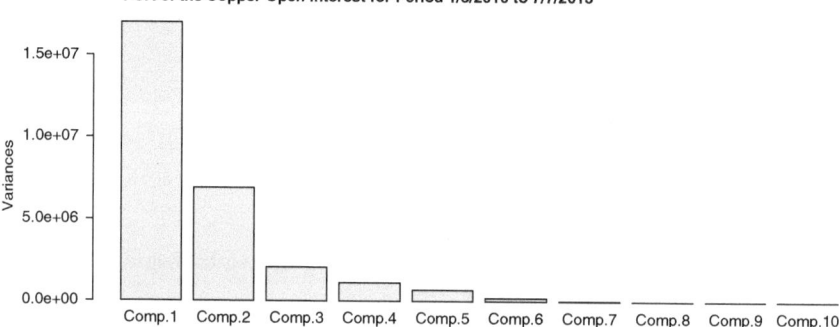

Fig. 12 Proportions of the variance explained by the loadings of the PCA of the open interest copper forward curves for the period January 3, 2000 to December 31, 2004 (*left*) and the period P_2 ranging from January 3, 2010 to July 7, 2013 (*right*)

Figure 11 gives the surface plots of the term structures of open interest for WTI crude oil and copper. The plot in the left pane clearly shows that the highest open interest is concentrated on the shortest available maturity (the variable Date being close to 0), and that for longer times to maturity, a secondary bump appears. However, the time evolution of the location of this bump shows a clear shift further down the curve in the mid 2000s. A similar phenomenon, though not as clean because of noise, can be observed in the right pane in the case of copper.

In order to provide one more graphical evidence of the open interest slide down the curve, we performed the PCA of the daily term structure of open interest over the two time periods considered so far. The results are reproduced in Fig. 12. Contrary to the daily changes in price, it appears that more factors are needed to explain the fluctuations over the second period. But looking at the loadings plotted in Fig. 13, we clearly see a shift to the right of the bumps representing where most of the open interest is expected.

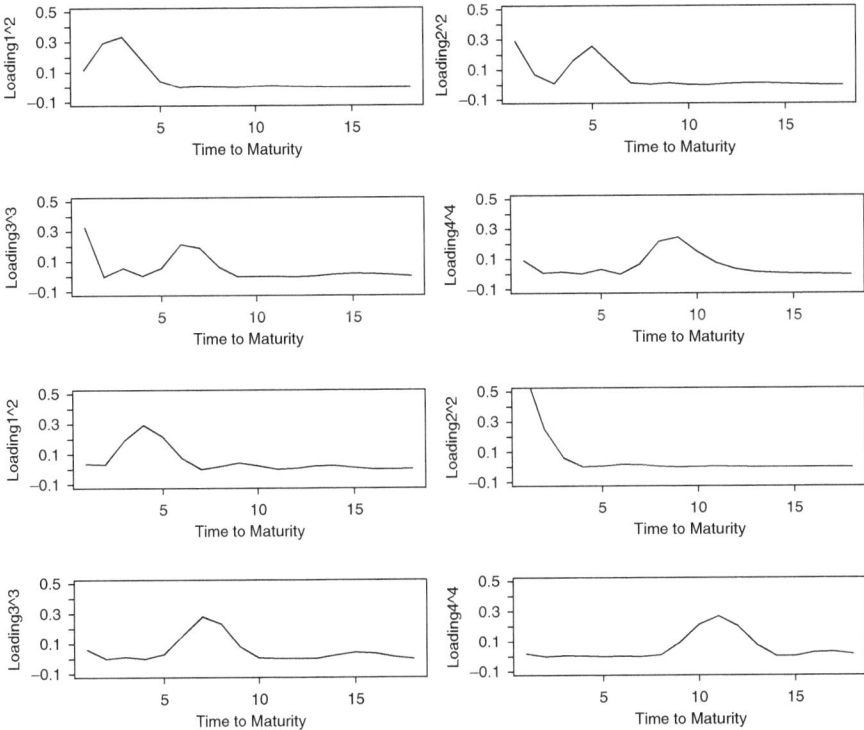

Fig. 13 Loadings of the PCA of the open interest forward curves of Copper for the period January 3, 2000 to December 31, 2004 (*left*) and the period ranging from January 3, 2010 to July 7, 2013 (*right*)

Acknowledgements The work of Prof. Carmona is partially supported by NSF - DMS 0806591. The author would like to thank the referee for a thorough reading of the first version of the manuscript, and for insightful suggestions which led to a more readable write up.

References

1. Basak, S., Pavlova, A.: A model of financialization of commodities. Technical Report, London Business School (2013)
2. Benth, J.S., Benth, F.E., Koekebakker, S.: Stochastic Modeling of Electricity and Related Markets. Advanced Series in Statistical Science & Applied Probability, vol. 11. World Scientific, Singapore (2008)
3. Bjerga, A.: Endless Appetites: How the Commodities Casino Creates Hunger and Unrest. Bloomberg Press, New Jersey (2011)
4. Buyuksahin, B., Harris, J.: Do speculators drive crude oil prices? Energy J. **32**, 167–202 (2011)
5. Buyuksahin, B., Robe, M.A.: Speculators, commodities and cross-market linkages. Technical Report, 2012

6. Carmona, R.: Statistical Analysis of Financial Data in R-. Springer Texts in Statistics, 2nd edn. Springer, New York (2014)

7. Carmona, R., Coulon, M.: A survey of commodity markets and structural approaches to modeling electricity. In: Benth, F. (ed.) Energy Markets, Proceedings of the WPI Special Year, pp. 1–42. Springer, New York (2013)

8. Carter, C.A., Rausser, G.C., Schmitz, A.: Efficient asset portfolios and the theory of normal backwardation. J. Polit. Econ. **91**, 319–331 (1983)

9. Cheng, I.H., Xiong, W.: The financialization of commodity markets. Ann. Rev. Financ. Econ. **6**, 419–441 (2014)

10. Dennis, P.J., Strickland, D.: Who blinks in volatile markets, individuals or institutions? J. Finance **57**, 1923–1949 (2002)

11. Food & Water Watch: Casino of Hunger: How Wall street speculators fueled the global food crisis. Technical Report (2009)

12. Gabaix, X., Gopikrishnan, P., Plerou, V., Stanley, H.E.: Institutional investors and stock market volatility. Q. J. Econ. **121**, 461–504 (2006)

13. Hamilton, J.D.: Causes and consequences of the oil shock of 2007–08. Brookings Papers on Economic Activity, pp. 215–259 (2009)

14. Hamilton, J.D., Wu, C.: Effects of index-fund investing on commodity futures prices. Technical Report, U.C. San Diego (2012)

15. Henderson, B., Pearson, N., Wang, L.: New evidence on the financialization of the commodity markets. Technical Report, George Washington University, July 22 (2013)

16. Irwin, S.H., Sanders, D.R.: Index funds, financialization, and commodity futures markets. Appl. Econ. Perspect. Pol. **33**, 1–31 (2011)

17. Keynes, J.M.: Some Aspects of Commodity Markets. Manchester Guardian Commercial: European Reconstruction Series, Sect. 13, March 29 (1923)

18. Keynes, J.M.: A Treatise of Money, vol. 2. Macmillan, London (1950)

19. Kilian, L.: Not all oil price shocks are alike: disentangling demand and supply shocks in the crude oil market. Am. Econ. Rev. **99**, 1053–1069 (2009)

20. Kilian, L., Murphy, D.P.: The role of inventories and speculative trading in the global market for crude oil. J. Appl. Econ. **29**(3) 454–478 (2013)

21. Kilian, L., Fattouh, B., Mahadeva, L.: The role of speculation in oil markets: what have we learned so far? Technical report, University of Michigan (2012)

22. Krugman, P.: Fuels on the hill. The New York Times, June 27, 2008

23. Kyle, A., Xiong, W.: Contagion as a wealth effect. J. Finance **56**, 1401–1440 (2001)

24. Litterman, R., Scheinkman, J.: Common factors affecting bond returns. J. Fixed Income **1**, 49–53 (1991)

25. Lombardi, M.J., Van Robays, I.: Do financial investors destabilize the oil price? Technical Report, European Central Bank (2011)

26. Morana, C.: Oil price dynamics, macro-finance interactions and the role of financial speculation. J. Bank. Finance **37**(1), 206–226 (2013)

27. Mou, Y.: Limits to arbitrage and commodity index investment: front-running the goldman roll. Technical Report, Columbia University (2011)

28. Sanders, D.R., Irwin, S.H.: A speculative bubble in commodity futures prices? Cross-sectional evidence. Agric. Econ. **41**, 25–32 (2010)

29. Sanders, D.R., Irwin, S.H.: The impact of index funds in commodity futures markets. J. Altern. Invest. **14**, 40–49 (2011)

30. Sanders, D.R., Irwin, S.H.: New evidence on the impact of index funds in US grain futures markets. Can. J. Agric. Econ. **59**(4), 519–532 (2011)

31. Singleton, K.: Investor flows and the 2008 boom/bust in oil prices. Technical Report, Stanford University (2012)

32. Tang, K., Xiong, W.: Index investment and financialization of commodities. Financ. Anal. J. **68**, 54–74 (2012)

33. UNCTAD: Price formation in financialized commodity markets: the role of information. Technical Report UNCTAD/GDS/2011/1, United Nations, New York and Geneva (2011)
34. UNCTAD: Don't blame the physical markets: financialization is the root cause of oil and commodity price volatility. Technical Report Policy Brief No. 25, United Nations, New York and Geneva (September 2012)
35. U.S. Commodity Futures Trading Commission: Disaggregated commitments of traders report, exploratory notes. Technical Report (2014)

Understanding the Tracking Errors of Commodity Leveraged ETFs

Kevin Guo and Tim Leung

Abstract Commodity exchange-traded funds (ETFs) are a significant part of the rapidly growing ETF market. They have become popular in recent years as they provide investors access to a great variety of commodities, ranging from precious metals to building materials, and from oil and gas to agricultural products. In this article, we analyze the tracking performance of commodity leveraged ETFs and discuss the associated trading strategies. It is known that leveraged ETF returns typically deviate from their tracking target over longer holding horizons due to the so-called volatility decay. This motivates us to construct a benchmark process that accounts for the volatility decay, and use it to examine the tracking performance of commodity leveraged ETFs. From empirical data, we find that many commodity leveraged ETFs underperform significantly against the benchmark, and we quantify such a discrepancy via the novel idea of *realized effective fee*. Finally, we consider a number of trading strategies and examine their performance by backtesting with historical price data.

1 Introduction

The advent of commodity exchange-traded funds (ETFs) has provided both institutional and retail investors with new ways to gain exposure to a wide array of commodities, including precious metals, agricultural products, and oil and gas. All commodity ETFs are traded on exchanges like stocks, and many have very high liquidity. For example, the SPDR Gold Trust ETF (GLD), which tracks the daily London gold spot price, is the most traded commodity ETF with an average trading volume of 8 million shares and market capitalization of US $31 billion in 2013.[1]

---

[1] According to ETF Database website (http://www.etfdb.com/compare/volume).

K. Guo • T. Leung (✉)
Industrial Engineering & Operations Research (IEOR) Department, Columbia University,
New York, NY 10027, USA
e-mail: klg2138@columbia.edu; tl2497@columbia.edu

© Springer Science+Business Media New York 2015
R. Aïd et al. (eds.), *Commodities, Energy and Environmental Finance*,
Fields Institute Communications 74, DOI 10.1007/978-1-4939-2733-3_2

Within the commodity ETF market, some funds are designed to track a constant multiple of the daily returns of a reference index or asset. These are called leveraged ETFs (LETFs). An LETF maintains a constant leverage ratio by holding a variable portfolio of assets and/or derivatives, such as futures and swaps, based on the reference index. For example, the Dow Jones U.S. Oil & Gas Index (DJUSEN) or the Dow Jones U.S. Basic Materials Index (DJUSBM) and their associated ETFs track the stocks of a basket of commodities producers, as opposed to the physical commodity prices. On the other hand, most LETFs are based on total return swaps and commodity futures. The most common leverage ratios are ± 2 and ± 3, and LETFs typically charge an expense fee. Major issuers include ProShares, iShares, VelocityShares and PowerShares (see Table 1). For example, the ProShares Ultra Long Gold (UGL) seeks to return 2x the daily return of the London gold spot price minus a small expense fee. One can also take a bearish position by buying shares of an LETF with a negative leverage ratio. The ProShares Ultra Short Gold (GLL) is an inverse LETF that tracks $-2x$ the daily return of the London gold fixing price.

Table 1 A summary of the 23 LETFs studied in this paper, arranged by commodity type and then leverage

LETF	Reference	Underlying	Issuer	β	Fee (%)	Inception
SLV	SLVRLN	Silver bullion	iShares	1	0.50	04/21/2006
AGQ	SLVRLN	Silver bullion	ProShares	2	0.95	12/01/2008
ZSL	SLVRLN	Silver bullion	ProShares	-2	0.95	12/01/2008
USLV	SPGSSIG	Silver bullion	VelocityShares	3	1.65	10/13/2011
DSLV	SPGSSIG	Silver bullion	VelocityShares	-3	1.65	10/14/2011
GLD	GOLDLNPM	Gold bullion	iShares	1	0.40	11/18/2004
UGL	GOLDLNPM	Gold bullion	ProShares	2	0.95	12/01/2008
GLL	GOLDLNPM	Gold bullion	ProShares	-2	0.95	12/01/2008
UGLD	SPGSGCP	Gold bullion	VelocityShares	3	1.35	10/13/2011
DGLD	SPGSGCP	Gold bullion	VelocityShares	-3	1.35	10/14/2011
IYE	DJUSEN	Oil & gas	iShares	1	0.48	06/12/2000
DDG	DJUSEN	Oil & gas	ProShares	-1	0.95	06/10/2008
DIG	DJUSEN	Oil & gas	ProShares	2	0.95	01/30/2007
DUG	DJUSEN	Oil & gas	ProShares	-2	0.95	01/30/2007
DBO	DBOLIX	WTI crude oil	PowerShares	1	0.75	01/05/2007
UCO	DJUBSCL	WTI crude oil	ProShares	2	0.95	11/24/2008
SCO	DJUBSCL	WTI crude oil	ProShares	-2	0.95	11/24/2008
UWTI	SPGSCLP	WTI crude oil	VelocityShares	3	1.35	02/06/2012
DWTI	SPGSCLP	WTI crude oil	VelocityShares	-3	1.35	02/06/2012
IYM	DJUSBM	Building materials	iShares	1	0.48	06/12/2000
SBM	DJUSBM	Building materials	ProShares	-1	0.95	03/16/2010
UYM	DJUSBM	Building materials	ProShares	2	0.95	01/30/2007
SMN	DJUSBM	Building materials	ProShares	-2	0.95	01/30/2007

Notice that the non-leveraged (1x) ETFs have the smallest expense fees, and LETFs with higher absolute leverage ratios, $|\beta| \in \{2, 3\}$, tend to have higher expense fees. Finally, notice that higher β LETFs are much more recent additions to the market

LETFs are a highly accessible and liquid instrument, thereby making them attractive instruments for traders who wish to gain leveraged exposure to a commodity without borrowing money or using derivatives.

For a long LETF, with a leverage ratio $\beta > 0$, the fund must add to a winning position in a bull market to maintain a constant leverage ratio. On the other hand, during a bear market, the fund must sell its losing positions to maintain the same leverage ratio. Similar arguments can be made for short (or inverse) LETFs ($\beta < 0$). As a consequence, LETFs can potentially outperform β times its reference during periods of market trending. However, should the LETF exhibit high volatility but no significant movement in price over a period of time, the constant daily re-balancing would cause the fund to decline in value. Therefore, LETFs can be viewed as long momentum but short volatility, and the value erosion due to realized variance of the reference is called *volatility decay* (see [2–4]). This raises the important question of how well do LETFs perform over a long horizon.

Since their introduction to the market, LETFs a number of criticisms from both practitioners and regulators.[2] Some are concerned that the returns of LETFs exhibit some discrepancies from the goals stated in their prospectuses. In fact, some issuers provide warnings that LETFs are unsuitable for long-term buy-and-hold investors.

Many existing studies focus on equity-based ETFs and their leveraged counterparts. For example, Avellaneda and Zhang [2] study the price behavior and discuss the volatility decay of equity LETFs in different sectors. They find minimal 1-day tracking errors among the most liquid equity ETFs. They explain that an equity LETF can replicate the leveraged returns of its reference through a dynamic portfolio consisting of the component equities.

In contrast, commodities are unique because the physical assets cannot be stored easily. As such, ETF issuers are required to replicate through either warehousing,[3] which is very costly, and thus uncommon except for precious metals such as silver and gold, or trading futures with multiple counterparties (see [5]). Since the reference indices may represent the spot prices of physical commodities, futures-based commodity ETFs may fail to track their reference indices perfectly and their tracking performance is subject to the fluctuation and term structure of futures prices. On top of that, most commodity LETFs use over-the-counter (OTC) total return swaps with multiple counterparties to generate the required leverage ratios. The lower liquidity of OTC contracts and counterparty risk can contribute to additional tracking errors. As we show in this paper, tracking errors can seriously affect the long-term fund performance of LETFs.

In a related work, Murphy and Wright [12] perform a *t*-test based on 1-day returns to determine if any commodity LETF has a non-zero tracking error. They conclude that all LETFs have a very good daily tracking performance. However,

---

[2]In 2009, the SEC and FINRA issued an alert on the risk of leveraged ETFs on http://www.sec.gov/investor/pubs/leveragedetfs-alert.htm.

[3]For more details on the issue of storage cost for commodity ETFs, we refer to the Morningstar Report: "An Ugly Side to Some Commodity ETFs" by Bradley Kay, August 19, 2009.

they do not conduct the analysis over a longer horizon, or account for the volatility decay. There is also no discussion of trading strategies there. On the other hand, Guedj et al. [5] discuss the difficulties faced by an ETF provider in replicating a commodity index using futures. In particular, they point out that the term structure of futures may lead to large deviations between the ETF price and the spot price of a commodity.

In this paper, we analyze the tracking performance of commodity leveraged ETFs. Through a series of regression analyses, we illustrate how the returns of commodity LETFs deviate from the reference returns multiplied by the leverage ratio over different holding periods. In particular, the average tracking error tends to turn more negative over a longer horizon and for higher leveraged ETFs. With in mind that realized variance of the reference can erode the LETF value, we examine the over/under-performance of LETFs with respect to a benchmark that incorporates the effect of volatility decay. From empirical data, we find that many commodity leveraged ETFs in our study underperform significantly against the benchmark, and we quantify such a discrepancy by introducing the *realized effective fee*. Finally, we consider a static trading strategy that involves shorting two LETFs with leverage ratios of different signs, and study its performance and dependence on the realized variance of the reference. We find that the resulting portfolio is always long realized variance both theoretically and empirically, but is also exposed to the tracking errors associated with the two LETFs. We also backtest the strategy through examining its empirical returns over rolling periods.

The rest of the paper is organized as follows. In Sect. 2, we analyze the returns of commodity LETFs over different holding periods and illustrate horizon dependence of tracking errors. In Sect. 3, we use a benchmark process that incorporates the realized variance of the reference to study the over/under-performance of each LETF. In Sect. 4, we discuss a static trading strategy and backtest using historical data. Section 5 concludes the paper and points out a number of directions for future research.

2 Analysis of Tracking Error

We first compare the returns of LETFs and their reference indices. For every ETF, we obtain its closing prices and reference index values from Bloomberg for the period Dec 2008–May 2013. We then calculate the n-day returns from $n = \{1, 2, \ldots, 30\}$ using disjoint successive periods (e.g. the return over days 1–30 then returns over days 31–60 for 30-day returns). Let L_t be the price of an LETF and S_t be the reference index value at time t. For a given leverage ratio β, we compare the log-returns of the LETF to β times the log-returns of the corresponding reference index. This leads us to define the n-day tracking error at time t by

$$Y_t^{(n)} = \ln \frac{L_{t+n\Delta t}}{L_t} - \beta \ln \frac{S_{t+n\Delta t}}{S_t}, \tag{1}$$

where Δt represents one trading day. We explore the empirical distribution of the n-day tracking error, and then analyze the effect of holding horizon on the magnitude of tracking errors. We remark there are alternative ways to define tracking errors for ETFs. For example, one can consider the difference in relative returns as opposed to log-returns, or the root mean square of the daily differences (see [10]).

2.1 Regression of Empirical Returns

We conduct a regression between log-returns of the LETF and its reference index based on the linear model:

$$\ln \frac{L_t}{L_0} = \hat{\beta} \ln \frac{S_t}{S_0} + \hat{c} + \epsilon, \tag{2}$$

where $\epsilon \sim N(0, \sigma^2)$ is independent of the reference index value S_t, $\forall t \geq 0$. In other words, we run an ordinary least square 1-variable regression between the log-returns for every fixed horizon of n days. Then, we increase the holding period from 1 to 30 days, and observe how the regression coefficients vary.

We display the regression results in Figs. 1, 2, 3, and 4 for log-returns over periods of 1, 5, 10, and 20 days. To avoid dependence among returns, we use disjoint time intervals to calculate returns. For example, we use $\frac{S_{20}}{S_0}, \frac{S_{40}}{S_{20}} \ldots$ and $\frac{L_{20}}{L_0}, \frac{L_{40}}{L_{20}} \ldots$ for 20-day log-returns as the inputs for the regression.

In Fig. 1, the regression coefficient $\hat{\beta}$ for DIG ($\beta = 2$, oil & gas) increases from 2 to 2.1 as the holding period lengthens from 1 to 20 days. Although the coefficient of determination R^2 is close to 99 % for up to 20 days, it is highest for 1-day returns. In Fig. 2 for DUG ($\beta = -2$, oil & gas), one again observes $\hat{\beta}$ increasing, and R^2 decreasing. For DUG ($\beta = -2$, oil & gas), as n varies from 1 to 20, $\hat{\beta}$ increases from -2 to -1.66. As a result, this implies that DIG ($\beta = 2$, oil & gas) effectively gains leverage as the holding time increases, while DUG ($\beta = -2$, oil & gas) loses leverage compared to the advertised fund β.

On the other hand, UGL ($\beta = 2$, gold) and GLL ($\beta = -2$, gold) exhibit very different return behaviors. In Fig. 3 the R^2 for UGL ($\beta = 2$, gold) is surprisingly worst for the shortest holding period of 1 day, whereas it increases to 95 % over a holding period of 20 days. In Fig. 4 for GLL ($\beta = -2$, gold), the R^2 increases from 35 % to 96 % when holding the fund from 1 to 20 days. Furthermore, the estimators $\hat{\beta}$ for UGL ($\beta = 2$, gold) and GLL ($\beta = -2$, gold) both slowly approach their advertised $\beta = \pm 2$. The variation of $\hat{\beta}$ for DIG ($\beta = 2$, oil & gas) and UGL ($\beta = 2$, gold) over different holding periods is summarized in Fig. 5.

We observe that LETFs that track an illiquid reference, such as the gold bullion index GOLDLNPM, tend to have more tracking errors than those tracking a liquid index, such as the oil & gas index DJUSEN. The oil & gas commodity LETFs

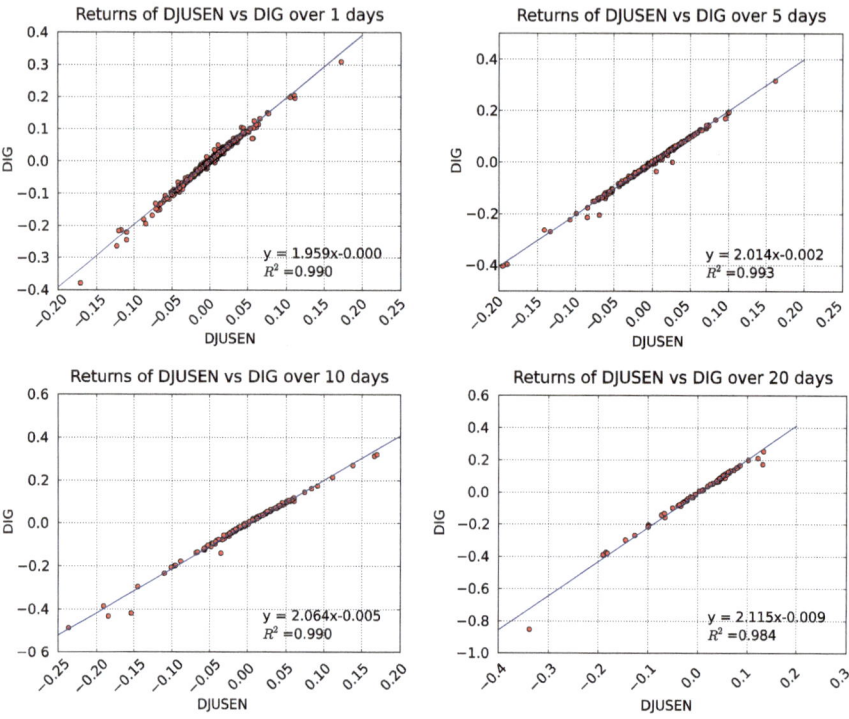

Fig. 1 *From top left to bottom right*: regression of DJUSEN-DIG ($\beta = 2$, oil & gas) 1, 5, 10, 20-day log-returns. We consider disjoint periods from Dec 2008 to May 2013

involve exchange-traded futures which are liquid proxy to the spot price. The gold and silver bullion LETFs consist of OTC total return swaps. The difficulty and higher costs replication using swaps, as well as infrequent (typically daily) update of the swaps' mark-to-market values can weaken the fund's tracking ability. For example, the 1-day regressions of UGL and GLL ($\beta = \pm 2$, gold) yield R^2 values less than 40 %, while DIG and DUG ($\beta = \pm 2$, oil & gas) have 1-day R^2 values of over 90 %. On the other hand, full physical replication yields the greatest R^2, with examples of the non-leveraged gold and silver ETFs, GLD and SLV, respectively. Hence, the replication strategy can significantly affect a fund's tracking errors. A more precise understanding of the effectiveness of swaps, futures, and other replication strategies requires the full holdings history from the ETF provider, which is not publicly available at all times.[4]

---

[4]For a detailed snapshot of the holdings for a proshares ETF, please see http://www.proshares.com/funds/XYZ_daily_holdings.html where {*XYZ*} is the ETF ticker.

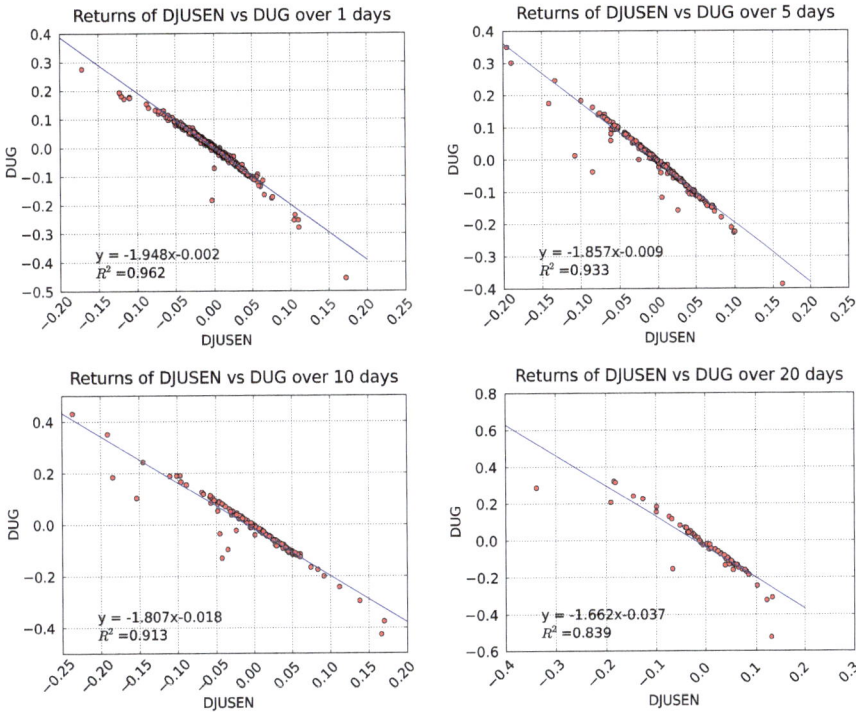

Fig. 2 *From top left to bottom right*: regression of DJUSEN-DUG ($\beta = -2$, oil & gas) 1, 5, 10, 20-day log-returns. We consider disjoint periods from Dec 2008 to May 2013

In addition, the LETFs we studied have an increasingly negative constant coefficient $\hat{c}$ as the holding time increases. For example, over a holding period of 20-days, DUG ($\beta = -2$, oil & gas) has a 3 % decay on returns compared to β times its reference index. We would expect this phenomenon, however, since the LETF would need to buy high and sell low, while the reference investor would simply hold his securities. Therefore, the longer the LETF is held, the more likely the fund will underperform against β times the reference index. As we will see in Sect. 3, the constant coefficient $\hat{c}$ depends on two factors, the expense fee charged by the issuer as well as the realized variance of the reference index.

Hence, with this simple linear model for LETF prices, we have observed that although LETFs safely replicate β times the reference over short holding periods, they begin to exhibit negative tracking error and deviations in their leverage ratios β as the holding time increases. Furthermore, we see that LETFs which attempt to track illiquid spot prices perform much more poorly than expected. We conclude that more factors must be considered when modeling LETF returns.

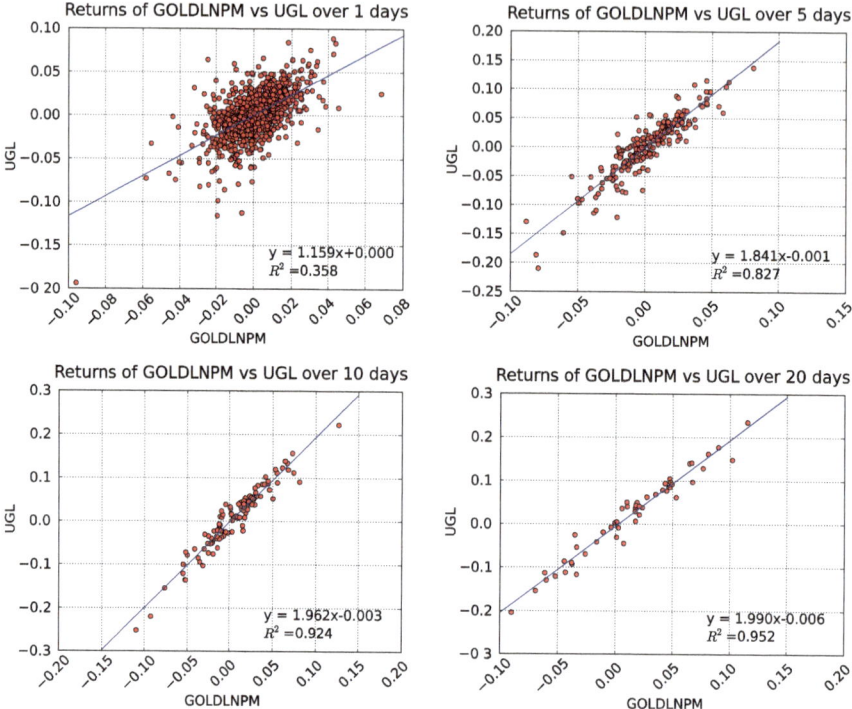

Fig. 3 *From top left to bottom right*: regression of GOLDLNPM-UGL ($\beta = 2$, gold) 1, 5, 10, 20-day log-returns. We consider disjoint periods from Dec 2008 to May 2013

2.2 Distribution of Tracking Errors

As defined in (1), the tracking error is the difference between the LETF's log-return and the corresponding multiple of its reference index's log-return. In this section, we examine the distribution of the tracking error. This provides a picture of the LETF's efficiency in its stated goal of replicating the leveraged return of a reference index.

For the 23 LETFs in Table 2, we compute the mean μ and standard deviation σ for the tracking errors using available price data during the period Dec 2008 to May 2013. For all these funds, the mean 1-day tracking error has $\mu \approx 0$, ranging from 0 % to −0.27 %. Therefore, all these LETFs on average successfully replicate the stated multiple β of the daily reference return, with a slight negative bias. In fact, many LETFs even continued to replicate returns over periods as long as 10 days. However, as the holding time increases, the average tracking error grows more negative, so that the LETF in fact underperforms its intended goal over longer holding periods (see Fig. 6).

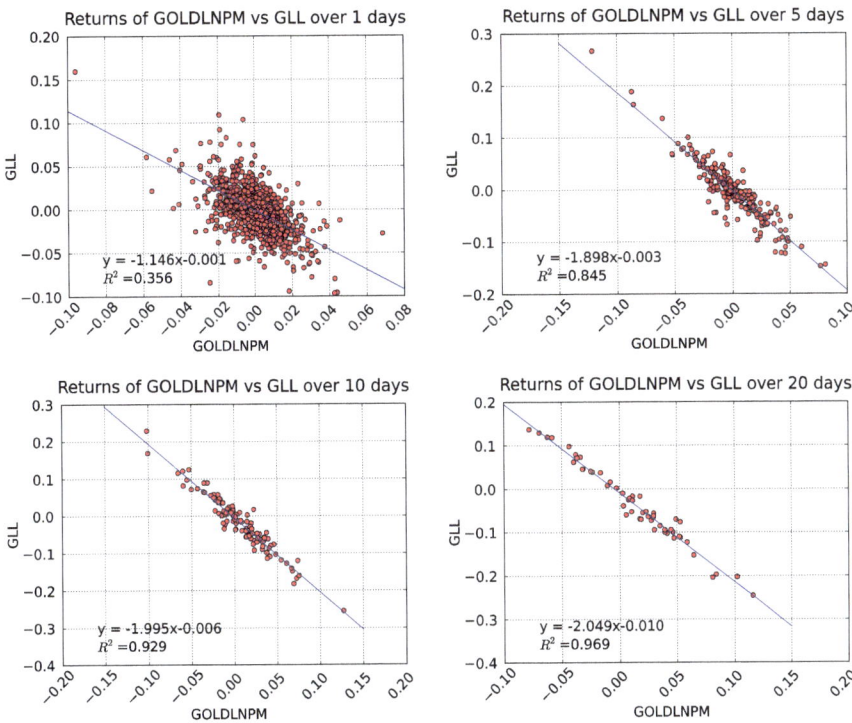

Fig. 4 *From top left to bottom right*: regression of GOLDLNPM-GLL ($\beta = -2$, gold) 1, 5, 10, 20-day log-returns. We consider disjoint periods from Dec 2008 to May 2013

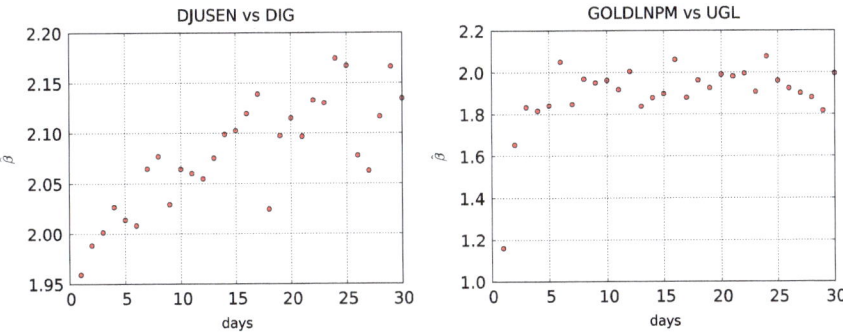

Fig. 5 The estimated $\hat{\beta}$ from the regressions for DJUSEN-DIG ($\beta = 2$, oil & gas), and GOLDLNPM-UGL ($\beta = 2$, gold)

Interestingly, the tracking errors for the silver and gold LETFs (AGQ, ZSL ($\beta = \pm2$, silver); UGL, GLL ($\beta = \pm2$, gold)) in Table 2 have σ several magnitudes higher than μ. For example, AGQ ($\beta = 2$, silver) has a tracking error σ of 5 % compared to a μ of 0.01 %. In other words, these four LETFs, while they might track

Table 2 Mean μ and standard deviation σ of the 1-day tracking error by commodity

LETF	Underlying	β	μ	σ
SLV	Silver bullion	1	0.0000	0.0302
AGQ	Silver bullion	2	−0.0009	0.0539
ZSL	Silver bullion	−2	−0.0022	0.0543
USLV	Silver bullion	3	−0.0014	0.0231
DSLV	Silver bullion	−3	−0.0027	0.0237
GLD	Gold bullion	1	0.0000	0.0128
UGL	Gold bullion	2	−0.0003	0.0221
GLL	Gold bullion	−2	−0.0005	0.0221
UGLD	Gold bullion	3	−0.0006	0.0134
DGLD	Gold bullion	−3	−0.0010	0.0139
IYE	Oil & gas	1	0.0000	0.0049
DDG	Oil & gas	−1	−0.0008	0.0118
DIG	Oil & gas	2	−0.0005	0.0044
DUG	Oil & gas	−2	−0.0018	0.0087
DBO	WTI crude oil	1	0.0000	0.0070
UCO	WTI crude oil	2	−0.0006	0.0135
SCO	WTI crude oil	−2	−0.0016	0.0132
UWTI	WTI crude oil	3	−0.0008	0.0147
DWTI	WTI crude oil	−3	−0.0017	0.0178
IYM	Building materials	1	0.0000	0.0020
SBM	Building materials	−1	−0.0004	0.0065
UYM	Building materials	2	−0.0005	0.0062
SMN	Building materials	−2	−0.0022	0.0149

their references well on average, may also exhibit positive and negative deviations over 1-day holding periods as well. These observations are consistent with the regressions in Figs. 3 and 4, where UGL and GLL ($\beta = \pm 2$, gold) show significant 1-day tracking errors. On the other hand, the non-leveraged gold and silver bullion ETFs, GLD and SLV, have almost no tracking error $\sigma \approx 0$, because they hold the underlying bullion according to their prospectuses. Since many investors use these ETFs to gain leveraged exposure to commodities, they should be aware of the large variance of the associated tracking errors.

In Fig. 6, we show the histogram for the tracking error for each ETF along with a quantile-quantile plot to illustrate the distribution. For DIG and DUG ($\beta = \pm 2$, oil & gas), the quantile-quantile plot shows that the tracking error distribution is not quite normal, and has a large negative tail, so that the commodity LETF tracking error is negatively biased even for the shortest possible holding period of 1 day. On the other hand, for UGL, GLL ($\beta = \pm 2$, gold) the distribution appears to be normal with R^2 close to 98 %. However, as noted in Table 2, the tracking errors for UGL and GLL ($\beta = \pm 2$, gold) also have a very large variance.

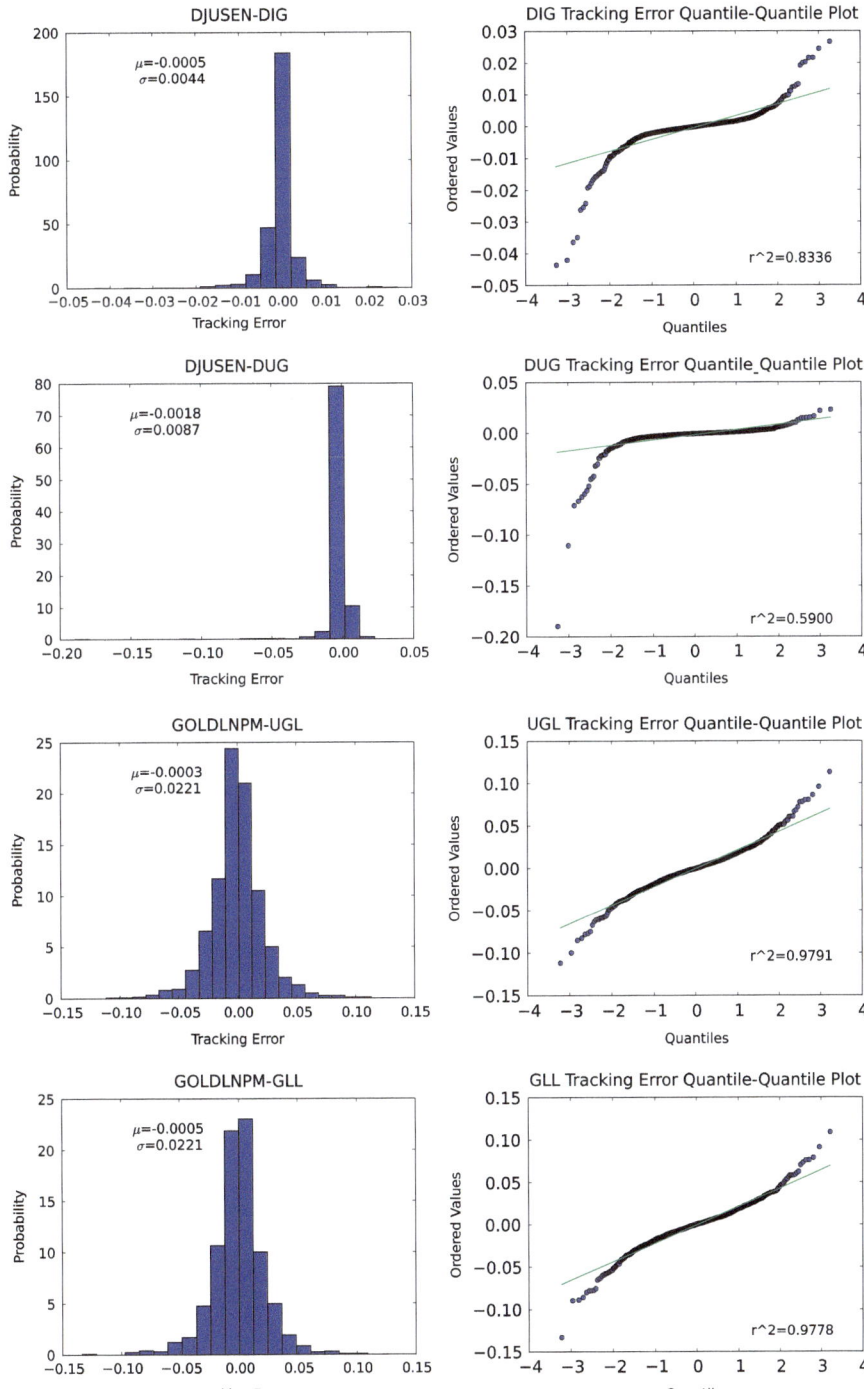

Fig. 6 Histograms and QQ plots of 1-day tracking errors for DIG, DUG ($\beta = \pm 2$, oil & gas); UGL, GLL ($\beta = \pm 2$, gold) *from top to bottom*

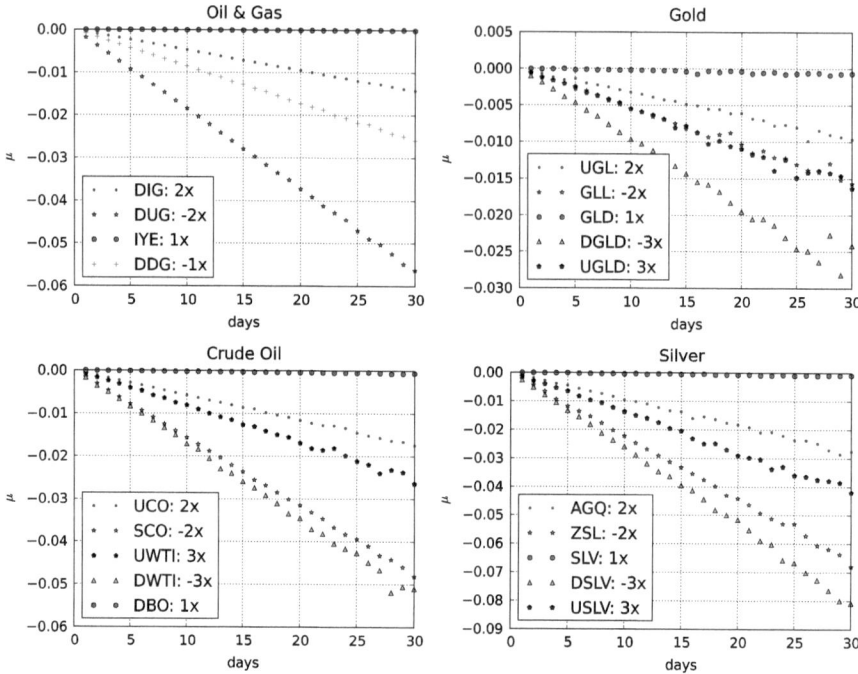

Fig. 7 A plot of no. of days vs the mean tracking error arranged by commodities tracked. *From top left to bottom right*: US Oil & Gas, Gold, Crude Oil,and Silver. As the holding period increases, the average tracking error becomes more negative as well

Next, we examine the horizon effect of tracking errors. Figure 7 indicates that higher leveraged ETFs tend to have more negative average tracking errors, which appear to be decreasing linearly over longer holding periods. In addition, negative leveraged LETFs have a more negative average tracking error than their positive counterparts. For example, in Fig. 7, GLL ($\beta = -2$, gold) has a lower slope than UGL ($\beta = 2$, gold) even though they have the same absolute value of leverage ratio $|\beta|$. Furthermore, with few exceptions, the average tracking error is most negative when $\beta = -3$ followed by $\beta = 3, -2, 2, -1, 1$. Thus, there is a higher holding horizon punishment for buying short than long LETFs.

Our analysis of the tracking error distribution reveals several characteristics of the tracking error defined in (1). Over a very short holding period, most LETFs perform close to their objectives stated in their prospectuses. Nevertheless, the realized tracking error varies over time, and can be positive or negative. For gold and silver LETFs, the tracking error is more volatile. Moreover, the magnitude of the mean tracking error depends heavily on the β of the LETF, with bear LETFs suffering a higher penalty than bull LETFs.

3   Incorporating Realized Variance into Tracking Error Measurement

As is well known in the industry (see [2, 3]), the price dynamics of an LETF depends on the realized variance of the reference index. This leads us to incorporate the realized variance in measuring the performance of an LETF. We run a regression analysis based on empirical LETF and reference prices that incorporates the realized variance as an independent variable. We then derive a realized effective fee associated with each LETF and analyze the realized price behavior relative to a theoretical benchmark to better quantify the over/under-performance.

3.1   Model for the LETF Price

Let S_t be the price of the reference index, and L_t be the price of the LETF at time t. Also denote f as the expense rate, r as the interest rate and β as the leverage ratio. Assume the reference asset follows the SDE

$$\frac{dS_t}{S_t} = \mu_t dt + \sigma_t dW_t, \quad t \geq 0, \tag{3}$$

with stochastic drift $(\mu_t)_{t \geq 0}$ and volatility $(\sigma_t)_{t \geq 0}$. For our analysis herein, we assume a general diffusion framework, but do not need to specify a parametric model. Many well-known models, including the CEV, Heston, and exponential Ornstein-Uhlenbeck models, fit within the above framework.

A long β-LETF L can be constructed through a dynamic portfolio. Specifically, the portfolio at time t consists of the cash amount $\$\beta L_t$ invested in the reference index S_t, while $\$(\beta - 1)L_t$ is borrowed at the positive risk free rate r. As a result, the LETF satisfies the SDE

$$dL_t = L_t \beta \frac{dS_t}{S_t} - L_t((\beta - 1)r + f)dt. \tag{4}$$

Solving the SDE, the log-return of the LETF is given by

$$\ln \frac{L_t}{L_0} = \beta \ln \frac{S_t}{S_0} + \frac{\beta - \beta^2}{2} V_t + ((1 - \beta)r - f)t, \tag{5}$$

where

$$V_t = \int_0^t \sigma_s^2 ds \tag{6}$$

is the realized variance of S accumulated up to time t. Therefore, under this general diffusion model, the log-return of the LETF is proportional to the log-return of the reference index by a factor of β, but also proportional to the variance by a factor of $\frac{\beta-\beta^2}{2}$. The latter factor is negative if $\beta \notin (0, 1)$, which is true for every LETF traded on the market. Also, the expense fee f reduces the return of the LETF.

Our regression analysis will focus on testing the functional form (5). We observe from (5) that the functional form of L_t in terms of S_t and V_t holds for any parametric model within the diffusion framework in (3). Considering the daily LETF returns, we set $\Delta t = \frac{1}{252}$ as one trading day. Let R_t^S be the daily return of the reference index at time t. At any time t, the n-day log-returns of an LETF follows

$$\ln \frac{L_{t+n\Delta t}}{L_t} = \beta \ln \frac{S_{t+n\Delta t}}{S_t} + \frac{\beta - \beta^2}{2} V_t^{(n)} + ((1 - \beta)r - f)n\Delta t, \tag{7}$$

$$V_t^{(n)} = \sum_{i=0}^{n-1}(R_{t+i\Delta t}^S - \bar{R}_t^{\,S})^2, \quad \bar{R}_t^{\,S} = \frac{1}{n}\sum_{i=0}^{n-1}R_{t+i\Delta t}^S. \tag{8}$$

This serves as a benchmark process for our subsequent analysis.

3.2   Regression of Empirical Returns

The log-return equation (7) suggests a regression with two predictors: the log-returns and the realized variance of the reference over n-days. This results in the linear model

$$\ln \frac{L_t}{L_0} = \hat{\beta} \ln \frac{S_t}{S_0} + \hat{\theta}V_t + \hat{c} + \epsilon, \tag{9}$$

where $\hat{c}$ is a constant intercept to be determined, and $\varepsilon \sim N(0, \sigma^2)$ is independent of $(S_t)_{t\geq 0}$.

In Table 3, we summarize the estimated $\hat{\theta}$ from our regression with holding periods of 30 days. Again, we use price data from disjoint periods to calculate returns. The realized variance is calculated using the inter-period returns (30 days). The choice of 30-day periods gives us sufficient points to compute the realized variance while providing enough disjoint periods during the period Dec 2008–May 2013 to perform a regression. A longer price history would certainly have helped in balancing this tradeoff, but all these commodity LETFs were introduced only in the past 5 years.

Our empirical analysis confirms several aspects of our theoretical model in (5) and provides explanations in cases where there is discrepancy. The theoretical value of θ according to (5) is given by $\frac{\beta-\beta^2}{2}$. Table 3 shows that the estimator $\hat{\theta}$ is typically in the neighborhood of θ, its theoretical value. For example, SCO ($\beta = -2$, crude oil) has $\hat{\theta} = 2.93$ versus a theoretical θ of 3. In addition, the non-leveraged ETFs all

Table 3 $\hat{\theta}$ vs. θ, estimated from 30-day multi-variable regression of returns, with a partial correlation table

LETF	Underlying	β	$\hat{\theta}$	θ	r^2	$r^2_{x|y}$	$r^2_{y|x}$
SLV	Silver bullion	1	0.11	0	0.9799	0.9503	0.0078
AGQ	Silver bullion	2	−1.31	−1	0.9885	0.9751	0.3892
ZSL	Silver bullion	−2	−3.27	−3	0.9995	0.9988	0.7514
USLV	Silver bullion	3	−2.24	−3	0.9995	0.9988	0.7514
DSLV	Silver bullion	−3	−6.94	−6	0.9994	0.9989	0.9654
GLD	Gold bullion	1	−0.14	0	0.9898	0.9791	0.0064
UGL	Gold bullion	2	−2.44	−1	0.9934	0.9867	0.2900
GLL	Gold bullion	−2	−0.96	−3	0.9914	0.9828	0.0417
UGLD	Gold bullion	3	−2.38	−3	0.9982	0.9955	0.6355
DGLD	Gold bullion	−3	−6.26	−6	0.9846	0.9685	0.0809
IYE	Oil & gas	1	−0.06	0	0.9988	0.9965	0.1905
DDG	Oil & gas	−1	−0.99	−1	0.8866	0.7662	0.2342
DIG	Oil & gas	2	−1.11	−1	0.9996	0.9989	0.9498
DUG	Oil & gas	−2	−3.31	−3	0.9884	0.9769	0.8873
DBO	WTI crude oil	1	−0.02	0	0.9992	0.9981	0.0035
UCO	WTI crude oil	2	−1.15	−1	0.9987	0.9972	0.7747
SCO	WTI crude oil	−2	−2.93	−3	0.9987	0.9975	0.9619
UWTI	WTI crude oil	3	−2.14	−3	0.9974	0.9939	0.6218
DWTI	WTI crude oil	−3	−7.25	−6	0.9974	0.9939	0.6218
IYM	Building materials	1	0.03	0	0.9996	0.9987	0.0495
SBM	Building materials	−1	−0.98	−1	0.9970	0.9920	0.5446
UYM	Building materials	2	−1.10	−1	0.9997	0.9993	0.9380
SMN	Building materials	−2	−3.59	−3	0.9613	0.9221	0.5301

$r^2_{y|x}$ stands for the marginal predictive power of adding the realized variance (y) into the model, holding constant the predictive power of the reference index returns (x). Similar definition for $r^2_{x|y}$. Data from Dec 2008–May 2013

have $\hat{\theta}$ close to 0, suggesting that realized variance does not play an important role in its price process, as predicted. However, some LETFs have $\hat{\theta}$ diverging significantly from θ. For example, the $\hat{\theta}$ for UGL ($\beta = 2$, gold) differs from its theoretical value by a factor of 114 % even with a regression R^2 of 99 %.

We attribute the deviation of $\hat{\theta}$ from θ in our regression to the collinearity effect of the two predictors ($\ln \frac{S_t}{S_0}$ and V_t). Of course $\ln \frac{S_t}{S_0}$ and V_t cannot be independent observations, since V_t depends on the price path process of S_t, the reference index. In general, the reference returns and the realized variance are negatively correlated. When the realized variance is high, it is likely the reference has suddenly dropped in value. When the realized variance is low, it usually implies a period of steady positive growth for the reference. Thus, the multi-collinearity effect is responsible for shifting predictive power among the different predictor variables. In order to measure the magnitude of the collinearity effect and the contribution of each correlated predictor variable, we compute the coefficients of partial determination for our regression model.

The factor $r^2_{y|x}$ which measures the marginal predictive power of adding the realized variance into the model. As $r^2_{y|x}$ increases, $\hat{\theta}$ becomes closer to θ, suggesting a larger dependence of LETF returns on realized variance during holding periods of high volatility. For example, for the 3 LETFs DIG ($\beta = 2$, oil & gas), SCO ($\beta = -2$, crude oil), and UYM ($\beta = 2$, building materials) all have $r^2_{y|x}$ over 90 %. Their estimated $\hat{\theta}$ is similarly very close to the theoretical θ, never differing by more than 10 %. However, for non-leveraged ETFs, the realized variance has minimal added predictive power in the model. For those ETFs, we observe $\hat{\theta} \approx 0$. For example, SLV ($\beta = 1$, silver), GLD ($\beta = 1$, gold), and DBO ($\beta = 1$, crude oil) all have $r^2_{y|x} \approx 0$, and they subsequently have $\hat{\theta} \approx 0$. In addition, $r^2_{x|y}$, which is the marginal predictive power of adding the log-returns of the reference into our regression model, is always very high, indicating that the log-returns of the reference affect the LETF prices the most, but that the realized variance is still important for predictive power, especially when leverage and the holding period is high.

3.3 Realized Effective Fee

In Fig. 8, we show three empirical price paths: the LETF log-returns, the benchmark process defined in (5), and β times the reference index log-returns. As we can see, the value erosion due to realized variance (volatility decay) starts to play a significant role in determining LETF prices as the holding time increases. The path associated with β times the reference log-returns dominates the LETF log-returns after about 1 month of holding. After about 1 year, the benchmark which incorporates volatility decay more closely models the empirical LETF log-returns. For example, after 6 months of holding, SCO ($\beta = -2$, crude oil) diverges from β times the reference, illustrating the effects of volatility decay.

However, there are also some strong deviations from the predictions given by the benchmark, which compound as the holding time increases. This causes the LETF to underperform even after the volatility decay is accounted for. For example, DUG's ($\beta = -2$, oil & gas) empirical returns begin to trail its benchmark significantly around 2009. Therefore, the volatility decay cannot explain all the LETF underperformance.

We are therefore motivated to quantify the over/under-performance of the LETFs after observing deviations from the benchmark in Fig. 8. We introduce the concept of *realized effective fee* (REF) as the effective deduction rate charged by the LETF provider over the frictionless dynamic portfolio from which the LETF is constructed in Sect. 3.1. For a holding interval $[0, t]$, the corresponding REF is defined by

$$\widehat{f}_t = (1 - \beta)r - \frac{\ln \frac{L_t}{L_0} - \beta \ln \frac{S_t}{S_0} - \frac{\beta - \beta^2}{2} V_t}{t}. \tag{10}$$

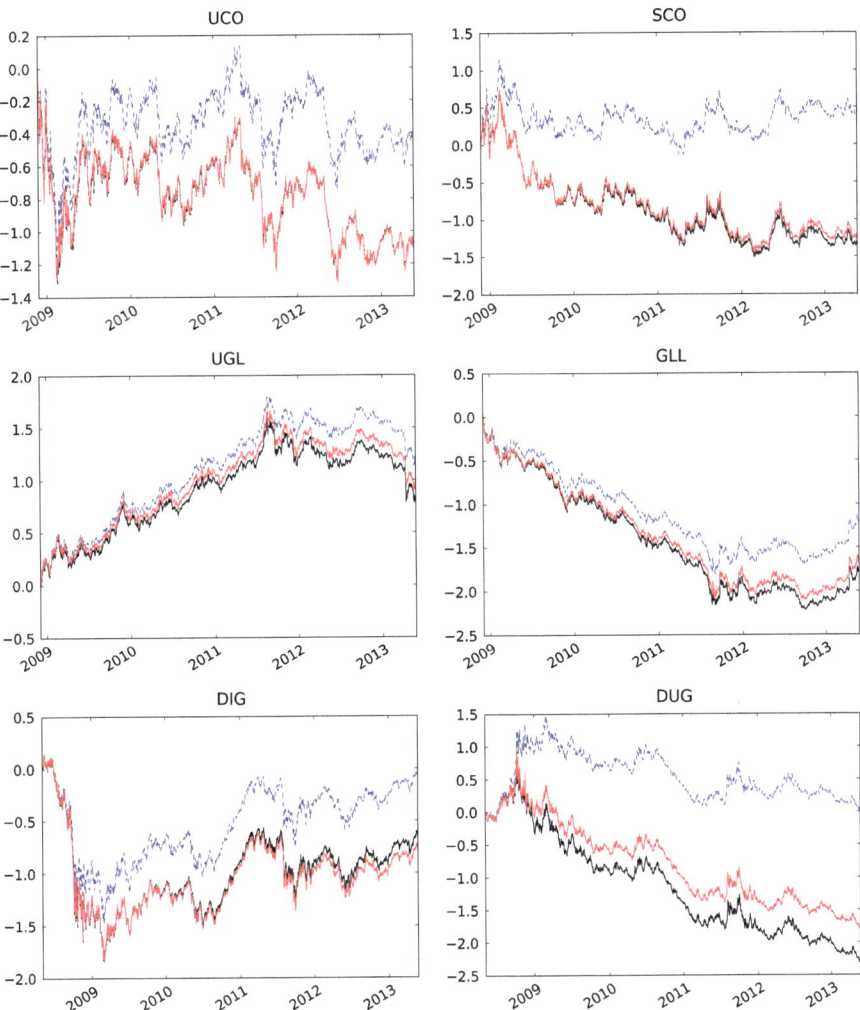

Fig. 8 Cumulative empirical log-returns of the LETF (*solid dark*) vs benchmark (*solid light*) and β times reference (*dashed light*), from Dec 2008–May 2013. *From top left to bottom right*: UCO, SCO (crude oil); UGL, GLL (gold); DIG, DUG (building materials). UCO, UGL, and DIG have $\beta = 2$ while SCO, GLL, and DUG have $\beta = -2$

Since for each LETF, L_t, S_t, V_t, β, and r are all known, we can calculate the REF $\widehat{f_t}$ for any LETF over a given holding period $[0, t]$ using historical prices. We remark that the REF, which is indexed by time t, depends on the selected holding horizon.

In many cases, the REF is seen to be much larger than the fund's advertised fee, indicating significant underperformance. Out of the 23 commodity LETFs, 2 have negative implied costs, so that the fund overperforms by the end of the 5 year period Dec 2008 to May 2013. If the REF exceeds the advertised fee, then the investor effectively pays an extra price for the opportunity to invest in the LETF. As a general trend, the bear LETFs tend to charge higher REFs than bull LETFs with the same magnitude of leverage $|\beta|$. For example, USLV ($\beta = 2$, silver) has a REF of 93 bps, while DSLV ($\beta = -2$, silver) has an REF of 504 bps over the period Dec 2008–May 2013. The two highest REFs correspond to DUG ($\beta = -2$, oil & gas) and SMN ($\beta = -2$, building materials), whose REFs are 1,134 bps and 1,625 bps respectively. Figure 8 illustrates that DUG ($\beta = -2$, oil & gas) drastically underperforms the benchmark, thereby realizing a high REF. Notice that in both cases, however, DUG and SMN's bull counterparts DIG ($\beta = 2$, oil & gas) and UYM ($\beta = 2$, building materials) respectively display a negative REF, indicating overperformance during the same period. It is possible that as the reference trends upwards for a long period of time, the bear LETF will underperform, while the bull LETF will overperform (Table 4).

Table 4 Comparison of the official fee for the LETF charged on the fund prospectus and the REF calculated using 5 years of price data (Dec 2008–May 2013) for the LETF and reference (see (10))

LETF	Underlying	β	Prospectus fee (bps)	Realized effective fee (bps)
SLV	Silver bullion	1	50	96
AGQ	Silver bullion	2	95	524
ZSL	Silver bullion	−2	95	567
USLV	Silver bullion	3	165	93
DSLV	Silver bullion	−3	165	504
GLD	Gold bullion	1	40	48
UGL	Gold bullion	2	95	343
GLL	Gold bullion	−2	95	406
UGLD	Gold bullion	3	135	139
DGLD	Gold bullion	−3	135	521
IYE	Oil & gas	1	48	50
DDG	Oil & gas	−1	95	953
DIG	Oil & gas	2	95	−142
DUG	Oil & gas	2	95	1,134
DBO	WTI crude oil	1	75	56
UCO	WTI crude oil	2	95	84
SCO	WTI crude oil	−2	95	321
UWTI	WTI crude oil	3	135	3
DWTI	WTI crude oil	−3	135	549
IYM	Building materials	1	48	11
SBM	Building materials	−1	95	456
UYM	Building materials	2	95	−204
SMN	Building materials	−2	95	1,625

We set $r = 69.1$ bps, the annualized LIBOR rate

4 A Static LETF Portfolio

Taking advantage of the volatility decay, a well-known trading strategy used by practitioners involves shorting a $\pm\beta$ pair of LETFs with the same reference, as discussed in [2, 7, 9, 11]. Since the LETFs have opposite daily returns on the same reference index, the portfolio has very little exposure to the reference as long as the holding period is sufficiently short. With this strategy, the volatility decay can help generate profit, which is the intuition of many practitioners. However, the portfolio is exposed to risk during periods of low volatility and high trending, as well as tracking errors. In this section, we describe an extension of this trading strategy by allowing the positive and negative leverage ratios to differ. We determine the portfolio weights to approximately eliminate the dependence on the reference. We show that the resulting portfolio is long volatility. For a number of LETF pairs, we find from empirical data that on average the strategy is profitable with enormous tail risk.

We now construct a weighted portfolio which is short the LETF with leverage ratio $\beta_+ > 0$ and short another LETF with leverage ratio $\beta_- < 0$. We emphasize that both LETFs having the same reference, but that β_+ and $|\beta_-|$ may differ. We hold fraction $\omega \in (0, 1)$ of the portfolio in the β_+-LETF and $(1 - \omega)$ of the portfolio in the β_--LETF. At time T, the normalized return from this strategy is

$$\mathscr{R}_T = 1 - \omega \frac{L_T^+}{L_0^+} - (1 - \omega)\frac{L_T^-}{L_0^-}. \tag{11}$$

Applying (5), $\mathscr{R}_T$ admits the expression

$$\mathscr{R}_T = 1 - \omega \left(\frac{S_T}{S_0}\right)^{\beta_+} \exp\left(\Gamma_T^+\right) - (1 - \omega)\left(\frac{S_T}{S_0}\right)^{\beta_-} \exp\left(\Gamma_T^-\right), \tag{12}$$

where

$$\Gamma_T^\pm = \frac{\beta_\pm - \beta_\pm^2}{2} V_T + ((1 - \beta_\pm)r - f_\pm)T, \tag{13}$$

Here, $\beta_\pm$ and $f_\pm$ are the respective leverage ratios and fees of the two LETFs in the portfolio defined in (11). Over a short holding period such that $\frac{L_T}{L_0} \approx 1$, one can pick an appropriate weight ω^* to approximately remove the dependence of $\mathscr{R}_T$ on S_T.

Proposition 1. *Select the portfolio weight $\omega^* = \frac{-\beta_-}{\beta_+ - \beta_-}$. For $\frac{L_T}{L_0} \approx 1$, the return from this strategy is given by*

$$\mathscr{R}_T = \frac{-\beta_- \beta_+}{2} V_T - \frac{\beta_-}{\beta_+ - \beta_-}(f_+ - f_-)T + (f_- - r)T. \tag{14}$$

Table 5 Table of (β_+, β_-) pairs vs ω^* the weight of the β_+ portfolio, and $\frac{-\beta_- - \beta_+}{2}$ the dependence of the strategy on V_t (see Proposition 1)

(β_+, β_-)	ω^*	$\frac{-\beta_- - \beta_+}{2}$
$(1, -1)$	$1/2$	$1/2$
$(1, -2)$	$2/3$	1
$(1, -3)$	$3/4$	$3/2$
$(2, -1)$	$1/3$	1
$(2, -2)$	$1/2$	2
$(2, -3)$	$3/5$	3
$(3, -1)$	$1/4$	$3/2$
$(3, -2)$	$2/5$	3
$(3, -3)$	$1/2$	$9/2$

Proof. For $\frac{L_T}{L_0} \approx 1$, we can substitute for $\frac{L_T}{L_0}$ with $\ln\frac{L_T}{L_0} + 1$ in (11). Then, we set $\omega = \omega^*$ and apply (5) to conclude (14).

The return (14) corresponding to portfolio weight ω^* reflects a linear dependence on the realized variance. In particular, the coefficient $\frac{-\beta_- - \beta_+}{2}$ is strictly positive, so the strategy is effectively long volatility (V_T). Also, as it does not depend on S_T, the ω^* portfolio is Δ-*neutral* as long as the reference does not move significantly. In Table 5, we summarize the coefficient of V_T and the weighted portfolio $(\omega^*, 1 - \omega^*)$ for different combinations of leverage ratios. Note that as long as $\beta_+ = -\beta_-$, we end up with the portfolio weight $\omega^* = \frac{1}{2}$. Also, the coefficient $\frac{-\beta_- - \beta_+}{2}$ exceeds or equals to 1 except for the pair $(\beta_+, \beta_-) = (1, -1)$, and it is largest for the pair $(\beta_+, \beta_-) = (3, -3)$.

We now backtest the ω^* strategy from Proposition 1 as follows. For each LETF pair, we short \$0.5 of the β_+-LETF and \$0.5 of the β_--LETF with $\beta_+ = -\beta_- = 2$ and hold the position for some time T. The normalized return $\mathcal{R}_T$ depends on the relative weights on the long/short-LETFs but not the absolute cash amounts. More generally, one can also test the strategy with different $\beta_\pm$ and ω^*.

Dividing the price data from Dec 2008 to May 2013 into n-day rolling (overlapping) periods, we calculate the returns from the strategy over each period. For every n-day return, we compare against the realized variance over the same period. This is illustrated in Fig. 9. As a theoretical benchmark, we also plot $\mathcal{R}_T$ in (14) as a linear function. Each point (dot) on the plots represents a 5-day return, but over rolling periods the returns are not independent. In other words, the lines in Fig. 9 are not generated by regression but taken from (14). We choose (14) as a benchmark because it is expected to hold *pathwise* as long as $\frac{L_T}{L_0} \approx 1$ with negligible tracking error.

We can observe from Fig. 9 that the returns exhibit positive dependence on the realized variance (V_T). In particular, for the energy pairs (DIG-DUG ($\beta = \pm 2$, oil & gas) and UCO-SCO ($\beta = \pm 2$, crude oil)), the returns tend to be very positive when the realized variance is high. This is because the strategy captures the volatility decay as profit. Nevertheless, there is also a visible amount of noise in the returns deviating from the linear dependence on V_T, especially for the gold and silver pairs

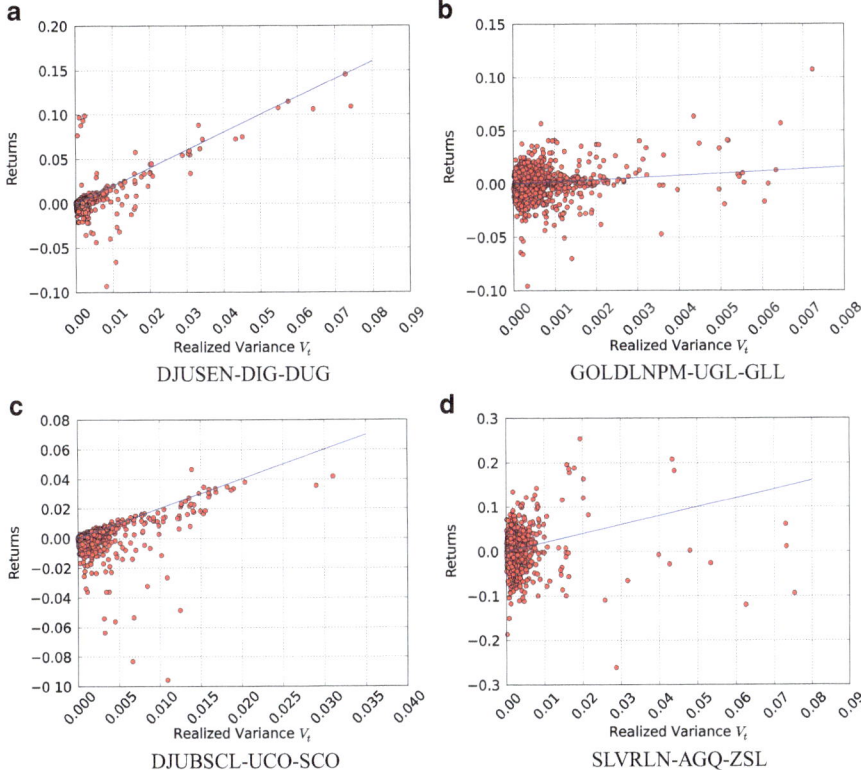

Fig. 9 Plot of trading returns vs realized variance for a double short strategy over 5-day rolling holding periods, with $\beta_\pm = \pm 2$ for each LETF pair. We compare with the empirical returns (*circle*) from the ω^* strategy with the predicted return (*solid line*) in Proposition 1. Trading pairs are DIG-DUG (oil & gas), UGL-GLL (gold), UCO-SCO (crude oil), AGQ-ZSL (silver)

(UGL-GLL ($\beta = \pm 2$, gold) and AGQ-ZSL ($\beta = \pm 2$, silver), respectively). This can be partly attributed to tracking errors from both LETFs in the portfolio. Also, the ω^*-strategy loses its Δ-neutrality if the reference moves significantly.

While this portfolio is expected to be Δ-neutral (with respect to the reference index) for small reference movements, in reality the strategy is also short-Γ. One way to see this is through Fig. 10 that plots the returns against the reference index returns. Common to all four LETF pairs, when the reference return is either very positive or negative, the return of the ω^*-strategy tends to be negative. As a theoretical benchmark, we also plot the normalized return equation (12) which applies even for large reference movements.

In contrast to the energy pairs, the gold and silver pairs yield very noisy returns. This is consistent with our earlier observations from our regressions in Figs. 3 and 4. For instance, both UGL and GLL ($\beta = \pm 2$, gold) show substantial tracking errors

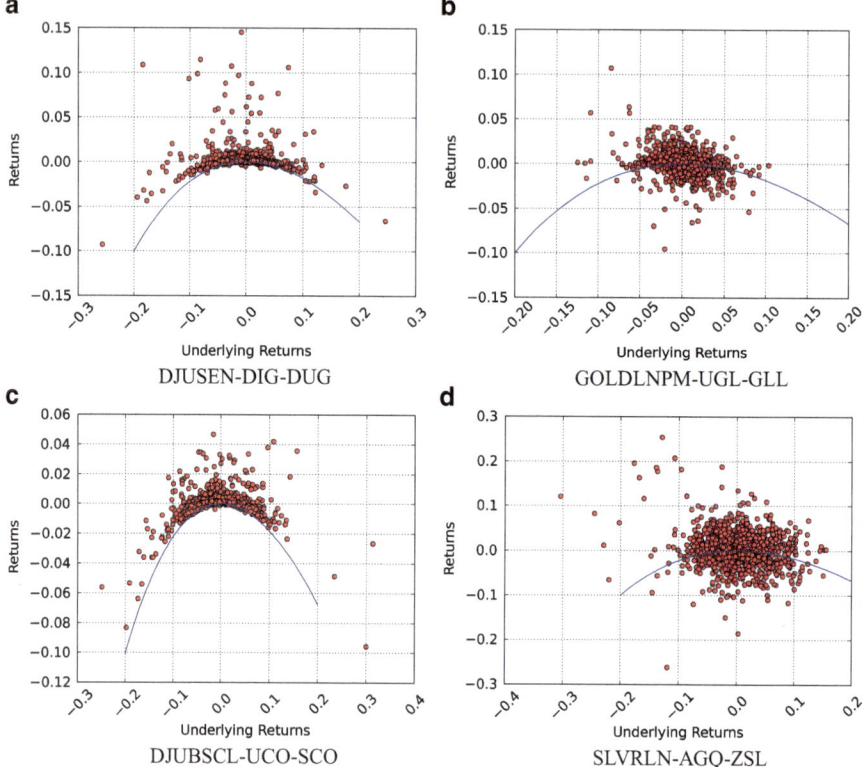

Fig. 10 Plot of returns of reference index vs trading returns for a double short strategy over 5-day rolling, holding periods. $\beta_{\pm} = \pm2$ for each LETF pair. We compare the empirical returns from our trading strategy (*dark solid circle*) with the predicted dependence on reference returns according to (12), using $\Gamma_T^{\pm} = 0$ (*light solid line*). Trading pairs are DIG-DUG (oil & gas), UGL-GLL (gold), UCO-SCO (crude oil), AGQ-ZSL (silver)

over short periods such as 5 days, and their regressed leverage ratios differ from the stated ones. On the other hand, the DIG and DUG ($\beta = \pm2$, oil & gas) regressions in Figs. 1 and 2 reflect much less tracking errors.

Furthermore, Fig. 11 shows that as the holding time increases, the returns from the ω^* strategy increases as well. The performance is best for the energy pairs UCO-SCO ($\beta = \pm2$, crude oil) and DIG-DUG ($\beta = \pm2$, oil & gas), but more subdued for the bullion pairs UGL-GLL ($\beta = \pm2$, gold) and AGQ-ZSL ($\beta = \pm2$, silver). However, over longer holding periods, the ω^* portfolio may lose its Δ-neutral status, thereby generating more risk as well. Although average returns from the ω^* strategy are positive, one is subject to enormous tail risk, which increases with the holding time of the static portfolio. In order to ensure that we do not subject ourselves to excessive tail risk, we should not only be sure of a high volatility environment, but we must also adjust the holding time to account for the extra risk associated with time horizon of returns.

Fig. 11 Average returns from a double short trading strategy by commodity pair over no. of days holding period. $\beta_{\pm} = \pm 2$ for each LETF pair. Trading pairs are DIG-DUG (oil & gas), UGL-GLL (gold), UCO-SCO (crude oil), AGQ-ZSL (silver)

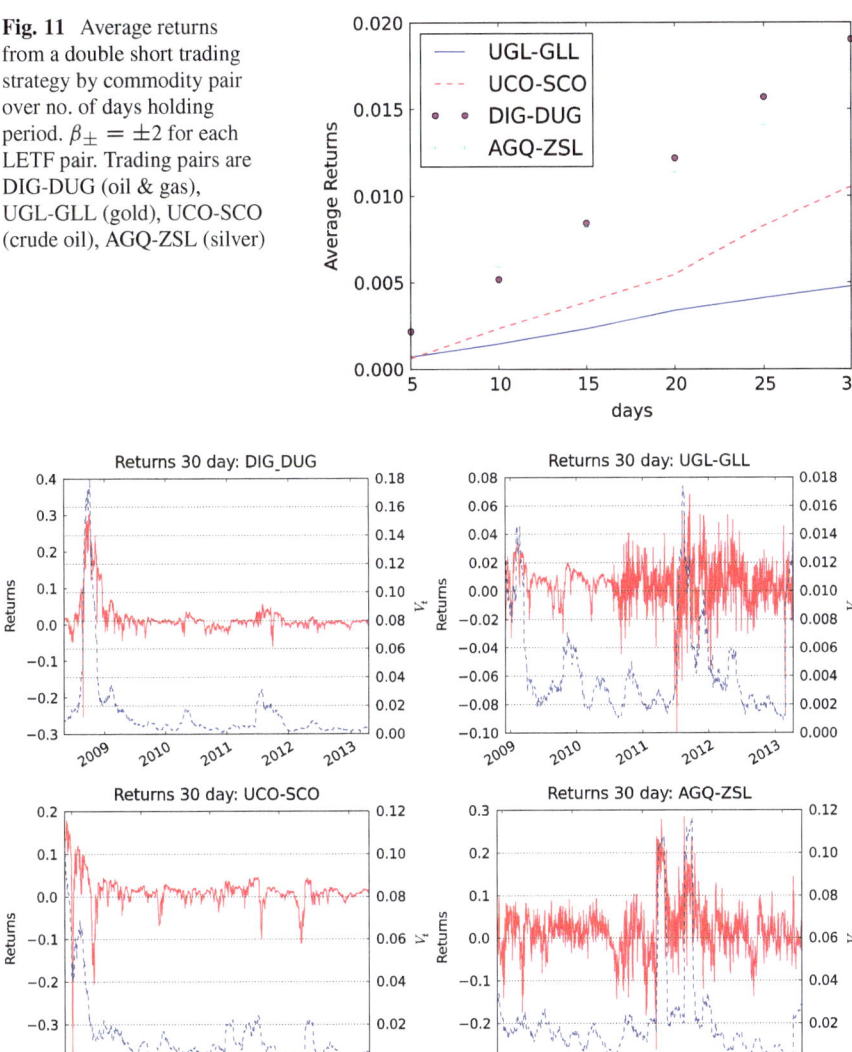

Fig. 12 Time series of returns for a double short strategy over 30-day rolling, holding periods, with $\beta_{\pm} = \pm 2$ for each LETF pair. Notice how during the periods of greatest volatility the double short strategy has the greatest return. Trading pairs are DIG-DUG (oil & gas), UGL-GLL (gold), UCO-SCO (crude oil), AGQ-ZSL (silver)

Figure 12 gives another perspective of the ω^* strategy's dependence on realized variance. It shows the time series of the 30-day rolling returns along with the realized variance of the reference index from Dec 2008 to May 2013. We see that when the realized variance increases sharply, the strategy returns also spike sharply. For example, when DJUSEN index realized variance spikes, the DIG-DUG

($\beta = \pm2$, oil & gas) trading pair accumulates a 30 % return over a single 30-day holding period. However, when realized variance is subdued over a period of time, the ω^* returns may turn quite negative as well.

In summary, the double-short trading strategy studied herein is profitable on average, but it is commodity specific and subject to enormous tail risk, as seen from empirical prices. The strategy's profitability depends strongly on a high volatility from the reference index. Although longer holding times tend to enhance the average return, they also enormously increase the horizon risk. According to these findings, this strategy appears to be appealing only during times of high volatility in the reference index.

5   Concluding Remarks

The ETF market has continued to grow in quantity and diversity, especially in the past 5 years. For both investors and regulators, it is very important to understand and quantify the risks involved with various ETFs. In this paper, we have focused on commodity ETFs and their leveraged counterparts. We find that the LETF returns tend to deviate significantly from the corresponding multiple of the reference returns as the holding horizon lengthens. To study the performance of an LETF, we have applied a new benchmark process that accounts for the realized variance of the underlying. We find that many commodity LETFs still diverge, typically negatively, from this benchmark over time. These empirical observations motivate us to illustrate the over/under-performance of an LETF via the concept of realized expense fee. Based on the funds and the time periods we have studied, most commodity LETFs effectively charge significantly higher expense fees than stated on their prospectuses.

In view of LETFs' common pattern of value erosion over time, one well-known trading strategy in the industry involves statically shorting both long and short LETFs in order to capture the volatility decay as profit. We systematically study an extension of this strategy that is applicable to LETF pairs with different asymmetric leverage ratios. We analytically derive the specific weights in the LETFs so that the resulting portfolio is approximately Δ-neutral, but short-Γ as well. This strategy can potentially be quite profitable but its return can be negatively impacted by tracking errors generated by the LETFs and large movements of the reference index. These two factors both depend on the holding horizon. This should motivate future research on the horizon risk for LETF strategies. To this end, Leung and Santoli [7] study the admissible holding horizon and leverage ratio given a risk constraint. The recent papers [6, 13, 14] examine the dynamics of price spreads between ETF pairs, for example, gold vs. silver.

Our analysis herein does not assume a parametric stochastic volatility model for the underlying. It is of practical interest to investigate the price behavior of LETF under a number of well-known stochastic volatility models, such as the Heston and SABR models. On top of LETFs, there are also options written on these funds.

This gives rise to the question of consistent pricing of LETF options across leverage ratios (see [1, 8]). Finally, models that capture the connection between LETFs and the broader financial market would be very useful for not only traders and investors, but also regulators.

Acknowledgements The authors would like to thank Scott Weiner of VelocityShares, and the participants of the 2013 Focus Program on Commodities, Energy and Environmental Finance held at Field's Institute and the 2014 Joint Mathematics Meetings in Baltimore for their helpful suggestions and comments.

References

1. Ahn, A., Haugh, M., Jain, A.: Consistent pricing of options on leveraged ETFs. Working Paper, Columbia University (2012)
2. Avellaneda, M., Zhang, S.: Path-dependence of leveraged ETF returns. SIAM J. Financ. Math. **1**, 586–603 (2010)
3. Cheng, M., Madhavan, A.: The dynamics of leveraged and inverse exchange-traded funds. J. Invest. Manag. **7**(4), 43–62 (2009)
4. Dobi, D., Avellaneda, M.: Structural slippage of leveraged ETFs. Working Paper (August 2012)
5. Guedj, I., Li, G., McCann, C.: Futures-based commodities ETFs. J. Index Invest. **2**(1), 14–24 (2011)
6. Leung, T., Li, X.: Optimal mean reversion trading with transaction costs and stop-loss exit. J. Theoretical & Applied Finance **18**(3), p. 1550020 (2015)
7. Leung, T., Santoli, M.: Leveraged exchange-traded funds: admissible leverage and risk horizon. J. Invest. Strateg. **2**, 39–61 (2013)
8. Leung, T., Sircar, R.: Implied volatility of leveraged ETF options. Applied Mathematical Finance, **22**(2), pp. 162–188 (2015). http://www.tandfonline.com/eprint/jt8MFtBFhkjMIPDYiS9E/full
9. Mackintosh, P., Lin, V.: Longer term plays on leveraged ETFs. Credit Suisse: Portfolio Strategy, pp. 1–6 (April 2010)
10. Mackintosh, P., Lin, V.: Tracking down the truth. Credit Suisse: Portfolio Strategy, pp. 1–10 (February 2010)
11. Mason, C., Omprakash, A., Arouna, B.: Few strategies around leveraged ETFs. BNP Paribas Equities Derivatives Strategy, pp. 1–6 (April 2010)
12. Murphy, R., Wright, C.: An empirical investigation of the performance of commodity-based leveraged ETFs. J. Index Invest. **1**(3), 14–23 (2010)
13. Naylor, M., Wongchoti, U., Gianotti, C.: Abnormal returns in gold and silver exchange traded funds. J. Index Invest. **2**(2), 1–34 (2011)
14. Triantafyllopoulos, K., Montana, G.: Dynamic modeling of mean-reverting spreads for statistical arbitrage. Comput. Manag. Sci. **8**, 23–49 (2009)

Integration of Commodity Derivative Markets: Has It Gone Too Far?

Delphine Lautier, Julien Ling, and Franck Raynaud

Abstract We examine the impact of two financial crises on commodity derivative markets: the subprime crisis and the bankruptcy of Lehman Brothers. These crises are "external" to the commodity markets because they occurred in the financial sphere. Still, because commodity markets are now highly integrated with each other and with other financial markets, such events could have had an impact. In order to fully comprehend this possible impact, we rely on tools inspired by the graph theory that allow for the study of large databases. We examine the daily price fluctuations recorded in 14 derivative markets from 2000 to 2009 in three dimensions: the observation time, the space dimension—the same underlying asset can be traded simultaneously in two different places—and the maturity of the transactions. We perform an event study in which we first focus on the efficiency of the price shock's transmission to the commodity markets during the crises. Then we concentrate on whether the paths of shock transmission are modified. Finally, relying on the measure proposed by Bonacich (Am J Sociol 92(5):1170–1182, 1987) for social networks, we focus on whether the centrality of the price system changes.

1 Introduction

In this paper, we examine the impact of two financial crises on the commodity derivative markets: the subprime crisis and the bankruptcy of Lehman Brothers. These crises are exogenous to the commodity markets because they occurred in the financial sphere. Still, such events could have propagated to the commodity markets because these markets are highly integrated with each other and with other financial markets (see [5–7, 12, 15, 22]).

D. Lautier • J. Ling (✉)
PSL, Université Paris-Dauphine, Paris, France
e-mail: delphine.lautier@dauphine.fr; julien.ling@dauphine.fr

F. Raynaud
EPFL, Lausanne, Switzerland
e-mail: franck.raynaud@epfl.ch

© Springer Science+Business Media New York 2015   65
R. Aïd et al. (eds.), *Commodities, Energy and Environmental Finance*,
Fields Institute Communications 74, DOI 10.1007/978-1-4939-2733-3_3

Specifically, in this paper, we analyze the shock transmission through the dynamic behavior of the correlations between price returns. Following [13], we consider that there is transmission if market co-movements increase significantly after a shock.

In order to fully comprehend the potential impact of such crises on the commodity derivative markets, we perform an event study in which we examine price fluctuations in three dimensions: the observation time, the space dimension—the same underlying asset can be traded in two exchanges simultaneously—and the maturity of the transactions. We focus on a time window of 1 month (i.e., ten trading days before and after the beginning of the crises). We situate the triggering event on August 9, 2007 for the subprime crisis and on September 15, 2008 for the Lehman Brothers bankruptcy (see sections "Some Important Events Around the Subprime Crisis" and "Some Important Events Around Lehman Brothers Bankruptcy" in Appendix 1 for more details on the chronology of the crises).

Such an analysis requires the use of high dimensional data. In this context, the tools of the graph theory have already proved to be very interesting in various fields of finance. First, they provide a way to synthesize the information contained in the data and to obtain meaningful visual representations, second they allow for the quantification of high dimensional information (see for instance [10, 17, 19]). In what follows, we rely mainly on the methodology proposed by Lautier et al. [17]. These authors provide a *long-term* analysis of the connections between 14 derivative markets between 2000 and 2009. They give evidence of an increasing integration along the time period under scrutiny, and they show that it is a condition for systemic risk to appear. Taking advantage of the fact that between 2000 and 2009 two main financial crises occurred, we perform an *event study* on the same markets. This study gives us the possibility to concretely assess the potential consequences of market integration. Moreover, we introduce a new method that was initially proposed by Bonacich [3] for social networks. This method allows us to better evaluate the organization of the graph. It gives insights into the localization of the center of the graph that, as far as systemic risk is concerned, is crucial.

Following [17], the nodes of the graphs correspond to price returns: there is one node per futures contract and per maturity. The link between each pair of nodes depends on the correlations between their returns. Relying on several measures, we provide a dynamic analysis of these graphs and their behavior around the crises. We also empirically compute how exceptional these events are compared to what can be observed in the whole period.

First, in order to filter the information contained in the graphs, we use Minimum Spanning Trees—MST—[18]. Because they capture the most important links between the markets, they are the most probable and the most efficient paths of price shock transmission. Taking into account the length of the MST, we can ask a first question: does the efficiency of the price shock transmission improve during crises? We then concentrate on the organization of the graph, namely the topology of the MST and ask a second question: do the paths of shock transmission change during crises and how? In order to answer these questions, several tools are used. First, we use survival ratios that indicate the number of links that change from one day to the other and give indications about large reorganizations of the graphs. Second, the allometric coefficients measure how far a tree stands from a linear or, on the contrary, a star-like organization. These two extreme configurations have radically opposite consequences from the systemic point of view: with a chain-like tree, a shock appearing at one extremity of the tree must spread through all nodes before reaching the other extremity. On the contrary, with a star-like tree, a shock arising at the center of the graph might rapidly affect all other nodes. Finally, we focus on the centrality of the price system: does it change? Does it increase? In a first approach, we simply identify the center of the price system as the most connected node. We then improve this analysis with the measure developed by Bonacich [3]: in a nutshell, instead of focusing on one single node, we take into account the whole organization of the network, that is, the number and proximity of *the direct as well as the indirect neighbors* of a node.

This paper is organized as follows. We first explain how to build a graph on the basis of our data. We then examine the efficiency of the shock transmission, the organization of the price system and its centrality. At each step, we compare the behavior of the price system in the whole period with what happened during the crises.

2   The Price System

After a short description of the data used for the study, we explain the way we build price graphs.

2.1   Data

For the empirical study, we examine 14 futures markets corresponding to three different sectors of activity: 6 energy markets that comprise 2 markets each of crude oil, natural gas and petroleum products; 4 agricultural markets (wheat, corn, soy oil and soy bean) and 4 financial assets (Mini S&P500 index, gold, USD/EUR exchange rate, and 3 month Eurodollar interest rate). We selected the contracts that were characterized by the largest transaction volumes over a long time period, thanks to the Futures Industry Association's monthly volume reports. We used Datastream in order to collect settlement prices on a daily basis.

In the absence of reliable spot data for most commodity markets, we approximated all spot prices with the nearest futures prices. Such an approximation is very common in finance. We also rearranged the futures prices in order to reconstitute the daily term structures, i.e., the relationships linking, at a specific date, several futures contracts with different delivery dates. We removed some maturities from the database because the price curves were shorter at the beginning of the period. The number of contract maturities indeed usually rises on a derivative market; the growth in the transaction volumes of existing contracts results in the introduction

Table 1 Characteristics of the collected data: nature of the underlying asset, trading location (CME stands for Chicago Mercantile Exchange, ICE for Inter Continental Exchange, US for United States and EU for Europe), longest maturity traded (in months), number of contracts (this number is added just after the name of the underlying asset on the figures)

Underlying asset	Exchange-Zone	Maturities	# contracts
Light crude oil	CME-US	Up to 84	33
Brent crude	ICE-EU	Up to 18	17
Heating oil	CME-US	Up to 18	18
Gasoil	ICE-EU	Up to 12	12
Natural gas (US)	CME-US	Up to 36	36
Natural gas (Eu)	ICE-EU	Up to 9	9
Wheat	CME-US	Up to 15	6
Soy bean	CME-US	Up to 14	7
Soy oil	CME-US	Up to 15	15
Corn	CME-US	Up to 25	4
Eurodollar	CME-US	Up to 120	40
Gold	CME-US	Up to 60	17
USD/EUR Exchange rate	CME-US	Up to 12	4
Mini S&P500	CME-US	Up to 6	2

of new delivery dates. Finally, when performing spatial and 3D analyses, we used the longest common time period for all of the underlying assets, from 2000/01/04 to 2009/08/12. Once these selections have been carried out, our database still contains more than 655,000 prices, that comprise 220 time series in the 3D analysis and a subset of 14 in the spatial one.

Table 1 summarizes the main characteristics of our database.

2.2 Building the Graphs

Our graphs are built on the basis of the correlations between the price returns. We use this measure in order to capture the synchronous price movements in the system. To obtain a graph, these correlations are transformed into distances.

2.2.1 Correlations of Price Returns

The first step towards the analysis of market integration is the computation of the synchronous correlation coefficients $\rho_{ij}(t)$ of the price returns, defined as follows:

$$\rho_{ij}(t) = \frac{\langle r_i r_j \rangle - \langle r_i \rangle \langle r_j \rangle}{\sqrt{\left(\langle r_i^2 \rangle - \langle r_i \rangle^2\right)\left(\langle r_j^2 \rangle - \langle r_j \rangle^2\right)}}, \tag{1}$$

In the spatial dimension, i and j stand for the nearby futures contracts of a pair of assets (crude oil or corn for example), whereas they stand for pairs of delivery dates in the maturity dimension. Both are present in the 3D analysis with the 220 time series. The daily logarithm price differential stands for the price returns r_i, with $r_i = (\ln F_i(t) - \ln F_i(t - \Delta t))/\Delta t$, where $F_i(t)$ is the price of the futures contract i at date t. The time interval is Δt and $\langle . \rangle$ denotes the statistical average performed over time, for the trading days of the study period.

For a given time period and a given set of data, we thus compute the matrix $\mathbf{C}$ of $N \times N$ correlation coefficients, for all of the pairs ij. $\mathbf{C}$ is symmetric with $\rho_{ij}(t)$ equal to one when $i = j$. Thus, it is characterized by $N(N-1)/2$ coefficients.

Performing dynamic studies on the basis of rolling windows requires the choice of a proper window length. On the one hand, we want it to represent typical economic periods (one semester, 1 year, 5 years...) and to be as short as possible in order to give evidence of sudden changes. On the other hand, we are confronted with a technical constraint: in order to ensure representative results, the number of observations has to be larger than the number of nodes. Having to deal with 220 series of price returns (i.e., 220 nodes), we thus use a rolling window of 1 year (252 trading days). We do the same in the spatial dimension for comparison purposes. As robustness checks, we also perform computations with 2-year windows, as illustrated in section "Robustness Checks" in Appendix 2. Further, we use rolling windows situated before the observation date. So when we look at what happens on August 9, 2007, the information used is situated 1 year before that event. Fortunately, because the two crises are separated by more than 1 year, there is no overlap between them.

2.2.2 From Correlations to Distances

In order to use the tools of the graph theory, we need to introduce a metric. The correlation coefficient ρ_{ij} cannot be used as a distance d_{ij} between i and j because it does not fulfill the three axioms that define a metric [14, p. 30]:

- $d_{ij} = 0$ if and only if $i = j$
- $d_{ij} = d_{ji}$
- $d_{ij} \le d_{ik} + d_{kj}$

However, a metric d_{ij} can be extracted from the correlation coefficients through a nonlinear transformation. This Euclidean distance is defined as follows[1]:

$$d_{ij}(t) = \sqrt{2\left(1 - \rho_{ij}(t)\right)}. \tag{2}$$

A distance matrix $\mathbf{D}$ is thus extracted from each correlation matrix $\mathbf{C}$ (at each date t) according to Eq. (2). The matrices $\mathbf{C}$ and $\mathbf{D}$ are both $N \times N$ dimensional.

---

[1] Taking the square of $\rho_{ij}(t)$ has no impact on the results (computations are available on request).

While the coefficients $\rho_{ij}(t)$ can be positive for the correlated returns or negative for the anti-correlated returns, the distance $d_{ij}(t)$ is always positive. The distance matrix corresponds to a fully connected graph; it represents all the possible connections in the price system.

3   The Efficiency of the Shock Transmission

Considering the dimensionality of our price system and the number of nodes in our graph, it is very difficult to visualize. We thus resort to a filtering technique which is especially suited to our context: the Minimum Spanning Tree (MST).

3.1   The Minimum Spanning Tree

In order to understand the organizing principles of a system through its representation as a graph, the latter needs to be spanned. However, there are a lot of paths that span a graph. For a weighted graph like ours, the MST divulges the most relevant connections of each element of the system and it reduces the information space from $N(N-1)/2$ to $N-1$.

The MST is the path spanning all the nodes of the graph without any loop. It has less weight than any other tree and is unique. The distance $d_{ij}(t)$ is more than just an Euclidean metric; it is the subdominant ultrametric that satisfies the triangular inequality: $d_{ij}(t) \leq \max \{d_{ik}(t); d_{kj}(t)\}$.

When the graph is weighted with distances, the latter corresponding to the correlations between the price returns, the MST is especially useful for the study of systemic risk. In an analogy with signal transmission, the ultrametric provides the shortest path between all of the nodes, that is, the path where the signal suffers the least losses and travels the fastest. We interpret this feature as the efficiency (in speed and in accuracy) in the transmission of the signal. Furthermore, if a price shock is assimilated to a signal and if transmission is appreciated through the analysis of the dynamic behavior of the correlations between the price returns, then the MST "can be assimilated into the shortest and most probable path for the propagation of price shocks" [17].

The visualization of the trees (which are plotted with the software Graphviz) addresses the meaningfulness of the taxonomy that emerges from the system. Because we are considering the links between markets and/or delivery dates belonging to the MST, if a link between two markets or maturities does not appear in the tree, it only means that this link does not correspond to a minimal distance. Note also that, in such an analysis, the results depend on the nature and the number of markets chosen for the study.

Figure 1 presents the MST obtained on the basis of our price system for the spatial dimension and over the whole period. It is scaled: the closest nodes

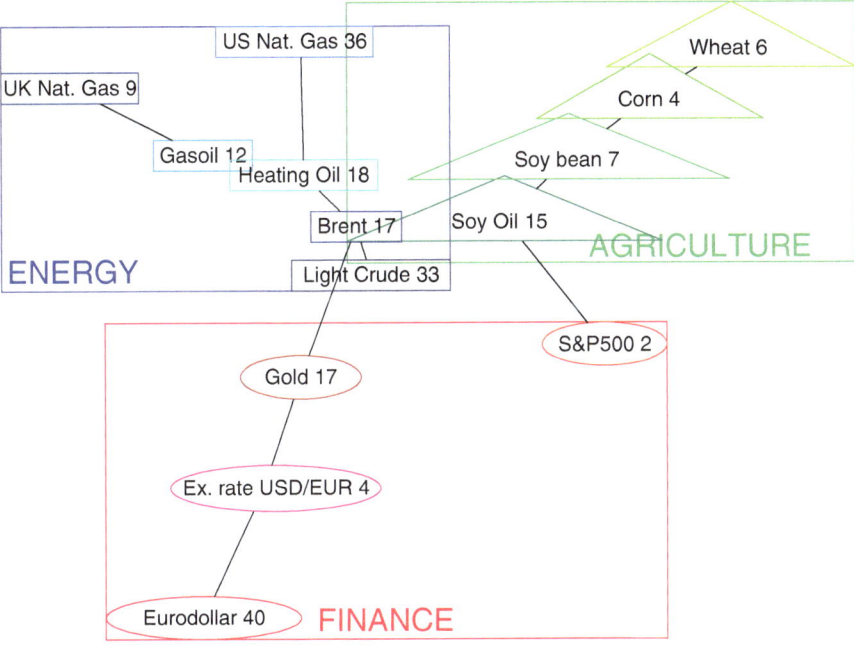

Fig. 1 Scaled MST in the spatial dimension, 2000–2009

correspond to the most correlated price series. Three sectors can be identified: energy is in the top left-hand. It gathers American as well as European markets and is situated between agriculture (on the right) and financial assets (at the bottom).

The link between the energy and agricultural products passes through soy oil. This is interesting because soy oil can be used for fuel. The link between commodities and financial assets passes through gold, which is also meaningful, because gold can be seen as a commodity as well as a reserve of value. The only surprise comes from the Mini S&P500 that is more correlated to soy oil than to financial assets. This connection between the Mini S&P500 and agricultural markets could be interpreted as evidence of the financialization of the commodity markets. However, in a dynamic analysis, this connection is very unstable. At least two reasons could explain such a result: first, Buyukşahin et al. [8] find that the correlations between grains and equities fluctuate a lot; and second, compared to all other contracts taken into account, the Mini S&P500 is the least actively traded.

At first glance (if we accept that counting the number of links allows for the identification of the center of the graph) the most connected node is the one corresponding to Brent crude oil, which makes it—a priori—the best candidate for the transmission of price fluctuations in the tree (actually, the same could be said for American crude oil—Light Crude—because the distance between these products is very short). Last but not least, the energy sector seems the most integrated, as the distances between the nodes are short.

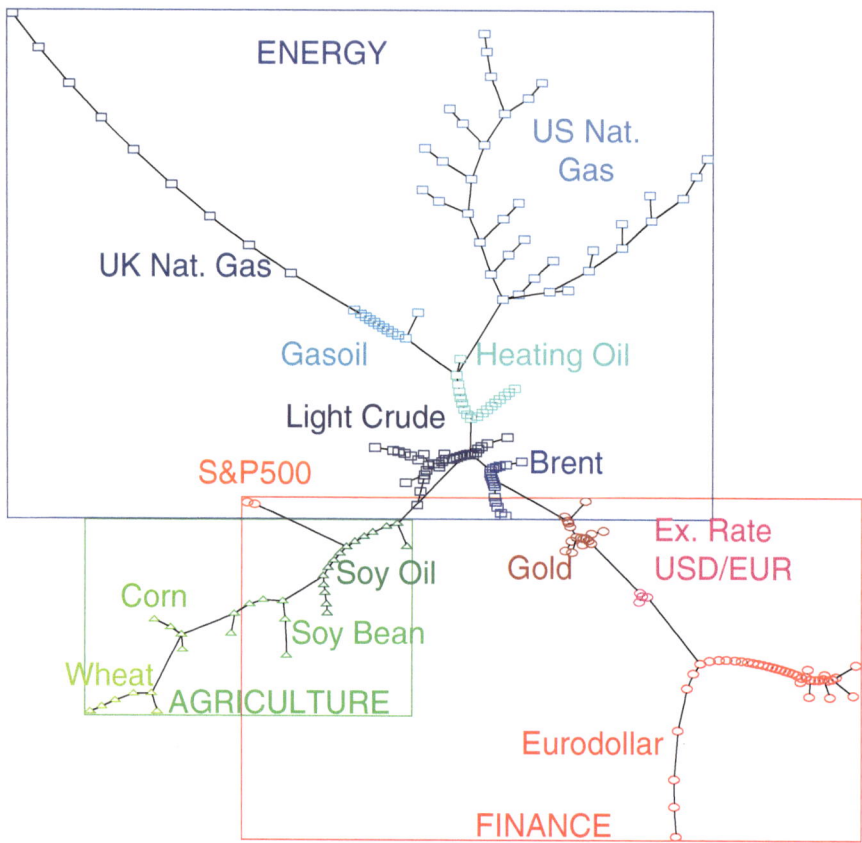

Fig. 2 Scaled MST in 3D, 2000–2009

Such a star-like organization leads to specific conclusions regarding systemic risk. A price movement appearing in the energy markets, situated at the heart of the price system, will have more impact than a fluctuation affecting the peripheral markets such as interest rates or wheat. This configuration explains why we consider the subprime and the Lehman Brothers crises as exogenous events in this study.

The 3D MST comprises 220 time series (nodes). Depicted by Fig. 2, it is less easy to read (this is why we removed the captions in the nodes), but it can be interpreted through the prism of the spatial tree. The same topology prevails, except that adding the maturities introduces linear branches in each market (with the noticeable exception of American natural gas). Moreover, this scaled representation shows that some markets are more integrated than others: clusters of maturities can be seen, at the center of the graph, for the energy sector (except for the two natural gas markets). Strong integration can also be observed in the financial branch; this is especially true for the Eurodollar contract after the eighth maturity.

Because these topologies are very stable over time [17], we use them as references in the remainder of this study.

3.2 How Does the Length of the MST Behave?

We first explain how this measure can be obtained and how it behaves on the whole sample. We then study it around crises.

3.2.1 The Measure

The normalized length of the tree can be defined as the average of the lengths of the edges belonging to the MST:

$$\mathscr{L}(t) = \frac{1}{N-1} \sum_{(i,j)\in MST} d_{ij}(t), \qquad (3)$$

where t denotes the date of the construction of the tree and $N-1$ is the number of edges in the MST. The length of a tree is higher when the distances increase and consequently when correlations are low. Thus, the more the length diminishes, the more integrated the system is.

Figure 3 represents the dynamic behavior of the normalized length of the MST in the spatial dimension over the whole period under consideration. The general pattern is that the length decreases, which reflects the increasing integration of the system. Thus the most efficient transmission path for price fluctuations becomes shorter as time goes by. This finding is consistent with e.g., [21] and [22]. A more in-depth examination of the graph also shows some very important moves at specific dates, one of them being around the Lehman Brothers bankruptcy.

3.2.2 The Length of the Trees Around the Crises

A first appraisal of the importance of the crises consists in measuring whether the changes in the length of the MST that occurred around the events were tail events or not.

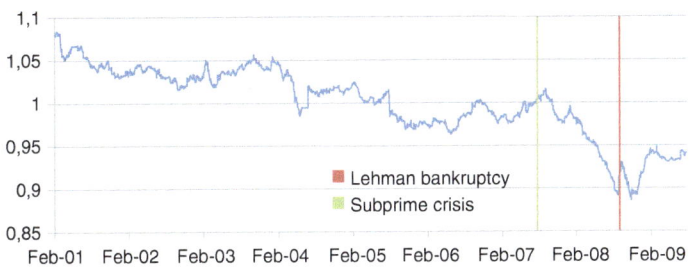

Fig. 3 Normalized tree's length in the spatial dimension, 2000–2009

We compute the empirical distribution of the length variations over the whole sample and examine the probability of the occurrence of fluctuations situated above (for increases) or below (for decreases) those observed around the crises. At 5 %, the changes recorded on August 16, 2007 (five trading days after the beginning of the subprime crisis) and on September 12, 2008 (one trading day before the bankruptcy of Lehman Brothers) are in the tail of the distribution, both in the spatial dimension and in 3D. In the spatial dimension only, we can add August 14, 2007, and in 3D only September 17, 2008. These last two events and the one recorded on September 12, 2008 have a probability of occurrence that is close to 1 %. Consequently, compared to what was observed between 2000 and 2009, the two crises have generated exceptional changes in the length of the MST.

A recurrent result in finance is the observation of an increase in the price correlations just after a crisis (see, e.g., [9] for an analysis of the equity market around Black Monday on October 19, 1987, [6] and [22] for commodity-equity markets, or [20] for a review of several studies on these topics). Figure 4a–d, which represent the evolution of the length of the trees on a 1-month time window around the crises under consideration both in the spatial dimension and in 3D, do not exhibit such behavior. On the contrary, in three cases (subsets b, c and d) out of four, we find an increase in the length of the MST.

For the subprime crisis, the peak appears on August 15, 2007, four trading days after the beginning of the crisis. For the Lehman Brothers bankruptcy, the change in the behavior of the tree arrives before the event, between September 11 and 12 of 2008. These dates correspond to the period when the difficulties

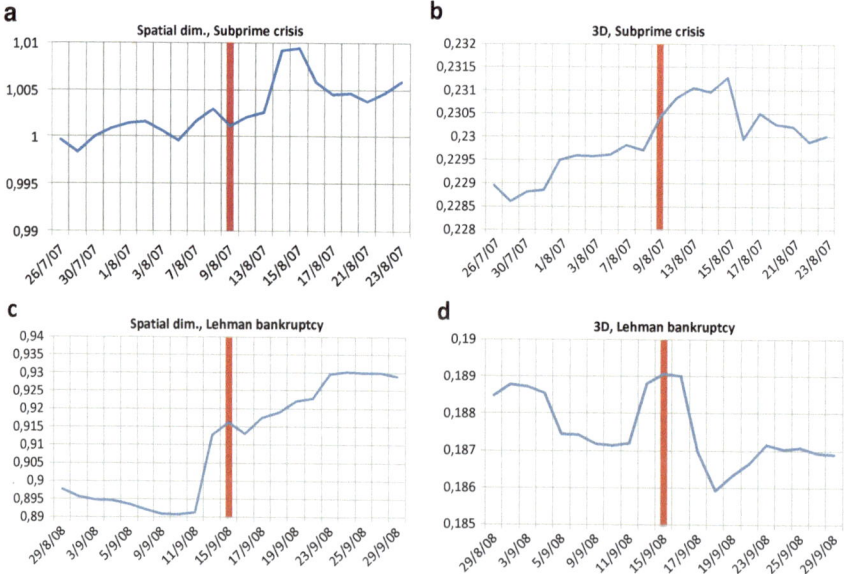

Fig. 4 Normalized tree's length in the spatial dimension and in 3D for each event

encountered by the bank became public knowledge (see sections "Some Important Events Around the Subprime Crisis" and "Some Important Events Around Lehman Brothers Bankruptcy" in Appendix 1).

However, this increase in the global length of the MST comes with a decrease in certain subsets of the trees. This is especially the case for the Eurodollar market around the Lehman Brothers bankruptcy as shown by the scaled MST in Fig. 5, where there clearly is a shrink in the trees around the two crises. Such a result

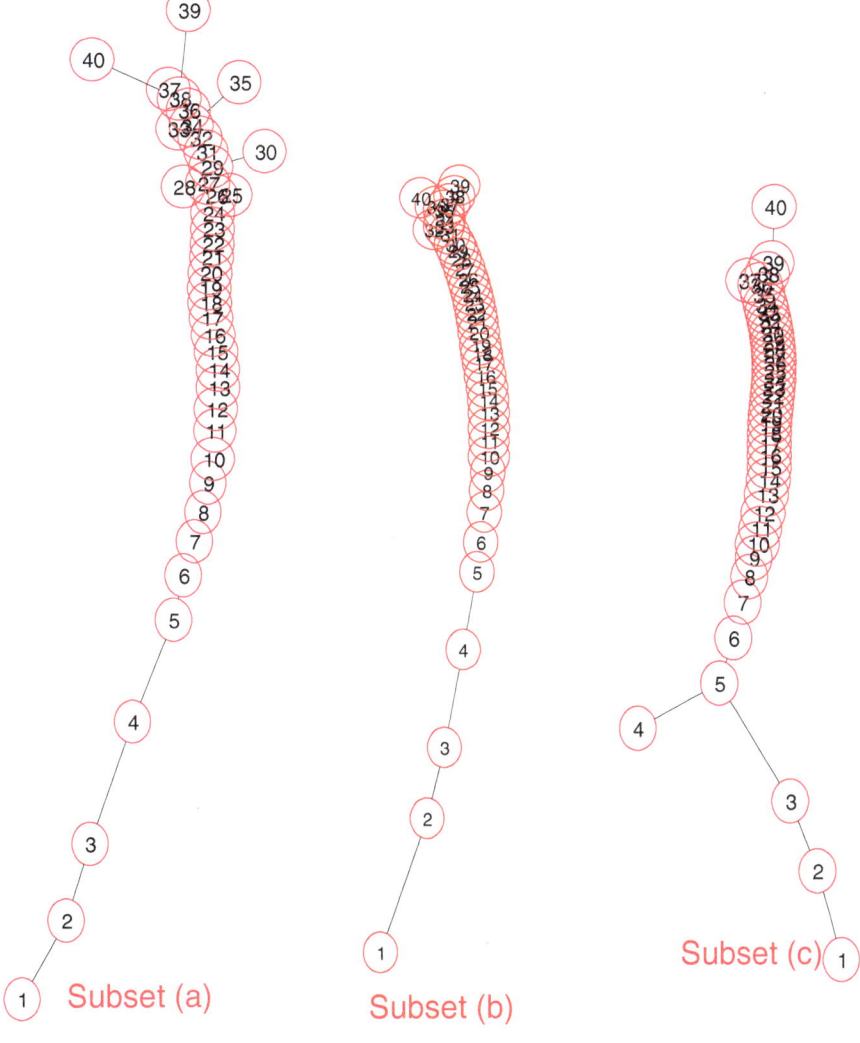

Fig. 5 Scaled MST in the maturity dimension, Eurodollar market. Subset (**a**) 2000–2009 ; subset (**b**) 1-month time window including the subprime crisis; subset (**c**) 1-month time window including the Lehman Brothers bankruptcy

is reasonable: first because the 3-month interest rate is a pure financial asset and second because what we observe here is a branch of the tree where only the maturity dimension is taken into account. As mentioned by Lautier and Raynaud [17], under the pressure of arbitrage operations, the markets are more integrated in the maturity dimension than in the spatial one.

The analysis of the length of the trees shows that, even if our price system becomes more and more integrated between 2000 and 2009, these two crises, born in the financial sphere, did not harm the commodity markets as a whole. This conclusion is consistent with the findings of Buyukşahin and Robe [5] who observed that the link between the equity index and the energy futures is weaker in times of crises or of Corsetti et al. [11] who find that correlations decrease in some episodes of crisis. As expected, these crises had an impact on the financial sphere: there is a local increase in the integration of the futures contracts written on the financial assets. However, as far as commodity markets are concerned, they became temporarily less connected with the financial assets.

4   The Organization of the Tree

Measuring the length of the MST does not give the possibility to ask whether or not the paths for shock transmission change during the crises. In order to answer this question, the graph theory provides several tools: first the survival ratios and second the allometric coefficients.

4.1   The Survival Ratios

This measure (S_R) indicates the fraction of links that survives, in the MST, between two consecutive trading days [9]:

$$S_R(t) = \frac{1}{N-1} |E(t) \cap E(t-1)|. \tag{4}$$

In this equation, $E(t)$ refers to the set of the tree edges at date t, $\cap$ is the intersection operator and $|\,.\,|$ gives the number of elements contained in the set. Due to the finite number of links, the ratios take discrete values.

The use of this measure naturally raises the same question as before: how exceptional are the values of the survival ratios observed around the crises? As before, we evaluate the probability of the occurrence of high reconfigurations in the graph. We find that only the changes recorded on September 18 and 19 of 2008 (the 17th is close) are below the 5 % probability of occurrence in the spatial dimension. In 3D, only September 17 and 24 of 2008 appear below the 5 % threshold. According to these figures, the subprime crisis shows nothing specific:

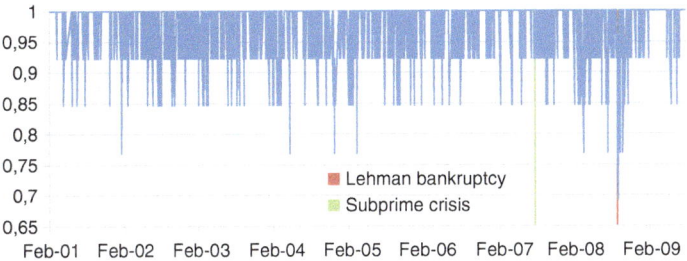

Fig. 6 Survival ratios in the spatial dimension, 2000–2009

even if, as shown by the length of the MST, the trees locally shrink in the financial sphere on this occasion, the path of the price shock transmission remains the same.

This result is confirmed by Fig. 6. The figure shows first that under normal circumstances, the topology of the trees is very stable between two dates, in the spatial dimension as well as in 3D: most of the time, between 2000 and 2009, more than 85 % of the links remain unchanged from one day to the next. Second, nothing special happens around the subprime crisis. This is far less obvious for the Lehman Brothers bankruptcy. In this case, the most important reorganizations appear in the spatial dimension, where more than 30 % of the graph is reorganized.

Finally, while some fluctuations of the survival ratios might be due to real changes in the behavior of the system, it is worth noting that others might simply be due to noise. This is why a deeper analysis is needed. We will perform it through the use of allometric coefficients.

4.2 The Allometric Coefficients

The computation of the allometric coefficients of a MST quantifies where this tree stands between two asymptotic topologies: star-like trees and chain-like trees. These two topologies have very different implications for systemic risk.

The first model of the allometric scaling on a spanning tree was developed by Banavar et al. [1]. In their method, the first step consists in assigning a value A_i equal to 1 to each node i. Then the root (also called the central node) of the graph must be identified. In what follows, the root is determined with Bonacich's measure defined in Sect. 5. As a robustness check, we perform the same tests with a root identified as the node with the highest number of links. The results remain qualitatively the same and are available on request.

Starting from the root, the second step of the method consists in updating the coefficients A_i and in assigning the coefficients B_i of each node i as follows:

$$A_i = \sum_j A_j + 1 \text{ and } B_i = \sum_j B_j + A_i, \tag{5}$$

Fig. 7 Star-like structure

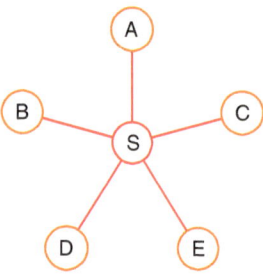

Fig. 8 Chain-like structure

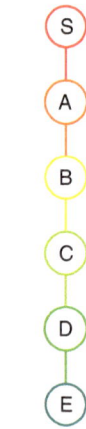

where j stands for all of the nodes connected to i in the MST. The allometric scaling relation is defined as the relationship between A_i and B_i:

$$B \sim A^\eta, \tag{6}$$

where η is the allometric exponent. It represents the degree or complexity of the tree and stands between two extreme values: 1^+ for star-like trees (Fig. 7) and 2^- for chain-like trees (Fig. 8).

A MST belonging to the first or to the second structure will not have the same implications in terms of shock transmission. One way to explain such an interpretation is to rely once again on the analogy with the transmission of a signal in a network. Let us assume that a signal is transmitted in each network represented by Figs. 7 and 8. In each case, the signal is transmitted from node S at time t and there is some latency in the transmission. In the star-like tree, all of the others nodes (A, B, C, D and E) will receive the transmission simultaneously at time $t + 1$. Comparatively, in the chain-like tree, the first receiver is node A, the second is node B, etc. In such a topology with N nodes, it takes $N - 1$ time periods (i.e., five in Fig. 8) before reaching the end of the network. Meanwhile, if there is noise in the transmission channel, the signal will suffer some losses. In our case, where the distances in the networks stand for correlations between price returns, a price

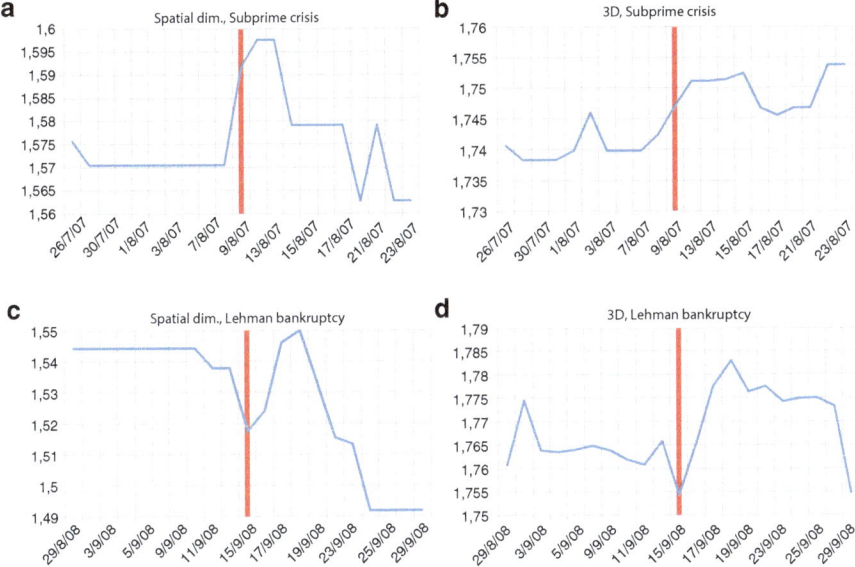

Fig. 9 Allometric coefficients, in the spatial dimension and in 3D, for each event

shock emerging at node S will spread more efficiently if the structure of the tree is star-like, because it will more quickly reach all of the other nodes. It is thus crucial to correctly identify the center of the graph.

Relying on the allometric coefficients, [17] show that: (1) the MST are almost linear in the maturity dimension of most markets, (2) they stand right in the middle of the two extreme configurations in the spatial dimension at 1.5, and (3) the allometric coefficients are around 1.75 in the 3D case. Around the crises, as shown by Fig. 9, the levels of the allometric coefficients remain the same. Moreover, their variations are *not* exceptional at 5 % except those recorded in 3D on September 2, 2008 and on September 29, 2008, around the bankruptcy of Lehman Brothers.

5 Examining the Centrality of the Graphs

When studying systemic risk, it is important to correctly detect the center of the trees. For regulatory authorities, such nodes can be assimilated to regions of higher fragility. Even though we examine exogenous events in this study, the question of centrality remains crucial. What if these events create shocks that reach the center of the graph? They would then spread rapidly to all of the other markets, as noted in the above subsection.

The most common way to identify the center of a graph is to assess the degree (i.e., the number of links) of each node in the trees. However, such an analysis might be insufficient: first because it does not take into consideration the distances between the nodes, and second because it only accounts for the direct neighbors of a node. It could be interesting, on the contrary, to be mindful of the overall configuration of the graph.

In what follows, we first give an example of an analysis based on degree only: we focus on the evolution of the trees in the spatial dimension around the Lehman Brothers bankruptcy. As noted on the basis of the survival ratios, there is indeed an important reconfiguration of the graph on this occasion. Then, we propose the use of a new measure of centrality that was introduced by Bonacich in 1987 for social networks and recently used by Bloch and Quérou [2], as well as, in finance, by Cohen-Cole et al. [10].

5.1 The Degree of the Nodes

The scaled MST in the spatial dimension at the Lehman Brothers bankruptcy is depicted by Fig. 10. If we compare this tree with the one computed for the whole period as illustrated by Fig. 1 (as shown in Sect. 4.1 the MST is very stable; the tree computed on the whole period can thus be taken as a reference) then we can see some changes: the Mini S&P500 is not linked to soy oil anymore, but now to wheat; the UK natural gas is not directly connected to the energy sector anymore; and, more importantly, gold now stands at the center of the graph. From an economic point of view, such a result is very reasonable. In a situation where high uncertainty affects the whole financial system, we indeed expect investors to consider gold as a reserve of value. Yet the story is not so simple.

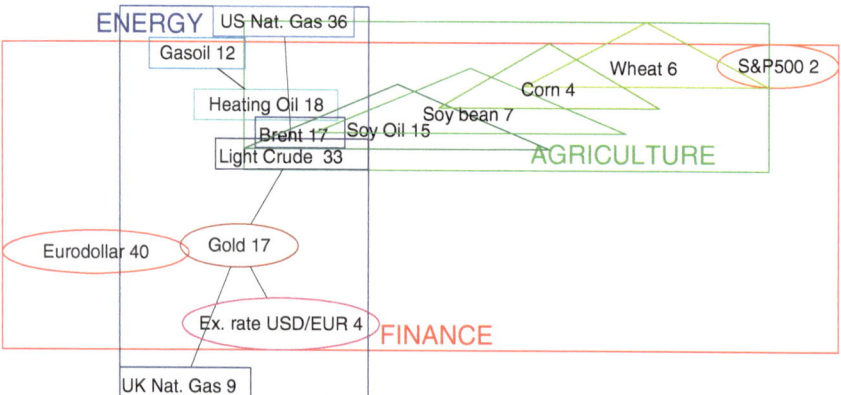

Fig. 10 Scaled MST in the spatial dimension at Lehman Brothers bankruptcy (September 15, 2008)

Fig. 11 Connectivity (A, B, or C) versus centrality (S)

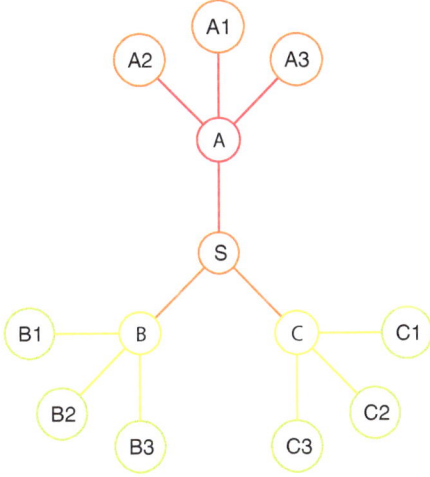

5.2   The Katz-Bonacich Centrality Measure

We first present the method and its advantages. Then we use it for the event study.

5.2.1   The Method

The Katz-Bonacich centrality measure aims at taking into consideration the whole configuration of a graph, that is, the direct as well as the indirect neighbors of a node. Looking only at the direct neighbors, as done when one relies on the degree, might be insufficient as illustrated by Fig. 11: the node labelled "A" (or B or C) exhibits the highest degree (four in this case). However, the "S" node is obviously the most central one.

The measure proposed by Bonacich [3] is an extension of the one developed by Katz [16]. This author was the first to pay attention to the indirect neighbors of a node. In addition, the measure developed by Bonacich [3] gives the possibility of taking into account the "negative" relations, i.e. the fact that, if the value of a node increases, then its neighbors' value decreases.

The centrality vector, which gives one value per node, is computed as follows:

$$c\,(\alpha, \beta) = \alpha\,(\mathbf{I} - \beta\mathbf{R})^{-1}\,\mathbf{R}\mathbf{1}$$

where $\mathbf{I}$ is the identity square matrix, $\mathbf{R}$ is the matrix of the weights of the graph, and $\mathbf{1}$ a vector of 1s. The coefficient α is a scale factor. According to Bonacich, the coefficient β can be interpreted in different ways: "the degree to which an individual's status is a function of the statuses of those to whom he or she is connected" or "a radius within which the researcher wishes to assess centrality". Note also that the centrality values are sensitive to both the weights of the graph and its topology. Since these values take into account infinitely far neighbors, a small change in the topology of the graph can result in large changes in the centrality values.

The use of this relationship matrix requires first a measure of similarity: the quantities in **R** must be such that, the higher the β, the easier the transmission. A second requirement is that all R_{ij} are positive. Third the R_{ii} must be equal to zero. To fulfill the first requirement, we use the correlation matrix for **R**. More precisely, because we are interested in the central node of the MST, we consider the price correlations in the MST, and we compute **R** as follows:

$$R_{ij}(t) = C_{ij}(t) * E_{ij}^{MST}(t),$$

where $\mathbf{C}(t)$ is the correlation matrix and $\mathbf{E}^{MST}(t)$ is the edge matrix of the tree; $E_{ij}^{MST}(t)$ equals to one if there is a link between i and j in the MST and zero otherwise. This matrix is symmetric, with $N - 1$ ones.

The use of the filtered correlation matrix for **R** simplifies the application of the method developed by Bonacich. This matrix can be directly identified to **R**, because it fits all of the requirements. Moreover, such a choice leads to more precise results, because it allows for taking into account the specific value of each link instead of averaging them into a β coefficient (which we thus drop).

5.2.2 Empirical Results

For comparison purposes, it is interesting to go back to the scaled MST in the spatial dimension commented on in Sect. 3 and represented by Fig. 1. When relying on the degrees of the nodes, the root of the tree corresponds to crude oil. However, taking into account the overall organization of the tree leads to a conclusion that is more nuanced. Table 2 presents the results of the method when it is applied in the spatial dimension between 2000 and 2009. Relying on the centrality measures leads to putting more emphasis on both heating oil and crude oil; the heating oil is ranked first. Moreover, a dynamic analysis shows that, especially after August 17, 2005, the agricultural markets play a more important role. This result calls for further analysis, but it is probably due to the introduction of the rules regarding bioethanol in the United States in 2005. Second, half of the markets under consideration in the spatial analysis never reach a centrality value above 1: this is true for the 3-month eurodollar, the USD/EUR exchange rate, the Mini S&P500 index, gold, gasoil and for the US and UK natural gases. These markets thus have a centrality that is unusually low and are hence less important.

The results associated with the centrality measures around the crises are depicted, for the spatial dimension, in sections "Ranking by Centrality Measure in the Spatial Dimension, Around the Subprime Crisis (August 9, 2007)" and "Rankings by Centrality Measure in the Spatial Dimension, Around the Lehman Brothers Bankruptcy (September 15, 2008)" in Appendix 2. Once again, the subprime crisis does not affect the organization of the trees, whereas the Lehman Brothers

Table 2 Bonacich's centrality measure in the spatial dimension, 2000–2009

Market	Centrality measure	Rank
Heating oil	1.148228	1
Brent	1.108484	2
Light crude	0.856703	3
Gasoil	0.591487	4
US Natural gas	0.364067	5
Gold	0.231502	6
USD/EUR Exchange rate	0.036973	7
UK Natural gas	0.034241	8
Eurodollar	−0.00875	9
Mini S&P500	−0.189855	10
Wheat	−1.144788	11
Soy oil	−1.159204	12
Corn	−1.890017	13
Soy bean	−1.979338	14

bankruptcy has an impact (mostly temporary, though). Around this event, the ranking of the nodes puts light crude oil first, gold second and heating oil third.

In 3D, the most central nodes are about the same as in the spatial dimension. Due to the large number of nodes (220), we cannot display the tables in this case but the results are available on request. As before, we do not find many changes around the subprime crisis and many more around the Lehman Brothers bankruptcy.

Finally, the most interesting phenomena appear in the maturity dimension around the Lehman Brothers bankruptcy. There are some changes in the direction of certain propagation paths. The most illustrative example of such behavior is that of light crude on September 10, 2008: before that date, many short-term maturities of light crude oil are among the most central nodes of the tree (they are situated above the rank of 20 according to the centrality measure), while most of the long-term maturities are among the least central (below the rank 200). From one day to the next, however, there is an inversion: the least central nodes become the most central ones (they even reach the rank 1) while the previously most central ones go as low as rank 220. Finally, things revert back to the initial state.

6   Conclusion

For a decade, commodity derivative markets have been experiencing a process of financialization due to managers seeking the diversification of their portfolios and to the arrival of new actors. This phenomenon has raised questions and worries about the eventuality of meaningless links, from an economic point of view, between commodities and more traditional financial markets like bonds and stocks. These fears have been largely confirmed by the acknowledgment of a growing integration between commodity markets as well as between commodities and other financial assets. One could wonder to what extent a shock originating from financial markets could propagate to commodities and strongly impact them. Investigating such a question is the purpose of this paper.

To this aim we examine the impact, on commodity markets of two recent financial crises: the Subprime crisis and the Lehman Brothers bankruptcy. Using the insightful tools of the graph theory, on the basis of several measures, we show that those shocks did not affect the commodity markets as hard as one might have expected.

Acknowledgements This paper is based on work supported by the Chair Finance and Sustainable Development and the FIME Research Initiative. Comments by the referees and the audience at the 7th Financial Risks International Forum and at the 31st AFFI conference are gratefully acknowledged. The same is true for fruitful remarks from the reviewer of this special volume and from Michel Robe.

Appendix 1: Timelines Around the Events

Some Important Events Around the Subprime Crisis

Based on [4], News feeds, Wikipedia (Table 3).

Table 3 Important events around the subprime crisis (S denotes the date of the trigger of the crisis, on August 9th, 2008)

Trading date	Calendar date	Events
S-10	2007-07-26	• Home sales declined and largest home builder reported loss
S-7	2007-07-31	• American Home Mortgage Investment Corp. faces difficulties
S-6 to S	2007-08-01–2007-08-09	• Quantitative hedge funds suffered losses that trigger margin calls, fire sales, and correlation across strategies
S-6	2007-08-01	• US Crude oil prices reach a new high due to declining stocks and decreased output
S-4	2007-08-03	• Officials state that the housing crisis should not spread
S-3	2007-08-06	• America Home Mortgage Investment Corp. goes bankrupt
S	2007-08-09	• BNP Paribas froze redemption of 3 of its investment funds due to inability to value structured products • Triggered the first illiquidity wave on the interbank market and support from Central Banks
S+1	2007-08-10	• Decreases propagate to Asian markets, triggering support from Central Banks
S+2 to S+8	2007-08-13–2007-08-21	• Central Banks increase their support and lower rates

Some Important Events Around Lehman Brothers Bankruptcy

Based on [4], News feeds, Wikipedia (Table 4).

Appendix 2: Additional Results

Robustness Checks

This section of the appendix is devoted to a sensitivity analysis. It provides the results obtained around the two events, with the different measures used in the analysis (length of the MST, survival ratios and allometric coefficients) when the rolling window is extended to 2 years instead of 1 year. The comparison shows that overall, the behavior remains qualitatively the same. As expected, compared with the 1-year rolling window, the 2-year window has a smoothing effect (Figs. 12, 13, and 14).

Table 4 Important events around Lehman Brothers bankruptcy (L denotes the date of Lehman Brothers default, on September 15, 2008)

Trading date	Calendar date	Events
L-6	2008-09-05	US Government's plan to bail out Fannie Mae and Freddie Mac leaks
L-3	2008-09-10	OPEC will cut oil production by 500,000 barrels a day
		Announcement of the worst losses of Lehman
L-1	2008-09-12	The Federal Reserve tries to find buyers for Lehman and warns CME of a potential default
L	2008-09-15	Lehman files for bankruptcy in the morning, because of lack of buyers and of bail out
		Merrill Lynch is sold to Bank of America
L+1	2008-09-16	AIG is bailed out
L+2	2008-09-17	Russia helps its biggest banks
L+3	2008-09-18	Russia extends help
		Lloyds TSB purchases HBOS, largely exposed to subprime mortgages
L+4	2008-09-19	The Troubled Asset Relief Program leaks
		US Treasury guarantees money market mutual funds up to $50 billion
		Nigerian oil production is cut by 280,000 barrels per day and a pipeline of Royal Dutch Shell was destroyed
L+5	2008-09-22	G7 commits to protect the financial system
L+9	2008-09-26	The Federal Deposit Insurance Corporation seizes Washington Mutual to sell it to JPMorgan Chase

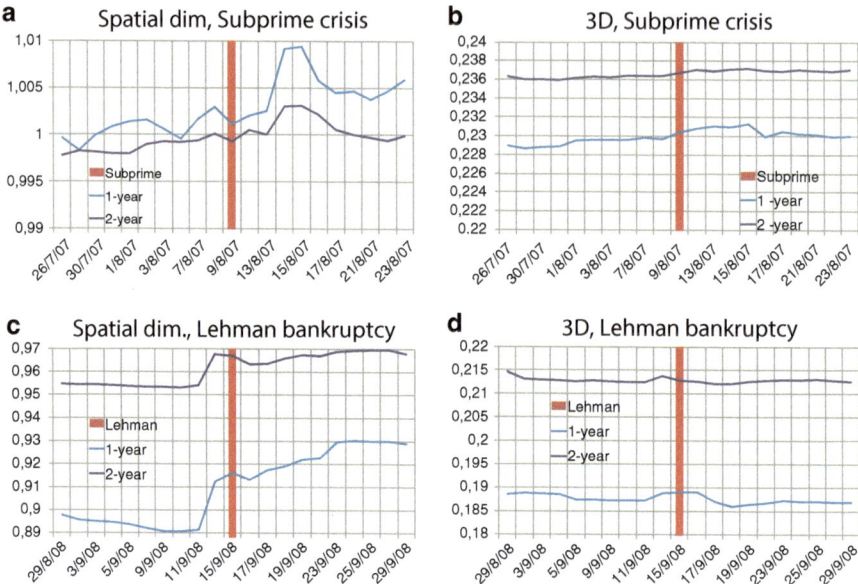

Fig. 12 Normalized length in the spatial dimension and in 3D, around the events

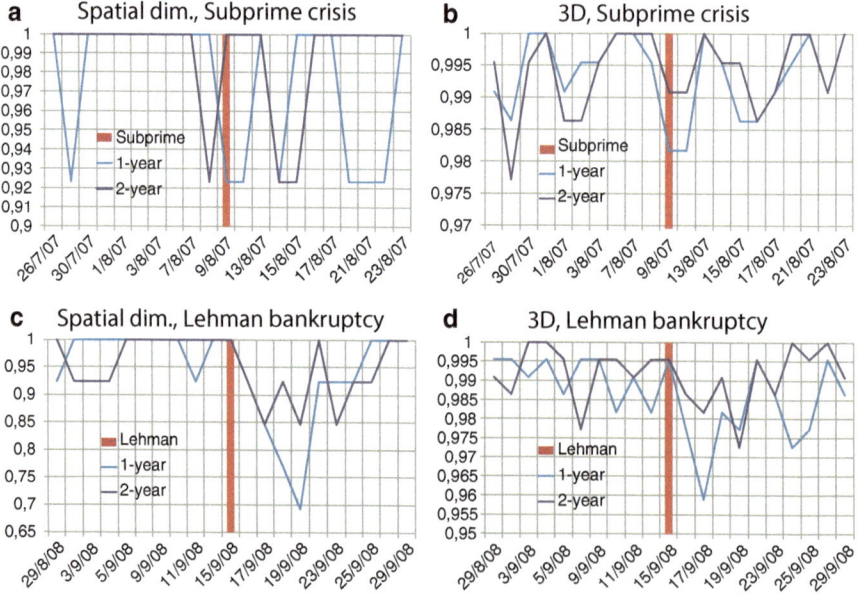

Fig. 13 Survival ratios in the spatial dimension and in 3D, around the events

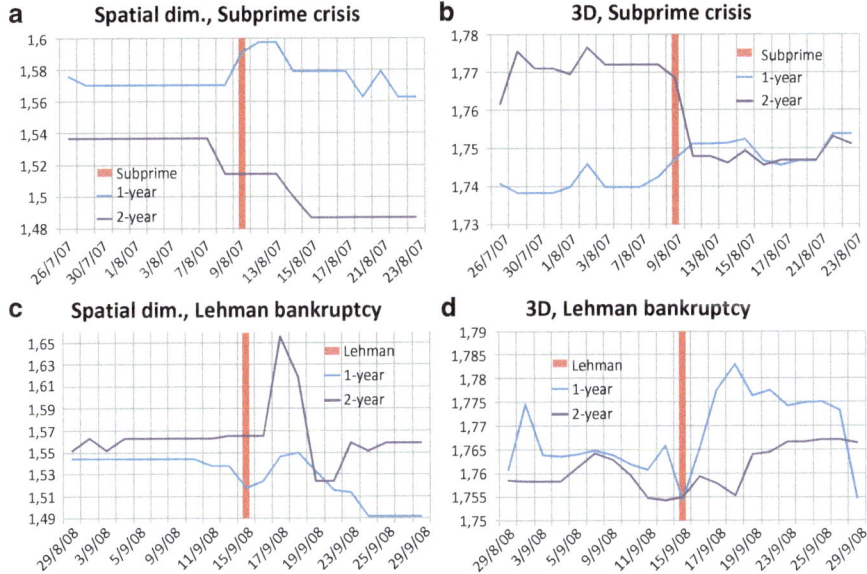

Fig. 14 Allometric coefficients in the spatial dimension and in 3D, around the events

Evolution of the Markets Rankings by Centrality, Around the Events and Sector by Sector

Ranking by Centrality Measure in the Spatial Dimension, Around the Subprime Crisis (August 9, 2007) (Fig. 15)

Rankings by Centrality Measure in the Spatial Dimension, Around the Lehman Brothers Bankruptcy (September 15, 2008) (Fig. 16)

Fig. 15 Evolution of the ranks in the spatial dimension around the Subprime crisis. Subset (**a**) is for agricultural markets, subset (**b**) is for only 4 energy markets (for readability) and subset (**c**) is for financial markets

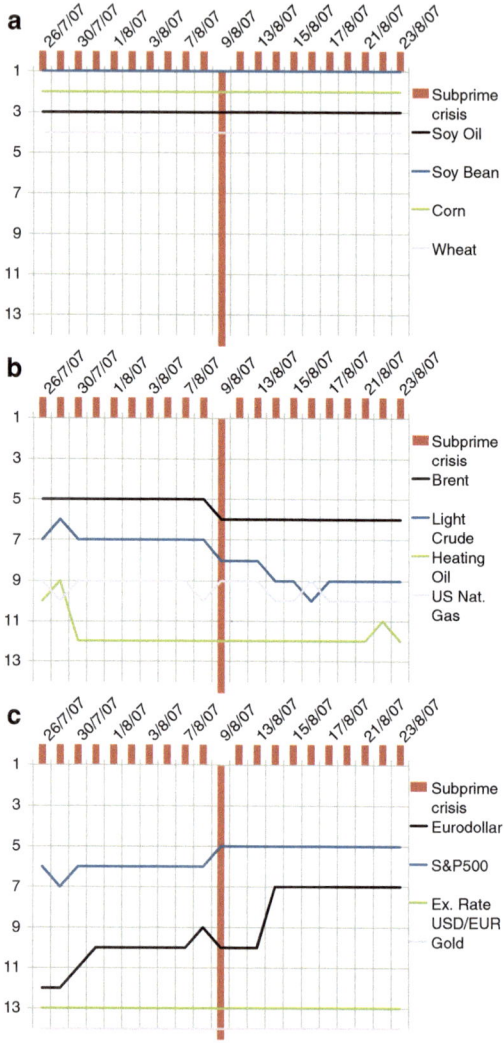

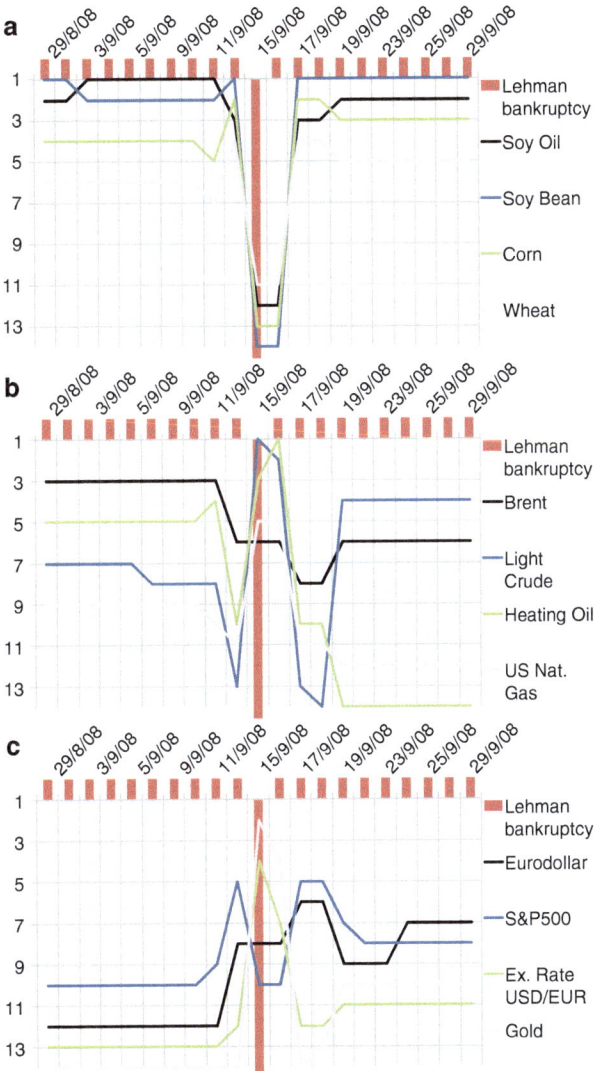

Fig. 16 Evolution of the ranks in the spatial dimension, around the Lehman Brothers bankruptcy. Subset (**a**) is for agricultural markets, subset (**b**) is for only 4 energy markets (for readability) and subset (**c**) is for financial markets

References

1. Banavar, J.R., Maritan, A., Rinaldo, A.: Size and form in efficient transportation networks. Nature **399**(6732), 130–132 (1999)
2. Bloch, F., Quérou, N.: Pricing in social networks. Games Econ. Behav. **80**, pp. 243–261 (2013)
3. Bonacich, P.: Power and centrality: a family of measures. Am. J. Sociol. **92**(5), 1170–1182 (1987)
4. Brunnermeier, M.K.: Deciphering the liquidity and credit crunch 2007–2008. J. Econ. Perspect. **23**(1), 77–100 (2009)
5. Buyukşahin, B., Robe, M.A.: Does paper oil matter? Energy markets' financialization and equity-commodity co-movements. Working Paper, American University, (Revised) July (2011)
6. Buyukşahin, B., Robe, M.A.: Speculators, commodities and cross-market linkages. J. Int. Money Finance **42**, 38–70 (2014)
7. Buyukşahin, B., Haigh, M.S., Robe, M.A.: Commodities and equities: ever a "market of one"? J. Altern. Invest. **12**(3), 76–95 (2010)
8. Buyukşahin, B., Robe, M.A., Bruno, V.G.: The financialization of food? Working Paper (2014)
9. Chakraborti, A., Kaski, K., Kertész, J., Onnela, J.P.: Dynamic asset trees and black Monday. Physica A **324**(1–2), 247–252 (2003)
10. Cohen-Cole, E., Kirilenko, A., Patacchini, E., Fouque, J., Langsam, J.: Strategic interactions on financial networks for the analysis of systemic risk. In: Handbook on Systemic Risk, Cambridge University Press, p. 306 (2012)
11. Corsetti, G., Pericoli, M., Sbracia, M.: Some contagion, some interdependence: more pitfalls in tests of financial contagion. J. Int. Money Finance **24**(8), 1177–1199 (2005)
12. Fattouh, B., Killian, L., Mahadeva, L.: The role of speculation in oil markets: what have we learned so far? Energy J. **34**(3), 7–33 (2013)
13. Forbes, K.J., Rigobon, R.: No contagion, only interdependence: measuring stock market comovements. J. Finance **57**(5), 2223–2261 (2002)
14. Fréchet, M.: Sur quelques points du calcul fonctionnel. Rendiconti del Circolo Mathematico di Palermo **22**, 1–74 (1906)
15. Irwin, S.H., Sanders, D.R.: Index funds, financialization, and commodity futures markets. Appl. Econ. Perspect. Policy **33**(1), 1–31 (2011)
16. Katz, L.: A new status index derived from sociometric analysis. Psychometrika **18**(1), 39–43 (1953)
17. Lautier, D., Raynaud, F.: Systemic risk in energy derivative markets: a graph-theory analysis. Energy J. **33**(3), 215–239 (2012)
18. Mantegna, R.N.: Hierarchical structure in financial markets. Eur. Phys. J. B **11**, 193–197 (1999)
19. Onnela, J.P., Chakraborti, A., Kaski, K., Kertesz, J., Kanto, A.: Dynamics of market correlations: taxonomy and portfolio analysis. Phys. Rev. E **68**(5), 056110 (2003)
20. Pesaran, M.H., Pick, A.: Econometric issues in the analysis of contagion. J. Econ. Dyn. Control **31**(4), 1245–1277 (2007)
21. Silvennoinen, A., Thorp, S.: Financialization, crisis and commodity correlation dynamics. J. Int. Financ. Mark. Inst. Money **24**, 42–65 (2013)
22. Tang, K., Xiong, W.: Index investing and the financialization of commodities. Financ. Anal. J. **68**(6), 54–74 (2012)

Margrabe Revisited

Hans J.H. Tuenter

Abstract We introduce a new representation of the bivariate normal distribution to first give a short derivation of the classic Margrabe exchange-option formula, using elementary integration methods. The second application is a new and simple technique to provide an accurate lower bound for the value of a spread option with a nonzero strike.

1 Introduction

Exchange options were introduced by William Margrabe in a seminal paper [7], published in 1978. This type of option allows the holder to exchange one asset for another at expiration. Such options are ubiquitous in foreign exchange markets, bond markets, stock markets, and commodity markets, among others. In energy markets, in particular, they have found applications in locational spreads, calendar spreads, crack spreads, and spark spreads. (See Clewlow and Strickland [4, pp. 80–81], Geman [5, pp. 287–294], and Pilipovic [8, pp. 361–374].) The survey by Carmona and Durrleman [2] provides a good introduction to the topic.

Margrabe studied European-style exchange options in a Black-Scholes framework, where the rate of return on each asset is given by

$$d S_i(t) = S_i(t) \left[r \, dt + \sigma_i \, dW_i(t) \right], \quad i = 0, 1, \tag{1}$$

with r the risk-free interest rate, σ_i the instantaneous volatilities, $W_i(t)$ Wiener processes, and ρ the correlation coefficient between the increments dW_0 and dW_1. The payoff on the option to exchange S_0 for S_1 at time T, is given by

$$\left(S_1(T) - S_0(T) \right)^+, \tag{2}$$

where $x^+ = \max\{x, 0\}$.

H.J.H. Tuenter (✉)
Mathematical Finance Program, University of Toronto, Toronto, ON, Canada
e-mail: hans.tuenter@utoronto.ca

© Springer Science+Business Media New York 2015
R. Aïd et al. (eds.), *Commodities, Energy and Environmental Finance*,
Fields Institute Communications 74, DOI 10.1007/978-1-4939-2733-3_4

Margrabe derived the risk-neutral value of this option as

$$e^{-rT} \mathbb{E}\left(S_1(T) - S_0(T)\right)^+ = S_1(0)\Phi(d_+) - S_0(0)\Phi(d_-), \tag{3}$$

where $\mathbb{E}$ denotes the expectation operator, Φ the cumulative density function of the standard normal distribution, and

$$d_\pm = \frac{\ln(S_1(0)/S_0(0))}{\sigma\sqrt{T}} \pm \frac{1}{2}\sigma\sqrt{T} \quad \text{and} \quad \sigma^2 = \sigma_0^2 + \sigma_1^2 - 2\rho\sigma_0\sigma_1.$$

He obtains his formula by deriving a partial differential equation for the price of the option, together with its initial and boundary conditions. He postulates a solution and shows that it is the unique solution by employing a change-of-numéraire approach that transforms the valuation into a Black-Scholes type problem.

The first objective of this article is to provide a brief and simple derivation of the Margrabe formula that is based on a new representation of the bivariate normal distribution. This approach reduces the derivation to an elementary integration, and improves on previous approaches using plain integration, such as the one used by Li et al. [6, Proposition 2]. The second and main objective is to showcase a new lower bound for the value of a spread option with a nonzero strike, similar to the one derived by Carmona and Durrleman [2, § 6.1], but arrived at by a much simpler technique.

2   Bivariate Normal Distribution

In the Black-Scholes framework the logarithms of the asset prices at maturity follow a bivariate normal distribution. So, it seems only natural to first study the expected value of $(X_1 - X_0)^+$, where $\ln X_0$ and $\ln X_1$ are correlated normal variables, without the distraction of the stochastic process that generates them.

We derive this expectation for a particular case that is readily evaluated, and then show that the general case can always be mapped to it. This implies that, within the Black-Scholes framework, the particular case can be interpreted as a canonical formulation for an exchange option.

2.1   The Particular Case

As the particular case, we take

$$\ln X_0 = \mu_0 + aY + bZ \quad \text{and} \quad \ln X_1 = \mu_1 + aY + cZ, \tag{4}$$

where $a \geq 0$, $b > c$, and Y and Z are independent, standard normal variables.

Lemma 1.

$$\mathbb{E}\,(X_1 - X_0)^+ = \mathbb{E}\,X_1\,\Phi\left(z^* - c\right) - \mathbb{E}\,X_0\,\Phi\left(z^* - b\right),\tag{5}$$

where $z^* = (\mu_1 - \mu_0)/(b - c)$.

Proof. Substitute the expressions for X_0 and X_1, separate the factor e^{aY}, and take the expectation over Y, to give

$$\mathbb{E}\,(X_1 - X_0)^+ = e^{\frac{1}{2}a^2}\mathbb{E}\left(e^{\mu_1 + cZ} - e^{\mu_0 + bZ}\right)^+.\tag{6}$$

The value for Z, that renders the expression within brackets equal to zero, is given by $z^* = (\mu_1 - \mu_0)/(b - c)$. The expression is positive for values of Z smaller than z^*, and negative for values of Z larger than z^*. Now integrate over Z, and simple algebra, combined with the expectations $\mathbb{E}\,X_1 = e^{\mu_1 + \frac{1}{2}a^2 + \frac{1}{2}c^2}$ and $\mathbb{E}\,X_0 = e^{\mu_0 + \frac{1}{2}a^2 + \frac{1}{2}b^2}$, will show the validity of (5), and proves the lemma. $\qquad\square$

Note that the constant z^* also has significance in that $\Phi(z^*)$ is the probability that X_0 is smaller than or equal to X_1, and the corresponding exchange option pays out.

- The condition $b > c$ is not really a restriction, as we can switch easily from the case $b < c$, by taking $-Z$ instead of Z in (4), using the fact that the standard normal distribution is symmetric. It was chosen for convenience in the proof of Lemma 1 to give the range of integration for Z as $(-\infty, z^*]$.
- The special case $b = c$ implies that $\ln X_0$ and $\ln X_1$ are the same random variable, except for a difference in their mean. The valuation in this case is simple as $\mathbb{E}\,(X_1 - X_0)^+ = \mathbb{E}\left(e^{\mu_1 + \sigma_0 Z} - e^{\mu_0 + \sigma_0 Z}\right)^+ = (e^{\mu_1} - e^{\mu_0})^+ e^{\frac{1}{2}\sigma_0^2}$, and is equal to $\mathbb{E}\,X_1 - \mathbb{E}\,X_0$, when $\mu_1 > \mu_0$, and zero otherwise. This is not a practical case that one would encounter in the setting of an exchange option. However, it is worth noting that Lemma 1 includes this as a boundary case and thus ensures continuity of solution. Taking the limit of $(\mu_1 - \mu_0)/(b - c)$, as b approaches c from above, gives $z^* = +\infty$ and $z^* = -\infty$, when $\mu_1 > \mu_0$ and $\mu_1 < \mu_0$, respectively, so that (5) gives the correct limit values.

2.2 The General Case

In the general case, we have a bivariate normal distribution where the distributions of the logarithm of X_0 and X_1 are normal with mean μ_0 and μ_1, standard deviation σ_0 and σ_1, and correlation coefficient ρ. We make the very mild assumption that $\sigma_0^2 + \sigma_1^2 \neq 2\rho\sigma_0\sigma_1$. These two variables can be represented in several ways as a linear combination of independent, standard normal variables. The linear combination that is of interest in our setting is the following:

$$\ln X_0 = \mu_0 + \frac{\sigma_0}{\sqrt{\sigma_0^2 + \sigma_1^2 - 2\rho\sigma_0\sigma_1}} \left(\sigma_1 \sqrt{1 - \rho^2}\, Y + (\sigma_0 - \rho\sigma_1)Z \right) \qquad (7)$$

and

$$\ln X_1 = \mu_1 + \frac{\sigma_1}{\sqrt{\sigma_0^2 + \sigma_1^2 - 2\rho\sigma_0\sigma_1}} \left(\sigma_0 \sqrt{1 - \rho^2}\, Y + (\rho\sigma_0 - \sigma_1)Z \right), \qquad (8)$$

where Y and Z are independent, standard normal variables. It is easy to verify that this construct gives two normal variables with the required means, standard deviations and correlation coefficient. Now note that the coefficients of Y in (7) and (8) are identical and nonnegative, and that the coefficient of Z in (7) is strictly larger than the coefficient of Z in (8). This implies that (7) and (8) are of the form (4), with

$$a = \frac{\sigma_0\sigma_1}{\sigma} \sqrt{1 - \rho^2}, \quad b = \frac{\sigma_0}{\sigma}(\sigma_0 - \rho\sigma_1), \quad \text{and} \quad c = \frac{\sigma_1}{\sigma}(\rho\sigma_0 - \sigma_1), \qquad (9)$$

where $\sigma^2 = \sigma_0^2 + \sigma_1^2 - 2\rho\sigma_0\sigma_1$. We note the following properties: $a^2 + b^2 = \sigma_0^2$, $a^2 + c^2 = \sigma_1^2$, and $b - c = \sigma$.

- The above shows that one can always cast the general bivariate normal distribution into the form (4); thus, they are equivalent and justifies referring to (4) as a canonical formulation. The representation of the bivariate normal distribution, given in (7) and (8), seems to be new. Although, it should be noted that, when $\sigma_0 = \sigma_1 = 1$ (and $\mu_0 = \mu_1 = 0$), it reduces to a well-known form that is used to generate correlated, standard normal variables. (See Tong [9, p. 11].)
- We imposed the condition $\sigma_0^2 + \sigma_1^2 \neq 2\rho\sigma_0\sigma_1$, but this is not restrictive. Equality holds if, and only if, $\rho = 1$ and $\sigma_1 = \sigma_0$. The implication is that $\ln X_0$ and $\ln X_1$ are the same random variable, except for a difference in their mean. This case was dealt with in the last bullet point of Sect. 2.1.

3   Margrabe's Formula

As noted in the introduction, the risk-neutral value of the exchange option within the Black-Scholes framework is given by $e^{-rT} \mathbb{E}\, (S_1(T) - S_0(T))^+$, where $S_0(T)$ and $S_1(T)$ are correlated lognormal variables. This means that we can apply Lemma 1, and, to do this, we take $X_1 = e^{-rT} S_1(T)$ and $X_0 = e^{-rT} S_0(T)$.

It is straightforward to show that this implies $\mathbb{E}\, X_i = S_i(0)$, $\mu_i = \ln S_i(0) - \frac{1}{2}\sigma_i^2 T$, $b = \frac{\sigma_0}{\sigma}(\sigma_0 - \rho\sigma_1)\sqrt{T}$ and $c = \frac{\sigma_1}{\sigma}(\rho\sigma_0 - \sigma_1)\sqrt{T}$, with $\sigma^2 = \sigma_0^2 + \sigma_1^2 - 2\rho\sigma_0\sigma_1$. Simple substitution of these expressions in (5) gives the Margrabe formula (3).

4   Technical Interlude

In the subsequent analysis, it turns out to be beneficial to distinguish between a few basic case types for the canonical formulation (4). These correspond to whether or not the volatility parameters b and c are each positive or negative. Since we imposed the condition $b > c$, this gives three cases, as listed in Table 1.

The breakdown into the different case types has a straightforward interpretation in the general formulation in terms of a bound for the correlation coefficient. Note that, all things being equal, type II is the one most likely to be encountered, as it covers negative, zero and small positive correlations. We also note that, when one is holding the asset with lower volatility, it corresponds to either type II or III, but never to type I. Conversely, when one is holding the asset with higher volatility, it corresponds to either type I or II, but never to type III.

4.1   Classification and Roots

The classification into different case types is not merely an exercise in taxonomy, but plays a key role in the subsequent section on spread options. To ease the notational burden, let us define the function

$$f(z) = e^{\mu_1 + cz} - e^{\mu_0 + bz}. \tag{10}$$

This function already appeared in the proof of Lemma 1, in particular Eq. (6), where we had to find the unique root of $f(z) = 0$, in order to determine the range of integration. The situation is different for the general equation $f(z) = k$, as it may have a unique solution, none or two. The basic shape of $f(z)$ depends only on the signs of b and c, as can be seen more clearly from $f(z + z^*) = e^{\mu_1 + cz^*} \left(e^{cz} - e^{bz} \right)$, where z^* is the unique solution to $f(z) = 0$, that already featured in Lemma 1.

The different shapes are graphed in Fig. 1, from which one can infer the location of the roots of $f(z) = k$.

Table 1 Characteristics of the different case types

Formulation	Type I	Type II	Type III
Canonical	$b > c > 0$	$b > 0 > c$	$0 > b > c$
General	$\rho > \sigma_1/\sigma_0$	$\rho < \min\{\sigma_0/\sigma_1, \sigma_1/\sigma_0\}$	$\rho > \sigma_0/\sigma_1$
$\sigma_0 < \sigma_1$	–	$\rho < \sigma_0/\sigma_1$	$\rho > \sigma_0/\sigma_1$
$\sigma_0 > \sigma_1$	$\rho > \sigma_1/\sigma_0$	$\rho < \sigma_1/\sigma_0$	–

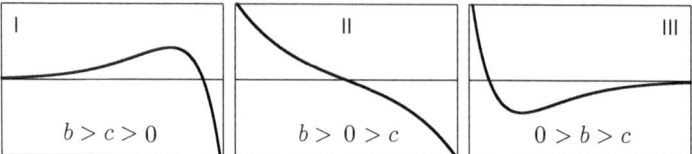

Fig. 1 The shape of $f(z)$ for the different case types

4.2 Case Type Analysis

To simplify the analysis, we will work with the normalized version $f_0(t) = e^{ct} - e^{bt}$, as the relevant properties of $f(z)$ are readily inferred from it. The former has exactly one root at zero, and, under the imposed condition $b > c$, is positive for $t < 0$, and negative for $t > 0$. The solvability of $f_0(t) = k$, depends on the case type. Since the basic shapes of $f_0(t)$ are the same as those of $f(z)$, the graphs in Fig. 1 will be helpful in visualizing the analysis of the different case types.

Type I When $b > c > 0$, then $f_0(t)$ has a unique extremum at

$$t^* = -\frac{\ln(b/c)}{b-c}, \quad \text{with value} \quad f_0^* = \left(\frac{b}{c}\right)^{\frac{b-c}{c}} - \left(\frac{b}{c}\right)^{\frac{b-c}{b}}, \tag{11}$$

and this extremum is positive and a maximum. The function $f_0(t)$ has zero as a left asymptote, is strictly increasing for $t < t^*$, and strictly decreasing for $t > t^*$. This implies that $f_0(t) = k$ has no solution when $k > f_0^*$, two solutions when $0 < k < f_0^*$, and one when $k = f_0^*$ or $k \leq 0$.

Type II When $b > 0 > c$, then $f_0(t)$ is a strictly decreasing function of t, and has no minimum or maximum. This implies that $f_0(t) = k$ has exactly one solution, for every value of k.

Type III When $0 > b > c$, we have the mirror image of case type I, and $f_0(t)$ has a unique extremum at t^* with value f_0^*, as given in (11), with the difference that this extremum is negative and a minimum. The function $f_0(t)$ is strictly decreasing for $t < t^*$, strictly increasing for $t > t^*$, and has zero as a right asymptote. This implies that $f_0(t) = k$ has no solution when $k < f_0^*$, two solutions when $f_0^* < k < 0$, and one when $k = f_0^*$ or $k \geq 0$.

4.3 Generalization of Lemma 1

In this section, we consider the expected value of $(X_1 - X_0 - K)^+$, where K is a constant. Using the canonical formulation (4), one can write

$$\mathbb{E}(X_1 - X_0 - K)^+ = \mathbb{E}\left(e^{aY}f(Z) - K\right)^+, \tag{12}$$

where Y and Z are standard normal and independent random variables, and $f(z)$ is as defined in (10). For $K = 0$, the expectation in (12) factorizes and results in (5). For general values of K, a simple and closed-form expression is not known. However, as noted by several researchers, expressions that provide good lower bounds do exist. Carmona and Durrleman [2, § 6.1], in the context of spread options, provide such an expression, although it involves solving a not-so-simple equation with quite a few trigonometric functions. An important feature of their lower bound is that it has a structure that is like that of the Margrabe formula, but with the addition of a term that is linear in K. It turns out that one can use our canonical representation to derive similar lower bounds with far less effort.

Lemma 2. *When the equation*

$$f(z) = Ke^{-\frac{1}{2}a^2}, \tag{13}$$

where K is nonnegative, has one or more solutions and z^* is the largest of them, then

$$\mathbb{E}(X_1 - X_0 - K)^+ \geq \mathbb{E} X_1 \Phi(z^* - c) - \mathbb{E} X_0 \Phi(z^* - b) - K\Phi(z^*). \tag{14}$$

Proof. We use representation (12). By conditioning on the value of Z, and then taking the expectation over Y, one can derive the lower bound

$$\mathbb{E}\left(e^{aY}f(Z) - K\right)^+ \geq \mathbb{E}\left(e^{\frac{1}{2}a^2}f(Z) - K\right)^+. \tag{15}$$

This follows easily from the fact that the function x^+ is convex and an application of Jensen's inequality. The problem is now reduced to that of determining the roots of (13). These roots provide the range of integration that allows the lower bound to be evaluated. From Sect. 4.2, we know that we have either one, none or two roots. When we are dealing with case type II, then Eq. (13) has a unique solution z^*, and the same applies when we are dealing with case type III, as K is nonnegative. This means that we can evaluate the right-hand side of (15) as

$$\int_{-\infty}^{z^*}\left(e^{\frac{1}{2}a^2}f(z) - K\right)d\Phi(z) = \mathbb{E} X_1 \Phi(z^* - c) - \mathbb{E} X_0 \Phi(z^* - b) - K\Phi(z^*).$$

When we are dealing with case type I, then Eq. (13) has roots $z_2^* \leq z_1^*$, as per condition of the lemma, and the right-hand side of (15) evaluates to

$$\int_{z_2^*}^{z_1^*}\left(e^{\frac{1}{2}a^2}f(z) - K\right)d\Phi(z) > \int_{-\infty}^{z_1^*}\left(e^{\frac{1}{2}a^2}f(z) - K\right)d\Phi(z), \tag{16}$$

with the latter integral evaluating to the same expression as for case types II and III, only with z_1^* instead of z^*. This shows, when (13) has a solution, that the largest root provides the lower bound (14) and proves the Lemma. □

- The condition in Lemma 2 that the constant K is nonnegative is not restrictive. Using the fact that $x^+ = x + (-x)^+$, gives the parity formula $\mathbb{E}(X_1 - X_0 - K)^+ = \mathbb{E}X_1 - \mathbb{E}X_0 - K + \mathbb{E}(X_0 - X_1 + K)^+$, so that one can easily convert from a negative value of K to a positive one.
- The condition in Lemma 2 that Eq. (13), where K is nonnegative, has one or more solutions is easily verified. Case types II and III always have exactly one solution. Case type I has one or two solutions if, and only if, $\max f(z) \geq Ke^{-\frac{1}{2}a^2}$. Using the analysis in Sect. 4.2, one can show that this is equivalent to

$$\frac{\mu_1 b - \mu_0 c}{b - c} + \ln f_0^* \geq \ln K - \frac{1}{2}a^2, \tag{17}$$

with f_0^* as defined in (11).
- The convexity-based lower bound (14) has a format that is similar to the lower bound derived by Carmona and Durrleman [2, Eqn. 6.3]. However, the arguments of and inputs to their lower bound are much more complex and their computation more involved. The computational effort to compute our lower bound is limited to a simple root-finding procedure that can be implemented efficiently by a binary search or a Newton-Raphson method, and has no convergence issues.
- The bound (15) holds with equality when either $K = 0$ or $a = 0$. This implies that the convexity bound (14) is exact for the case $K = 0$, and thus that Lemma 2 is indeed a generalization of Lemma 1. When a is zero, this corresponds to X_0 and X_1 being perfectly correlated and $\rho = \pm 1$.

5 Spread Options

In many practical settings it is not really the intent to exchange one asset for another, but to lock in a price differential. In these settings, one has to overcome a fixed cost or pay a price to exercise (or strike) the option. In the set-up of Sect. 2, this means that we are considering the expected value of $(X_1 - X_0 - K)^+$, where K denotes the strike price. The exchange option can thus be seen as a spread option with a zero strike.

5.1 Validation

To validate Lemma 2 and get a sense of its applicability and usefulness in deriving a lower bound for the value of an exchange option, we use the test case from Carmona and Durrleman [3, Table 1], and compare against their results. They take

$S_0(0) = 100, q_0 = 2\%, \sigma_0 = 15\%$, and $S_1(0) = 110, q_1 = 3\%, \sigma_1 = 10\%$, where the q_i represent the continuous dividend yields. The time to maturity is taken as $T = 1$ year and the risk-free rate as $r = 5\%$. For the strike and the correlation coefficient, all combinations of $K \in \{-20, -10, 0, 5, 15, 25\}$ and $\rho \in \{-1, -0.5, 0, 0.3, 0.8, 1\}$ are considered.

Although Margrabe's original formulation did not include dividends (he only considered capital gains), the introduction of the continuous dividend yields q_i does not fundamentally change things. The spot prices are now assumed to follow the stochastic differential equation

$$d\,S_i(t) = S_i(t)\,[(r - q_i)\,dt + \sigma_i\,dW_i(t)]\,, \quad i = 0, 1. \tag{18}$$

The risk-neutral value of the spread option is given by $e^{-rT}\mathbb{E}\,(S_1(T) - S_0(T) - K)^+$. This means that we can apply Lemma 2, with $X_1 = e^{-rT}S_1(T)$ and $X_0 = e^{-rT}S_0(T)$, and a strike price of $e^{-rT}K$. It is straightforward to show that $\mathbb{E}X_i = e^{-q_iT}S_i(0)$, $\mu_i = \ln S_i(0) - (q_i + \frac{1}{2}\sigma_i^2)T$, $a = \frac{\sigma_0\sigma_1}{\sigma}\sqrt{1 - \rho^2}\sqrt{T}$, $b = \frac{\sigma_0}{\sigma}(\sigma_0 - \rho\sigma_1)\sqrt{T}$, and $c = \frac{\sigma_1}{\sigma}(\rho\sigma_0 - \sigma_1)\sqrt{T}$, with $\sigma^2 = \sigma_0^2 + \sigma_1^2 - 2\rho\sigma_0\sigma_1$. Substitution of these expressions in (14) gives the following lower bound for the value of this spread option:

$$e^{-q_1T}S_1(0)\Phi(z^* - c) - e^{-q_0T}S_0(0)\Phi(z^* - b) - e^{-rT}K\Phi(z^*), \tag{19}$$

where z^* solves $f(z) = e^{-rT}Ke^{-\frac{1}{2}a^2}$, and $f(z)$ is as defined in (10).

The numerical values of the three volatility parameters a, b and c, under each of the six correlation coefficients considered, are given in Table 2. The results of the lower bound (19) for the value of the spread option are listed in Table 3, together with the lower bound, as derived by the Carmona-Durrleman method [3, p. 24].

As can be seen, the two methods give virtually the same results, with the Carmona-Durrleman bound being a little bit better for large, positive values of the strike.

5.2 Relevance of the Lower Bounds

Using a Monte Carlo simulation with 100,000 trials, Carmona and Durrleman [3, p. 24] showed that these lower bounds are extremely close to the true value. This finding is corroborated in a study by van der Hoek and Korolkiewicz [10] who use a different valuation technique. Their approach is based on a perturbation

Table 2 Volatility parameters for the Carmona-Durrleman test case under the canonical formulation (4)

ρ	-1	-0.5	0	0.3	0.8	1
a	0	0.05960	0.08321	0.09334	0.09762	0
b	0.15	0.13765	0.12481	0.11742	0.11389	0.15
c	-0.1	-0.08030	-0.05547	-0.03588	0.02169	0.1

Table 3 Comparison of the lower bounds

K	ρ					
	-1	-0.5	0	0.3	0.8	1
-20	29.656	28.994	28.381	28.070	27.770	27.754
	29.656	28.990	28.373	28.062	27.769	27.754
-10	21.868	20.904	19.888	19.270	18.381	18.244
	21.869	20.903	19.885	19.265	18.379	18.244
0	15.133	13.917	12.523	11.561	9.632	8.821
	15.133	13.918	12.524	11.562	9.633	8.821
5	12.244	10.956	9.445	8.367	5.967	4.454
	12.244	10.956	9.444	8.365	5.963	4.454
15	7.521	6.242	4.744	3.679	1.342	0.049
	7.522	6.236	4.729	3.657	1.303	0.049
25	4.201	3.129	1.961	1.219	0.103	0
	4.201	3.115	1.930	1.178	0.076	0

For every strike value, the first row gives the bound by the Carmona-Durrleman method, and the second row the bound (19) by our method

expansion of the solution to the differential equation that the price of the spread-option satisfies, and they use this expansion to derive analytic formulae for second-order approximations. They employ the same test case and validate their bounds with a recombining binomial tree model. Bjerksund and Stensland [1] study a modified version of the Kirk approximation, and verify their results with a quasi-Monte Carlo method using a two-dimensional Halton sequence with 100,000 pairs, combined with a variance reduction technique. They also use the test case from Carmona and Durrleman, and show that the lower bounds are extremely accurate. Their methodology appears simpler and as accurate as the Carmona-Durrleman approach. The conclusion in each of these studies is the same: these lower bounds provide extremely accurate approximations.

6   Improving the Lower Bounds

To obtain better lower bounds for $\mathbb{E}\,(X_1 - X_0 - K)^+ = \mathbb{E}\,\left(e^{aY}f(Z) - K\right)^+$, we can partition the range for Y, from the one interval, covering the totality of the real line, into n intervals: $I_i = [y_i, y_{i+1})$, $i = 0, \ldots n - 1$, with $y_0 = -\infty$ and $y_n = +\infty$. This gives

$$\mathbb{E}\,\left(e^{aY}f(Z) - K\right)^+ = \sum_{i=0}^{n-1} \mathbb{E}\,\left[\left(e^{aY}f(Z) - K\right)^+ \mid Y \in I_i\right] \mathsf{Prob}\,(Y \in I_i) \geq$$

$$\sum_{i=0}^{n-1} \mathbb{E}\,\left(\left(\Phi(y_{i+1} - a) - \Phi(y_i - a)\right) e^{\frac{1}{2}a^2}f(Z) - \left(\Phi(y_{i+1}) - \Phi(y_i)\right)K\right)^+,$$

where the lower bound follows from applying Jensen's inequality to the conditional expectation. For this lower bound, similar to Lemma 2, let z_i^* be the largest value for Z that equates the argument of the expectation in the ith summand to zero. This expectation is then evaluated by integrating Z over the interval $(-\infty, z_i^*]$. To simplify the notation, we define $p_i = \text{Prob}\,(Y \in I_i)$ and $\tilde{p}_i = \text{Prob}\,(Y + a \in I_i)$, and derive the following generalization of the lower bound (14):

$$\sum_{i=0}^{n-1} \tilde{p}_i \left[\mathbb{E}\,X_1 \Phi(z_i^* - c) - \mathbb{E}\,X_0 \Phi(z_i^* - b) - K\Phi(z_i^*)p_i/\tilde{p}_i \right], \tag{20}$$

where z_i^* is the largest of the values that solves $f(z) = Ke^{-\frac{1}{2}a^2} p_i/\tilde{p}_i$.

This leaves us with a multitude of ways to choose the partition. An appealing choice is the construction where the intervals $[y_i - a, y_{i+1} - a)$ are equally probable, so that $\tilde{p}_i = 1/n$. This implies $y_i = \Phi^{-1}(i/n) + a$, and gives the lower bound as

$$\mathbb{E}\,\left(e^{aY}f(Z) - K\right)^+ \geq \frac{1}{n} \sum_{i=0}^{n-1} \mathbb{E}\,\left(e^{\frac{1}{2}a^2}f(Z) - np_i K\right)^+ \tag{21}$$

$$\geq \frac{1}{n} \sum_{i=0}^{n-1} \left[\mathbb{E}\,X_1\,\Phi(z_i^* - c) - \mathbb{E}\,X_0\,\Phi(z_i^* - b) - np_i K\Phi(z_i^*) \right]. \tag{22}$$

We note that these bounds might also give some insight into the structure of the spread option formula. As the number of partitions goes to infinity, the limit of (21) will converge to the true value of the spread option. For case type II, this evaluates to $\frac{1}{n} \sum_{i=0}^{n-1} \int_{-\infty}^{z_i^*} e^{\frac{1}{2}a^2}f(z)\,dz - \sum_{i=0}^{n-1} p_i K\Phi(z_i^*)$. This structure implies that there is a ζ_n, such that the first expression is the same as the integral $\int_{-\infty}^{\zeta_n} e^{\frac{1}{2}a^2}f(z)\,dz$, and that there is a ξ_n, such that the second expression is the same as $K\Phi(\xi_n)$. These aggregations imply that, for case type II, the value of $\mathbb{E}\,(X_1 - X_0 - K)^+$ can be given by a formula of the type:

$$\mathbb{E}\,X_1\,\Phi(d_1) - \mathbb{E}\,X_0\,\Phi(d_0) - K\Phi(d_2),$$

where $d_1 - d_0 = b - c = \sqrt{\sigma_0^2 - 2\rho\sigma_0\sigma_1 + \sigma_1^2}$, which has a pleasant resemblance to the Black-Scholes formula.

6.1   Test Results

To measure the effect of increasing the number of partitions, we implemented lower bound (22) and again used the Carmona-Durrleman test case. The results for a selected number of partitions are given in Table 4. We note that our benchmark values agree perfectly with those given by Bjerksund and Stensland [1, Table 1].

For negative values of the strike K, the choice of $n = 1$, corresponding to the lower bound (14) from Lemma 2, already gives excellent results. This could already have been surmised from Table 3, but the test results show that the approximations

Table 4 The effect of the number of partitions n on the accuracy of lower bound (22)

K	n	ρ					
		-1	-0.5	0	0.3	0.8	1
−20	1	29.656138	28.989825	28.372967	28.062162	27.768572	27.753786
	2		28.992999	28.378177	28.067242	27.769537	
	5		28.994296	28.380299	28.069303	27.769931	
	10		28.994605	28.380802	28.069789	27.770024	
	20		28.994726	28.380997	28.069977	27.770061	
	50		28.994782	28.381088	28.070064	27.770078	
	100		28.994797	28.381112	28.070086	27.770082	
	1,000		28.994808	28.381129	28.070102	27.770086	
-10	1	21.868637	20.902992	19.884984	19.265409	18.378517	18.243872
	2		20.904239	19.887451	19.268381	18.380153	
	5		20.904750	19.888463	19.269599	18.380818	
	10		20.904873	19.888705	19.269891	18.380976	
	20		20.904921	19.888801	19.270005	18.381036	
	50		20.904943	19.888846	19.270059	18.381065	
	100		20.904949	19.888858	19.270073	18.381072	
	1,000		20.904954	19.888866	19.270083	18.381077	
0	1	15.133217	13.917957	12.523665	11.561761	9.632542	8.821249
5	1	12.244123	10.955506	9.443681	8.364984	*5.962979*	4.454214
	2		10.955956	9.444728	8.366512	5.965534	
	5		10.956141	9.445161	8.367146	5.966597	
	10		10.956185	9.445266	8.367300	5.966857	
	20		10.956203	9.445307	8.367361	5.966961	
	50		10.956211	9.445327	8.367390	5.967011	
	100		10.956213	9.445332	8.367398	5.967025	
	1,000		10.956215	9.445336	8.367404	5.967035	
15	1	7.521812	*6.235713*	*4.729195*	***3.657248***	***1.302566***	0.048825
	2		6.239841	*4.738887*	*3.671555*	***1.328113***	
	5		6.241535	4.742874	*3.677442*	*1.338485*	
	10		6.241941	4.743833	3.678856	*1.340930*	
	20		6.242101	4.744211	3.679413	1.341874	
	50		6.242176	4.744390	3.679677	1.342310	
	100		6.242196	4.744438	3.679748	1.342423	
	1,000		6.242210	4.744472	3.679798	1.342500	

(continued)

Table 4 (continued)

K	n	ρ					
		-1	-0.5	0	0.3	0.8	1
25	1	4.201368	*3.114953*	*1.929799*	*1.177618*	**0.076431**	0.000000
	2		*3.124543*	*1.950431*	*1.204798*	**0.094212**	
	5		3.128465	*1.958832*	*1.215786*	**0.101371**	
	10		3.129400	*1.960822*	*1.218362*	**0.103045**	
	20		3.129765	1.961594	*1.219351*	0.103687	
	50		3.129936	1.961954	*1.219806*	0.103983	
	100		3.129981	1.962048	*1.219923*	0.104059	
	1,000		3.130013	1.962113	1.220002	0.104112	

For the cases with $K = 0$ or $\rho = \pm 1$, our method is exact and attained for $n = 1$, so these numbers are not replicated further. For all other cases, we have taken the value for $n = 1,000$ as the "true" and benchmark value. The numerical results that deviate more than 0.05 % from their benchmark are typeset in italics, and those that deviate more than 0.5 % are typeset in bold-italics

are all within 0.05 % of their benchmark. For positive values of K, the results are still very good, but decrease in accuracy with increasing value of the strike K, and increasing value of the correlation coefficient ρ. However, the choice of $n = 5$ partitions does give an approximation within 0.5 % of the benchmark, for all cases, except for those with the highest correlation coefficient of 0.8.

7 Discussion

In this section we add a few comments and observations that would have obstructed the flow of the discussion had they been incorporated into the main section.

Why Are the Lower Bounds So Accurate? One naturally wonders why the analytical lower bounds for the price of the spread option are so accurate. All the approaches that rely on using convexity arguments and replacing a random variable by its expectation reduce the problem from a two-dimensional to a one-dimensional one. The approximation being so accurate must mean that the problem is, in some sense, close to a one-dimensional problem. The traditional formulation does not show this, but the canonical formulation provides some insight. The convexity argument uses the approximation $\mathbb{E}\left[e^{aY}f(Z) - K\right]^{+} \approx \mathbb{E}\left[\left(\mathbb{E}\,e^{aY}\right)f(Z) - K\right]^{+}$, so that, the lower the variability of e^{aY}, the better the approximation is likely to be. For the Carmona-Durrleman test case, the numerical value of a is close to zero, and Table 5 shows that e^{aY} is close to one, as measured by its expectation and standard deviation.

For new test cases, with larger values of a, one would hazard a guess that the lower bounds are likely to be less accurate. It would be interesting to compare the

Table 5 Characteristics of the volatility parameter a

ρ	-1	-0.5	0	0.3	0.8	1
a	0	0.05960	0.08321	0.09334	0.09762	0
$\mathbb{E}\,e^{aY}$	1	1.00178	1.00347	1.00437	1.00478	1
SDev e^{aY}	0	0.09763	0.08364	0.09395	0.09832	0

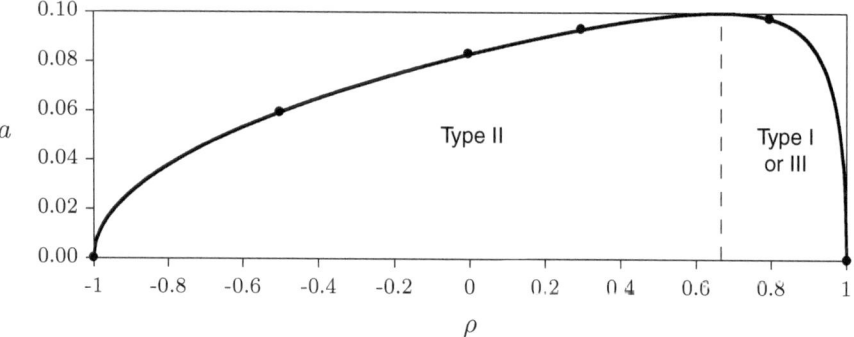

Fig. 2 The effect of the correlation on the volatility parameter a for the Carmona-Durrleman test case. The maximum occurs at $\rho = 2/3$ with value 0.1. The *bullets* correspond to the values for $\rho \in \{-1, -0.5, 0, 0.3, 0.8, 1\}$

results from the Carmona-Durrleman and the Bjerksund-Stensland approaches to ours for a wider range of parameters.

The Effect of Correlation When the correlation is perfect, that is, $\rho = \pm 1$, the value of a is zero, and our lower bound gives the true value of the spread option. Since larger values of a imply a larger standard deviation for e^{aY}, the behavior of a and its maximum, as a function of the correlation coefficient, is of interest. It is easier to look at a^2 and differentiate this with respect to ρ:

$$\frac{\partial a^2}{\partial \rho} = \frac{2\sigma_0^2\sigma_1^2(\sigma_1\rho - \sigma_0)(\rho\sigma_0 - \sigma_1)}{(\sigma_0^2 + \sigma_1^2 - 2\rho\sigma_0\sigma_1)^2}.$$

This derivative is zero for either $\rho = \sigma_0/\sigma_1$ or $\rho = \sigma_1/\sigma_0$. As ρ lies within the interval $[-1, 1]$, it is not difficult to show that a^2 is strictly increasing for $\rho \leq r$, where $r = \min\{\sigma_0/\sigma_1, \sigma_1/\sigma_0\}$, and strictly decreasing for $\rho \geq r$. The unique maximum of a^2 is at $\rho = r$ with value $\min\{\sigma_0^2, \sigma_1^2\}$, so that $a \leq \min\{\sigma_0, \sigma_1\}$. Note that the upper bound also follows from the fact that $a^2 + b^2 = \sigma_0^2$ and $a^2 + c^2 = \sigma_1^2$. For the Carmona-Durrleman test case, the value of a, as a function of ρ, is graphed in Fig. 2. Typically, for smaller values of a, the lower bound (14) is more accurate.

Acknowledgements The simplified proof of the Margrabe formula, using elementary integration techniques, was developed in the context of a course on energy commodities that the author teaches in the Mathematical Finance Program at the University of Toronto. I thank the anonymous referee for comments and suggestions that helped to improve the presentation, and my wife Marguerite Martindale for a professional line edit.

References

1. Bjerksund, P., Stensland, G.: Closed form spread option valuation. Quant. Finance **14**(10), 1785–1794 (2014)
2. Carmona, R., Durrleman, V.: Pricing and hedging spread options. SIAM Rev. **45**(4), 627–685 (2003)
3. Carmona, R., Durrleman, V.: Pricing and hedging spread options in a log-normal model. Technical Report, Department of Operations Research and Financial Engineering. Princeton University, Princeton, NJ, 16 March (2003)
4. Clewlow, L., Strickland, C.: Energy Derivatives: Pricing and Risk Management. Lacima Publications, London (2000)
5. Geman, H.: Commodities and Commodity Derivatives: Modeling and Pricing for Agriculturals, Metals and Energy. Wiley, London (2005)
6. Li, M., Deng, S.-J., Zhou, J.: Closed-form approximations for spread option prices and Greeks. J. Deriv. **15**(3), 58–80 (Spring 2008)
7. Margrabe, W.: The value of an option to exchange one asset for another. J. Finance **33**(1), 177–186 (1978)
8. Pilipovic, D.: Energy Risk: Valuing and Managing Energy Derivatives, 2nd edn. McGraw-Hill, New York (2007)
9. Tong, Y.L.: The Multivariate Normal Distribution. Springer Series in Statistics. Springer, New York (1990)
10. van der Hoek, J., Korolkiewicz, M.W.: New analytic approximations for pricing spread options. In: Cohen, S.N., Madan, D., Siu, T.K.,Yang, H. (eds.) Stochastic Processes, Finance and Control: A Festschrift in Honor of Robert J. Elliott. Advances in Statistics, Probability and Actuarial Science, vol. 1, pp. 259–284. World Scientific, Singapore (2012)

Part II
Electricity and Related Markets

Cross-Commodity Modelling by Multivariate Ambit Fields

Ole E. Barndorff-Nielsen, Fred Espen Benth, and Almut E.D. Veraart

Abstract This paper proposes a multivariate model for commodity forward curves which is based on multivariate ambit fields. We show how a multivariate ambit field can be used to describe complex dependencies between commodities while staying in a tractable multivariate martingale framework. Moreover, we study in detail how spread options can be priced in our new ambit framework. Here we consider both calendar spreads written on one commodity as well as spread options on different commodity futures.

1 Introduction

Modelling forward and futures prices is of key importance in commodity markets in general, and in energy markets in particular. This paper introduces a new framework for modelling various commodity forward curves simultaneously in a multivariate framework.

Following the famous Heath-Jarrow-Morton methodology, see [23], we will model forward prices directly rather than deriving them from the corresponding spot prices. A forward price, denoted by $F(t, T)$, where $t \geq 0$ denotes the current time and $T \geq t$ denotes the time of maturity, can be regarded as a multiparameter process, i.e. a random field, depending on two parameters t and T. This finding has

O.E. Barndorff-Nielsen
Thiele Center, Department of Mathematics CREATES, Department of Economics and Business
Aarhus University, Aarhus C, Denmark
e-mail: oebn@math.au.dk

F.E. Benth
Department of Mathematics, University of Oslo, Oslo, Norway
e-mail: fredb@math.uio.no

A.E.D. Veraart (✉)
Department of Mathematics, Imperial College London, London, UK
e-mail: a.veraart@imperial.ac.uk

© Springer Science+Business Media New York 2015   109
R. Aïd et al. (eds.), *Commodities, Energy and Environmental Finance*,
Fields Institute Communications 74, DOI 10.1007/978-1-4939-2733-3_5

triggered research on modelling the term structure of interest rates, but also more general forward prices, by random fields, following the influential work by Kennedy [26, 27].

In a recent article, Barndorff-Nielsen et al. [7], suggested to use so-called *ambit fields* to model energy forward and futures prices in a univariate framework. Ambit fields are special types of random fields, which have been developed in the context of modelling turbulence in physics by Barndorff-Nielsen and Schmiegel [3]. Their key building block is a stochastic integral of the form

$$X(t, T) = \int_{A(t,T)} g(t, T; s, \xi)\sigma(s, \xi)L(ds, d\xi),$$

where L is a so-called Lévy basis, g is a deterministic kernel function and σ is a stochastic volatility/intermittency component. In addition to the stochastic integral, a shift term is often included which allows for rich model specification including the possibility of modelling skewness in a flexible way.

The parameter space for T could be multivariate as well, however, here we will focus on the case when $t, T \in \mathbb{R}$. The name *ambit field* comes from the fact that the stochastic integral is computed over a so-called *ambit set* $A(t, T)$ which denotes the sphere of influence of the random noise. A rigorous mathematical definition as well as suitable integrability conditions which ensure that the stochastic integral is well-defined will be given in the next section.

This paper builds upon the work of Barndorff-Nielsen et al. [7] and extends it in the following directions: First of all, we extend the univariate ambit fields to a multivariate set-up in order to be able to model various commodity forward curves simultaneously. Second, we allow for the possibility of unbounded ambit sets $A(t, T)$ which allows to incorporate *stationary* processes in the modelling framework—an aspect we will discuss in more detail later in the paper. Third, we build a two-factor model, which allows for a Gaussian part with stochastic volatility as well as a pure jump part.

The main challenge in the multivariate set-up is to understand and model cross-commodity dependencies. We will discuss how serial and cross-sectional dependence can be introduced in our ambit framework.

Having a multivariate set-up at hands, naturally raises the question of how derivatives written on various commodity forward/futures can be priced in the new model. In that context, pricing of so-called *spread options* is of key importance in energy markets. Carmona and Durrleman [17] provide an excellent review of the relevant literature on spread options pricing and discuss the most relevant cases in the context of energy markets: Here we need to distinguish between both the *temporal spread* and the *location spread*. While the former considers energy commodities with different times of maturity, e.g. a *calendar spread*, a location spread considers commodity futures for commodity at different locations with potentially the same time of maturity.

Two of the most widely studied energy spread option are the *crack spread* options and *spark spread* options. While the crack spread depends on daily futures prices of crude oil, unleaded gasoline and heating oil, there are also different variants, such as the *gasoline crack* spread, which depends on crude oil and unleaded gasoline, or the *heating oil crack* spread, which depends on crude oil and heating oil. Moreover, the spark spread depends on the difference of electricity and natural gas and can be considered as the cost of converting gas into electricity.

In order to find a fair price of such options, we need a multivariate modelling framework. Here we follow the key idea of Margrabe [28] and link a spread option with strike price zero to a classical European option. In the context of jump-diffusion models, extensions of the Margrabe-formula have been studied in e.g. [18] and [12]. In this paper, we will go a step further and extend the Margrabe formula to the ambit set-up. In addition, we study the case when the strike price is not equal to zero and the Margrabe formula is no longer applicable.

The remaining part of this article is structured as follows. Section 2 lays out the background of ambit fields and presents the model assumptions including in particular the martingale condition which ensures that forward prices are martingales under the risk neutral probability measure. Section 3 presents further important model properties including a detailed survey on how we model dependence in the ambit framework and how the forward price is linked to the corresponding spot price. Next, Sect. 4 discusses in detail how spread options can be priced in our new model, where we present both the case of temporal spreads as well as location spread options. A further extension of our modelling framework is introduced in Sect. 5. Finally Sect. 6 concludes. The proofs of our main results have been relegated to the Appendix.

2   The Model

Throughout the paper, we denote by $(\Omega, \mathscr{F}, \mathbb{P})$ a complete probability space. Let us first recall the basic traits of multivariate Lévy bases which have been introduced by [5]. We will later use such multivariate Lévy bases to construct multivariate ambit fields.

2.1   Preliminaries

Let $(S, \mathscr{S}, Leb)$ denote a Lebesgue-Borel space where S denotes a Borel set in $\mathbb{R}^k$ for $k \in \mathbb{N}$. In this paper, we will choose $k = 2$, i.e. one parameter will represent the *time* parameter and the other one the *time of maturity* parameter. We denote by $\mathscr{S} = \mathscr{B}(S)$ the Borel σ-algebra on S and by Leb the Lebesgue measure. Note that the Lebesgue measure on matrix or vector spaces should be understood as the product of the coordinate-wise Lebesgue measures throughout the paper. In addition, we define $\mathscr{B}_b(S) = \{A \in \mathscr{S} : Leb(A) < \infty\}$, which is the subset of $\mathscr{S}$ that contains sets which have bounded Lebesgue measure. Note that the set $\mathscr{B}_b(S)$ is closed under finite union, relative complementation, and countable intersection and is therefore a δ-ring.

Definition 1. Let $n \in \mathbb{N}$. A family of $\mathbb{R}^n$-valued random variables denoted by $L = \{L(A) : A \in \mathscr{B}_b(S)\}$ is called an $\mathbb{R}^n$-valued Lévy basis on S, if it is a random measure which is independently scattered and infinitely divisible, i.e. if it satisfies the following three conditions:

1. For any pairwise disjoint sets $(A_i)_{i \in \mathbb{N}} \subset \mathscr{B}_b(S)$ satisfying $\cup_{i \in \mathbb{N}} A_i \in \mathscr{B}_b(S)$, the infinite series $\sum_{i \in \mathbb{N}} L(A_i)$ converges almost surely and $L(\cup_{i \in \mathbb{N}} A_i) = \sum_{i \in \mathbb{N}} L(A_i)$ a.s. (i.e. L is a random measure);
2. for any $d \in \mathbb{N}$ and for any pairwise disjoint sets $A_1, \ldots, A_d \in S$ the $\mathbb{R}^n$-valued random variables $L(A_1), \ldots, L(A_d)$ are independent (i.e. L is independently scattered);
3. the distribution of $L(A)$ is infinitely divisible for all $A \in \mathscr{B}_b(S)$ (i.e. L is infinitely divisible).

Throughout this paper, we will restrict our attention to Lévy bases, which are *homogeneous* and *factorisable*, in the sense that their characteristic function can be expressed as

$$\mathbb{E}(\exp(i\theta^\top L(A))) = \exp(\phi(\theta)Leb(A)),$$

for all $\theta \in \mathbb{R}^n$ and $A \in \mathscr{B}_b(S)$, where

$$\phi(\theta) = i\theta^\top \gamma - \frac{1}{2}\theta^\top \Sigma^* \theta + \int_{\mathbb{R}^n} \left(e^{i\theta^\top z} - 1 - i\theta^\top z \mathbb{I}_{\{||z|| \leq 1\}}(z)\right) \nu(dz), \qquad (1)$$

where $\gamma \in \mathbb{R}^n$ denotes a constant, Σ denotes a symmetric, positive semidefinite $n \times n$-dimensional matrix and ν denotes a Lévy measure on $\mathbb{R}^n$. Here $|| \cdot ||$ denotes the Euclidean norm. We call $(\gamma, \Sigma^*, \nu, Leb)$ the *characteristic quadruplet* (CQ) associated with the Lévy basis L which determines the Lévy basis uniquely, see [8, 30] for details. Note that we call the infinitely divisible $\mathbb{R}^n$-valued random variable L' having cumulant function given by (1) the *Lévy seed* associated with L. Note that one can define the Lévy process $(L'_t)_{t \geq 0}$ associated with L' by setting $L'_1 := L'$.

From [5, Theorem 2.2] we know that the homogeneous, factorisable Lévy basis L with characteristic quadruplet given by $(\gamma, \Sigma^*, \nu, Leb)$ has a Lévy-Itô decomposition. That is there exists a modification $\tilde{L}$ of L with identical CQ which satisfies

$$\tilde{L}(A) = \gamma Leb(A) + \tilde{L}^G(A) + \int_A \int_{||z|| \leq 1} z(N(dz, dy) - \nu(dz)dy)$$

$$+ \int_A \int_{||z|| > 1} zN(dz, dy),$$

for all $A \in \mathcal{B}_b(S)$, and for all $\omega \in \Omega$, where $\tilde{L}^G$ denotes an $\mathbb{R}^n$-valued Lévy basis on S with CQ $(0, \Sigma^*, 0, Leb)$ and N denotes an independent Poisson random measure on $(\mathbb{R}^n \times S, \mathcal{B}(\mathbb{R}^n \times S))$ having intensity measure $\nu \otimes Leb$. We typically write $\tilde{N} = N - \nu \otimes Leb$ for the compensated Poisson random measure.

2.2 Model Assumptions

Inspired by the success of jump-diffusion models, we will work with two types of stochastic noise given by two $\mathbb{R}^n$-valued Lévy bases. That is we build a two-factor model, where the first factor is a mixed Gaussian random field and the second factor is driven by a pure jump Lévy basis.

More precisely, suppose that $W = (W_1, \dots, W_n)^\top$ is a multivariate homogeneous and factorisable Gaussian Lévy basis with CQ $(0, \Sigma^*, 0, Leb)$. By definition, W has zero mean and its second moment exists. Moreover, we assume that the diagonal elements of Σ^* are all standardised to 1, so that we are dealing with unit variances throughout.

Further, let $L = (L_1, \dots, L_n)^\top$ denote a multivariate homogeneous and factorisable Lévy basis with CQ $(\gamma, 0, \nu, Leb)$. We assume that L has zero mean and finite variance. The zero mean assumption implies that

$$\gamma = -\int_{||z||>1} z\nu(dz).$$

We know that there exists a modification $\tilde{L}$ of L such that under the zero-mean assumption

$$\tilde{L}(A) = \gamma Leb(A) + \int_A \int_{||z|| \leq 1} z(N(dz, dy) - \nu(dz)dy) + \int_A \int_{||z||>1} zN(dz, dy)$$

$$= \int_A \int_{\mathbb{R}^n} z(N(dz, dy) - \nu(dz)dy).$$

Note that we write

$$J_{L_j}(dz_j, dy) = N(\mathbb{R}, \dots, \mathbb{R}, dz_j, \mathbb{R}, \dots, \mathbb{R}, dy),$$

for the Poisson random measure associated with the jth jump component, which is obtained by integrating out the remaining components. Also the corresponding univariate Lévy measure is given by

$$\nu_{L_j}(dz_j) = \nu(\mathbb{R}, \dots, \mathbb{R}, dz_j, \mathbb{R}, \dots, \mathbb{R}).$$

Then $\tilde{J}_{L_j}$ denotes the compensated Poisson random measure associated with L_j. Note further that W and L are assumed to be independent of each other.

We will use W and L as stochastic integrators, where we define stochastic integrals in the sense of Walsh [32], see [8] for a recent review of this integration concept. The integration concept of Walsh [32] follows the idea of splitting the integration domain into time and space and of working with an Itô-type integration theory with respect to the time variable.

In particular, we define the (two-sided) stochastic process $(W(t, A))_{t \in \mathbb{R}}$ for fixed $A \in \mathscr{B}_b(\mathbb{R})$ as follows: For $t \geq 0$, we set $W(t, A) = W([0, t] \times A)$. For $t < 0$ we take an independent copy of W denoted by W^*, say, having the same characteristic quadruplet as W. Then we define $W(t, A) = -W^*([0, -t])$ for $t < 0$. This is analogue to the construction of a two-sided Brownian motion. Similarly, we define the process $(L(t, A))_{t \in \mathbb{R}}$.

Next, we need to define a suitable filtration. We assume that $(\mathscr{F}_t)_{t \in \mathbb{R}}$ denotes a filtration satisfying the usual conditions of right-continuity and completeness such that for fixed $A \in \mathscr{B}_b(\mathbb{R})$ both $(W(t, A))_{t \in \mathbb{R}}$ and $(L(t, A))_{t \in \mathbb{R}}$ are martingales with respect to that filtration.

Example 1. Let us briefly illustrate how a suitable filtration can be constructed. To this end, we can define the filtration generated by the increments of the Gaussian Lévy basis W by

$$\mathscr{F}_t^{W, \text{incr}} = \cap_{n=1}^{\infty} \mathscr{F}_{t+1/n}^0, \qquad \text{where}$$

$$\mathscr{F}_t^0 = \sigma\{W(u, A) - W(s, A), -\infty < s \leq u \leq t, A \in \mathscr{B}_b(\mathbb{R})\} \vee \mathscr{N},$$

where $\mathscr{N}$ denotes the $\mathbb{P}$-null sets of $\mathscr{F}$. Clearly, the filtration is right-continuous and complete, and for fixed $A \in \mathscr{B}_b(\mathbb{R})$ the process $(W(t, A))_{t \in \mathbb{R}}$ is a martingale with respect to the filtration $(\mathscr{F}_t^{W, \text{incr}})_{t \in \mathbb{R}}$.

However, the filtration $(\mathscr{F}_t)_{t \in \mathbb{R}}$ is assumed to be bigger than just the one generated from suitable increments of W and L in that it can also support drift and stochastic volatility processes which are independent of L and W. That is let $\mu : \mathbb{R}^3 \times \Omega \mapsto \mathbb{R}^n$ and $\sigma : \mathbb{R}^2 \times \Omega \mapsto \mathbb{R}_+^n$ denote the stochastic drift and volatility processes, respectively. We assume that they are adapted in the temporal component and that the (μ_j, σ_j) are assumed to be independent of (W_j, L_j) for $j = 1, \ldots, n$.

Throughout the paper, we will work with stochastic processes of the form

$$Y_j(t, T) = \int_{A_j(t)} \mu_j(T; s, \xi) ds d\xi + \int_{A_j(t)} g_j(T; s, \xi) \sigma_j(s, \xi) W_j(ds, d\xi)$$

$$+ \int_{A_j(t)} h_j(T; s, \xi) L_j(ds, d\xi), \qquad (2)$$

for $j = 1, \ldots, n$, where $g_j, h_j : \mathbb{R}^3 \to \mathbb{R}$ are measurable deterministic functions.

The range of integration is determined by the so-called *ambit sets* $A_j(t) \subset (-\infty, T^*] \times [0, \infty)$, where $T^* > 0$ denotes the finite time horizon we consider throughout the paper. That is we will have $0 \leq t \leq T \leq T^*$.

In order to ensure that the integrals in (2) are well-defined, we require that the σ_j are predictable for all $j = 1, \ldots, n$ and that

$$\|\mu_j\|_{Leb} := \mathbb{E}\left[\int_{(-\infty,T^*]\times[0,\infty)} \mu_j(\boldsymbol{\xi},s)dsd\boldsymbol{\xi}\right] < \infty, \tag{3}$$

$$\|g_j\sigma_j\|_W^2 := \mathbb{E}\left[\int_{(-\infty,T^*]\times[0,\infty)} g_j^2(T;\boldsymbol{\xi},s)\sigma_j^2(s,\xi)\,Var(W_j')dsd\boldsymbol{\xi}\right] < \infty, \tag{4}$$

$$\|h_j\|_L^2 := \mathbb{E}\left[\int_{(-\infty,T^*]\times[0,\infty)} h_j^2(\boldsymbol{\xi},s)\,Var(L_j')dsd\boldsymbol{\xi}\right] < \infty. \tag{5}$$

Then, the integrals in (2) are well-defined in the L^2-sense for $0 \le t \le T \le T^*$.

Remark 1. Note that for purely deterministic integrands, the integration theory of Rajput and Rosinski [30] can be employed, which has been formulated for the multivariate case in [5]. For stochastic integrands, the L^2-theory can be further extended, see in particular [13] and also [10, 19] for recent work on the stochastic integration theory for ambit processes and fields beyond the L^2-framework.

2.3 The Geometric Model

We will model forward rates directly under the risk-neutral measure, which we will denote by $\mathbb{P}$ throughout the paper. Under the modelling assumptions introduced in the previous section, the random field $Y_j(t,T)$ defined in (2) can be expressed as

$$Y_j(t,T) = \int_{A_j(t)} \mu_j(T;s,\xi)dsd\xi + \int_{A_j(t)} g_j(T;s,\xi)\sigma_j(s,\xi)W_j(ds,d\xi)$$

$$+ \int_{A_j(t)}\int_{\mathbb{R}} zh_j(T;s,\xi)\tilde{J}_{L_j}(dz,ds,d\xi), \tag{6}$$

for $j = 1, \ldots, n$ and $0 \le t \le T \le T^*$.

In our modelling framework we allow for stochastic volatility in the second component, where we have a Gaussian Lévy basis as integrator. In the third term, which reflects the jumps, we do not allow for stochastic volatility to keep the exposition simple. However, from a mathematical point of view, it would be no problem also to include stochastic volatility in the second factor. Note that stochastic volatility could also be included via extended subordination by meta times, as described in detail in [2].

In the definition of the ambit field, the so-called ambit sets $A_j(t)$ occur, which determine the relevant range for the integration. In the following, we choose $A_j(t) = (-\infty,t] \times [0,\infty)$. The reason for this choice is that the first parameter reflects the time parameter. Here we would like to be able to account for the entire past up to the current time t and hence we choose the time interval $(-\infty, t]$. While we typically consider modelling the forward price for $t \geq 0$, we require the infinite past in the time domain to potentially obtain stationary processes. The second parameter relates to the time of maturity of the contract, which could potentially be any positive number. Hence we choose the interval $[0, \infty)$. Note that we are interested in the case when $t \leq T$. Given that the variable s corresponds to the time-parameter and the variable ξ to the time of maturity parameter, one could think of choosing kernel functions g_j which satisfy $g_j(T; s, \xi) = 0$ whenever $s \leq \xi$.

The forward price $F_j(t, T)$ at time t with delivery $T \geq t$ of the jth commodity is defined by a geometric model of the form

$$F_j(t, T) = \exp(Y_j(t, T)), \text{ for } 0 \leq t \leq T \leq T^*, \tag{7}$$

where $\mu_j(T; s, \xi)$ is chosen such that $F_j(t, T)$ is a martingale.

Remark 2. In energy markets, forward prices typically do not just depend on a time of maturity T, but typically on a *delivery interval*. It is well-known, how a model for $F_j(t, T)$ can be extended to allow for a delivery period, see e.g. [7, Section 3.2]. In the following, however, we will ignore any dependence on a delivery period to keep our exposition as simple as possible when we work with the new ambit framework.

2.4 Martingale Condition

Since we model directly under the risk-neutral probability measure $\mathbb{P}$, we need to ensure that forward prices are martingales under $\mathbb{P}$. This entails in a drift condition on $\mu_j(T; s, \xi)$, as we shall see. First, notice that using the definition $A_j^s(t) = [s, t] \times [0, \infty)$ for $s \leq t$, we have

$$Y_j(t, T) = Y_j(s, T) + \int_{A_j^s(t)} \mu_j(T; u, \xi)dud\xi + \int_{A_j^s(t)} g_j(T; u, \xi)\sigma_j(u, \xi)W_j(du, d\xi)$$

$$+ \int_{A_j^s(t)} \int_{\mathbb{R}} zh_j(T; u, \xi)\tilde{J}_{L_j}(dz, du, d\xi).$$

Therefore,

$$F_j(t, T) = F_j(s, T) \exp\left(\int_{A_j^s(t)} \mu_j(T; u, \xi)dud\xi + \int_{A_j^s(t)} g_j(T; u, \xi)\sigma_j(u, \xi)W_j(du, d\xi) \right.$$

$$\left. + \int_{A_j^s(t)} \int_{\mathbb{R}} zh_j(T; u, \xi)\tilde{J}_{L_j}(dz, du, d\xi) \right),$$

with $s \leq t$. We have the following condition on the model for F_j ensuring the martingale property.

Proposition 1. *If Y_j has an exponential moment and*

$$\mu_j(T;t,\xi) + \frac{1}{2}g_j^2(T;t,\xi)\sigma_j^2(t,\xi) + \int_{\mathbb{R}} \left(e^{zh_j(T;t,\xi)} - 1 - zh_j(T;t,\xi)\right) v_{L_j}(dz) = 0,$$

(8)

then $t \mapsto F_j(t,T)$ is a martingale for $t \leq T$. Under this martingale condition, we have

$$Y_j(t,T) = -\int_{A_j(t)} \frac{1}{2}g_j^2(T;s,\xi)\sigma_j^2(s,\xi)dsd\xi + \int_{A_j(t)} g_j(T;s,\xi)\sigma_j(s,\xi)W_j(ds,d\xi)$$

$$- \int_{A_j(t)} \int_{\mathbb{R}} \left(e^{zh_j(T;s,\xi)} - 1 - zh_j(T;s,\xi)\right) v_{L_j}(dz)dsd\xi$$

$$+ \int_{A_j(t)} \int_{\mathbb{R}} zh_j(T;s,\xi)\tilde{J}_{L_j}(ds,d\xi).$$

The proof is given in the Appendix.

Proposition 2. *Under the martingale condition, the dynamics of F_j can be expressed as follows.*

$$\frac{d_t F_j(t,T)}{F_j(t-,T)} = \int_{\mathbb{R}} g_j(T;t,\xi)\sigma_j(t,\xi)W_j(dt,d\xi) + \int_{\mathbb{R}^2} \left(e^{zh_j(T;t,\xi)} - 1\right)\tilde{J}_{L_j}(dz,dt,d\xi),$$

(9)

where d_t indicates that we look at the differential with respect to the time parameter, which is here denoted by t.

The proof is given in the Appendix.

Remark 3. Note that the integration with respect to Lévy bases should always be understood as joint integration and not as iterative integration. Hence, Eq. (9) should only be understood as a formal representation rather than an iterative integral.

Remark 4. Note that in this paper we are interested in ambit fields which are indeed martingales since forward contracts are tradeable and hence need to satisfy the martingale condition under a risk-neutral probability measure. So throughout the paper, we will stick to our martingale set-up. However, when we compare our model with the original model specification of ambit fields in the context of modelling turbulence as discussed by Barndorff-Nielsen and Schmiegel [3], we see that ambit fields are not in general (semi-) martingales and there is a lot of interest in further developing the non-semimartingale case. This being particularly the case in turbulence, but also in finance, as we discussed in more detail in the context of modelling energy spot prices in [6], which do not necessarily need to be modelled as (semi-) martingales. For example electricity spot prices are not directly tradeable, hence a non-semimartingale modelling framework could be applied.

2.5   Examples

In the following, we give some relevant examples how we can specify a fully parametric model within our ambit framework. To this end, we need to specify the weight functions g_j and h_j, the stochastic volatility processes σ_j and the Lévy basis L_j.

2.5.1   Specifying the Lévy Basis

Let us start with the Lévy basis. In the following we consider a homogeneous Lévy basis L_j with characteristic quadruplet $(\gamma_j, 0, \nu_{L_j}, Leb)$.

Example 2 (Poisson Lévy Basis). When we choose $\nu_{L_j}(dz) = \lambda_j \delta_1(dz)$, where δ_1 denotes the Dirac measure with point mass at 1 and $\lambda_j > 0$ is the intensity rate, then L constitutes a Poisson Lévy basis.

Example 3 (Generalised Hyperbolic Lévy Basis). Suppose the Lévy measure is of the form

$$\nu_{L_j}(dz) = \frac{\gamma_j^{r_j} \alpha_j^{1-2r_j}}{\sqrt{2\pi} K_{r_j}(\delta_j \gamma_j)} \overline{K}_{r_j-1/2}\left(\alpha_j \sqrt{\delta_j^2 + (z-\mu_j)^2}\right) \exp(\beta_j(z-\mu_j))dz, \quad z \in \mathbb{R},$$

i.e. we are dealing with the generalised hyperbolic (GH) distribution with parameters $(r_j, \alpha_j, \beta_j, \mu_j, \delta_j)$, where $\alpha_j = \sqrt{\beta_j^2 + \gamma_j^2}$. Note that K_r is the modified Bessel function of the third kind satisfying $K_r(z) = K_{-r}(z)$. Also, we set $\overline{K}_r(z) = z^r K_r(z)$. Note that $\alpha_j, \delta_j > 0$, $\mu_j, r_j \in \mathbb{R}$ and β_j is such that $\alpha_j^2 - \beta_j^2 > 0$. The GH distribution contains many well-known distributions as special cases. In our context, the normal inverse Gaussian (NIG) distribution is particularly relevant, which we obtain when setting $r_j = 1/2$.

Example 4. In the above examples, we have specified the marginal Lévy measures. The joint Lévy measure of $L = (L_1, \dots, L_n)$ can be constructed using Lévy copulas, which allow for a great variability in terms of modelling the dependence structure between the individual components. Alternatively, one could directly specify a parametric model for the multivariate Lévy basis L. For example one could assume that the multivariate Lévy measure of L follows a multivariate generalised hyperbolic distribution.

2.5.2 Specifying the Weight Functions

Next, one might ask how one could fully parameterise the weight functions g_j and h_j. This has been discussed in [7] and we briefly review some relevant specifications for g_j, similar specifications could be used for h_j.

Suppose the weight function factorises as

$$g_j(T, s, \xi) = g_j^{(1)}(T, s)g_j^{(2)}(\xi), \tag{10}$$

for suitable functions $g_j^{(1)}, g_j^{(2)}$.

Example 5 (Choices for $g_j^{(1)}$). Extending the classical models for forward prices in the literature, we can consider choosing

$$g_j^{(1)}(T, s) = \exp(-\alpha_j(T - s)), \text{ for } \alpha_j > 0,$$

which is motivated from Ornstein-Uhlenbeck processes. Extensions to weight functions mimicking continuous-time ARMA (CARMA) processes could also be considered.

A different specification, which is motivated by the work by Bjerksund et al. [14] suggesting that it can model the Samuelson effect well, is given by

$$g_j^{(1)}(T, s) = \frac{\alpha_j}{T - s + \beta_j}, \text{ for } \alpha_j, \beta_j > 0,$$

Example 6 (Choices for $g_j^{(2)}$). While choices for $g_j^{(1)}$ are rather straightforward to come up with since there is a wide empirical literature on the time series properties of forward and futures prices, modelling the dependence in the spatial variable appears to be slightly trickier. This is mainly due to the fact that the data in the time-to-maturity direction are much more sparse than in the time direction. While choosing $g_j^{(2)}(\xi) \equiv 1$ would essentially bring us back to classical one-parameter models of forward rates, empirical work by Audet et al. [1] suggests that choosing

$$g_j^{(2)}(\xi) = \exp(-\beta_j\xi), \text{ for } \beta_j > 0,$$

could be a good starting point to allow for decreasing dependence when time to maturity/date of maturity for different contracts are far apart. In future research, it will be interesting to investigate empirically which functional forms are the most suitable choices for commodity futures.

2.5.3 Specifying the Stochastic Volatility Field

In the mixed-Gaussian part of the model specification we allow for a stochastic volatility component denoted by $\sigma_j(s, \xi)$. That is in principle we allow for the volatility to depend both on the time as well as on a time to/of maturity parameter.

Let us consider the case when the volatility factorises as $\sigma_j(s, \xi) = \sigma_j^{(1)}(s)\sigma_j^{(2)}(\xi)$. Also, consider the situation when the weight function factorises as in (10). Then we have that

$$\int_{(-\infty,t]\times[0,\infty)} g_j^{(1)}(T, s)g_j^{(2)}(\xi)\sigma_j(s)W_j(ds, d\xi)$$

is conditional on $\mathscr{F}^{\sigma_j}$ normally distributed with zero mean and conditional variance given by

$$\int_{-\infty}^t \left(g_j^{(1)}(T, s)\right)^2 \left(\sigma_j^{(1)}(s)\right)^2 ds \cdot \int_0^\infty \left(g_j^{(2)}(\xi)\right)^2 \left(\sigma_j^{(2)}(\xi)\right)^2 d\xi.$$

This is an interesting finding: The first factor resembles the accumulated stochastic volatility in time—up to time t, which is very much in-line with classical time series models for forward prices. However, the second factor can be regarded as the accumulated stochastic volatility in the *spatial* direction, i.e. the time to delivery direction. This is an important generalisation beyond the time-series framework.

Further (fully parametric) examples for the stochastic volatility can be found in [7] and [8].

3 Properties of the Ambit Model

3.1 Autocorrelation and Cross-Correlation

Let us briefly investigate which types of dependence are implied by our multivariate ambit model. Here we want to study both serial dependence as well as cross-commodity dependence.

To simplify the exposition, we focus on two key building blocks in the model: For $j \in \{1, \ldots, n\}$, let

$$Y_j^c(t, T) := \int_{A_j(t)} g_j(T; s, \xi)\sigma_j(s, \xi)W_j(ds, d\xi),$$

$$Y_j^d(t, T) = \int_{A_j(t)} zh_j(T; s, \xi)\tilde{J}_{L_j}(dz, ds, d\xi).$$

Recall that the correlation between the Gaussian Lévy bases is governed by the covariance matrix Σ^* in the CQ. More precisely, let us assume that

$$Cov\left(W^{(i)}(d\xi, ds), W^{(j)}(d\xi, ds)\right) = \rho_{i,j}d\xi ds,$$

for $-1 \leq \rho_{i,j} \leq 1$ and $i, j \in \{1, \ldots, n\}$.

For the jump part, we introduce the bivariate Lévy measure $v_{i,j}$ describing the dependence between the i and jth component.

So we see that dependence between the two Lévy bases can be incorporated either through the Gaussian part or the jump part or a combination of both. In addition, the stochastic volatility component introduces an additional level of cross-sectional dependence.

In order to shorten the exposition slightly, we focus on the Gaussian and pure-jump cases separately.

Proposition 3. *Under the assumptions above, we get the following covariation functions:*

$$Cov\left(Y_i^c(t,T), Y_j^c(\tilde{t},\tilde{T})\middle| \mathscr{F}^{\sigma_i} \vee \mathscr{F}^{\sigma_j}\right)$$
$$= \rho_{i,j} \int_{A_i(t)\cap A_j(\tilde{t})} g_i(T;s,\xi)g_j(\tilde{T};s,\xi)\sigma_i(s,\xi)\sigma_j(s,\xi)dsd\xi.$$

The unconditional covariation is given by

$$Cov\left(Y_i^c(t,T), Y_j^c(\tilde{t},\tilde{T})\right)$$
$$= \rho_{i,j} \int_{A_i(t)\cap A_j(\tilde{t})} g_i(T;s,\xi)g_j(\tilde{T};s,\xi)\mathbb{E}(\sigma_i(s,\xi)\sigma_j(s,\xi))dsd\xi.$$

The above results can be easily verified by computing the joint characteristic function of $Y_i^c(t,T)$ and $Y_j^c(\tilde{t},\tilde{T})$ by first conditioning on $\mathscr{F}^{\sigma_i} \vee \mathscr{F}^{\sigma_j}$, where $\mathscr{F}^{\sigma_i}$ and $\mathscr{F}^{\sigma_j}$ denote the filtration generated by σ_i and σ_j, respectively. That is

$$\mathscr{F}^{\sigma_i} = \sigma\{\sigma_i(s,\xi) : (s,\xi) \in A_i(T^*)\}, \quad \mathscr{F}^{\sigma_j} = \sigma\{\sigma_j(s,\xi) : (s,\xi) \in A_j(T^*)\}.$$

The unconditional result follows from the well-known law of total (co)variance. Likewise, we get the following result in the pure jump case.

Proposition 4. *Under the assumptions above, we get the following covariation functions:*

$$Cov\left(Y_i^d(t,T), Y_j^d(\tilde{t},\tilde{T})\right)$$
$$= \int_{A_i(t)\cap A_j(\tilde{t})} h_i(T;s,\xi)h_j(\tilde{T};s,\xi)dsd\xi \int_{\mathbb{R}^2} z_iz_j v_{i,j}(dz_i,dz_j).$$

Again, this result can be verified by computing the joint characteristic function of $Y_i^d(t,T)$ and $Y_j^d(\tilde{t},\tilde{T})$.

Note that from a modelling point of view there are many possibilities in modelling the joint Lévy measure $v_{i,j}$. For example one could work with classical multivariate Lévy measures. Another possibility would be to apply Lévy copulas, see e.g. [20], to model the dependence structure.

Note that in the formulas above, we see that the intersection of the corresponding ambit sets is an important ingredient in determining the correlation structure for general ambit fields. In our modelling context, however, we work with a rather simple structure of the ambit field, which is motivated by the fact that we want to have a martingale structure, where we simply obtain $A_i(t) \cap A_j(\tilde{t}) = (-\infty, \min(t, \tilde{t})] \times [0, \infty)$.

Note that the intersections of more general ambit sets typically lead to more interesting shapes as the one obtained here, see e.g. the recent investigations by Barndorff-Nielsen et al. [9].

3.2 Relation Between Forward and Spot

Let us briefly study the spot model which is implied by our forward model. At time $t = T$, the spot price of commodity j is given by

$$X_j(T) = \exp(Y_j(T, T)),$$

where

$$Y_j(T, T) = \int_{A_j(T)} \mu_j(T; s, \xi) ds d\xi + \int_{A_j(T)} g_j(T; s, \xi) \sigma_j(s, \xi) W_j(ds, d\xi)$$

$$+ \int_{A_j(T)} h_j(T; s, \xi) L_j(ds, d\xi). \tag{11}$$

In the case when $A_j(T) = (-\infty, T] \times [0, \infty)$, we observe that $Y_j(T, T)$ can be regarded as a superposition of volatility modulated Lévy-driven Volterra (VMLV) processes of the form

$$Z(T) = \int_{-\infty}^{T} \mu_j(T; s) ds + \int_{-\infty}^{T} g_j(T; s) \sigma_j(s) W_j'(ds) + \int_{-\infty}^{T} h_j(T; s) L_j'(ds), \tag{12}$$

where W_j' and L_j' denote the Lévy processes associated with the Lévy seeds of W and L. Note that as soon as the $g_j(T; s) = g(T-s)$ and $h_j(T; s) = h(T-s)$ we would call the corresponding stochastic integrals *Lévy semistationary (LSS)* processes, see [6].

Barndorff-Nielsen et al. [6] have shown that arithmetic LSS processes can describe the empirical behaviour of energy spot prices very well. The corresponding result for the geometric model can be found in an earlier version of that paper. Hence if we model forward prices by ambit fields, the implied spot price model is indeed a rather realistic one, which is an encouraging result.

Also note that the relation between spot and forward prices in an ambit framework has been studied in detail in the univariate set-up by Barndorff-Nielsen et al. [7] and we refer to that paper for more details since the results carry over directly to the multivariate case. The extension of LSS processes to the multivariate case has been studied by Veraart and Veraart [31] in the context of modelling multivariate energy spot prices.

4 Application to Spread Options

Next we show how spread options prices can be computed in our multivariate ambit framework. Spread options are popular derivatives in energy markets and their payoff typically depends on the price of at least two commodities.

Carmona and Durrleman [17] provide a very detailed review of the spread option market and survey the methods for pricing and hedging of spread options. One result which is of particular importance is the so-called Margrabe formula, see [28], which we will extend to our ambit framework in the following.

As mentioned before, we fix a finite time horizon $T^* > 0$ and assume that all options and futures expire before that point in time.

4.1 Spread Option Set-Up

Consider two asset prices $S_j(t)$ for $j = 1, 2$ and $t \geq 0$. Throughout this paper we only consider asset prices which are indeed forward prices, meaning that the $S_j(t)$ are martingales. Hence the following exposition will differ slightly from the classical Margrabe result.

We denote the time of maturity of the spread option by T_o and the payoff is given by

$$\max(S_2(T_o) - S_1(T_o) - K, 0),$$

for strike price $K \geq 0$. For now, we focus on the case that $K = 0$ (which is in fact the case of an exchange option). Then the price of the spread option is given by the risk-neutral expectation formula for the time 0 price of the option

$$\mathrm{Spr}((S_1(0), S_2(0)); T_o, K) = e^{-rT_o}\mathbb{E}_0[\max(S_2(T_o) - S_1(T_o) - K, 0)],$$

see e.g. [25]. Here $\mathbb{E}_0$ is the short-hand notation for the conditional expectation given $\mathscr{F}_0$. For $K = 0$ we can rewrite the price as follows.

$$\text{Spr}((S_1(0), S_2(0)); T_o, 0) = e^{-rT_o} \mathbb{E}_0[\max(S_2(T_o) - S_1(T_o), 0)]$$

$$= e^{-rT_o} \mathbb{E}_0 \left[\max \left(\frac{S_2(T_o)}{S_1(T_o)} - 1, 0 \right) S_1(T_o) \right]$$

$$= e^{-rT_o} S_1(0) \mathbb{E}_0^{\tilde{\mathbb{P}}} \left[\max \left(\frac{S_2(T_o)}{S_1(T_o)} - 1, 0 \right) \right],$$

where we used $S_1(t)/S_1(0)$ as numeraire in the generalised Bayes' formula, see e.g. [4, p. 124]. Moreover, as the numeraire in the cases we are studying becomes a martingale with mean equal to 1, we defined a new probability measure by $\frac{d\tilde{\mathbb{P}}}{d\mathbb{P}}\Big|_{\mathscr{F}_t} = \frac{S_1(t)}{S_1(0)}$.

In the context when both S_2 and S_1 are geometric Brownian motions, Margrabe [28] used the above result to link the pricing of an exchange option to the classical Black-Scholes option pricing problem, see [15].

4.2 Calendar Spreads

First of all, we study the case of calendar spreads, which are written on *one* commodity only. W.l.o.g. we study a calendar spread option written on $F_1(t, T)$. We fix two different dates of maturity satisfying $T_o < T_2 < T_1 < T^*$ and set

$$S_1(t) = \exp(Y_1(t, T_1)), \qquad\qquad S_2(t) = \exp(Y_1(t, T_2)).$$

First, we observe that from the martingale property of $F_1(t, T_1)$, it holds for $t \geq s \geq 0$,

$$\mathbb{E} \left[\frac{S_1(t)}{S_1(0)} \,\middle|\, \mathscr{F}_s \right] = \frac{1}{S_1(0)} \mathbb{E}\left[S_1(t) \mid \mathscr{F}_s\right] = \frac{S_1(s)}{S_1(0)}.$$

Here, we applied the $\mathscr{F}_0$-measurability of $S_1(0) = \exp(Y_1(0, T_1))$. Hence, $S_1(t)/S_1(0)$ is a martingale for $0 \leq t \leq T_1$. But, as the expectation of a martingale is equal to its expected value at zero, we find

$$\mathbb{E} \left[\frac{S_1(t)}{S_1(0)} \right] = 1.$$

Hence, $S_1(t)/S_1(0)$ is a density process and $\tilde{\mathbb{P}}$ is a probability.

We need to find the dynamics of the ratio S_2/S_1 under the new probability measure $\tilde{\mathbb{P}}$. In order to derive the dynamics, we proceed in three steps: First, we formulate a Girsanov theorem which is relevant for our ambit set-up. Next, we use an extended Itô's formula, to derive the dynamics of S_2/S_1 under the risk-neutral probability measure $\mathbb{P}$. Finally, we combine the two results, to find the dynamics of S_2/S_1 under the probability measure $\tilde{\mathbb{P}}$.

In the following, we will state the results without proofs since we will present the proofs for the more general cases covered in the next subsection in the Appendix.

Lemma 1. *Define the probability measure* $\left.\frac{d\tilde{\mathbb{P}}}{d\mathbb{P}}\right|_{\mathscr{F}_t} = \frac{S_1(t)}{S_1(0)}$, *for* $0 \leq t \leq T_1$, *where*

$$
\begin{aligned}
\frac{S_1(t)}{S_1(0)} = \exp\bigg(&- \int_{A_1^0(t)} \frac{1}{2} g_1^2(T_1; s, \xi) \sigma_1^2(s, \xi) ds d\xi \\
&+ \int_{A_1^0(t)} g_1(T_1; s, \xi) \sigma_1(s, \xi) W_1(ds, d\xi) \\
&- \int_{A_1^0(t)} \int_{\mathbb{R}} \left(e^{zh_1(T_1; s, \xi)} - 1 - zh_1(T_1; s, \xi) \right) \nu_{L_1}(dz) ds d\xi \\
&+ \int_{A_1^0(t)} \int_{\mathbb{R}} zh_1(T_1; s, \xi) \tilde{J}_{L_1}(dz, ds, d\xi) \bigg).
\end{aligned}
$$

Then,

$$
\hat{W}_1(ds, d\xi) = W_1(ds, d\xi) - g_1(T; s, \xi) \sigma_1(s, \xi) ds d\xi
$$

is a Gaussian homogeneous and factorisable Lévy bases under $\tilde{\mathbb{P}}$. *Also,*

$$
\hat{N}_1(dz_1, ds, d\xi) = J_{L_1}(dz_1, ds, d\xi) - e^{z_1 h_1(T_1; s, \xi)} \nu_{L_1}(dz_1) ds d\xi,
$$

is a compensated Poisson random measure under $\tilde{\mathbb{P}}$.

Lemma 2. *Under* $\mathbb{P}$, *the dynamics of* $S_2(t)/S_1(t)$ *are given by*

$$
\begin{aligned}
\frac{d\left(\frac{S_2(t)}{S_1(t)}\right)}{S_2(t-)/S_1(t-)} = &\int_0^\infty (-g_1(T_1; t, \xi) + g_1(T_2; t, \xi)) \sigma_1(t, \xi) W_1(dt, d\xi) \\
&+ \int_0^\infty (g_1^2(T_1; t, \xi) - g_1(T_1; t, \xi) g_1(T_2; t, \xi)) \sigma_1^2(t, \xi) dt d\xi \\
&+ \int_0^\infty \int_{\mathbb{R}} \left(e^{z_1 h_1(T_2; t, \xi)} - e^{z_1 h_1(T_1; t, \xi)} \right) \tilde{J}_{L_1}(dz_1, dt, d\xi) \\
&+ \int_0^\infty \int_{\mathbb{R}} \left(e^{z_1 h_1(T_2; t, \xi) - z_1 h_1(T_1; t, \xi)} + e^{z_1 h_1(T_1; t, \xi)} - e^{z_1 h_1(T_2; t, \xi)} - 1 \right) \nu_{L_1}(dz_1) dt d\xi.
\end{aligned}
$$

Proposition 5. *Using the same notation as in Lemma 1, the dynamics of S_2/S_1 under $\tilde{\mathbb{P}}$ are given by*

$$\frac{d\left(\frac{S_2(t)}{S_1(t)}\right)}{S_2(t-)/S_1(t-)} = \int_0^\infty (-g_1(T_1;t,\xi) + g_1(T_2;t,\xi))\sigma_1(t,\xi)\hat{W}_1(dt,d\xi)$$

$$+ \int_0^\infty \int_{\mathbb{R}} \left(e^{z_1(h_1(T_2;t,\xi)-h_1(T_1;t,\xi))} - 1\right) \hat{N}_1(dz_1,dt,d\xi).$$

Then the solution is given by

$$\frac{S_2(t)}{S_1(t)} = \frac{S_2(0)}{S_1(0)} \exp(Z(t)),$$

where

$$Z(t) = \int_{A_1^0(t)} (-g_1(T_1;s,\xi) + g_1(T_2;s,\xi))\sigma_1(s,\xi)\hat{W}_1(ds,d\xi)$$

$$-\frac{1}{2}\int_{A_1^0(t)} (-g_1(T_1;s,\xi) + g_1(T_2;s,\xi))^2\sigma_1^2(s,\xi)dsd\xi$$

$$+ \int_{A_1^0(t)}\int_{\mathbb{R}} (z_1h_1(T_2;s,\xi) - z_1h_1(T_1;s,\xi))\hat{N}_1(dz_1,ds,d\xi)$$

$$+ \int_{A_1^0(t)}\int_{\mathbb{R}} (z_1h_1(T_2;s,\xi) - z_1h_1(T_1;s,\xi)) + 1 - \exp(z_1h_1(T_2;s,\xi)$$

$$-z_1h_1(T_1;s,\xi))v_{L_1,\tilde{\mathbb{P}}}(dz_1)dsd\xi, \tag{13}$$

where $v_{L_1,\tilde{\mathbb{P}}}$ denotes the Lévy measure of L_1 under $\tilde{\mathbb{P}}$.

Based on the preceding results, we can now formulate the Margrabe formula for a calendar spread option.

Proposition 6. *The time 0 price of the calendar spread option is given by*

$$Spr((S_1(0), S_2(0)); T_o, 0) = e^{-rT_o}S_1(0)\mathbb{E}_0^{\tilde{\mathbb{P}}}\left[\max\left(\frac{S_2(T_o)}{S_1(T_o)} - 1, 0\right)\right], \tag{14}$$

where $S_2(T_o)/S_1(T_o) = S_2(0)/S_1(0)\exp(Z(T_o))$, where Z is defined as in (13).

Remark 5. Note that the option price at time 0 is actually stochastic since it depends on the random variables $S_1(0)$ and $S_2(0)$. In practice, however, one could use the observations at time 0, which can be viewed as realisations of $S_1(0)$ and $S_2(0)$, to pin down a price. By taking the expected value of $Spr((S_1(0), S_2(0)); T_o, 0)$, we find the best predicted option price from the model of the forwards. We may also provide insight into the bid-ask spread of prices for this option based on the distribution of $\mathrm{Spr}((S_1(0), S_2(0)); T_o, 0)$ (computing its standard deviation, say). In illiquid markets as electricity, it may be informative to have a random variable describing the distribution of prices rather than one price.

Clearly, the key question is: How can we compute the expectation given in (14)? Since

$$\mathbb{E}_0^{\tilde{\mathbb{P}}}\left[\max\left(\frac{S_2(T_o)}{S_1(T_o)} - 1, 0\right)\right] = \mathbb{E}_0^{\tilde{\mathbb{P}}}\left[\max\left(\frac{S_2(0)}{S_1(0)}\exp(Z(T_o)) - 1, 0\right)\right]$$
$$= \frac{S_2(0)}{S_1(0)}\mathbb{E}_0^{\tilde{\mathbb{P}}}\left[\max\left(\exp(Z(T_o)) - \frac{S_1(0)}{S_2(0)}, 0\right)\right],$$

and since $S_1(0)/S_2(0)$ is $\mathscr{F}_0$-measurable, we know that, as soon as we can compute $\mathbb{E}_0^{\tilde{\mathbb{P}}}\left[\max(\exp(Z(T_o)) - \kappa, 0)\right]$, for a positive constant κ, we can derive the corresponding price of the spread option when setting $\kappa = S_1(0)/S_2(0)$.

In the absence of jumps and stochastic volatility, this essentially boils down to the classical Margrabe formula, which is based on the Black-Scholes formula. In the more general case we have here, we can for instance follow the approach of using Fourier methods, see e.g. [25].

4.2.1 The Gaussian Case

Consider the case when $h_1 \equiv 0$, i.e. there is no jump component and we have

$$Z(t) = \int_{A_1^0(t)} (-g_1(T_1; s, \xi) + g_1(T_2; s, \xi))\sigma_1(s, \xi)\hat{W}_1(ds, d\xi)$$
$$- \frac{1}{2}\int_{A_1^0(t)} (-g_1(T_1; s, \xi) + g_1(T_2; s, \xi))^2\sigma_1^2(s, \xi)dsd\xi.$$

Conditional on $\mathscr{F}^{\sigma_1} \vee \mathscr{F}_0$, $Z(t)$ is normally distributed with mean

$$m(t) = -\frac{1}{2}\int_{A_1^0(t)} (-g_1(T_1; s, \xi) + g_1(T_2; s, \xi))^2\sigma_1^2(s, \xi)dsd\xi,$$

and variance

$$v^2(t) = \int_{A_1^0(t)} (-g_1(T_1; s, \xi) + g_1(T_2; s, \xi))^2\sigma_1^2(s, \xi)dsd\xi.$$

Proposition 7. *Using the above notation, we get*

$$\mathbb{E}_0^{\tilde{\mathbb{P}}} \left[\max \left(\exp(Z(T_o)) \right) - \kappa, 0 \right] | \mathscr{F}^{\sigma_1} \right)$$

$$= \kappa \Phi \left(\frac{m(T_o) - \log(\kappa)}{v(T_o)} \right) - \Phi \left(\frac{m(T_o) - \log(\kappa)}{v(T_o)} + v(T_o) \right), \qquad (15)$$

where Φ denotes the cumulative distribution function of the standard normal distribution.

Note that the proof of the above results follows the classical arguments of computing the call price associated with a lognormally distributed random variable and is hence omitted.

Clearly, when σ_1 is just a deterministic function, we are done since the results from Propositions 6 and 7 specify an extended version of the Margrabe formula.

However, in the case when σ_1 is truly stochastic, we need to compute the conditional expectation of (15) given $\mathscr{F}_0$. This is in line with the approach advocated in [24]. However, we are not able to find an analytic formula for this expectation and hence need a different methodology for the case of stochastic volatility (and also when we have a general jump term).

4.2.2   A Fourier Approach for the General Case

In the general case, we can work with the Fourier approach where the corresponding integral needs to be computed numerically.

Proposition 8. *For any $R > 1$ such that $\mathbb{E}^{\tilde{\mathbb{P}}} \left[\exp(i\theta Z(T_o)) \right] < \infty$, for $\theta :=$ $\theta(R, u) := -(u + iR)$, we have*

$$\mathbb{E}_0^{\tilde{\mathbb{P}}} \left[\max \left(\exp(Z(T_o)) \right) - \kappa, 0 \right) \right] = \frac{1}{2\pi} \int_{\mathbb{R}} \mathbb{E}_0^{\tilde{\mathbb{P}}} \left[\exp(i\theta(R, u)Z(T_o)) \right] \hat{f}(-\theta(R, u)) du,$$

$$(16)$$

and $\hat{f}(z) = \kappa^{1+iz} / (iz(1 + iz))$ and

$$\mathbb{E}_0^{\tilde{\mathbb{P}}} \left[\exp(i\theta Z(T_o)) \right]$$

$$= \mathbb{E}_0^{\tilde{\mathbb{P}}} \left[\exp \left(\left(-\frac{\theta^2}{2} - \frac{i\theta}{2} \right) \int_{A_1^0(T_o)} (-g_1(T_1; s, \xi) + g_1(T_2; s, \xi))^2 \sigma_1^2(s, \xi) ds d\xi \right) \right.$$

$$\left. \cdot \exp \left((1 - i\theta) \int_{A_1^0(T_o)} \int_{\mathbb{R}} \left(e^{z_1(h_1(T_2; s, \xi) - h_1(T_1; s, \xi))} - 1 \right) v_{L_1, \tilde{\mathbb{P}}}(dz_1) ds d\xi \right) \right].$$

In the above result, there is still an expectation which needs to be computed if we have a stochastic volatility component. In order to get an analytic expression for this expectation, we introduce a modelling assumption for the volatility process. Suppose that

$$\sigma_1^2(s, \xi) = \varpi(s-, \xi), \quad \text{where} \quad \varpi(s, \xi) = \int_{A_1(s)} l(s, \xi; x, y) L_\sigma(dx, dy),$$

where l is assumed to be a non-negative continuous function, which is integrable with respect to the homogeneous and factorisable Lévy basis L_σ, which has CQ given by $(0, 0, \nu_\sigma, Leb)$ under $\mathbb{P}$, where ν_σ corresponds to a Lévy measure of a subordinator. An application of Fubini's theorem (provided the corresponding integrals exist) leads to

$$\mathbb{E}_0^{\tilde{\mathbb{P}}}\left[\exp\left(\left(-\theta^2 - \frac{i\theta}{2}\right)\int_{A_1^0(T_o)}(-g_1(T_1; s, \xi) + g_1(T_2; s, \xi))^2\sigma_1^2(s, \xi)dsd\xi\right)\right]$$

$$= \exp\left(\int_{(-\infty, 0]\times[0,\infty)} iK(x, y)L_\sigma(dx, dy) + \int_{A_1^0(T_o)} \phi_{L_\sigma}(K(x, y))dxdy\right),$$

where ϕ_{L_σ} is the cumulant function, i.e. the distinguished logarithm of the characteristic function, associated with L_σ and

$$K(x, y) = \left(i\theta^2 - \frac{\theta}{2}\right)\int_{[x,T_o]\times[0,\infty)}(-g_1(T_1; s, \xi) + g_1(T_2; s, \xi))^2 l(s, \xi; x, y)dsd\xi.$$

4.3 Spreads Between Different Forward Contracts

Now we turn to spread options written on *two* commodities, where we need to use our multivariate modelling framework. Hence we set the number of assets to $n = 2$ and consider assets labelled with subscripts 1 and 2. That is for $i = 1, 2$ we write

$$S_i(t) = \exp(Y_i(t, T)).$$

That is we consider the spread between forward prices for two different commodities having the same times of maturity T.

In the following, we will denote by $\rho = \rho_{1,2} \in [-1, 1]$ the correlation coefficient determined by $\mathbb{E}(W_1(dt, d\xi)W_2(dt, d\xi)) = \rho dt d\xi$.

Before we present the dynamics of the ratio S_2/S_1, we derive two lemmas. The first lemma is a variant of the Girsanov theorem.

Lemma 3. *Define the probability measure by* $\frac{d\tilde{\mathbb{P}}}{d\mathbb{P}}\Big|_{\mathscr{F}_t} = \frac{S_1(t)}{S_1(0)}$, *for* $0 \leq t \leq T$, *where*

$$
\frac{S_1(t)}{S_1(0)} = \exp\left(-\int_{A_1^0(t)} \frac{1}{2} g_1^2(T; s, \xi) \sigma_1^2(s, \xi) ds d\xi \right.
$$

$$
+ \int_{A_1^0(t)} g_1(T; s, \xi) \sigma_1(s, \xi) W_1(ds, d\xi)
$$

$$
- \int_{A_1^0(t)} \int_{\mathbb{R}} \left(e^{z h_1(T; s, \xi)} - 1 - z h_1(T; s, \xi) \right) \nu_{L_1}(dz) ds d\xi
$$

$$
\left. + \int_{A_1^0(t)} \int_{\mathbb{R}} z h_1(T; s, \xi) \tilde{J}_{L_1}(dz, ds, d\xi) \right).
$$

Then,

$$
\hat{W}_1(ds, d\xi) = W_1(ds, d\xi) - g_1(T; s, \xi) \sigma_1(s, \xi) ds d\xi,
$$

$$
\hat{W}_2(ds, d\xi) = W_2(ds, d\xi) - \rho g_1(T; s, \xi) \sigma_1(s, \xi) ds d\xi
$$

are Gaussian homogeneous and factorisable Lévy bases under $\tilde{\mathbb{P}}$ *satisfying*

$$
\hat{W}_1(ds, d\xi) \hat{W}_2(ds, d\xi) = \rho ds d\xi.
$$

Also,

$$
\hat{N}_{1,2}(dz_1, dz_2, ds, d\xi) = J_{(L_1, L_2)}(dz_1, dz_2, ds, d\xi) - e^{z_1 h_1(T; s, \xi)} \nu_{(L_1, L_2)}(dz_1, dz_2) ds d\xi,
$$

is a compensated Poisson random measure under $\tilde{\mathbb{P}}$.

The proof is given in the Appendix.

Next, we would like to find the dynamics of the ratio S_2/S_1 under the new probability measure $\tilde{\mathbb{P}}$. In a first step, we derive the dynamics under $\mathbb{P}$ and then change measure.

Lemma 4. *Under* $\mathbb{P}$ *the dynamics of* $S_2(t)/S_1(t)$ *are given by*

$$
\frac{d\left(\frac{S_2(t)}{S_1(t)}\right)}{S_2(t-)/S_1(t-)}
$$

$$
= -\int_0^\infty g_1(T; t, \xi) \sigma_1(t, \xi) W_1(dt, d\xi) + \int_0^\infty g_2(T; t, \xi) \sigma_2(t, \xi) W_2(dt, d\xi)
$$

$$
+ \int_0^\infty g_1^2(T; t, \xi) \sigma_1^2(t, \xi) dt d\xi - \rho \int_0^\infty g_1(T; t, \xi) g_2(T; t, \xi) \sigma_1(t, \xi) \sigma_2(t, \xi) dt d\xi
$$

$$+ \int_0^\infty \int_{\mathbb{R}^2} \left(e^{z_2 h_2(T;t,\xi)} - e^{z_1 h_1(T;t,\xi)} \right) \tilde{J}_{(L_1,L_2)}(dz_1, dz_2, dt, d\xi)$$

$$+ \int_0^\infty \int_{\mathbb{R}^2} \left(e^{z_2 h_2(T;t,\xi) - z_1 h_1(T;t,\xi)} + e^{z_1 h_1(T;t,\xi)} - e^{z_2 h_2(T;t,\xi)} - 1 \right)$$

$$\cdot \nu_{(L_1,L_2)}(dz_1, dz_2) dt d\xi.$$

The proof is given in the Appendix.

Proposition 9. *Using the same notation as in Lemma 3, the dynamics of* S_2/S_1 *under* $\tilde{\mathbb{P}}$ *are given by*

$$\frac{d\left(\frac{S_2(t)}{S_1(t)}\right)}{S_2(t-)/S_1(t-)}$$

$$= - \int_0^\infty g_1(T;t,\xi)\sigma_1(t,\xi)\hat{W}_1(dt,d\xi) + \int_0^\infty g_2(T;t,\xi)\sigma_2(t,\xi)\hat{W}_2(dt,d\xi)$$

$$+ \int_0^\infty \int_{\mathbb{R}^2} \left(e^{z_2 h_2(T;t,\xi) - z_1 h_1(T;t,\xi)} - 1 \right) \hat{N}_{1,2}(dz_1, dz_2, dt, d\xi),$$

which is a geometric ambit field. Then the solution is given by

$$\frac{S_2(t)}{S_1(t)} = \frac{S_2(0)}{S_1(0)} \exp(Z(t)),$$

where

$$Z(t) = - \int_{A_j^0(t)} g_1(T;s,\xi)\sigma_1(s,\xi)\hat{W}_1(ds,d\xi) + \int_{A_j^0(t)} g_2(T;s,\xi)\sigma_2(s,\xi)\hat{W}_2(ds,d\xi)$$

$$- \frac{1}{2} \int_{A_j^0(t)} g_1^2(T;s,\xi)\sigma_1^2(s,\xi)dsd\xi - \frac{1}{2} \int_{A_j^0(t)} g_2^2(T;s,\xi)\sigma_2^2(s,\xi)dsd\xi$$

$$+ \rho \int_{A_j^0(t)} g_1(T;s,\xi)g_2(T;s,\xi)\sigma_1(s,\xi)\sigma_2(s,\xi)dsd\xi$$

$$+ \int_{A_j^0(t)} \int_{\mathbb{R}^2} (z_2 h_2(T;s,\xi) - z_1 h_1(T;s,\xi))\hat{N}_{1,2}(dz_1, dz_2, ds, d\xi)$$

$$+ \int_{A_j^0(t)} \int_{\mathbb{R}^2} (z_2 h_2(T;s,\xi) - z_1 h_1(T;s,\xi)) + 1 - \exp(z_2 h_2(T;s,\xi)$$

$$- z_1 h_1(T;s,\xi))\nu_{(L_1,L_2),\tilde{\mathbb{P}}}(dz_1, dz_2)dsd\xi, \tag{17}$$

where $\nu_{(L_1,L_2),\tilde{\mathbb{P}}}$ *denotes the Lévy measure of* (L_1, L_2) *under* $\tilde{\mathbb{P}}$.

The proof is given in the Appendix.

As in the case of calendar spread options, we get the following Margrabe-type formula for spread options written on different commodities.

Proposition 10. *The time 0 price of the spread option is given by*

$$Spr((S_1(0), S_2(0)); T_o, 0) = e^{-rT_o} S_1(0) \mathbb{E}_0^{\tilde{\mathbb{P}}} \left[\max \left(\frac{S_2(T_o)}{S_1(T_o)} - 1, 0 \right) \right], \qquad (18)$$

where $S_2(T_o)/S_1(T_o) = S_2(0)/S_1(0) \exp(Z(T_o))$, *where* Z *is defined as in* (17).

As before we note that as soon as we can compute $\mathbb{E}_0^{\tilde{\mathbb{P}}} [\max (\exp(Z(T_o)) - \kappa, 0)]$, for a positive constant κ, we can derive the corresponding price of the spread option.

4.3.1 The Gaussian Case

First we consider the case when $h_1 \equiv h_2 \equiv 0$, i.e. there is no jump component and we have

$$Z(t) = - \int_{A_j^0(t)} g_1(T; s, \xi) \sigma_1(s, \xi) \hat{W}_1(ds, d\xi) + \int_{A_j^0(t)} g_2(T; s, \xi) \sigma_2(s, \xi) \hat{W}_2(ds, d\xi)$$

$$- \frac{1}{2} \int_{A_j^0(t)} g_1^2(T; s, \xi) \sigma_1^2(s, \xi) ds d\xi - \frac{1}{2} \int_{A_j^0(t)} g_2^2(T; s, \xi) \sigma_2^2(s, \xi) ds d\xi$$

$$+ \rho \int_{A_j^0(t)} g_1(T; s, \xi) g_2(T; s, \xi) \sigma_1(s, \xi) \sigma_2(s, \xi) ds d\xi.$$

Conditional on the $\mathscr{F}^{\sigma_1} \vee \mathscr{F}^{\sigma_2} \vee \mathscr{F}_0$, $Z(t)$ is normally distributed with mean

$$m(t) = -\frac{1}{2} \int_{A_j^0(t)} g_1^2(T; s, \xi) \sigma_1^2(s, \xi) ds d\xi - \frac{1}{2} \int_{A_j^0(t)} g_2^2(T; s, \xi) \sigma_2^2(s, \xi) ds d\xi$$

$$+ \rho \int_{A_j^0(t)} g_1(T; s, \xi) g_2(T; s, \xi) \sigma_1(s, \xi) \sigma_2(s, \xi) ds d\xi,$$

and variance

$$v^2(t) = \int_{A_j^0(t)} g_1^2(T; s, \xi) \sigma_1^2(s, \xi) ds d\xi + \int_{A_j^0(t)} g_2^2(T; s, \xi) \sigma_2^2(s, \xi) ds d\xi$$

$$- 2\rho \int_{A_j^0(t)} g_1(T; s, \xi) g_2(T; s, \xi) \sigma_1(s, \xi) \sigma_2(s, \xi) ds d\xi.$$

Proposition 11. *Using the same notation as above, we get*

$$\mathbb{E}_0^{\tilde{\mathbb{P}}}\left[\max\left(\exp(Z(T_o)) - \kappa, 0\right)\mid \mathscr{F}^{\sigma_1} \vee \mathscr{F}^{\sigma_2}\right)$$

$$= K\Phi\left(\frac{m(T_o) - \log(\kappa)}{v(T_o)}\right) - \Phi\left(\frac{m(T_o) - \log(\kappa)}{v(T_o)} + v(T_o)\right), \qquad (19)$$

where Φ denotes the cumulative distribution function of the standard normal distribution.

As in the case of calendar spread options, we note that if σ_1 and σ_2 are just deterministic functions, then we get a Margrabe-type formula from Propositions 10 and 11.

4.3.2 A Fourier Approach for the General Case

In order to find the option price in the general modelling framework, we apply the Fourier approach again.

Proposition 12. *For any $R > 1$ such that $\mathbb{E}^{\tilde{\mathbb{P}}}\left[\exp(i\theta Z(T_o))\right] < \infty$ for $\theta :=$ $\theta(R, u) := -(u + iR)$, we have*

$$\mathbb{E}_0^{\tilde{\mathbb{P}}}\left[\max\left(\exp(Z(T_o)) - \kappa, 0\right)\right] = \frac{1}{2\pi}\int_{\mathbb{R}}\mathbb{E}_0^{\tilde{\mathbb{P}}}\left[\exp(i\theta(R, u)Z(T_o))\right]\hat{f}(-\theta(R, u))du,$$

$$(20)$$

and $\hat{f}(z) = \kappa^{1+iz}/(iz(1 + iz))$ and

$$\mathbb{E}_0^{\tilde{\mathbb{P}}}\left[\exp(i\theta Z(T_o))\right]$$

$$= \mathbb{E}_0^{\tilde{\mathbb{P}}}\left[\exp\left(\left(-\frac{\theta^2}{2} - \frac{i\theta}{2}\right)\int_{A_1^0(T_o)}(g_1^2(T; s, \xi)\sigma_1^2(s, \xi)\right.\right.$$

$$\left.-2\rho g_1(T; s, \xi)g_2(T; s, \xi)\sigma_1(s, \xi)\sigma_2(s, \xi) + g_2^2(T; s, \xi)\sigma_2^2(s, \xi))dsd\xi)\right]$$

$$\left.\cdot\exp\left((1 - i\theta)\int_{A_1^0(T_o)}\int_{\mathbb{R}^2}\left(e^{z_2 h_2(T; s, \xi) - z_1 h_1(T; s, \xi)} - 1\right)v_{(L_1, L_2), \tilde{\mathbb{P}}}(dz_1, dz_2)dsd\xi\right)\right].$$

4.3.3 Spreads with $K \neq 0$

In the case when the strike price equals $K = 0$, a spread option is in fact an exchange option and we have discussed both an extension of the Margrabe formula and a Fourier approach for determining the price of an exchange option in an ambit framework.

Let us now focus on the case when $K \neq 0$. Let

$$f(K) := \mathbb{E}_0[\max(S_2(T_o) - S_1(T_o) - K, 0)].$$

A first order Taylor approximation could be used for small K. In that case we would have

$$f(K) \approx f(0) + f'(0)K,$$

where the derivative of f can be computed as follows. Suppose that the random variable $S(T_o) := S_2(T_o) - S_1(T_o)$ conditional on $\mathscr{F}_0$ has a probability density denoted by $f_{S(T_o)}$. Then

$$f'(K) = \frac{d}{dK} \int_K^\infty (x - K) f_{S(T_o)}(x) dx = - \int_K^\infty f_{S(T_o)}(x) dx = -\mathbb{P}_0(S(T_o) > K).$$

Hence we have

$$f(K) \approx f(0) - K\mathbb{P}_0(S_2(T_o) - S_1(T_o) > 0).$$

Note that

$$\mathbb{P}_0(S_2(T_o) - S_1(T_o) > 0) = \mathbb{P}_0(Y_2(T_o, T) > Y_1(T_o, T)). \tag{21}$$

In the Gaussian case (i.e. in the absence of stochastic volatility and jumps), this probability can be computed explicitly. Since this goes along the lines of our previous computations, we skip this and focus directly on the general case.

Note that [11] have carried out an analysis of the Taylor approximation for spread options on Gaussian spot price models.

In the general case, we need to find the density of $Y_2(T_o, T) - Y_1(T_o, T)$ given $\mathscr{F}_0$ which we denote by $f_{Y_2(T_o,T)-Y_1(T_o,T)|\mathscr{F}_0}$, then the probability in (21) can be computed numerically.

It is well-known, see e.g. [22] for a survey, that the density can be obtained from the inverse Fourier transform by

$$f_{Y_2(T_o,T)-Y_1(T_o,T)|\mathscr{F}_0}(x) = \frac{1}{2\pi} \int_{-\infty}^\infty \phi_{Y_2(T_o,T)-Y_1(T_o,T)|\mathscr{F}_0}(t) e^{-itx} dt,$$

where $\phi_{Y_2(T_o,T)-Y_1(T_o,T)|\mathscr{F}_0}$ denotes the characteristic function of $Y_2(T_o, T) - Y_1(T_o, T)$ given $\mathscr{F}_0$. The latter can be computed in our model. Hence, we can numerically approximate the correction term $f'(0)K$ which is needed in the first order Taylor approximation.

5   Extensions

So far, we have studied a multivariate model consisting of univariate ambit fields which are linked together by multivariate Lévy bases. A generalisation of this approach would be to work with a more general multivariate model of the type

$$Y(t,T) = \int_{A(t)} \mu(T;s,\xi)dsd\xi + \int_{A(t)} G(T;s,\xi)\Sigma(s,\xi)W(ds,d\xi)$$

$$+ \int_{A(t)} H(T;s,\xi)L(ds,d\xi),$$

where $A(t) = (-\infty,t] \times [0,\infty)$ and the integrals should be understood componentwise (for each element in the vector). Here μ denotes a possibly stochastic vector taking values in $\mathbb{R}^n$. G and H denote deterministic kernel functions taking values in the $n \times n$-dimensional matrices, and Σ denotes a stochastic process taking values in the $n \times n$-dimensional positive semi-definite matrices. As in the univariate case, we assume that the kernel function satisfy suitable integrability conditions and that Σ is predictable in the time component. The Lévy bases $W = (W_1,\ldots,W_n)^\top$ and $L = (L_1,\ldots,L_n)^\top$ are the same as before. W and L are assumed to be independent, and (μ,Σ) is assumed to be independent of (W,L).

Note that for $j \in \{1,\ldots,n\}$ the jth component of Y is given by

$$Y_j(t,T) = \int_{A(t)} \mu_j(T;s,\xi)dsd\xi + \sum_{j=1}^n \sum_{l=1}^n \int_{A(t)} G_{j,k}(T;s,\xi)\Sigma_{k,l}(s,\xi)W_l(ds,d\xi)$$

$$+ \sum_{l=1}^n \int_{A(t)} H_{j,l}(T;s,\xi)L_l(ds,d\xi).$$

Example 7.   Consider the special case where

$$G(T;s,\xi) = \mathrm{diag}(g_1(T;s,\xi),\ldots,g_n(T;s,\xi)),$$

$$\Sigma(s,\xi) = \mathrm{diag}(\sigma_1(s,\xi),\ldots,\sigma_n(s,\xi)),$$

$$H(T;s,\xi) = \mathrm{diag}(h_1(T;s,\xi),\ldots,h_n(T;s,\xi)).$$

Then we obtain the model we studied throughout the paper. Note that in that case, we allow for stochastic volatility, but not for stochastic correlation. Such a modelling assumption has often been used in the literature to obtain parsimonious multivariate models, cf. the famous CCC model introduced by Bollerslev [16].

Clearly, a full multivariate model is more general and it can introduce dependence between different commodities in an even more flexible way. So far, we have introduced dependence through multivariate Lévy bases W and L. In addition, in a full multivariate model, dependence can furthermore be modelled via the kernel function and the stochastic volatility matrix, which in particular makes it possible to allow for stochastic correlation.

6   Conclusion

We have proposed a multivariate model based on two-factor multivariate geometric ambit fields to model forward curves of various commodities simultaneously. We have shown that our model is highly analytical tractable and accounts for key stylised facts found in energy forward prices.

Moreover, we show how spread options can be priced in our new modelling framework. Here we have considered both calendar spread options as well as spread options written on two different commodities. First we considered the case when the strike price of the spread option is equal to zero, meaning that we are dealing with exchange options. We have obtained two key results: In the absence of stochastic volatility and jumps, we have extended the Margrabe formula to Gaussian ambit fields. In the general case, we have derived a pricing formula based on Fourier techniques and the characteristic function of ambit fields, which is available analytically. In the case when the strike price of the spread option is not equal to zero, we have presented an approximation method for the option price based on a first-order Taylor approximation.

Acknowledgements We wish to thank the editors and the referee for constructive comments on an earlier draft of this article.

F.E. Benth acknowledges financial support by the Norwegian Research Council within the project "Managing Weather Risk in Electricity Markets" (MAWREM).

A.E.D. Veraart acknowledges financial support by a Marie Curie FP7 Career Integration Grant within the 7th European Union Framework Programme.

Appendix: Proofs

In the following, we provide the proofs of our main results.

Proof of Proposition 1

As $Y_j(t, T)$ is assumed to have an exponential moment, $F_j(t, T)$ is integrable. We have

$$F_j(t, T) = F_j(s, T) \exp\left(\int_{A_j^s(t)} \mu_j(T; u, \xi) du d\xi + \int_{A_j^s(t)} g_j(T; u, \xi) \sigma_j(u, \xi) W_j(du, d\xi) \right)$$

$$+ \int_{A_j^s(t)} \int_{\mathbb{R}} z h_j(T; u, \xi) \tilde{J}_{L_j}(dz, du, d\xi) \Bigg),$$

for $s \leq t$. Hence

$$\mathbb{E}\left[F_j(t, T) \mid \mathscr{F}_s\right] = F_j(s, T) \mathbb{E}\left[\exp(Z_j(s, t, T)) \mid \mathscr{F}_s\right],$$

with

$$Z_j(s, t, T) = \int_{A_j^s(t)} \mu_j(T; u, \xi) du d\xi + \int_{A_j^s(t)} g_j(T; u, \xi) \sigma_j(u, \xi) W_j(du, d\xi)$$

$$+ \int_{A_j^s(t)} \int_{\mathbb{R}} z h_j(T; u, \xi) \tilde{J}_{L_j}(dz, du, d\xi).$$

The martingale property follows if $\mathbb{E}[\exp(Z_j(s, t, T)) \mid \mathscr{F}_s] = 1$ for all $s \leq t \leq T$. But by double conditioning using $\mathscr{F}^{\sigma_j} \vee \mathscr{F}^{\mu_j} \vee \mathscr{F}_s$ we have, by independence of the two stochastic integrals,

$$\mathbb{E}\left[\exp(Z_j(s, t, T)) \mid \mathscr{F}_s\right]$$

$$= \mathbb{E}\Bigg[\exp\Bigg(\int_{A_j^s(t)} \mu_j(T; u, \xi) du d\xi + \frac{1}{2} \int_{A_j^s(t)} g_j^2(T; u, \xi) \sigma_j^2(u, \xi) du d\xi$$

$$+ \int_{A_j^s(t)} \int_{\mathbb{R}} e^{z h_j(T; u, \xi)} - 1 - z h_j(T; u, \xi) v_{L_j}(dz) du d\xi\Bigg) \Bigg| \mathscr{F}_s\Bigg].$$

But this is equal to one by assumption. Then the result follows.

Proof of Proposition 2

Using Itô's formula, we get

$$F_j(t, T) = F_j(0, T) + \int_0^t F_j(s-, T) d_s Y_j(s, T) + \frac{1}{2} \int_0^t F_j(s-, T) d_s [Y]_j^c(s, T)$$

$$+ \sum_{0 \leq s \leq t} \Delta_s F_j(s, T) - F_j(s-, T) \Delta_s Y_j(s, T)$$

$$= F_j(0, T) + \int_0^t F_j(s-, T) d_s Y_j(s, T) + \frac{1}{2} \int_0^t F_j(s-, T) d_s [Y]_j^c(s, T)$$

$$+ \sum_{0 \leq s \leq t} F_j(s-, T) \left(\exp(\Delta_s Y_j(s, T)) - 1 - \Delta_s Y_j(s, T)\right)$$

$$=F_j(0,T) + \int_{A_j^0(t)} F_j(s-,T)g_j(T;s,\xi)\sigma_j(s,\xi)W_j(ds,d\xi)$$

$$+ \int_{A_j^0(t)} \int_{\mathbb{R}} F_j(s-,T)\left(e^{zh_j(T;s,\xi)} - 1\right) \tilde{J}_{L_j}(dz,ds,d\xi).$$

Note that we have applied the Itô formula to the process $(F_j(t,T))_{0\le t\le T}$ for fixed T. From Proposition 1 we deduce that this process is indeed a martingale and hence we can apply the classical Itô formula when we fix the parameter T. The corresponding results on the quadratic variation of Y are direct consequences of the Walsh-integration theory.

Proof of Proposition 8

Formula (16) is a direct consequence of [21, Theorem 2.2] and the functional form of $\hat{f}$ has been derived in [21, Example 5.1]. Finally, we need to compute the extended characteristic function $\mathbb{E}_0^{\tilde{\mathbb{P}}}[\exp(i\theta(R,u)Z(t))]$, where we set $\theta = \theta(R,u) \in \mathbb{C}$ in the proof to simplify the exposition.

For fixed t define the two random variables

$$Z_1(t) := \int_{A_1^0(t)} (-g_1(T_1;s,\xi) + g_1(T_2;s,\xi))\sigma_1(s,\xi)\hat{W}_1(ds,d\xi)$$

$$- \frac{1}{2}\int_{A_1^0(t)} (-g_1(T_1;s,\xi) + g_1(T_2;s,\xi))^2\sigma_1^2(s,\xi)dsd\xi$$

$$Z_2(t) := \int_{A_1^0(t)}\int_{\mathbb{R}} (z_1 h_1(T_2;s,\xi) - z_1 h_1(T_1;s,\xi))\hat{N}_1(dz_1,ds,d\xi)$$

$$- \int_{A_1^0(t)}\int_{\mathbb{R}} (z_1 h_1(T_2;s,\xi) - z_1 h_1(T_1;s,\xi)) + 1 - \exp(z_1 h_1(T_2;s,\xi)$$

$$- z_1 h_1(T_1;s,\xi))v_{L_1,\tilde{\mathbb{P}}}(dz_1)dsd\xi.$$

Note that the σ-algebra $\sigma(Z_2(t))$ is independent of $\sigma(\mathscr{F}_0, Z_1(t))$ (assuming a suitable choice of the underlying filtration $(\mathscr{F}_t)$). Hence

$$\mathbb{E}_0^{\tilde{\mathbb{P}}}[\exp(i\theta Z(t))] = \mathbb{E}_0^{\tilde{\mathbb{P}}}[\exp(i\theta Z_1(t))]\,\mathbb{E}^{\tilde{\mathbb{P}}}[\exp(i\theta Z_2(t))],$$

where

$$\mathbb{E}_0^{\tilde{\mathbb{P}}}[\exp(i\theta Z_1(t))]$$

$$= \mathbb{E}_0^{\tilde{\mathbb{P}}} \left[\exp\left(\left(-\frac{\theta^2}{2} - \frac{i\theta}{2} \right) \int_{A_1^0(t)} (-g_1(T_1; s, \xi) + g_1(T_2; s, \xi))^2 \sigma_1^2(s, \xi) ds d\xi \right) \right],$$

where we conditioned on $\mathscr{F}^{\sigma_1}$. For the jump part, we have

$$\mathbb{E}^{\tilde{\mathbb{P}}} \left[\exp(i\theta Z_2(t)) \right]$$

$$= \exp\left((1 - i\theta) \int_{A_1^0(t)} \int_{\mathbb{R}} \left(e^{z_1 (h_1(T_2; s, \xi) - h_1(T_1; s, \xi))} - 1 \right) \nu_{L_1, \tilde{\mathbb{P}}}(dz_1) ds d\xi \right).$$

Proof of Lemma 3

This follows directly from the martingale condition derived in Proposition 1 and a straightforward extension of the Girsanov theorem for Itô-Lévy processes, see e.g. [29, Theorem 1.33]. For the Poisson random measure, note that

$$\hat{N}_{1,2}(dz_1, dz_2, ds, d\xi)$$

$$= \tilde{J}_{(L_1, L_2)}(dz_1, dz_2, ds, d\xi) + (1 - e^{z_1 h_1(T; s, \xi)}) \nu_{(L_1, L_2)}(dz_1, dz_2) ds d\xi$$

$$= J_{(L_1, L_2)}(dz_1, dz_2, ds, d\xi) - \nu_{(L_1, L_2)}(dz_1, dz_2) ds d\xi$$

$$+ (1 - e^{z_1 h_1(T; s, \xi)}) \nu_{(L_1, L_2)}(dz_1, dz_2) ds d\xi$$

$$= J_{(L_1, L_2)}(dz_1, dz_2, ds, d\xi) - e^{z_1 h_1(T; s, \xi)} \nu_{(L_1, L_2)}(dz_1, dz_2) ds d\xi,$$

denotes the compensated Poisson random measure under $\tilde{\mathbb{P}}$.

Proof of Lemma 4

First of all, we apply the Itô formula for higher dimensions and get

$$\frac{d\left(\frac{S_2(t)}{S_1(t)} \right)}{S_2(t-)/S_1(t-)} = -\frac{dS_1(t)}{S_1(t-)} + \frac{dS_2(t)}{S_2(t-)} + \frac{d[S_1]^c(t)}{S_1^2(t-)} - \frac{1}{S_1(t-)S_2(t-)} d[S_1, S_2]^c(t)$$

$$+ \frac{S_1(t-)}{S_2(t-)} \left(\frac{S_2(t)}{S_1(t)} - \frac{S_2(t-)}{S_1(t-)} - \Delta S_1(t) \left(-\frac{S_2(t-)}{S_1^2(t-)} \right) \right.$$

$$\left. - \Delta S_2(t) \left(\frac{1}{S_1(t-)} \right) \right)$$

$$= -\frac{dS_1(t)}{S_1(t-)} + \frac{dS_2(t)}{S_2(t-)} + \frac{d[S_1]^c(t)}{S_1^2(t-)} - \frac{1}{S_1(t-)S_2(t-)}d[S_1,S_2]^c(t)$$

$$+ \frac{S_1(t-)}{S_2(t-)}\Delta\left(\frac{S_2(t)}{S_1(t)}\right) - \Delta S_1(t)\left(-\frac{S_2(t-)}{S_1^2(t-)}\right)\frac{S_1(t-)}{S_2(t-)}$$

$$- \Delta S_2(t)\left(\frac{1}{S_1(t-)}\right)\frac{S_1(t-)}{S_2(t-)}$$

$$= -\frac{dS_1(t)}{S_1(t-)} + \frac{dS_2(t)}{S_2(t-)} + \frac{d[S_1]^c(t)}{S_1^2(t-)} - \frac{1}{S_1(t-)S_2(t-)}d[S_1,S_2]^c(t)$$

$$+ \frac{1}{S_2(t-)/S_1(t-)}\Delta\left(\frac{S_2(t)}{S_1(t)}\right) + \frac{\Delta S_1(t)}{S_1(t-)} - \frac{\Delta S_2(t)}{S_2(t-)}.$$

Since

$$\frac{d[S_1]^c(t)}{S_1^2(t-)} = \int_0^\infty g_1^2(T;t,\xi)\sigma_1^2(t,\xi)dtd\xi,$$

and

$$\frac{d[S_1,S_2]^c(t)}{S_1(t-)S_2(t-)} = \int_0^\infty g_1(T;t,\xi)g_2(T;t,\xi)\sigma_1(t,\xi)\sigma_2(t,\xi)\rho dtd\xi,$$

we have

$$\frac{d\left(\frac{S_2(t)}{S_1(t)}\right)}{S_2(t-)/S_1(t-)}$$

$$= -\int_0^\infty g_1(T;t,\xi)\sigma_1(t,\xi)W_1(dt,d\xi) - \int_0^\infty\int_{\mathbb{R}}\left(e^{zh_1(T;t,\xi)}-1\right)\tilde{J}_{L_1}(dz,dt,d\xi)$$

$$+ \int_0^\infty g_2(T;t,\xi)\sigma_2(t,\xi)W_2(dt,d\xi) + \int_0^\infty\int_{\mathbb{R}}\left(e^{zh_2(T;t,\xi)}-1\right)\tilde{J}_{L_2}(dz,dt,d\xi)$$

$$+ \int_0^\infty g_1^2(T;t,\xi)\sigma_1^2(s,\xi)dtd\xi - \rho\int_0^\infty g_1(T;t,\xi)g_2(T;t,\xi)\sigma_1(t,\xi)\sigma_2(t,\xi)dtd\xi$$

$$+ \frac{1}{S_2(t-)/S_1(t-)}\Delta\left(\frac{S_2(t)}{S_1(t)}\right) + \int_0^\infty\int_{\mathbb{R}}\left(e^{zh_1(T;t,\xi)}-1\right)\tilde{J}_{L_1}(dz,dt,d\xi)$$

$$- \int_0^\infty\int_{\mathbb{R}}\left(e^{zh_2(T;t,\xi)}-1\right)\tilde{J}_{L_2}(dz,dt,d\xi).$$

In the case of jumps of finite variation, we can further simplify and get

$$
\frac{d\left(\frac{S_2(t)}{S_1(t)}\right)}{S_2(t-)/S_1(t-)}
$$

$$
= -\int_0^\infty g_1(T;t,\xi)\sigma_1(t,\xi)W_1(dt,d\xi) + \int_0^\infty g_2(T;t,\xi)\sigma_2(t,\xi)W_2(dt,d\xi)
$$

$$
+ \int_0^\infty g_1^2(T;t,\xi)\sigma_1^2(t,\xi)dtd\xi - \rho\int_0^\infty g_1(T;t,\xi)g_2(T;t,\xi)\sigma_1(t,\xi)\sigma_2(t,\xi)dtd\xi
$$

$$
+ \frac{1}{S_2(t-)/S_1(t-)}\Delta\left(\frac{S_2(t)}{S_1(t)}\right).
$$

In the general case—writing formally—we have

$$
\frac{1}{S_2(t-)/S_1(t-)}\Delta\left(\frac{S_2(t)}{S_1(t)}\right)
$$

$$
= \frac{S_1(t-)}{S_2(t-)}\Delta\left(\frac{S_2(t)}{S_1(t)}\right) = \frac{\exp(Y_1(t-,T))}{\exp(Y_2(t-,T))}\Delta\left(\frac{\exp(Y_2(t,T))}{\exp(Y_1(t,T))}\right)
$$

$$
= \frac{\exp(Y_1(t-,T))}{\exp(Y_2(t-,T))}\left(\frac{\exp(Y_2(t,T))}{\exp(Y_1(t,T))} - \frac{\exp(Y_2(t-,T))}{\exp(Y_1(t-,T))}\right)
$$

$$
= \exp(Y_1(t-,T) - Y_2(t-,T))\,(\exp(Y_2(t,T) - Y_1(t,T))
$$

$$
\qquad - \exp(Y_2(t-,T) - Y_1(t-,T)))
$$

$$
= \exp(\Delta Y_2(t) - \Delta Y_1(t)) - 1.
$$

So altogether the finite variation jump term, has the form

$$
\frac{1}{S_2(t-)/S_1(t-)}\Delta\left(\frac{S_2(t)}{S_1(t)}\right) + \frac{\Delta S_1(t)}{S_1(t-)} - \frac{\Delta S_2(t)}{S_2(t-)} = e^{\Delta Y_2(t)-\Delta Y_1(t)} + e^{\Delta Y_1(t)} - e^{\Delta Y_2(t)} - 1.
$$

Summing up and using the joint Poisson random measure, we have

$$
\sum_{0\le s\le t}\left(\frac{1}{S_2(s-)/S_1(s-)}\Delta\left(\frac{S_2(s)}{S_1(s)}\right) + \frac{\Delta S_1(s)}{S_1(s-)} - \frac{\Delta S_2(s)}{S_2(s-)}\right)
$$

$$
= \int_{A_j(t)}\int_{\mathbb{R}^2}\left(e^{z_2 h_2(T;s,\xi)-z_1 h_1(T;s,\xi)} + e^{z_1 h_1(T;s,\xi)} - e^{z_2 h_2(T;s,\xi)} - 1\right)
$$

$$
\cdot J_{(L_1,L_2)}(dz_1,dz_2,ds,d\xi).
$$

Also,

$$-\int_{A_j(t)}\int_{\mathbb{R}}\left(e^{z_1 h_1(T;s,\xi)}-1\right)\tilde{J}_{L_1}(dz_1,ds,d\xi)$$

$$+\int_{A_j(t)}\int_{\mathbb{R}}\left(e^{z_2 h_2(T;s,\xi)}-1\right)\tilde{J}_{L_2}(dz_2,ds,d\xi)$$

$$=\int_{A_j(t)}\int_{\mathbb{R}^2}\left(e^{z_2 h_2(T;s,\xi)}-e^{z_1 h_1(T;s,\xi)}\right)\tilde{J}_{(L_1,L_2)}(dz_1,dz_2,ds,d\xi).$$

Adding the jump terms, we get

$$\int_{A_j(t)}\int_{\mathbb{R}^2}\left(e^{z_2 h_2(T;s,\xi)}-e^{z_1 h_1(T;s,\xi)}\right)\tilde{J}_{(L_1,L_2)}(dz_1,dz_2,ds,d\xi)$$

$$+\int_{A_j(t)}\int_{\mathbb{R}^2}\left(e^{z_2 h_2(T;s,\xi)-z_1 h_1(T;s,\xi)}+e^{z_1 h_1(T;s,\xi)}-e^{z_2 h_2(T;s,\xi)}-1\right)$$

$$\cdot J_{(L_1,L_2)}(dz_1,dz_2,ds,d\xi)$$

$$=\int_{A_j(t)}\int_{\mathbb{R}^2}\left(e^{z_2 h_2(T;s,\xi)-z_1 h_1(T;s,\xi)}-1\right)\tilde{J}_{(L_1,L_2)}(dz_1,dz_2,ds,d\xi)$$

$$+\int_{A_j(t)}\int_{\mathbb{R}^2}\left(e^{z_2 h_2(T;s,\xi)-z_1 h_1(T;s,\xi)}-1+e^{z_1 h_1(T;s,\xi)}-e^{z_2 h_2(T;s,\xi)}\right)$$

$$\cdot \nu_{(L_1,L_2)}(dz_1,dz_2)dsd\xi.$$

Proof of Proposition 9

This is a direct consequence of the preceding two lemmas. Note in particular, that for the Gaussian part, we have that W_1 and W_2 are Gaussian bases with correlation coefficient ρ, meaning that there exists a homogeneous, factorisable Gaussian basis W_3 (under $\mathbb{P}$) such that $W_2 = \rho W_1 + \sqrt{1-\rho^2}W_3$. Under the new measure $\tilde{\mathbb{P}}$, we have that W_3 does not change and

$$\hat{W}_1(ds,d\xi) = W_1(ds,d\xi) - g_1(T;s,\xi)\sigma(s,\xi)dsd\xi,$$

$$\hat{W}_3(ds,d\xi) = W_3(ds,d\xi),$$

$$\hat{W}_2(ds,d\xi) = \rho\hat{W}_1(ds,d\xi) + \sqrt{1-\rho^2}\hat{W}_3(ds,d\xi)$$

$$= W_2(ds,d\xi) - \rho g_1(T;s,\xi)\sigma(s,\xi)dsd\xi,$$

are homogeneous, factorisable Gaussian bases under $\tilde{\mathbb{P}}$, where $\hat{W}_1$ and $\hat{W}_3$ are independent and $\hat{W}_1$ and $\hat{W}_2$ have correlation coefficient ρ. So for the mixed Gaussian part, we have

$$
-\int_0^\infty g_1(T;t,\xi)\sigma_1(t,\xi)W_1(dt,d\xi) + \int_0^\infty g_2(T;t,\xi)\sigma_2(t,\xi)W_2(dt,d\xi)
$$

$$
+\int_0^\infty g_1^2(T;t,\xi)\sigma_1^2(t,\xi)dtd\xi - \rho\int_0^\infty g_1(T;t,\xi)g_2(T;t,\xi)\sigma_1(t,\xi)\sigma_2(t,\xi)dtd\xi
$$

$$
= -\int_0^\infty g_1(T;t,\xi)\sigma_1(t,\xi)(W_1(dt,d\xi) - g_1(T;t,\xi)\sigma_1(t,\xi)dtd\xi)
$$

$$
+\int_0^\infty g_2(T;t,\xi)\sigma_2(t,\xi)(W_2(dt,d\xi) - \rho g_1(T;t,\xi)\sigma_1(t,\xi)dtd\xi)
$$

$$
= -\int_0^\infty g_1(T;t,\xi)\sigma_1(t,\xi)\hat{W}_1(dt,d\xi) + \int_0^\infty g_2(T;t,\xi)\sigma_2(t,\xi)\hat{W}_2(dt,d\xi).
$$

For the jump part, we have

$$
\int_0^\infty \int_{\mathbb{R}^2} \left(e^{z_2 h_2(T;t,\xi)-z_1 h_1(T;t,\xi)} - 1\right)\tilde{J}_{(L_1,L_2)}(dz_1,dz_2,dt,d\xi)
$$

$$
+\int_0^\infty \int_{\mathbb{R}^2} \left(e^{z_2 h_2(T;t,\xi)-z_1 h_1(T;t,\xi)} - 1 + e^{z_1 h_1(T;t,\xi)} - e^{z_2 h_2(T;t,\xi)}\right)
$$

$$
\cdot \nu_{(L_1,L_2)}(dz_1,dz_2)dtd\xi
$$

$$
= \int_0^\infty \int_{\mathbb{R}^2} \left(e^{z_2 h_2(T;t,\xi)-z_1 h_1(T;t,\xi)} - 1\right)(\hat{N}_{1,2}(dz_1,dz_2,dt,d\xi) + (e^{z_1 h_1(T;t,\xi)} - 1)
$$

$$
\cdot \nu_{(L_1,L_2)}(dz_1,dz_2)dtd\xi)
$$

$$
+\int_0^\infty \int_{\mathbb{R}^2} \left(e^{z_2 h_2(T;t,\xi)-z_1 h_1(T;t,\xi)} - 1 + e^{z_1 h_1(T;t,\xi)} - e^{z_2 h_2(T;t,\xi)}\right)
$$

$$
\cdot \nu_{(L_1,L_2)}(dz_1,dz_2)dtd\xi
$$

$$
= \int_0^\infty \int_{\mathbb{R}^2} \left(e^{z_2 h_2(T;t,\xi)-z_1 h_1(T;t,\xi)} - 1\right)\hat{N}_{1,2}(dz_1,dz_2,dt,d\xi)
$$

$$
+\int_0^\infty \int_{\mathbb{R}^2} \left(e^{z_2 h_2(T;t,\xi)-z_1 h_1(T;t,\xi)} - 1\right)(e^{z_1 h_1(T;t,\xi)} - 1)\nu_{(L_1,L_2)}(dz_1,dz_2)dtd\xi
$$

$$
+\int_0^\infty \int_{\mathbb{R}^2} \left(e^{z_2 h_2(T;t,\xi)-z_1 h_1(T;t,\xi)} - 1 + e^{z_1 h_1(T;t,\xi)} - e^{z_2 h_2(T;t,\xi)}\right)
$$

$$
\cdot \nu_{(L_1,L_2)}(dz_1,dz_2)dtd\xi
$$

$$
= \int_0^\infty \int_{\mathbb{R}^2} \left(e^{z_2 h_2(T;t,\xi)-z_1 h_1(T;t,\xi)} - 1\right)\hat{N}_{1,2}(dz_1,dz_2,dt,d\xi).
$$

Note in particular, that

$$
\begin{aligned}
\hat{N}_{1,2}&(dz_1, dz_2, dt, d\xi) \\
&= \tilde{J}_{(L_1, L_2)}(dz_1, dz_2, dt, d\xi) + (1 - e^{z_1 h_1 (T;t,\xi)}) \nu_{(L_1, L_2)}(dz_1, dz_2) dt d\xi \\
&= J_{(L_1, L_2)}(dz_1, dz_2, dt, d\xi) - \nu_{(L_1, L_2)}(dz_1, dz_2) dt d\xi + (1 - e^{z_1 h_1 (T;t,\xi)}) \\
&\quad \cdot \nu_{(L_1, L_2)}(dz_1, dz_2) dt d\xi \\
&= J_{(L_1, L_2)}(dz_1, dz_2, dt, d\xi) - e^{z_1 h_1 (T;t,\xi)} \nu_{(L_1, L_2)}(dz_1, dz_2) dt d\xi,
\end{aligned}
$$

i.e. $\nu_{(L_1, L_2)}(dz_1, dz_2)$ has become $e^{z_1 h_1 (T;t,\xi)} \nu_{(L_1, L_2)}(dz_1, dz_2)$ under the new measure $\tilde{\mathbb{P}}$. Overall, we get

$$
\frac{d\left(\frac{S_2(t)}{S_1(t)}\right)}{S_2(t-)/S_1(t-)} = d\Xi(t),
$$

with

$$
\begin{aligned}
d\Xi(t) = &-\int_0^\infty g_1(T; t, \xi) \sigma_1(t, \xi) \hat{W}_1(dt, d\xi) + \int_0^\infty g_2(T; t, \xi) \sigma_2(t, \xi) \hat{W}_2(dt, d\xi) \\
&+ \int_0^\infty \int_{\mathbb{R}^2} \left(e^{z_2 h_2 (T;t,\xi) - z_1 h_1 (T;t,\xi)} - 1 \right) \hat{N}_{1,2}(dz_1, dz_2, dt, d\xi).
\end{aligned}
$$

Note that this differential notation should be understood in the sense that we need to consider stochastic integration over the ambit sets $A_j^0(t) = [0, t] \times [0, \infty)$. This implies that $\Xi(0) = 0$. Also, the initial value is given by $\frac{S_2(0)}{S_1(0)} = \exp(Y_2(0, T) - Y_1(0, T))$, which depends on the corresponding stochastic integrals when we integrate over the range $(-\infty, 0] \times [0, \infty)$. We can solve this stochastic differential equation using the stochastic exponential. More precisely, we have

$$
\begin{aligned}
\frac{S_2(t)}{S_1(t)} &= \frac{S_2(0)}{S_1(0)} \exp\left(\Xi(t) - \Xi(0) - \frac{1}{2}[\Xi](t) \right) \\
&\quad \cdot \prod_{0 \le s \le t} (1 + \Delta\Xi_s) \exp\left(-\Delta\Xi_s + \frac{1}{2}(\Delta\Xi_s)^2 \right) \\
&= \frac{S_2(0)}{S_1(0)} \exp\left(\Xi(t) - \Xi(0) - \frac{1}{2}\langle\Xi\rangle^c(t) \right) \prod_{0 \le s \le t} (1 + \Delta\Xi_s) \exp\left(-\Delta\Xi_s \right).
\end{aligned}
$$

Note that

$$
\log\left(\prod_{0\leq s\leq t}(1+\Delta\Xi_s)\exp(-\Delta\Xi_s)\right) = \sum_{0\leq s\leq t}(\log(1+\Delta\Xi_s)-\Delta\Xi_s)
$$

$$
= \int_{A_j^0(t)}\int_{\mathbb{R}^2}(z_2h_2(T;s,\xi)-z_1h_1(T;s,\xi))+1-\exp(z_2h_2(T;s,\xi)
$$
$$
-z_1h_1(T;s,\xi))J_{(L_1,L_2)}(dz_1,dz_2,ds,d\xi)
$$

$$
= \int_{A_j^0(t)}\int_{\mathbb{R}^2}(z_2h_2(T;s,\xi)-z_1h_1(T;s,\xi))+1-\exp(z_2h_2(T;s,\xi)
$$
$$
-z_1h_1(T;s,\xi))\hat{N}_{1,2}(dz_1,dz_2,ds,d\xi)
$$

$$
+\int_{A_j^0(t)}\int_{\mathbb{R}^2}(z_2h_2(T;s,\xi)-z_1h_1(T;s,\xi))+1-\exp(z_2h_2(T;s,\xi)
$$
$$
-z_1h_1(T;s,\xi))\hat{v}_{(L_1,L_2),\tilde{\mathbb{P}}}(dz_1,dz_2)dsd\xi.
$$

From the jump terms we get the following overall contribution:

$$
\Xi(t)^d+\log\left(\prod_{0\leq s\leq t}(1+\Delta\Xi_s)\exp(-\Delta\Xi_s)\right)
$$

$$
= \int_{A_j^0(t)}\int_{\mathbb{R}^2}(z_2h_2(T;s,\xi)-z_1h_1(T;s,\xi))\hat{N}_{1,2}(dz_1,dz_2,ds,d\xi)
$$

$$
+\int_{A_j^0(t)}\int_{\mathbb{R}^2}(z_2h_2(T;s,\xi)-z_1h_1(T;s,\xi))+1-\exp(z_2h_2(T;s,\xi)
$$
$$
-z_1h_1(T;s,\xi))\hat{v}_{(L_1,L_2),\tilde{\mathbb{P}}}(dz_1,dz_2)dsd\xi.
$$

Overall we have

$$
\frac{S_2(t)}{S_1(t)} = \frac{S_2(0)}{S_1(0)}\exp(Z(t)),
$$

where

$$
Z(t) = -\int_{A_j^0(t)}g_1(T;s,\xi)\sigma_1(s,\xi)\hat{W}_1(ds,d\xi)+\int_{A_j^0(t)}g_2(T;s,\xi)\sigma_2(s,\xi)\hat{W}_2(ds,d\xi)
$$

$$
-\frac{1}{2}\int_{A_j^0(t)}g_1^2(T;s,\xi)\sigma_1^2(s,\xi)dsd\xi-\frac{1}{2}\int_{A_j^0(t)}g_2^2(T;s,\xi)\sigma_2^2(s,\xi)dsd\xi
$$

$$+ \rho \int_{A_j^0(t)} g_1(T; s, \xi) g_2(T; s, \xi) \sigma_1(s, \xi) \sigma_2(s, \xi) ds d\xi$$

$$+ \int_{A_j^0(t)} \int_{\mathbb{R}^2} (z_2 h_2(T; s, \xi) - z_1 h_1(T; s, \xi)) \hat{N}_{1,2}(dz_1, dz_2, ds, d\xi)$$

$$+ \int_{A_j^0(t)} \int_{\mathbb{R}^2} (z_2 h_2(T; s, \xi) - z_1 h_1(T; s, \xi)) + 1 - \exp(z_2 h_2(T; s, \xi)$$

$$- z_1 h_1(T; s, \xi)) v_{(L_1, L_2), \tilde{\mathbb{P}}}(dz_1, dz_2) ds d\xi.$$

Proof of Proposition 12

The first part of the proof is analogue to the proof of Proposition 8. Hence we only need to compute the extended characteristic function $\mathbb{E}^{\tilde{\mathbb{P}}}[\exp(i\theta(R, u)Z(t))]$, where we again set $\theta = \theta(R, u) \in \mathbb{C}$. Then

$$\mathbb{E}_0^{\tilde{\mathbb{P}}}[\exp(i\theta Z(t))] = \mathbb{E}_0^{\tilde{\mathbb{P}}}[\exp(i\theta Z_1(t))] \, \mathbb{E}^{\tilde{\mathbb{P}}}[\exp(i\theta Z_2(t))],$$

where

$$\mathbb{E}_0^{\tilde{\mathbb{P}}}[\exp(i\theta Z_1(t))]$$

$$:= \mathbb{E}_0^{\tilde{\mathbb{P}}}\left[\exp\left(-i\theta \int_{A_j^0(t)} g_1(T; s, \xi) \sigma_1(s, \xi) \hat{W}_1(ds, d\xi) \right.\right.$$

$$+ i\theta \int_{A_j^0(t)} g_2(T; s, \xi) \sigma_2(s, \xi) \hat{W}_2(ds, d\xi)$$

$$- \frac{1}{2} i\theta \int_{A_j^0(t)} g_1^2(T; s, \xi) \sigma_1^2(s, \xi) ds d\xi - \frac{1}{2} i\theta \int_{A_j^0(t)} g_2^2(T; s, \xi) \sigma_2^2(s, \xi) ds d\xi$$

$$\left.\left. + \rho i\theta \int_{A_j^0(t)} g_1(T; s, \xi) g_2(T; s, \xi) \sigma_1(s, \xi) \sigma_2(s, \xi) ds d\xi \right) \right]$$

$$= \mathbb{E}_0^{\tilde{\mathbb{P}}}\left[\exp\left(\left(-\frac{\theta^2}{2} - \frac{i\theta}{2} \right) \int_{A_1^0(t)} (g_1^2(T; s, \xi) \sigma_1^2(s, \xi) \right.\right.$$

$$\left.\left. - 2\rho g_1(T; s, \xi) g_2(T; s, \xi) \sigma_1(s, \xi) \sigma_2(s, \xi) + g_2^2(T; s, \xi) \sigma_2^2(s, \xi)) ds d\xi \right) \right],$$

where we conditioned on $\mathscr{F}^{\sigma_1} \vee \mathscr{F}^{\sigma_2}$. For the jump part, we have

$$\mathbb{E}^{\tilde{\mathbb{P}}}\left[\exp(i\theta Z_2(t))\right]$$

$$:= \mathbb{E}^{\tilde{\mathbb{P}}}\left[\exp\left(i\theta \int_{A_1^0(t)} \int_{\mathbb{R}^2} (z_2 h_2(T;s,\xi) - z_1 h_1(T;s,\xi)) \hat{N}_{1,2}(dz_1, dz_2, ds, d\xi)\right.\right.$$

$$+ i\theta \int_{A_1^0(t)} \int_{\mathbb{R}^2} (z_2 h_2(T;s,\xi) - z_1 h_1(T;s,\xi)) + 1 - \exp(z_2 h_2(T;s,\xi)$$

$$\left.\left. - z_1 h_1(T;s,\xi)) \nu_{(L_1,L_2),\tilde{\mathbb{P}}}(dz_1, dz_2) ds d\xi\right)\right]$$

$$= \exp\left((1 - i\theta) \int_{A_1^0(t)} \int_{\mathbb{R}^2} \left(e^{z_2 h_2(T;s,\xi) - z_1 h_1(T;s,\xi)} - 1\right) \nu_{(L_1,L_2),\tilde{\mathbb{P}}}(dz_1, dz_2) ds d\xi\right).$$

References

1. Audet, N., Heiskanen, P., Keppo, J., Vehviläinen, I.: Modeling electricity forward curve dynamics in the Nordic Market. In: Bunn, D.W. (ed.) Modelling Prices in Competitive Electricity Markets, pp. 251–265. Wiley, New York (2004)
2. Barndorff-Nielsen, O.E., Pedersen, J.: Meta-times and extended subordination. Theory Probab. Appl. **56**(2), 319–327 (2012)
3. Barndorff-Nielsen, O.E., Schmiegel, J.: Lévy-based tempo-spatial modelling; with applications to turbulence. Uspekhi Mat. NAUK **59**, 65–91 (2004)
4. Barndorff-Nielsen, O.E., Shiryaev, A.: Change of Time and Change of Measure. Advanced Series on Statistical Science and Applied Probability, vol. 13. World Scientific, Singapore (2010)
5. Barndorff-Nielsen, O.E., Stelzer, R.: Multivariate SupOU processes. Ann. Appl. Probab. **21**(1), 140–182 (2011)
6. Barndorff-Nielsen, O.E., Benth, F.E., Veraart, A.E.D.: Modelling energy spot prices by volatility modulated Lévy-driven Volterra processes. Bernoulli **19**(3), 803–845 (2013)
7. Barndorff-Nielsen, O.E., Benth, F.E., Veraart, A.E.D.: Modelling electricity futures by ambit fields. Adv. Appl. Probab. **46**(3), 719–745 (2014)
8. Barndorff-Nielsen, O.E., Benth, F.E., Veraart, A.E.D.: Recent advances in ambit stochastics with a view towards tempo-spatial stochastic volatility/intermittency. Banach Center Publications **104**, 25–60 (2015)
9. Barndorff-Nielsen, O.E., Lunde, A., Shephard, N., Veraart, A.E.D.: Integer-valued trawl processes: a class of stationary infinitely divisible processes. Scand. J. Stat. **41**, 693–724 (2014)
10. Basse-O'Connor, A., Graversen, S.-E., Pedersen, J.: A unified approach to stochastic integration on the real line. Theory Probab. Appl. **58**, 355–380 (2013)
11. Benth, F.E., Zdanowicz, H.: Pricing energy spread options. In: Roncoroni, A. Fusai, G., Cummins, M. (eds.) Handbook of Multi-Commodity Markets and Products: Structuring, Trading and Risk Management. Wiley, New York (2014)
12. Benth, F.E., Di Nunno, G., Khedher, A., Schmeck, M.D.: Pricing of spread options on a bivariate jump market and stability to model risk. Appl. Math. Finance **22**(1), 28–62 (2015)

13. Bichteler, K., Jacod, J.: Random measures and stochastic integration. In: Kallianpur, G. (ed.) Theory and Application of Random Fields. Lecture Notes in Control and Information Sciences, vol. 49, pp. 1–18. Springer, Berlin/Heidelberg (1983)
14. Bjerksund, P., Rasmussen, H., Stensland, G.: Valuation and risk management in the Nordic electricity market. In: Bjørndal, P.M.P.E., Bjørndal, M., Pardalos, P., Rönnqvist, M. (eds.) Energy, Natural Resources and Environmental Economics, pp. 167–185. Springer, Berlin (2010)
15. Black, F., Scholes, M.: The pricing of options and corporate liabilities. J. Polit. Econ. **81**(3), 637–654 (1973)
16. Bollerslev, T.: Modelling the coherence in short-run nominal exchange rates: a multivariate generalized ARCH model. Rev. Econ. Stat. **72**(3), 498–505 (1990)
17. Carmona, R., Durrleman, V.: Pricing and hedging spread options. SIAM Rev. **45**(4), 627–685 (2003)
18. Cheang, G.H.L., Chiarella, C.: Exchange options under jump-diffusion dynamics. Appl. Math. Finance **18**(3), 245–276 (2011)
19. Chong, C., Klüppelberg, C.: Integrability conditions for space-time stochastic integrals: theory and applications. Bernoulli (2015, in press)
20. Cont, R., Tankov, P.: Financial Modelling with Jump Processes. Financial Mathematics Series. Chapman & Hall, Boca Raton (2004)
21. Eberlein, E., Glau, K., Papapantoleon, A.: Analysis of Fourier transform valuation formulas and applications. Appl. Math. Finance **17**(3), 211–240 (2010)
22. Goutis, C., Casella, G.: Explaining the saddlepoint approximation. Am. Stat. **53**(3), 216–224 (1999)
23. Heath, D., Jarrow, R., Morton, A.: Bond pricing and the term structure of interest rates: a new methodology for contingent claims valuation. Econometrica **60**(1), 77–105 (1992)
24. Hull, J., White, A.: The pricing of options on assets with stochastic volatilities. J. Finance **42**(2), 281–300 (1987)
25. Hurd, T.R., Zhou, Z.: A Fourier transform method for spread option pricing. SIAM J. Financ. Math. **1**(1), 142–157 (2010)
26. Kennedy, D.P.: The term structure of interest rates as a Gaussian random field. Math. Finance **4**(3), 247–258 (1994)
27. Kennedy, D.P.: Characterizing Gaussian models of the term structure of interest rates. Math. Finance **7**(2), 107–116 (1997)
28. Margrabe, W.: The value of an option to exchange one asset for another. J. Finance **33**(1), 177–186 (1978)
29. Øksendal, B., Sulem, A.: Applied Stochastic Control of Jump Diffusions, 3rd edn. Universitext. Springer, Berlin (2009)
30. Rajput, B., Rosinski, J.: Spectral representation of infinitely divisible distributions. Probab. Theory Relat. Fields **82**, 451–487 (1989)
31. Veraart, A.E.D., Veraart, L.A.M.: Modelling electricity day-ahead prices by multivariate Lévy semi-stationary processes. In: Benth, F.E., Kholodnyi, V., Laurence, P. (eds.) Wolfgang Pauli Proceedings, pp. 157–188. Springer, New York (2014)
32. Walsh, J.: An introduction to stochastic partial differential equations. In: Carmona, R., Kesten, H., Walsh, J. (eds.) Lecture Notes in Mathematics, vol. 1180. Ecole d'Eté de Probabilités de Saint–Flour XIV—1984. Springer, New York (1986)

Hedging Expected Losses on Derivatives in Electricity Futures Markets

Adrien Nguyen Huu and Nadia Oudjane

Abstract We investigate the problem of pricing and hedging derivatives of Electricity Futures contract when the underlying asset is not available. We propose to use a cross hedging strategy based on the Futures contract covering the larger delivery period. For that purpose we formulate the pricing problem in a stochastic target form along the lines of Bouchard et al. (SIAM J. Control Optim. 48:3123–3150, 2009), with a moment loss function. Following the same techniques as in the latter, we avoid to demonstrate the uniqueness of the value function by comparison arguments and explore convex duality methods to provide a semi-explicit solution to the problem. We then propose numerical results to support the new hedging strategy and compare our method to a Black–Scholes benchmark.

1 Introduction

We propose in this contribution a method involving numerical implementation for partially hedging financial derivatives on electricity futures contracts. Electricity futures markets present specific features. As a non-storable commodity, electrical energy is delivered as a power over time periods. Similarly, futures contracts exchange a present power price for delivery over a fixed future period against the future quoted price on that period: they refer explicitly to swap contracts, see [4]. Electricity being non-storable, arbitrage arguments do not hold, preventing anyone to construct a term structure via usual tools. As a natural result of liquidity restriction and uncertainty in the future, a limited number of contracts are quoted and the length of their delivery period increases with their time to maturity. This phenomenon we call the *granularity of the term structure*, or cascading rule [2, 23], implies that the most desired and flexible contracts (weekly or monthly-period contracts) are only

A. Nguyen Huu (✉)
ENPC, Marne-la-Vallée, France
e-mail: adrien.nguyen-huu@enpc.fr

N. Oudjane
EDF R&D, Clamart, France
e-mail: nadia.oudjane@edf.fr

R. Aïd et al. (eds.), *Commodities, Energy and Environmental Finance*,
Fields Institute Communications 74, DOI 10.1007/978-1-4939-2733-3_6
149

quoted a short time before their maturity. If one desires a fixed price for power delivered on a given distant month in the future, she shall get a contract covering a wider period, e.g. a quarter or a year contract. However, the risk remains if one is endowed with a derivative upon such a short period contract, before this contract is even quoted.

This is the explored situation: we consider here an agent endowed with a derivative upon a non-yet quoted asset. In practice, the unhedgeable risk of such a position is managed by deploying a cross-hedging strategy with a correlated quoted asset, see [13, 19] and [1]. The risk cannot be completely eliminated without prohibitive cost: the market is incomplete and the methodology requires a pricing criterion, as proposed in the above references. Lindell and Raab [19] use monthly forward contracts to minimize the variance when hedging hourly forward derivatives while [13] use additional hedging instruments such as power plants to target a mean-risk criterion for hedging spot prices on the futures period. This latter situation is closer to generation management than to financial hedging. In [1], another approach based on utility indifference pricing is developed to hedge derivatives on nontradable underlyings with correlated instruments. In what follows, we develop a partial financial hedging procedure in order to satisfy a loss constraint in expectation. One major interest of this approach is that the loss constraint is easily understood. Unlike quadratic hedging our criterion differentiates losses from gains. Besides, the expected loss threshold characterizing our criterion is easier to interpret than the risk aversion coefficient of the utility function. Finally, we are able to provide a ready-to-implement method to attain such objectives. This necessitates recent tools of stochastic control [7] and numerical approximations of coupled forward backward SDEs. However, we relate constantly the obtained results to well-known formulas and concepts, so that the method is easy to assimilate. We also provide strong and sufficient assumptions in order to avoid involved proofs in difficult cases. We believe that the specific method proposed hereafter can be understood and applied without much effort, and be profitable to a numerous variety of hedging problems.

The approach we adopt has been originated by Föllmer and Leukert [14, 15] but we develop the problem in the framework of control theory. The minimal initial portfolio value needed to satisfy the constraint of expectation of losses can be expressed as a value function of a stochastic target problem. The stochastic target approach has been introduced by Soner and Touzi [21, 22] to formulate the pricing problem in a control fashion. It has been extended to expectation criteria by Bouchard et al. [7], and applied for quantile hedging [7], loss constraints in liquidation problems [5], or loss constraints with small transaction costs [9]. However, the general approach is rather technical and necessitates to solve a non-linear Partial Differential Equations (PDE), when one has first proved a comparison theorem to resolve the uniqueness problem for the value function.

Here, we use the application of [7], where nice results can be provided in complete market without appealing to comparison arguments. However, our initial problem cannot be tackled directly with the stochastic target formulation of

[7] or [20]. The unhedgeable risk coming from the apparition of the desired asset price generates a non-trivial extension to the usual framework. Instead, we proceed in three steps in a backward fashion:

1. We first formulate in Sect. 3 the stochastic target problem in complete market, using an easy extension of the application in [7]. Our approach relies strongly on the convexity of the loss function. By convex duality arguments, the expected loss target can be expressed in order to provide a formula which is very close to a usual risk-neutral pricing formulation. This allows to express the value function V at the precise moment of the apparition of the missing contract.
2. To treat the random apparition of a non-yet traded asset, we use a face-lifting procedure that provides a new (intermediary) target for the period before quotation of the underlying asset. This is done in Sect. 4, in coherence with the initial hedging criterion, and allows to retrieve a complete market setting where results of step 1 can be partially used.
3. When the complete market setting cannot provide analytical formulas, as it happens after step 2, we shall make use of numerical approximations. We propose such an algorithm in Sect. 5, and illustrate the efficiency of the method in Sect. 6.

As it will be understood, this method can be applied recursively by proceeding repeatedly to steps 1 and 2 along the numerical algorithm provided in step 3. The remaining of the contribution follows the above order, preluded by the introduction of the problem in Sect. 2, where we develop the archetypal situation encountered by a financial agent on the market.

2   Description of the Control Problem

Let $0 < T < T^* < \infty$. We consider an agent endowed with a financial option with expiration date T^*, payoff g on a futures covering a monthly period. However, this asset is not yet quoted at initial time $t = 0$, and appears on the market at time T. The month is covered by a futures with delivery over a larger period, e.g. a quarter futures. We denote by $X := (X_t)_{t \in [0,T^*]}$ the discounted price of this instrument, avoiding the introduction of an interest rate hereafter. We assume it is available over the whole period of interest $[0, T^*]$, whereas the monthly-period futures upon which the agent has a position is only available on the interval $[T, T^*]$.

The heart of our approach is to assume a structural relation between the two instruments above, namely that the return of the two assets are perfectly correlated. More precisely, the monthly futures price is supposed to be given as the product, ΛX_t, of the quarterly futures price X_t and a given *shaping factor* $\Lambda > 0$, assumed to be a bounded random variable revealed at time T, completely independent of the asset price X. This model is a simple generalization of the *profile coefficients* used to construct a refined (for instance monthly or hourly) forward curve as described in [17]. This profile carries many types of information such as prices seasonality. For a given quarterly futures contract, the monthly coefficients provide monthly futures prices for every month inside the quarter by multiplying the quarterly futures price with the associated monthly weight. In our approach, since the monthly futures prices are not yet observed, the monthly weights are naturally modelled by stochastic weights Λ. This model can also be related to [19, 23] and [2] which exhibits strong co-integration between two available contracts of different length. Here, considering the unavailability of the month contract, this model appears as a simple convenient assumption. The boundedness condition follows from a structural relation between the two futures contracts, assuming implicitly that the underlying spot price is non-negative.

We consider a probability space $(\Omega, \mathscr{F}, \mathbb{P})$ supporting a Brownian motion W and the variable Λ. The filtration $\mathbb{F}$ is given by $\mathscr{F}_t := \sigma(W_s, \ 0 \le s \le t)$ for $t < T$, and by $\mathscr{F}_t := \sigma(W_s, \ 0 \le s \le t) \wedge \sigma(\Lambda)$ for $t \ge T$.

Assumption 1. *We denote by L the support of* Λ, *which is supposed to be a bounded subset of* $\mathbb{R}^+$, *and ρ its probability measure on L.*

On the period $[0, T]$, the agent trades with the asset $X^{t,x}$, which is assumed to be solution to the SDE:

$$dX^{t,x}_s = \mu(s, X^{t,x}_s)ds + \sigma(s, X^{t,x}_s)dW_s, \quad \text{for } s \ge t, \text{ and } X^{t,x}_t = x. \tag{1}$$

We assume the following for $X^{t,x}$ to be well and uniquely defined.

Assumption 2. *The functions* $(\mu, \sigma) \ : \ [0, T^*] \times \mathbb{R}_+ \rightarrow \mathbb{R} \times \mathbb{R}^*_+$ *are assumed to be such that the following properties are verified:*

1. *μ and σ are uniformly Lipschitz and verify $|\mu(t, x)| + |\sigma(t, x)| \le K(1 + |x|)$ for any $(t, x) \in [0, T^*] \times \mathbb{R}_+$.*
2. *for any $x > 0$, we have $X^{t,x}_s > 0 \ \mathbb{P} - a.s.$ for all $s \in [0, T^*]$*
3. *$\sigma(t, x) > 0$ for any $(t, x) \in [0, T^*] \times \mathbb{R}^*_+$;*
4. *if $x = 0$ then $\sigma(t, x) = 0$ for all $t \in [0, T^*]$;*
5. *σ is continuous in the time variable on $[0, T^*] \times \mathbb{R}_+$;*
6. *μ and σ are such that*

$$\theta(t, x) := \frac{\mu(t, x)}{\sigma(t, x)} < \infty \quad \text{uniformly in } (t, x) \in [0, T^*] \times \mathbb{R}^*_+. \tag{2}$$

Equation (2) implies the so-called Novikov condition. The submarket composed of only $(X_t)_{t \in [0, T^*]}$ is then a complete market associated to the Brownian subfiltration. The filtration $\mathbb{F}$ relates to an incomplete market because of Λ, which is unknown on the interval $[0, T)$ and cannot be hedged by a self-financed admissible portfolio defined as follows. In that manner, the market can be labelled as semi-complete in the sense of Becherer [3].

Definition 1. An admissible self-financed portfolio is a $\mathbb{F}$-adapted process $Y^{t,x,y,v}$ defined by $Y^{t,x,y,v}_t = y \ge -\kappa$ and

$$Y^{t,x,y,v}_s := y + \int_t^s v_u dX^{t,x}_u, \quad s \in [t, T^*], \tag{3}$$

where $v \in \mathcal{U}_{t,y}$ denotes a strategy and $\mathcal{U}_{t,y}$ the set of admissible strategies at time t, which is the set of $\mathbb{R}$-valued $\mathbb{F}$-progressively measurable, square-integrable processes such that $Y_s^{t,x,y,v} \geq -\kappa$ $\mathbb{P}$ $-$ a.s. for all $s \in [t, T^*]$ and some $\kappa \geq 0$ representing a finite credit line for the agent.

According to the asset model (1), it is strictly equivalent to consider a hedging portfolio with $X^{t,x}$ on $[0, T^*]$ or a switch at any time $r \in [T, T^*]$ for the newly appeared asset $\Lambda X^{t,x}$:

$$Y_s^{t,x,y,v} = y + \int_t^{s \wedge r} v_u dX_u^{t,x} + \int_{s \wedge r}^t v_u' d(\Lambda X_u^{t,x}) \, , r \geq T \text{ and } t \geq 0 \, ,$$

if $v' = v/\Lambda$ for $u \geq r$. We thus assume that the agent trades with X on $[0, T^*]$.

Assumption 3. *The final payoff of the option is given by $g(\Lambda X_{T^*})$, where the function g is assumed to be Lipschitz continuous.*

As it is well-known in the literature, the superhedging price of such an option can be prohibitive, even with Assumption 1 where Λ is bounded. To circumvent this problem, we propose to control the expected losses on partial hedging. That means that the agent gives herself a threshold $p < 0$ and a loss function ℓ to evaluate the loss over her terminal position $g(\Lambda X_{T^*}^{t,x}) - Y_{T^*}^{t,x,y,v}$.

Definition 2. *A loss function ℓ : $\mathbb{R}_+ \to \mathbb{R}_+$ is assumed to be continuous, strictly convex and strictly increasing on $\mathbb{R}_+$ with polynomial growth. We normalize the function so that $\ell(0) = 0$.*

The agent's objective at time t is to find the minimal value y and a portfolio strategy $v \in \mathcal{U}_{t,y}$ such that

$$\mathbb{E}\left[-\ell\left(\left(g\left(\Lambda X_{T^*}^{t,x}\right) - Y_{T^*}^{t,x,y,v}\right)^+\right)\right] \geq p \, . \tag{4}$$

More generally, it will be useful to measure the deviation between the payoff and the hedging portfolio through a general map Ψ : $\mathbb{R}_+ \times \mathbb{R} \to \mathbb{R}_-$. The previous example corresponds to the specific case where

$$\Psi(x, y) = -\ell((g(x) - y)^+). \tag{5}$$

Assumption 3 and Definition 2 imply that Ψ defined above satisfies the following assumption:

Assumption 4. *The function Ψ : $\mathbb{R}_+ \times \mathbb{R} \to \mathbb{R}_-$ is assumed to be*

- *continuous with polynomial growth in (x, y).*
- *concave and increasing in y on cl $\{y \in [-\kappa, \infty) \, : \, \Psi(x, y) < 0\}$ for any $x \in \mathbb{R}_+$.*

Notice that $\Psi(x, y) = 0$ in (5) for any $y \geq g(x)$, which makes Ψ not invertible on the domain $\mathbb{R}$ for any fixed x. Under Assumption 4, we can define the y inverse $\Psi^{-1}(x, p)$ on $\mathbb{R}_+ \times \mathbb{R}_-$ as a convex increasing function of p, where

$$\Psi^{-1}(x, 0) = \inf\{y \geq -\kappa \ : \ \Psi(x, y) = 0\} \ . \tag{6}$$

As explained above, the valuation approach of (4) has been introduced in [15, 16]. It can be written under the stochastic control form of [7] allowed by the Markovian framework (1)–(3).

Definition 3. Let $(t, x, p) \in \mathbf{S} := [0, T^*] \times \mathbb{R}_+^* \times \mathbb{R}_-^*$ be given. Then we define the value function V on $\mathbf{S}$ as

$$V(t, x, p) := \inf\left\{y \geq -\kappa \ : \ \mathbb{E}\left[\Psi(\Lambda X_{T^*}^{t,x}, Y_{T^*}^{t,x,y,\nu})\right] \geq p \text{ for } \nu \in \mathcal{U}_{t,y}\right\} \ . \tag{7}$$

Advanced technicalities and details on the general setting for the stochastic target problem with controlled loss can be found in [7] and [20]. In particular, we introduce the value function V only on the open domain of (x, p), and avoid treating the specific case $x = 0$ or $p = 0$. Furthermore, the function V is implicitly bounded by the superhedging price of $g(\Lambda X_{T^*}^{t,x})$ as we will see in the next section. Notice that the expectation in (7) involves an integration w.r.t. the law of Λ for $t \in [0, T)$, before the shaping factor Λ is revealed. Thus, the present problem is not standard due to the presence of Λ: the filtration $\mathbb{F}$ is not only due to the Brownian motion, and dynamic programming arguments of [7] do not apply directly. The approach we undertake is to separate the complete and incomplete market intervals in a piecewise problem.

Example 1. A particular example of loss function we will use in Sects. 5 and 6 is the special case of lower partial moment

$$\ell(x) := x^k \mathbb{1}_{\{x \geq 0\}}/k \ , \quad \text{with} \quad k > 1 \ . \tag{8}$$

In the case of $k = 1$, (which is not considered here) we obtain a criterion close to the expected shortfall, independently studied in [12], whereas allowing $k = 0$ allows to retrieve precisely the quantile hedging problem [7], although ℓ is not a loss function as in Definition 2 in that case. The case $k = 2$ gets closer to the quadratic hedging (or mean-variance hedging) objective, but with the advantage of considering only losses, and not gains. Notice that ℓ as in Eq. (8) for $k > 1$ follows Definition 2 and that Assumption 4 holds with (5) if Assumption 3 holds.

We finish this section with additional notations. In the sequel, we will denote $\mathbf{S} := \mathbf{S}_1 \cup \mathbf{S}_2$ where $\mathbf{S}_1 := [0, T] \times \mathbb{R}_+^* \times \mathbb{R}_-^*$ and $\mathbf{S}_2 := [T, T^*] \times \mathbb{R}_+^* \times \mathbb{R}_-^*$ are the two domains of the value function. For any real valued function φ, defined on $\mathbf{S}$, we denote by φ_t or φ_x the partial derivatives with respect to t or x. Partial derivatives of other variables, or second order partial derivatives are written in the same manner.

3   Solution in Complete Market

3.1   General Solution and Risk-Neutral Expectation

On the interval $[T, T^*]$, the variable Λ is known and the asset ΛX is tradable. We assume that Λ takes the value $\lambda \in L$. On this interval, λ can be seen as a coefficient affecting the loss function Ψ, and the problem (7) follows the standard formulation of [7]: the filtration is generated by the paths of the Brownian motion. In Sect. 4, we reduce the incomplete market setting of period $[0, T)$ to complete market problem of the form (7) on $[T, T^*]$, in a similar Brownian framework, with a function Ψ which is no more derived from a loss function as in (5), motivating the introduction of the general function Ψ.

Consequently and without loss of generality, we temporarily assume $L = \{1\}$, and omit λ in the notation. On $[0, T^*]$ then, we can work on problem (7) in the Brownian filtration and apply most results of [7]. The filtration $\mathbb{F}$ for $t \geq T$ in that case is given by $\mathscr{F}_t := \sigma(W_s \; : \; T \leq s \leq t)$ on a complete probability space $(\Omega, \mathscr{F}, \mathbb{P})$ with $\mathbb{P}[\Lambda = 1] = 1$.

Proposition 1. *Under Assumptions 2 and 4, the function V is given on $\mathbf{S}$ by*

$$V(t, x, p) = \mathbb{E}^{\mathbb{Q}_t}\left[\Psi^{-1}\left(X_{T^*}^{t,x}, J\left(X_{T^*}^{t,x}, Q_{T^*}^{t,q}\right)\right)\right] \tag{9}$$

where $J(x, q) := \mathrm{argsup}\left\{pq - \Psi^{-1}(x, p) \; : \; p \leq 0\right\}$ and

$$X_s^{t,x} = x + \int_t^s \sigma(u, X_u^{t,x})dW_u^{\mathbb{Q}_t} \quad \text{and} \quad Q_s^{t,q} = q + \int_t^s Q_u^{t,q}\theta(u, X_u^{t,x})dW_u^{\mathbb{Q}_t} , \tag{10}$$

with $W^{\mathbb{Q}_t}$ being the Brownian motion under the probability $\mathbb{Q}_t$, the latter being defined by $d\mathbb{P}_t/d\mathbb{Q}_t = Q^{t,1}$ and used for expectation (9). Finally q is given such that $\mathbb{E}\left[J\left(X_{T^*}^{t,x}, Q_{T^*}^{t,q}\right)\right] = p$.

To obtain such a result, we proceed in several steps. We first use Proposition 3.1 in [7] to express problem (7) in the standard form of [21], see Lemma 1 below. We then apply the Geometric Dynamic Programming Principle (GDP) Principle of [21] in order to obtain a PDE characterization of V in the viscosity sense. These results are recalled in Appendix. We then use properties of the convex conjugate of V to obtain V as the risk-neutral expectation (9) of Proposition 1.

Lemma 1. *Let $P^{t,p,\alpha}$ be a $\mathbb{F}$-adapted stochastic process defined by*

$$P_s^{t,p,\alpha} = p + \int_t^s \alpha_u P_u^{t,p,\alpha}dW_u , \quad 0 \leq t \leq s \leq T^* , \tag{11}$$

where α is a $\mathbb{F}$-progressively measurable process taking values in $\mathbb{R}$. Let us denote $\mathscr{A}_{t,p}$ the set of such processes such that $P^{t,p,\alpha}$ and $\alpha P^{t,p,\alpha}$ are square-integrable processes. Let Assumptions 2 and 4 hold. Then on $\mathbf{S}$

$$V(t, x, p) = \inf \left\{ y \geq -\kappa \; : \; Y_{T*}^{t,x,y,\nu} \geq \Psi^{-1}(X_{T*}^{t,x}, P_{T*}^{t,p,\alpha}) \text{ for } (\nu, \alpha) \in \mathcal{U}_{t,y} \times \mathcal{A}_{t,p} \right\} .$$
(12)

Moreover, for a given triplet (t, y, p), if there exists $\nu \in \mathcal{U}_{t,y}$ and a $\mathbb{F}$-progressively measurable process α taking values in $\mathbb{R}$ such that

$$Y_{T*}^{t,x,y,\nu} \geq \Psi^{-1}(X_{T*}^{t,x}, P_{T*}^{t,p,\alpha}) \; \mathbb{P} - a.s. \; ,$$
(13)

then there exists $\alpha' \in \mathcal{A}_{t,p}$ such that $Y_{T*}^{t,x,y,\nu} \geq \Psi^{-1}(X_{T*}^{t,x}, P_{T*}^{t,p,\alpha'}) \; \mathbb{P} - a.s.$

Proof. Under Assumption 4, $\Psi^{-1}(x, p)$ is well defined with (6). According to the polynomial growth of Ψ combined with Assumption 2, the stochastic integral representation theorem can be applied. The first result (12) is then a reformulation of Proposition 3.1 in [7].

The second result echoes Assumption 4 and Remark 6 in [5], as the statement is missing in [7]. According to (11), for $p < 0$, the process $P^{t,p,\alpha}$ is a negative local martingale, and therefore a bounded submartingale. Thus, $\mathbb{E}\left[\Psi(X_{T*}^{t,x}, Y_{T*}^{t,x,y,\nu})\right] \geq p$. According to the growth condition of Ψ, the martingale representation theorem implies the existence of a square-integrable martingale $P^{t,p,\alpha'}$, with $P_t^{t,p,\alpha'} = p$ and according to (13),

$$\mathbb{E}\left[\Psi(X_{T*}^{t,x}, Y_{T*}^{t,x,y,\nu}) - P_{T*}^{t,p,\alpha'}\right] = \mathbb{E}\left[\Psi(X_{T*}^{t,x}, Y_{T*}^{t,x,y,\nu})\right] - p \geq 0 .$$

Since $\Psi(x, y) \in \mathbb{R}_-$ for any $(x, y) \in \mathbb{R}_+ \times \mathbb{R}$, we can choose $P^{t,p,\alpha'}$ to follow dynamics (11). This implies that α' is a real-valued $(\mathcal{F}_t)$-progressively measurable process such that $(\alpha' P^{t,p,\alpha'}) \in L^2([0, T^*] \times \Omega)$, and $\alpha' \in \mathcal{A}_{t,p}$.   □

We can now turn to the proof of Proposition 1 by using the GDP recalled in Appendix. The proof follows closely developments of section 4 in [7], and is given for the sake of clarity, as for the introduction of important objects for Sect. 4.3.

Proof (of Proposition 1). We divide the proof in three steps

1. *We introduce conjugate and local test functions.* Let V_* be the lower semi-continuous version of V, as defined in Appendix. According to Assumption 4, dynamics (3) and definition (7), V and V_* are increasing functions of p on $\mathbb{R}_-$. The boundedness of V implies also the finiteness of V_*. For $(t, x, q) \in [0, T^*] \times \mathbb{R}_+^* \times \mathbb{R}_+^*$, we introduce the convex conjugate of V_* in $p < 0$, i.e.,

$$\tilde{V}(t, x, q) := \sup \left\{ pq - V_*(t, x, p) \; : \; p \leq 0 \right\} .$$

The map $q \mapsto \tilde{V}(., q)$ is convex and upper-semi-continuous on $\mathbb{R}_+^*$.

Let $\tilde{\varphi}$ be a smooth function with bounded derivatives, such that $(t_0, x_0, q_0) \in [0, T^*) \times \mathbb{R}_+^* \times \mathbb{R}_+^*$ is a local maximizer of $\tilde{V} - \tilde{\varphi}$ with $(\tilde{V} - \tilde{\varphi})(t_0, x_0, q_0) = 0$. The map $q \mapsto \tilde{\varphi}(., q)$ is convex. Without loss of generality, we can assume that $\tilde{\varphi}$ is strictly convex with quadratic growth in q, see the proof in Section 4 of [7].

The convex conjugate of $\tilde{\varphi}$ with respect to q is a strictly convex function of p defined by $\varphi(t,x,p) := \sup\{qp - \tilde{\varphi}(t,x,q) : q \geq 0\}$. We can then properly define the map $(t,x,q) \mapsto (\varphi_p)^{-1}(t,x,q)$ on $[0,T^*] \times \mathbb{R}_+^* \times \mathbb{R}_+^*$, where the inverse is taken in the p variable. According to the definition of $\tilde{V}$ and the quadratic growth of $\tilde{\varphi}$, there exists $p_0 \leq 0$ such that for the fixed q_0,

$$p_0 q_0 - V_*(t_0, x_0, p_0) = \tilde{V}(t_0, x_0, q_0) = \tilde{\varphi}(t_0, x_0, q_0) = \sup_{p \leq 0}\{pq_0 - \varphi(t_0, x_0, p)\}$$

which, by taking the left and right sides of the above equation, implies that (t_0, x_0, p_0) is a local minimizer of $V_* - \varphi$ and $(V_* - \varphi)(t_0, x_0, p_0) = 0$. The first order condition in the definition of φ implies that $p_0 = (\varphi_p)^{-1}(t_0, x_0, q_0)$.

2. *We prove that $\tilde{V}$ is subsolution to a linear PDE.* In our case, the control v takes unbounded values by definition of $\mathcal{U}_{t,y}$. It implies, together with Assumption 2, that

$$\mathcal{N}_0(t,x,p,d_x,d_p) := \{(u,a) \in \mathbb{R}^2 : |\sigma(t,x)(u-d_x) - apd_p| = 0\} \neq \emptyset \tag{14}$$

for any $(t,x,p,d_x,d_p) \in [0,T^*] \times \mathbb{R}_+ \times \mathbb{R}_- \times \mathbb{R}^2$. This holds in particular for the set $\mathcal{N}_0(t_0, x_0, p_0, \varphi_x(t_0, x_0, p_0), \varphi_p(t_0, x_0, p_0))$ which is then composed of elements of the form $((\varphi_x + a\varphi_p/\sigma)(t_0, x_0, p_0), a)$ for $a \in \mathbb{R}$. According to Theorem 2 in Appendix and changing v for its new expression, φ in (t_0, x_0, p_0) verifies

$$-\varphi_t - \frac{1}{2}\sigma^2(t_0, x_0)\varphi_{xx} - \inf_{a \in \mathbb{R}}\left\{\frac{1}{2}(ap_0)^2\varphi_{pp} - ap_0(\theta(t_0, x_0)\varphi_p - \sigma\varphi_{xp})\right\} \geq 0 . \tag{15}$$

Since $\varphi_{pp}(t_0, x_0, p_0) > 0$, the infimum in the above equation is reached for

$$a := -\left(\frac{\sigma\varphi_{xp} - \theta\varphi_p}{p_0\varphi_{pp}}\right)(t_0, x_0, p_0) \in \mathbb{R} , \tag{16}$$

which can be plugged back into (15) to obtain a new inequality at (t_0, x_0, p_0):

$$-\varphi_t - \frac{1}{2}\sigma^2(t_0, x_0)\varphi_{xx} + \frac{1}{2}(\varphi_{pp})^{-1}\left(\theta(t_0, x_0)\varphi_p - \sigma(t_0, x_0)\varphi_{xp}\right)^2 \geq 0 . \tag{17}$$

Recall that $p_0 = (\varphi_p)^{-1}(t_0, x_0, q_0)$. According to the Fenchel–Moreau theorem, $\tilde{\varphi}$ is its own biconjugate, $\tilde{\varphi}(t,x,q) = \sup\{pq - \varphi(t,x,p) : p \leq 0\}$, and $\tilde{\varphi}(t_0, x_0, q_0) = p_0 q_0 - \varphi(t_0, x_0, p_0)$. By differentiating in p, we get $\varphi_p(t_0, x_0, p_0) = q_0$. It follows by differentiating again that at point (t_0, x_0, p_0) we have the following correspondence:

$$(\varphi_t, \varphi_x, \varphi_{xx}, \varphi_{pp}, \varphi_{xp}) = \left(-\tilde{\varphi}_t, -\tilde{\varphi}_x, -\tilde{\varphi}_{xx} + \frac{\tilde{\varphi}_{xq}^2}{\tilde{\varphi}_{qq}}, \frac{1}{\tilde{\varphi}_{qq}}, -\frac{\tilde{\varphi}_{xq}}{\tilde{\varphi}_{qq}}\right) . \tag{18}$$

Plugging (18) into (17), we get that $\tilde{\varphi}$ satisfies at (t_0, x_0, q_0)

$$-\tilde{\varphi}_t - \frac{1}{2}\left(\sigma^2 \tilde{\varphi}_{xx} + |\theta|^2 q_0^2 \tilde{\varphi}_{qq} + 2\mu \tilde{\varphi}_{xq}\right)(t_0, x_0, q_0) \leq 0 . \tag{19}$$

By arbitrariness of $(t_0, x_0, q_0) \in [0, T^*) \times \mathbb{R}_+^* \times \mathbb{R}_+^*$, this implies that $\tilde{V}$ is a subsolution of (19) on $[0, T^*) \times \mathbb{R}_+^* \times \mathbb{R}_+^*$. The terminal condition is given by the definition of $\tilde{V}$ and Theorem 2 in Appendix:

$$\tilde{V}(T^*, x, q) = \sup_{p \leq 0}\{pq - V_*(T^*, x, p)\} = \sup_{p \leq 0}\{pq - \Psi^{-1}(x, p)\} . \tag{20}$$

3. *We compare V to a conditional expectation.* Let $\bar{V}$ be the function defined by $\bar{V}(t, x, q) = \mathbb{E}^{\mathbb{Q}_t}\left[\tilde{V}(T^*, X_{T^*}^{t,x}, Q_{T^*}^{t,q})\right]$ for $(t, x, q) \in [0, T^*] \times (0, \infty)^2$, with dynamics for $s \in [0, T^*]$ given by

$$X_s^{t,x} = x + \int_t^s \sigma(u, X_u^{t,x})dW_u^{\mathbb{Q}_t} \quad \text{and} \quad Q_s^{t,q} = q + \int_t^s Q_u^{t,q}\theta(u, X_u^{t,x})dW_u^{\mathbb{Q}_t} ,$$

where $\mathbb{Q}_t$ is a $\mathbb{P}$-equivalent measure such that $d\mathbb{P}/d\mathbb{Q}_t = Q^{t,1}$. According to the Feynman–Kac formula, $\bar{V}$ is a supersolution to Eq. (19), and thus $\bar{V} \geq \tilde{V}$. According to Assumption 4, $p \mapsto \Psi^{-1}(., p)$ is convex increasing on $\mathbb{R}_-$. Thus, for sufficiently large values of q, $J(x, q) := \arg\sup\{pq - \Psi^{-1}(x, p) : p \leq 0\}$ is well-defined and can take any value in $\mathbb{R}_-$. By the implicit function theorem, there exists a function q of (t, x, p) such that $\mathbb{E}^{\mathbb{Q}_t}\left[Q_{T^*}^{t,1}J\left(X_{T^*}^{t,x}, Q_{T^*}^{t,q(t,x,p)}\right)\right] = p$. Therefore,

$$\begin{aligned}
V(t, x, p) &\geq V_*(t, x, p) \geq \sup\{qp - \bar{V}(t, x, q) : q \geq 0\} \\
&\geq pq(t, x, p) - \mathbb{E}^{\mathbb{Q}_t}\left[\tilde{V}\left(T^*, X_{T^*}^{t,x}, Q_{T^*}^{t,q(t,x,p)}\right)\right] \\
&\geq q(t, x, p)\left(p - \mathbb{E}^{\mathbb{Q}_t}\left[Q_{T^*}^{t,1}J\left(X_{T^*}^{t,x}, Q_{T^*}^{t,q(t,x,p)}\right)\right]\right) \\
&\quad + \mathbb{E}^{\mathbb{Q}_x}\left[\Psi^{-1}\left(X_{T^*}^{t,x}, J\left(X_{T^*}^{t,x}, Q_{T^*}^{t,q(t,x,p)}\right)\right)\right] \\
&\geq \mathbb{E}^{\mathbb{Q}_t}\left[\Psi^{-1}\left(X_{T^*}^{t,x}, J\left(X_{T^*}^{t,x}, Q_{T^*}^{t,q(t,x,p)}\right)\right)\right] =: y(t, x, p) .
\end{aligned}$$

By the martingale representation theorem, there exists $\nu \in \mathscr{U}_{t,y}$ such that

$$Y_{T^*}^{t,y(t,x,p),\nu} = \Psi^{-1}\left(X_{T^*}^{t,x}, J\left(X_{T^*}^{t,x}, Q_{T^*}^{t,q(t,x,p)}\right)\right)$$

which implies that for $p \leq 0$

$$\mathbb{E}\left[\Psi\left(X_{T^*}^{t,x}, Y_{T^*}^{t,y(t,x,p),\nu}\right)\right] \geq \mathbb{E}\left[J\left(X_{T^*}^{t,x}, Q_{T^*}^{t,q(t,x,p)}\right)\right] = \mathbb{E}^{\mathbb{Q}_t}\left[Q_{T^*}^{t,1}J\left(X_{T^*}^{t,x}, Q_{T^*}^{t,q(t,x,p)}\right)\right]$$
$$\geq p$$

and therefore, by definition of the value function $y(t, x, p) \geq V(t, x, p)$ which provides the equality and (9) for $(t, x, p) \in \mathbf{S}$. □

3.2 Application to the Interval $[T, T^*]$

We now come back to $[T, T^*]$, where the *shaping factor* Λ, is assumed to take the known value λ at time T. To highlight the effect of λ as a given state parameter for $t \geq T$, we denote by $V(t, x, p, \lambda)$ the value function defined similarly as (7) where Ψ is given by (5).

Definition 4. Let $(t, x, p, \lambda) \in \mathbf{S}_2 \times L$. We can define the value function on $\mathbf{S}_2 \times L$ as

$$V(t, x, p, \lambda) := \inf\{y \geq -\kappa \ : \ \mathbb{E}\left[\ell\left(g\left(\lambda X_{T^*}^{t,x}\right) - Y_{T^*}^{t,x,y,\nu}\right)^+\right] \leq -p \ \text{ for } \nu \in \mathscr{U}_{t,y}\}, \tag{21}$$

Notice that if X is an exponential process, then we can explicitly change $V(t, x, p, \lambda)$ for $V(t, \lambda x, p, 1)$, recalling the assumption $\mathbb{P}[\Lambda = 1] = 1$ of the previous section. Let us recall the standard pricing concepts in complete market. Under Assumption 2, we can define $\mathbb{Q}$ the $\mathbb{P}$-equivalent martingale measure defined by

$$\left.\frac{d\mathbb{Q}}{d\mathbb{P}}\right|_{\mathscr{F}_t} = \exp\left\{-\int_T^t \theta(s, X_s)dW_s - \frac{1}{2}\int_T^t |\theta(s, X_s)|^2 ds\right\} \ , t \geq T \ . \tag{22}$$

In the present setting, $\mathbb{Q}$ is the unique risk-neutral measure with the drifted Brownian motion $W_t^{\mathbb{Q}} = W_t + \int_T^t \theta(s, X_s)ds$, and we can provide a unique no-arbitrage price for $g(\lambda X_{T^*})$.

Definition 5. We define for $(t, x, \lambda) \in [T, T^*] \times \mathbb{R}_+^* \times L$ the function

$$C(t, x, \lambda) := \mathbb{E}^{\mathbb{Q}}[g(\lambda X_{T^*}^{t,x})] \quad \text{for } t \geq T. \tag{23}$$

According to Assumption 2 and Assumption 3, we can apply Proposition 6.2, in [18], which implies that for any $\lambda \in L$, $(t, x) \mapsto C(t, x, \lambda)$ is Lipschitz continuous in the spatial variable with the same Lipschitz constant K as g. Moreover, $C(\cdot, \cdot, \lambda)$ is the unique classical solution of polynomial growth to the Black–Scholes equation

$$- C_t - \frac{1}{2}\sigma^2(t, x)C_{xx} = 0 \ \text{ on } [T, T^*) \times (0, +\infty) \tag{24}$$

with terminal condition $C(T^*, x, \lambda) = g(\lambda x)$. Now Proposition 1 applies to provide the following when we use a loss function as in Definition 2.

Corollary 1. *Let Assumptions 2 and 3 hold. Then V is given on $\mathbf{S}_2 \times L$ by*

$$V(t, x, p, \lambda) = \mathbb{E}^{\mathbb{Q}}[g(\lambda X_{T^*}^{t,x}) - \ell^{-1}(-P_{T^*}^{t,p})] \tag{25}$$

where $P_{T^*}^{t,p}$ is a $\mathscr{F}_{T^*}$-measurable random variable defined by

$$P_{T^*}^{t,p} = j \left(q \exp \left(\int_t^{T^*} \theta(s, X_s^{t,x}) dW_s^{\mathbb{Q}} - \frac{1}{2} \int_t^{T^*} \theta^2(s, X_s^{t,x}) ds \right) \right) \tag{26}$$

with $j(q) := -((\ell^{-1})')^{-1}(q)$ and q in (26) such that $\mathbb{E}^{\mathbb{Q}} \left[P_{T^*}^{t,p} \right] = p$. Moreover, V is convex and increasing in p and is $\mathscr{C}^{1,2}$ in (t, x) on $[T, T^*] \times \mathbb{R}_+^*$.

The solution (25)–(26) can be explicitly computed in simple cases, see Sect. 5 below. Notice that V is bounded on $[T, T^*] \times \mathbb{R}_+^* \times \mathbb{R}_-^* \times L$ by C. Since ℓ is convex increasing on $\mathbb{R}_+$, $-\ell^{-1}$ is convex. A look at (25)–(26) then convinces that V is continuous in p on $\mathbb{R}_-^*$, but Proposition 3.3 in [7] can be used in our setting to assert that V is also convex in p. According to what was said about the function C, V is $\mathscr{C}^{1,2}$ in (t, x) on $[T, T^*] \times \mathbb{R}_+^*$.

Remark 1. It is noticeable that the value function (25) is composed of the Black–Scholes price of the claim $g(\lambda X_{T^*}^{t,x})$ minus a term that corresponds to a penalty in the dual expression of the acceptance set if ℓ is a risk measure, see [16]. The hedging strategy is to be modified in consequence. First note that (25) is a conditional expectation of a function of two Markov processes $(X^{t,x}, Q^{t,q})$, for q well chosen, so that we can write $Y_s^{t,V(t,x,p),v} := y(X_s^{t,x}, Q_s^{t,q}) := V(s, X_s^{t,x}, P_s^{t,p})$. According to Corollary 1, y is a regular function, and $X^{t,x}, Q^{t,q}$ are martingales under the probability $\mathbb{Q}$, as well as $Y^{t,V(t,x,p),v}$. Therefore Itô formula provides

$$dY_s^{t,V(t,x,p),v} = y_x(X_s^{t,x}, Q_s^{t,q}) dX_s^{t,x} + y_q(X_s^{t,x}, Q_s^{t,q}) dQ_s^{t,q} .$$

Now expressing $dW_s^{\mathbb{Q}}$ with respect to $dX_s^{t,x}$, we obtain

$$dY_s^{t,V(t,x,p),v} = \left(y_x(X_s^{t,x}, Q_s^{t,q}) + \mu(s, X_s^{t,x}) Q_s^{t,q} y_q(X_s^{t,x}, Q_s^{t,q}) \right) dX_s^{t,x} \tag{27}$$

which allows to deduce the optimal dynamic strategy v.

Corollary 1 retrieves solutions of [15]. The original problem in the latter is to minimize the expected loss given an initial portfolio value, but the authors prove that our version of the problem is equivalent. The stochastic target problem of Definition 3 is actually linked to an optimal control problem in a similar way, as noticed in the introduction of [7] and developed in [6]. We can introduce the value function giving the minimal loss that can be achieved at time T^*, with at time t, the initial capital y, $X_t^{t,x} = x$ and the shaping factor $\Lambda = \lambda$.

Definition 6. For $(t, x, y, \lambda) \in [T, T^*] \times \mathbb{R}_+^* \times \mathbb{R} \times L$ we define

$$U(t, x, y, \lambda) := \sup \left\{ \mathbb{E} \left[-\ell \left(\left(g \left(\lambda X_{T^*}^{t,x} \right) - Y_{T^*}^{t,y,v} \right)^+ \right) \right] : v \in \mathscr{U}_{t,y} \right\} . \tag{28}$$

This corresponds to the problem of finding the best reachable threshold p if the initial portfolio value is given by y at time t. This result is of great use in the forthcoming resolution of the problem before T, by the following connection with V.

Lemma 2. *For* $(t, x, y, \lambda) \in [T, T^*] \times \mathbb{R}_+^* \times [-\kappa, \infty) \times L$ *we define* $V^{-1}(t, x, y, \lambda) :=$ $\sup \{p \leq 0 : V(t, x, p, \lambda) \leq y\}$. *Then we have*

$$U(t, x, y, \lambda) = V^{-1}(t, x, y, \lambda) . \tag{29}$$

Moreover the function U *is a concave increasing function of* y *bounded by* 0 *from above.*

Equality (29) is a direct application of Lemma 2.1 in [6], whereas the properties of U fall from the properties of V and Definition 6.

4 Solution in Incomplete Market

We now turn to the solution of the problem before T, i.e., when Λ is unknown. We first provide the face-lifting procedure that allows to reduce the new situation to the one handled in Sect. 3. This procedure can be done according to two model paradigms:

1. the law ρ of Λ is supposed to be known;
2. only the support L of the law of Λ is supposed to be known.

The first approach is a *probabilistic approach* with a prior distribution, whereas the second one is a *robust approach*, in connection with robust finance with parameter uncertainty, see [8] for a control theory version. Finally, the numerical complexity of the face-lifting procedure pushes us to give up explicit formulas for a numerical approach. We thus modify results of Sect. 3 to provide a convenient formulation of the problem to be numerically approximated in Sect. 5.

4.1 Faced-Lifted Intermediary Condition with Prior on the Shaping Factor

When $t < T$, the problem (7) cannot be treated with the methodology developed in [7]. To solve the problem, we are guided by the following argument. Considering $(t, y) \in [0, T) \times [-\kappa, \infty)$ and a strategy $v \in \mathcal{U}_{t,y}$, we arrive at time T to the wealth $Y_T^{t,x,y,v} \geq -\kappa$, at the apparition of the exogenous risk factor Λ. Assume that the agent wants to control the expected level of risk $p < 0$, at time T, by using the portfolio $Y_T^{t,x,y,v}$. It is obviously not possible with certainty if Λ takes a value λ such that $Y_T^{t,y,v} < V(T, X_T^{t,x}, p, \lambda)$. However, the optimal strategy after T consists in optimizing the portfolio by trying to achieve the optimal expected level of loss $U(T, X_T^{t,x}, Y_T^{t,y,v}, \lambda)$ given by Definition 6. In the complete market setting, this achievement is possible.

Lemma 3. *Under Assumptions 2 and 4, there exists a map $(x, y, \lambda) \mapsto v(x, y, \lambda) \in \mathcal{U}_{T,y}$ on $\mathbb{R}_+^* \times [-\kappa, \infty) \times L$ such that*

$$\mathbb{E}\left[\Psi\left(\lambda X_{T^*}^{T,x}, Y_{T^*}^{T,y,v(x,y,\lambda)} \right) \right] \geq U(T, x, y, \lambda) . \tag{30}$$

Proof. Fix $(x, y, \lambda) \in \mathbb{R}_+^* \times [-\kappa, \infty) \times L$. Let $p := U(T, x, y, \lambda)$. According to Lemma 2, $y \geq V(T, x, p, \lambda)$. Following Corollary 1 and Remark 1, and since Ψ is increasing in y, there exists v such that $\mathbb{E}\left[\Psi\left(\lambda X_{T^*}^{T,x}, Y_{T^*}^{T,y,v} \right) \right] \geq p$ for (x, y, λ) arbitrarily fixed.   □

Then the expected loss at T resulting from this strategy is averaged among the realizations of Λ for fixed $X_T^{t,x}$ and $Y_T^{t,y,v}$. The expectation is thus done also with regard to Λ just before T.

Definition 7. We define $\Xi : \mathbb{R}_+ \times [-\kappa, \infty) \to \mathbb{R}_-$ by

$$\Xi(x, y) := \int_L U(T, x, y, r)\rho(dr) . \tag{31}$$

The above function Ξ represents the expected optimal level of loss the agent can reach if she attains the wealth y at time T with a state $X_T^{T,x} = x$.

Lemma 4. *The function Ξ takes non-positive values, is $\mathscr{C}^2$ in $x \in \mathbb{R}_+^*$ and concave increasing in y.*

This is a direct consequence of Lemma 2 and corollary 1. This property ensures that parts of Assumption 4 hold in the new following problem, in order to apply Theorem 2 of Appendix. In particular, Lemma 4 allows to define a terminal condition with the generalized inverse Ξ^{-1}.

Definition 8. For $(t, x, p) \in \mathbf{S}_1$, we define

$$\bar{V}(t, x, p) := \inf\left\{ y \geq -\kappa \ : \ \mathbb{E}\left[\Xi(X_T^{t,x}, Y_T^{t,x,y,v}) \right] \geq p \text{ for } v \in \mathcal{U}_{t,y} \right\} . \tag{32}$$

We now prove that this new problem coincides with the one of Definition 3 on $\mathbf{S}_1$.

Proposition 2. *Let Assumptions 2 and 4 hold. Then $\bar{V}(t, x, p) = V(t, x, p)$ on $\mathbf{S}_1$.*

Proof. **1.** Fix $(t, x, p) \in \mathbf{S}_1$ and take $y > V(t, x, p)$. Then by definition there exists $v \in \mathcal{U}_{t,y}$ such that $\mathbb{E}\left[\Psi\left(X_{T^*}^{t,x}, Y_{T^*}^{t,x,y,v} \right) \right] \geq p$. Since $t < T$, the control can be written $v_t = v_t \mathbb{1}_{\{t \in [0,T)\}} + v_t(\Lambda)\mathbb{1}_{\{t \in [T,T^*]\}}$ where $v_t(.)$ follows from the canonical construction of $\mathbb{F}$: it is a measurable map from L to the set of square integrable control processes on $[T, T^*]$ which are adapted to the Brownian filtration $\sigma(W_s, T \leq s \leq .)$. From the flow property of Markov processes $X^{t,x}$ and $Y^{t,x,y,v}$ (see [21]) and the tower property of expectation,

$$\mathbb{E}\left[\Psi\left(X_{T*}^{t,x}, Y_{T*}^{t,x,y,\nu}\right)\right] =$$
$$\mathbb{E}\left[\int_L \mathbb{E}\left[\Psi\left(X_{T*}^{T,X_T^{t,x}}, Y_{T*}^{T,X_T^{t,x},Y_T^{t,x,y,\nu},\nu(\lambda)}\right)\,\bigg|\,\left(X_T^{t,x}, Y_T^{t,x,y,\nu}, \lambda\right)\right]\rho(d\lambda)\right]. \tag{33}$$

By taking the supremum over all possible maps $\nu(\lambda)$ and by Definition 6, we obtain

$$p \le \mathbb{E}\left[\Psi\left(X_{T*}^{T,X_T^{t,x}}, Y_{T*}^{T,Y_T^{t,x,y,\nu},\nu(\lambda)}\right)\,\bigg|\,\left(X_T^{t,x}, Y_T^{t,x,y,\nu}, \lambda\right)\right] \le U\left(T, X_T^{t,x}, Y_T^{t,x,y,\nu}, \lambda\right). \tag{34}$$

By integrating in λ over L, and then take the expectation, we immediately get that $y \ge \bar{V}(t,x,p)$. By arbitrariness of y, $V(t,x,p) \ge \bar{V}(t,x,p)$.

2. Take $y > \bar{V}(t,x,p)$. There exists a control $\nu \in \mathcal{U}_{t,y}$ on $[t,T]$ such that

$$\mathbb{E}\left[\int_L U\left(T, X_T^{t,x}, Y_T^{t,x,y,\nu}, \lambda\right)\rho(d\lambda)\right] \ge p.$$

Now Lemma 3 allows to assert the existence of a control $\nu^*(\lambda)$ on $[T,T^*]$ such that for any $\lambda \in L$

$$\mathbb{E}\left[\Psi\left(X_{T*}^{T,X_T^{t,x}}, Y_T^{T,Y_T^{t,x,y,\nu},\nu^*(\lambda)}\right)\right] = U\left(T, X_T^{t,x}, Y_T^{t,x,y,\nu}, \lambda\right).$$

By choosing the new admissible control $\nu' \in \mathcal{U}_{t,y}$ defined by the concatenation $\nu_t' = \nu_t \mathbb{1}_{\{t \in [0,T)\}} + \nu_t^*(\Lambda)\mathbb{1}_{\{t \in [T,T^*]\}}$, we obtain $y \ge V(t,x,p)$ for an arbitrary $y > \bar{V}(t,x,p)$. We thus have equality of V and $\bar{V}$ on $\mathbf{S}_1$.

$\square$

In the present context when the law of Λ is known, the face-lifting procedure of (31) allows to retrieve a stochastic target problem in expectation of [7] on the interval $[0,T]$. However, it is improbable that the terminal condition (31) is explicit for non-trivial models.

4.2 Variation to a Robust Approach

Due to the lack of data, or some non stability of the problem, it might be interesting to adopt an approach where the agent wants to control the expected level of loss without assuming the law ρ on L. Under Assumption 1, the robust approach is easy to undertake and is the following. Since the law of Λ is not known, we have to consider the worst case scenario, i.e., the supremum of (31) over a set of probability measures on L.

Definition 9. Let $\mathcal{P}(L)$ be the set of all probability measures over L including singular measures. Then we define for any $(x, y) \in \mathbb{R}_+^* \times [-\kappa, \infty)$ the function

$$\xi(x, y) := \sup_{\rho \in \mathcal{P}(L)} \Xi(x, y) . \tag{35}$$

It is straightforward that by considering singular measures in $\mathcal{P}(L)$ we shall have

$$\xi(x, y) = \max_{\lambda \in L} U(T, x, y, \lambda) , \tag{36}$$

which is finite for all $(x, y) \in \mathbb{R}_+^* \times [-\kappa, \infty)$ following Lemma 2. The reasoning of the previous section can be applied here with the new intermediary condition $\mathbb{E}\left[\xi(X_T^{t,x}, Y_T^{t,x,y,\nu})\right] \geq p$. We thus introduce the other stochastic target problem with controlled loss if $t < T$:

$$\bar{V}^R(t, x, p) := \inf\left\{y > -\kappa \; : \; \mathbb{E}\left[\xi(X_T^{t,x}, Y_T^{t,x,y,\nu})\right] \geq p \text{ for } \nu \in \mathscr{U}_{t,y}\right\} . \tag{37}$$

The monotonicity of U with respect to λ, provided by some additional assumption on g, leads to a direct solution to Eq. (36). In general, the resolution of problem (37) is similar to (32) with a terminal condition that is more tractable. In the sequel, we thus focus on problem (32).

4.3 From the Control Problem to an Expectation Formulation

In this section, we want to emphasize formally the link between the solution of the nonlinear PDE (15) (in the proof of Proposition 1) and a simple conditional expectation, in *sufficiently regular settings*. This idea will be used to propose a numerical scheme to approximate the solution of our partial hedging problem. Assume of starting that the nonlinear PDE (15) has a classical solution $\bar{V}$ on $\mathbf{S}_1$ such that it also verifies

$$\begin{cases} \bar{V}_t + \dfrac{\sigma^2(t, x)}{2} \bar{V}_{xx} + a^* p \left(\sigma(t, x)\bar{V}_{xp} - \theta(t, x)\bar{V}_p\right) + \dfrac{1}{2}(a^* p)^2 \bar{V}_{pp} = 0 , \text{ for } t < T \\ \bar{V}(T, x, p) = \Xi^{-1}(x, p) , \end{cases}$$
$$\tag{38}$$

where we recall the specific form of the control (16):

$$a^* := \left(p\bar{V}_{pp}\right)^{-1} \left(\theta(t, x)\bar{V}_p - \sigma(t, x)\bar{V}_{xp}\right) , \tag{39}$$

and the terminal condition is

$$\Xi^{-1}(x, p) := \inf\{y \geq -\kappa \; : \; \Xi(x, y) \geq p\} . \tag{40}$$

In particular if a^* in (39) is well defined (by strict convexity of $\bar{V}$ in p) and corresponds to the optimal value in (15), then formulations (15) and (38) are equivalent. Now let us assume that the function $\hat{V}$ such that

$$\hat{V}(t,x,p) = \mathbb{E}^{\mathbb{Q}}[\Xi^{-1}(X_T^{t,x}, P_T^{t,p,\alpha^*})], \quad 0 \le t \le T, \tag{41}$$

is well defined, with dynamics under $\mathbb{Q}$ given by

$$X_s^{t,x} = x + \int_t^s \sigma(u, X_u^{t,x})dW_u^{\mathbb{Q}} \; ; \; P_s^{t,p,\alpha^*} = p + \int_t^s P_u^{t,p,\alpha^*}\alpha_u^*\left(dW_u^{\mathbb{Q}} - \theta(u, X_u^{t,x})\right)du \tag{42}$$

and where the feedback control α^* is defined for $s \in [t, T]$ by

$$\alpha_s^* = \left(P_s^{t,p,\alpha^*}\bar{V}_{pp}\right)^{-1}\left(\theta(s, X_s^{t,x})\bar{V}_p - \sigma(s, X_s^{t,x})\bar{V}_{xp}\right)\left(s, X_s^{t,x}, P_s^{t,p,\alpha^*}\right). \tag{43}$$

Assume moreover that $\hat{V}$ is sufficiently regular to be a classical solution to the related linear Feynman–Kac PDE

$$\begin{cases} \varphi_t + \dfrac{\sigma^2(t,x)}{2}\varphi_{xx} + a^*p\left(\sigma(t,x)\varphi_{xp} - \theta(t,x)\varphi_p\right) + \dfrac{1}{2}(a^*p)^2\varphi_{pp} = 0 \,, \text{ for } t < T \\ \varphi(T,x,p) = \Xi^{-1}(x,p) \,, \end{cases} \tag{44}$$

where we recall the specific form of the control (39) given as a function of $\bar{V}$ and its derivatives. Now observe that $\bar{V}$ is also a classical solution to this linear PDE. Hence, if the classical solution to (44) is unique then we can conclude that $\hat{V} = \bar{V}$. In the following section, we propose a numerical scheme to solve (38) which relies, in some sense on this relation between the non linear PDE (38) and the conditional expectation (41) with dynamic (42)–(43).

5   Numerical Approximation Scheme

In the present section, we propose a numerical algorithm dedicated to the specific problem (41)–(43). The problem is specific for two reasons. First, the terminal condition at T is given by Ξ^{-1} which may be unexplicit and has to be numerically studied. Second, the optimal control is explicitly given by (43) so that the non-linear operator can be replaced by a proper expectation approximation. At this stage, the algorithm is presented as the result of a series of standard approximations. However, we do not provide any analysis of the approximation error induced by this algorithm so that it can only be considered as an heuristic. Nevertheless, some numerical simulations are provided in the next section and emphasize the practical interest of such numerical scheme.

5.1 Specification and Hedging in Complete Market

In this section, we will retrieve all the regularity assumptions by specifying the model. The question of the generalization of the presented procedure is a matter that is not treated in this paper. We assume that X is described by a geometrical Brownian motion:

$$\mu(t, x) = \mu x, \quad \text{and} \quad \sigma(t, x) = \sigma x, \tag{45}$$

with $(\mu, \sigma) \in \mathbb{R} \times \mathbb{R}_+^*$. The loss function is given as in Example 1, with $k > 1$. For the partial lower moment function $\ell(x) = x^k \mathbb{1}_{\{x \geq 0\}}/k$ for $k > 1$, Corollary 1 has an explicit solution, given for $t \geq T$ by

$$V(t, x, p, \lambda) = C(t, x, \lambda) - (-kp)^{1/k} \mathbb{E}^{\mathbb{Q}} \left[\exp \left\{ \frac{1}{2(k-1)} \int_t^{T^*} |\theta(u, X_u^{t,x})|^2 du \right\} \right]$$

$$= C(t, x, \lambda) - (-kp)^{1/k} \exp \left\{ \frac{\theta^2}{2(k-1)} (T^* - t) \right\} , \quad \text{where } \theta = \frac{\mu}{\sigma} ,$$
$$\tag{46}$$

where in this precise case, $C(t, x, \lambda)$ is given by the Black–Scholes price of the option with payoff $x \mapsto g(\lambda x)$, as in Definition 5. Following Sect. 3, $(t, x) \mapsto C(t, x, \lambda) \in \mathscr{C}^{1,2}([T, T^*) \times \mathbb{R}_+^*)$ for any $\lambda \in L$, so that according to (46) all the required partial derivatives of V exist. Note also that V is strictly convex in p since $k > 1$.

Consequently, we can explicit the strategy to hedge the expected loss constraint. If a^* is given by (39), then

$$v^*(t, x, p) = \left(V_x + \frac{a^* p V_p}{x\sigma} \right) (t, x, p) . \tag{47}$$

All the required derivatives are given by

$$\begin{cases} V_t(t, x, p, \lambda) &= C_t(t, x, \lambda) + \frac{\theta^2}{2(k-1)} \exp \left\{ \frac{\theta^2}{2(k-1)} (T - t) \right\} \\ V_x(t, x, p, \lambda) &= C_x(t, x, \lambda) \\ V_{xx}(t, x, p, \lambda) &= C_{xx}(t, x, \lambda) \\ V_p(t, x, p, \lambda) &= \exp \left\{ \frac{1-k}{k} \log(-kp) + \frac{\theta^2}{2(k-1)} (T - t) \right\} . \end{cases}$$

As anticipated in Remark 1, the strategy consists in hedging the claim $g(\lambda X_{T^*}^{t,x})$ plus a correcting term corresponding in hedging the constraint $P_{T^*}^{t,p,\alpha^*}$.

5.2 The Intermediary Target

In order to initiate the numerical procedure for $0 \leq t \leq T$, we need to compute the intermediary condition and its partial derivatives intervening in (38). According to (29) and (46),

$$U(t,x,y,\lambda) = \frac{-1}{k}(C(t,x,\lambda)-y)^k \exp\left\{-\frac{k\theta^2}{2(k-1)}(T^* - t)\right\} \mathbb{1}_{\{C(t,x,\lambda)\geq y\}}, \quad (48)$$

which provides the value of $\Xi(x,y)$ by integration according to the law ρ of Λ:

$$\Xi(x,y) = \frac{-1}{k}\exp\left\{-\frac{k\theta^2}{2(k-1)}(T^* - T)\right\}\int_L (C(T,x,\lambda)-y)^k \mathbb{1}_{\{C(T,x,\lambda)\geq y\}}\rho(d\lambda). \quad (49)$$

A numerical computation of the integral in (49) can be proceeded via numerical integration or Monte-Carlo expectation w.r.t. the law ρ. This in turn allows to obtain the desired function Ξ^{-1}, since the latter is monotonous in p.

For fixed $(x,p) \in \mathbb{R}_+^* \times \mathbb{R}_-^*$, define

$$M := \left\{\lambda \in L : C(T,x,\lambda) - \Xi^{-1}(x,p) \geq 0\right\}.$$

Let us introduce four real-valued functions $(f_{k-1}, \tilde{f}_{k-1}, f_{k-2}, \tilde{f}_{k-2})$ of $(x,p) \in \mathbb{R}_+^* \times \mathbb{R}_-^*$ defined by

$$\begin{cases} f_n(x,p) = \int_M (C(T,x,\lambda) - \Xi^{-1}(x,p))^n \rho(d\lambda) \\ \tilde{f}_n(x,p) = \int_M \lambda C_x(T,x,\lambda)(C(T,x,\lambda) - \Xi^{-1}(x,p))^n \rho(d\lambda), \end{cases} \quad (50)$$

for $n = k-1, k-2$, recalling that k denotes the exponent parameter determining the loss function (8). Then by a straightforward calculus we derive the partial derivative of Ξ^{-1} as follows

$$\begin{cases} \Xi_x^{-1}(x,p) = \frac{\tilde{f}_{k-1}}{f_{k-1}}(x,p) \\ \Xi_p^{-1}(x,p) = \exp\left\{\frac{\theta^2 k}{2(k-1)}(T^* - T)\right\}\frac{1}{f_{k-1}}(x,p) \\ \Xi_{pp}^{-1}(x,p) = \left[(k-1)\frac{f_{k-2}}{f_{k-1}}\left(\Xi_p^{-1}\right)^2\right](x,p) \\ \Xi_{xp}^{-1}(x,p) = \left[(k-1)\Xi_p^{-1}\frac{(\tilde{f}_{k-1}f_{k-2}-\tilde{f}_{k-2}f_{k-1})}{(f_{k-1})^2}\right](x,p). \end{cases} \quad (51)$$

The functions $(f_{k-1}, \tilde{f}_{k-1}, f_{k-2}, \tilde{f}_{k-2})$ can be computed numerically with the same methods as Ξ^{-1}. Having these derivatives, it is thus possible to obtain the values of controls at time T, $(v^*(T,x,p), a^*(T,x,p))$, given by (47) and (39).

5.3 Discrete Time Approximation and Regression Scheme

The approximation scheme that is proposed here is first based on a time discretization of the forward–backward dynamic determined by the system (41)–(43).

5.3.1 Time Discretization

Let us define a deterministic time grid $\pi := \{0 = t_0 < \ldots < t_N := T\}$
with regular mesh $|t_{i+1} - t_i| = T/N =: \Delta t$. We consider in this paragraph a
discrete time approximation of the process $(X^{0,x_0}, P^{0,p_0,\alpha^*})$ solution of (42) with
initial condition at time 0, $(X_0^{0,x_0}, P_0^{0,p_0,\alpha^*}) = (x_0, p_0)$. The process X^{0,x_0} possesses
an exact discretization at times $(t_i)_{i=0..N}$ which is denoted $(X_i)_{i=0..N}$, with increments
of the $\mathbb{Q}$-Brownian motion given by

$$W_{t_{i+1}} - W_{t_i} := \sqrt{\Delta t}\varepsilon_i \, , \tag{52}$$

$(\varepsilon_i)_{i=0,\cdots,N-1}$ being a sequence of i.i.d. centered and standard Gaussian random
variables. We introduce the sequence of random variables $(P_i^{\alpha^*})_{i=0\cdots N}$ obtained by
taking the exponential of the Euler approximation of $\log(P^{0,p_0,\alpha^*})$ on the mesh π.
Then, we can approximate the solution of (42), at the mesh instants π by the Markov
chain $(X_i, P_i^{\alpha^*})_{i=0,\cdots,N}$ satisfying the following dynamic for $i = 0, \ldots, N-1$:

$$\begin{cases} X_{i+1} = X_i \exp\left\{\sigma\sqrt{\Delta t}\varepsilon_i - (\sigma^2\Delta t)/2\right\} \\ P_{i+1}^{\alpha^*} = P_i^{\alpha^*} \exp\left\{-a_i^*(X_i, P_i^{\alpha^*})\left((\theta + \tfrac{1}{2}a_i^*(X_i, P_i^{\alpha^*}))\Delta t + \sqrt{\Delta t}\varepsilon_i\right)\right\} \end{cases} \tag{53}$$

with the initial condition, $X_0 = x_0$ and $P_0^{\alpha^*} = p_0$ and where at each time step i, a_i^*
is actually the function given by (39) at time t_i

$$a_i^*(x, p) := a^*(t_i, x, p) = \frac{\theta p \bar{V}_p(t_i, x, p) - \sigma x p \bar{V}_{xp}(t_i, x, p)}{pp \bar{V}_{pp}(t_i, x, p)} \tag{54}$$

In the sequel, we will denote $X_{i+1}^{i,x}$ and P_{i+1}^{i,x,p,a_i} the random variables satisfying
Eq. (53) with $X_i = x$, $P_i^{\alpha^*} = p$ and the function $a_i^* = a_i$.

5.3.2 Piecewise Constant Approximation of a_i^* and Tangent
Process Formula

Assume that at the discrete time t_i, for $i \in \{0, \cdots, N-1\}$, a piecewise constant
approximation of a_i^* is available, such that for any positive reals x and p, we have
the approximation $\hat{a}_i$ defined as follows

$$\hat{a}_i(x, p) = \sum_{r=1}^{R} a_{i,r} \mathbf{1}_{C_{i,r}}(x, p) , \tag{55}$$

where $(C_{i,r})_{r=1,\cdots R}$ is a partition of $\mathbb{R}_+^* \times \mathbb{R}_-^*$ and $(a_{i,r})_{r=1,\cdots R}$ is a sequence of reals. By the expectation formula (41), we obtain that the solution of problem (32) and equivalently (7) satisfies the following backward dynamic for $i \in \{0, \cdots, N-1\}$

$$\hat{V}(t_i, x, p) = \mathbb{E}^{\mathbb{Q}}[\hat{V}(t_{i+1}, X_{t_{i+1}}^{t_i, x}, P_{t_{i+1}}^{t_i, p, \alpha^*})] \quad \text{for } (x, p) \in \mathbb{R}_+^* \times \mathbb{R}_-^* .$$

Then, one can approximate $\hat{V}$ at the discrete instants of the mesh π, by injecting in the above formula two approximations consisting in:

1. replacing $(X_{t_{i+1}}^{t_i, x}, P_{t_{i+1}}^{t_i, p, x, \alpha^*})$ by the Markov chain approximation $(X_{i+1}^{i,x}, P_{i+1}^{i,p,x,a_i^*})$, obtained by the Euler scheme (53);
2. replacing the function a_i^* by the piecewise constant approximation $\hat{a}_i$ (55);

For $i \in \{0, \cdots, N-1\}$, we define $\hat{V}^i$ the resulting approximation of $\hat{V}(t_i, \cdot, \cdot)$ satisfying the following backward approximation scheme

$$\hat{V}^i(x, p) = \mathbb{E}[\hat{V}^{i+1}(X_{i+1}^{i,x}, P_{i+1}^{i,p,x,\hat{a}_i})] . \tag{56}$$

Let us assume, at this stage, that $\hat{V}^{i+1}$ is a given approximation of $\hat{V}(t_{i+1}, \cdot, \cdot)$, which is two times continuously differentiable w.r.t. both variables. Now recall that $\hat{a}_i$ is supposed to be constant on $C_{i,r}$, for any $r \in \{1, \cdots, R\}$. Then, for any $(x, p) \in Int(C_{i,r})$, $(X_{i+1}^{i,x}, P_{i+1}^{i,x,p,\hat{a}_i})$ follows a log-normal distribution (53) and we can apply tangent process approach [10] on (56) to obtain that $\hat{V}^i$ is two times continuously differentiable and a backward formula for the derivatives

$$\begin{cases} \hat{V}_p^i(x, p) = \frac{1}{p}\mathbb{E}\left[P_{i+1}^{i,x,p,\hat{a}_i} \hat{V}_p^{i+1}(X_{i+1}^{i,x}, P_{i+1}^{i,x,p,\hat{a}_i}) \right] \\ \hat{V}_{xp}^i(x, p) = \frac{1}{xp}\mathbb{E}\left[X_{i+1}^{i,x} P_{i+1}^{i,x,p,\hat{a}_i} \hat{V}_{xp}^{i+1}(X_{i+1}^{i,x}, P_{i+1}^{i,x,p,\hat{a}_i}) \right] \\ \hat{V}_p^i(x, p) = \frac{1}{pp}\mathbb{E}\left[(P_{i+1}^{i,x,p,\hat{a}_i})^2 \hat{V}_{pp}^{i+1}(X_{i+1}^{i,x}, P_{i+1}^{i,x,p,\hat{a}_i}) \right] . \end{cases} \tag{57}$$

5.3.3 Piecewise Constant Regression and Fixed Point Algorithm

Besides, recall that a_i^* is defined as a function of $\bar{V}_p(t_i, \cdot, \cdot)$, $\bar{V}_{px}(t_i, \cdot, \cdot)$ and $\bar{V}_{pp}(t_i, \cdot, \cdot)$ according to Eq. (54). Similarly, we want to impose the same relation between $\hat{a}_i$ and $\hat{V}_p^i$, $\hat{V}_{px}^i$ and $\hat{V}_{pp}^i$. For this purpose, let us define the map $f \mapsto T_i(f)$ such that for any real valued function f defined on $\mathbb{R}_+^* \times \mathbb{R}_-^*$,

$$T_i(f)(x, p) := \frac{\theta \mathbb{E}[P_{i+1}^{i,x,p,f} \hat{V}_p^{i+1}(X_{i+1}^{i,x}, P_{i+1}^{i,x,p,f})] - \sigma \mathbb{E}[X_{i+1}^{i,x} P_{i+1}^{i,x,p,f} \hat{V}_{xp}^{i+1}(X_{i+1}^{i,x}, P_{i+1}^{i,x,p,f})]}{\mathbb{E}[(P_{i+1}^{i,x,p,f})^2 \hat{V}_{pp}^{i+1}(X_{i+1}^{i,x}, P_{i+1}^{i,x,p,f})]} , \tag{58}$$

for all $(x, p) \in \mathbb{R}_+^* \times \mathbb{R}_-^*$. Notice that the map T_i, defined above, depends implicitly on the previous approximations $\hat{V}_p^{i+1}$, $\hat{V}_{px}^{i+1}$ and $\hat{V}_{pp}^{i+1}$. Then $\hat{a}_i$ could be obtained as a piecewise constant approximation of a fixed point of T_i.

One way to do this, is to approximate T_i by $\hat{T}_i$, obtained in replacing the conditional expectation, in (58), by a regression operator, $\hat{\mathbb{E}}_i$, on a set of regression functions which are piecewise constant. Consider for instance the following set of regression functions $(\mathbb{1}_{C_{i,r}})_{r=1,\cdots,R}$. We introduce

$$\hat{T}_i(f)(x, p) := \frac{\theta \hat{\mathbb{E}}_i[P_{i+1}^{i,x,p,f} \hat{V}_p^{i+1}(X_{i+1}^{i,x}, P_{i+1}^{i,x,p,f})] - \sigma \hat{\mathbb{E}}_i[X_{i+1}^{i,x} P_{i+1}^{i,x,p,f} \hat{V}_{xp}^{i+1}(X_{i+1}^{i,x}, P_{i+1}^{i,x,p,f})]}{\hat{\mathbb{E}}_i[(P_{i+1}^{i,x,p,f})^2 \hat{V}_{pp}^{i+1}(X_{i+1}^{i,x}, P_{i+1}^{i,x,p,f})]}$$

(59)

so that $\hat{T}_i(f)$ is automatically piecewise constant on the partition $(C_{i,r})_{r=1,\cdots,R}$.

Adding up all these approximations, we finally obtain, at each point, t_i, of the mesh grid, π, an approximation $(\hat{V}^i, \hat{V}_p^i, \hat{V}_{px}^i, \hat{V}_{pp}^i, \hat{a}_i)$ of

$$\left(\bar{V}(t_i, \cdot, \cdot), \bar{V}_p(t_i, \cdot, \cdot), \bar{V}_{px}(t_i, \cdot, \cdot), \bar{V}_{pp}(t_i, \cdot, \cdot), a^*(t_i, \cdot, \cdot) \right)$$

by the following algorithm applied with a fixed tolerance parameter $\varepsilon > 0$:

Initialization

$$\begin{cases} \hat{V}^N(x, p) = \Xi^{-1}(x, p) \\ \hat{V}_p^N(x, p) = \Xi_p^{-1}(x, p) \\ \hat{V}_{px}^N(x, p) = \Xi_{px}^{-1}(x, p) \\ \hat{V}_{pp}^N(x, p) = \Xi_{pp}^{-1}(x, p) \\ \hat{a}_N(x, p) = \frac{\theta p \hat{V}_p^N(x,p) - \sigma x p \hat{V}_{xp}^N(x,p)}{pp \hat{V}_{pp}^N(x,p)} . \end{cases}$$

(60)

From step A$(N-1)$ to A(0) :

A(i):   SET $a := \hat{a}_{i+1}$; GOTO B(i, a);

B(i, a):

1. SET $a' := \hat{T}_i(a)$ (recall that $\hat{T}_i$ depends on $\hat{V}_p^{i+1}$, $\hat{V}_{px}^{i+1}$ and $\hat{V}_{pp}^{i+1}$);
2. IF $|a' - a| \leq \varepsilon$

    – THEN SET

$$\begin{cases} \hat{a}_i & = a' \\ \hat{V}^i(x, p) & = \hat{\mathbb{E}}_i[\hat{V}^{i+1}(X_{i+1}^{i,x}, P_{i+1}^{i,p,x,\hat{a}_i})] \\ \hat{V}_p^i(x, p) & = \frac{1}{p} \hat{\mathbb{E}}_i \left[P_{i+1}^{i,x,p,\hat{a}_i} \hat{V}_p^{i+1}(X_{i+1}^{i,x}, P_{i+1}^{i,x,p,\hat{a}_i}) \right] \\ \hat{V}_{xp}^i(x, p) & = \frac{1}{xp} \hat{\mathbb{E}}_i \left[X_{i+1}^{i,x} P_{i+1}^{i,x,p,\hat{a}_i} \hat{V}_{xp}^{i+1}(X_{i+1}^{i,x}, P_{i+1}^{i,x,p,\hat{a}_i}) \right] \\ \hat{V}_p^i(x, p) & = \frac{1}{pp} \hat{\mathbb{E}}_i \left[(P_{i+1}^{i,x,p,\hat{a}_i})^2 \hat{V}_{pp}^{i+1}(X_{i+1}^{i,x}, P_{i+1}^{i,x,p,\hat{a}_i}) \right] . \end{cases}$$

```
        ·  IF i = 0 THEN STOP;
        ·  ELSE GOTO A(i − 1);

    −  ELSE GOTO B(i, a′);
```

Notice that limiting the previous algorithm to one fixed point iteration (in $B(i, a)$) reduces to an explicit scheme for which $\hat{a}_i$ is given as a function of the derivatives at the next time step t_{i+1}, $(\hat{V}^{i+1}, \hat{V}_p^{i+1}, \hat{V}_{px}^{i+1}, \hat{V}_{pp}^{i+1})$ and the control $\hat{a}_{i+1}$. However, in practice at most three iterations are sufficient to obtain reasonable convergence to the fixed point. In theory, the contraction of $\hat{T}_i$ should be proved for a sufficiently small time step Δt. But this is left for future works. If we don't proceed to a convergence analysis of the scheme with N, for a general diffusion and general loss function, we can still provide a numerical confirmation of the relevance of the method. To validate the algorithm we proceed in Sect. 6 to a comparison between the explicit formula of Corollary 1 and the value provided by the algorithm.

6 Numerical Tests

The present section is devoted to tests on real and simulated data meant to illustrate the interest of the partial hedging strategy developed in this article and to validate the numerical scheme introduced in the previous section. We proceed into four steps. First, we fit the parameters of the exponential model (1) on real data. Then, we point out the importance of the risk induced by the random shaping factor, by evaluating the hedging error implied, on real data, by the naive Black–Scholes hedging strategy based on a prediction of the shaping factor (without taking into account its randomness). This naive hedging approach will constitute our benchmark. To validate the numerical approximation scheme introduced in Sect. 5, we analyse its performance on the explicit case of Sect. 3. We finally compare the partial hedging procedure, on simulations, to the benchmark that shall be introduced right away.

6.1 Black–Scholes Benchmark

We consider the following Black–Scholes strategy for a naive agent. The naive agent assumes that the set L reduces to a singleton $\{\lambda_0\}$. This belief is accepted for example as a raw approximation of the expected value of Λ. In this situation, the previous setting reduces to the case of Sect. 3. However, since the market is complete, the naive agent desires to put in place a complete hedging strategy allowed by the Black–Scholes framework. The naive benchmark is thus given by

1. an initial value provided by the Black–Scholes price of the contingent claim $g(\lambda_0 X_{T*}^{t,x})$, given by Eq. (23): $C(t, x, \lambda_0)$.

2. A hedging strategy, associated to that belief, and given by the delta-hedging procedure $v_s = C_x(s, X_s^{t,x}, \lambda_0)$ on $[0, T]$. After the apparition at T of the asset ΛX, the option price is impacted immediately by the real value taken by Λ, different from λ_0, but the portfolio stays self-financed, and continuous. We assume here that the naive agent continues with the delta-hedging strategy until T^*.
3. A terminal hedging error that spreads from time T with value $\varepsilon_T := C(T, x, \lambda_0) - C(T, x, \Lambda)$. In the Black–Scholes setting, by a simple no-arbitrage argument and zero interest rate, this error remains constant until T^*.

The motivation of such a strategy is to average the losses by averaging the possible values taken by Λ. This is however wrong as the price of the derivative is mostly a non-linear function of the underlying price. In the studied example of the Call option below, if λ is fixed, we obtain a non-linear function of the strike:

$$C_{BS}(t, \lambda_0 x, K) := \mathbb{E}^{\mathbb{Q}}[(\lambda_0 X_{T^*}^{t,x} - K)^+] = \lambda_0 \mathbb{E}^{\mathbb{Q}}[(X_{T^*}^{t,x} - K/\lambda_0)^+]$$
$$= \lambda_0 C_{BS}(t, x, K/\lambda_0).$$

6.2 Analysis on Real Data

To provide a realistic framework, we refer to historical data. This allows to propose a model for L and Λ, and values for parameters (μ, σ) of the exponential dynamics (1). The available data designates daily quotations of futures prices on the French Power Market, provided by EEX. We consider a delivery period covering the period from October 2004 to March 2011, i.e., 78 Month delivery Futures during their whole quotation period and the respective Quarter delivery futures contracts covering them. Two estimations are made out of it.

1. This provides 78 observations for a supposedly repeated realisation of the random variable Λ. The average is $\hat{\Lambda} = 1.0012$ and its variance $V(\Lambda) = 0.081$. We then assume that Λ follows a scaled beta law with these characteristics: $\Lambda \sim 3\beta(114, 227)$. This is justified by the fact that Λ shall have a bounded support, which is assumed here to be the interval $[0, 3]$.
2. The parameters μ and σ in the exponential dynamics (1) are computed on the aggregated returns of month futures and quarter futures. Here, μ contains the discount rate (since we assumed that the interest rate is null by omitting it). The obtained drift $\hat{\mu}$ is null, and the obtained (yearly) volatility $\hat{\sigma} = 28\%$.

To quantify the impact of neglecting the uncertainty on the shaping factor $\Lambda = \lambda^*$, on the performance of the hedging strategy, we have implemented, on real data, the naive Black–Scholes hedging strategy supposing different given parameters λ varying around the real observed value λ^* with an amplitude of error of 50%. In our tests, we have considered call options on our 78 Month delivery futures with various maturities and strikes as indicated on Fig. 1. The resulting hedging error can be decomposed into four sources:

Fig. 1 Standard deviation of the hedging error as a function of the ratio between the shaping factor used in the hedging strategy λ and the real shaping factor λ^* impacting the historical scenarios. (**a**) $K = 0, 8X_T$. (**b**) $K = X_T$. (**c**) $K = 1.2X_T$

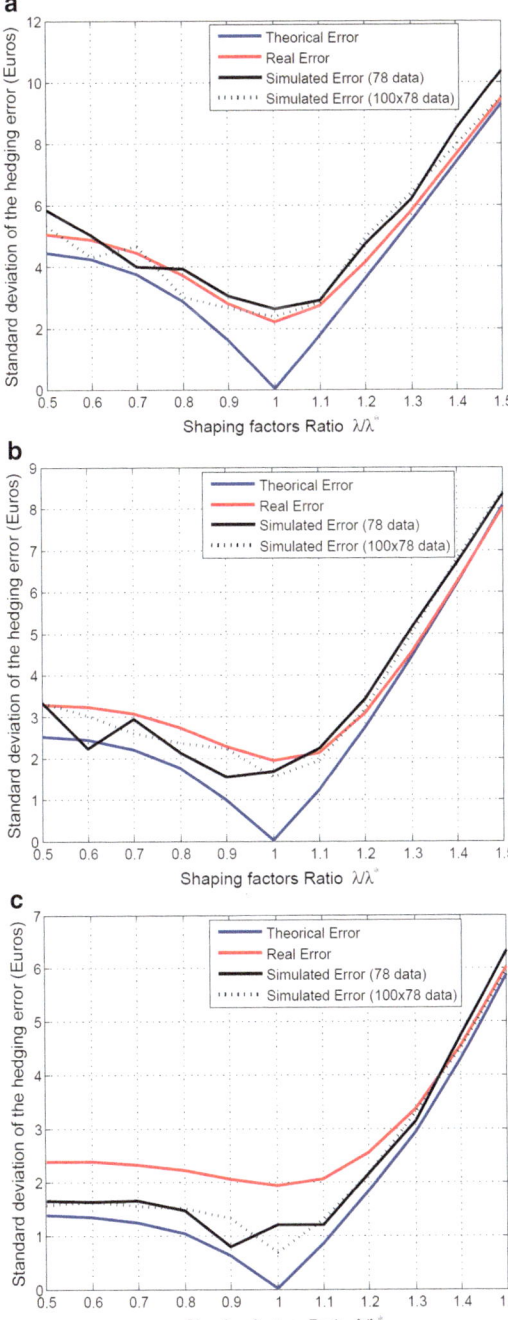

1. hedging at discrete times (the Delta hedging strategy is indeed implemented daily);
2. errors on the dynamical model or on the parameters (the hedging instrument may not have i.i.d. log-returns with log-normal distributions, $\hat{\mu}$ and $\hat{\sigma}$ are only estimations);
3. the limited number of hedging scenario inducing a statistical error;
4. error on the shaping factor value.

The scope of this paper focuses specifically on the latter source of error. Hence, to distinguish the contribution of each error and to separate the fourth one, we have represented in Fig. 1 four quantities:

1. *Real error* We evaluate the hedging error using the naive strategy on real data.
2. *Simulated Error (78 data)* We then do the same on simulated data on the same time grid, and with the same number of trajectories (78). This allows to quantify the error due to the model and the parameters estimation, by comparing it to the previous error.
3. *Simulated Error (78 × 100 data)* We repeat this procedure with a greater number of simulations (78 × 100) in order to confirm that the previous error is not erroneous because of the low number of studied trajectories.
4. *Theoretical error* We represent the error induced by hedging in the Black–Scholes framework (in continuous time) with the wrong shaping factor. This represents exclusively the hedging error due to the error on the shaping factor. Notice that for $t \geq T$, λ^* is known, hence on $[T, T^*]$ the naive Black–Scholes strategy is equal to the complete market replication strategy that would have been implemented from time 0 if λ^* was known. Hence the theoretical hedging error reduces to the difference between the values at time T of the naive hedging portfolio (with the wrong value of λ) and the perfect hedging portfolio (with the right value of $\lambda = \lambda^*$): $C_{BS}(t, \lambda x, K) - C_{BS}(t, \lambda^* x, K)$.

Altogether, the results presented in Fig. 1 push to the following temporary conclusions. An error on the value of Λ can significantly impact the error. After that, the main error is due to the discretization of the hedging strategy. Hence, it is worth developing a specific methodology to take into account the uncertainty of the shaping factor in the hedging strategy.

6.3 Convergence of the Approximation Scheme to Explicit Solution

We shall test the efficiency of the algorithm presented in Sect. 5. To do so, we compute the hedging strategy and the value function in the specific case of Sect. 3.2. Assume without loss of generality that $\Lambda = 1$. Recall that we obtain the following explicit expression for the $\mathbb{P}$-martingale $P^{t,p,\alpha}$ initialized at time $t \geq T$ and the function v for any $s \in [t, T^*]$,

$$\begin{cases} v(s, X_s^{t,x}, P_s^{t,p,\alpha}, 1) = C(s, X_s^{t,x}, 1) - (-kP_s^{t,p,\alpha})^{1/k} \exp\left\{\frac{\theta^2}{2(k-1)}(T^* - s)\right\} \\ P_s^{t,p,\alpha} = p\left(\frac{X_s^{t,x}}{x}\right)^{-\frac{k}{k-1}\frac{\mu}{\sigma^2}} \exp\left\{\frac{k^2(\theta^2-\mu)}{2(k-1)}(s - t)\right\} , . \end{cases} \quad (61)$$

Observe that $P_s^{t,p,\alpha}$ can be expressed as a function of $X_s^{t,x}$ i.e. $P_s^{t,p,\alpha} = p(t, s, X_s^{t,x})$. Hence we analyse the performance of our algorithm by observing its ability to approximate the one dimensional real valued function u_s such that

$$u_s(x) = V(s, x, p(t, s, x)) . \quad (62)$$

In our simulations, we consider the following parameters.

1. The model parameters are slightly modified with values $(\hat{\mu}, \hat{\sigma}) = (0.1, 0.28)$. The initial asset price value is fixed at $x = 50.89$.
2. The convexity parameter of the loss function is $k = 2$ and the level p takes the value 0.1 Euro^2.
3. The option is a call option with a strike K and a maturity of 20 trading days, i.e., $T = 20/250$.
4. We have performed our algorithm with $M = 10^5$ particles to estimate at each step of time the conditional expectations and a time discretization mesh $t_0 = 0, \cdots t_i, \cdots t_N = T$ with a time step $\Delta t = 1/250$.

In our tests, the fixed point algorithm was limited to three iterations. We have represented on Fig. 2 the value of $u_s(x)$ with respect to x computed by the explicit formula and the numerical algorithm. We also provide the value of the control v at the initial date to illustrate the convergence of derivatives too.

6.4 Performance with a Call Option

We now compare the loss approach (hereafter denoted *shortfall risk*, or SR) and the benchmark strategy (hereafter *Black–Scholes*, or BS) upon a call option. For each approach, we implement the associated hedging strategy on i.i.d. $M_{hedge} = 10,000$ simulated price paths. For each path we compute both hedging errors. Then we compute by Monte Carlo approximation (on these i.i.d. $M_{hedge} = 10,000$ simulations) the expected loss associated to the Black–Scholes approach and the shortfall risk hedge. Recalling Sect. 6.2, the trading strategies are not implemented continuously and the resulting hedging errors may differ from the theoretical time continuous setting.

1. The naive Black–Scholes strategy is settled with the value $\lambda_0 = \mathbb{E}[\Lambda] = 1.0012$. The variable Λ is given by a law $\beta(114, 227)$, and the price model as in Sect. 6.3.
2. For the option, we compare several strike possibilities for the Call option: $K = \gamma \lambda_0 x$ with γ taking values in the set $\{0.85; 0.9; 0.95; 1; 1.05, 1.1; 1.15; 1.2\}$.

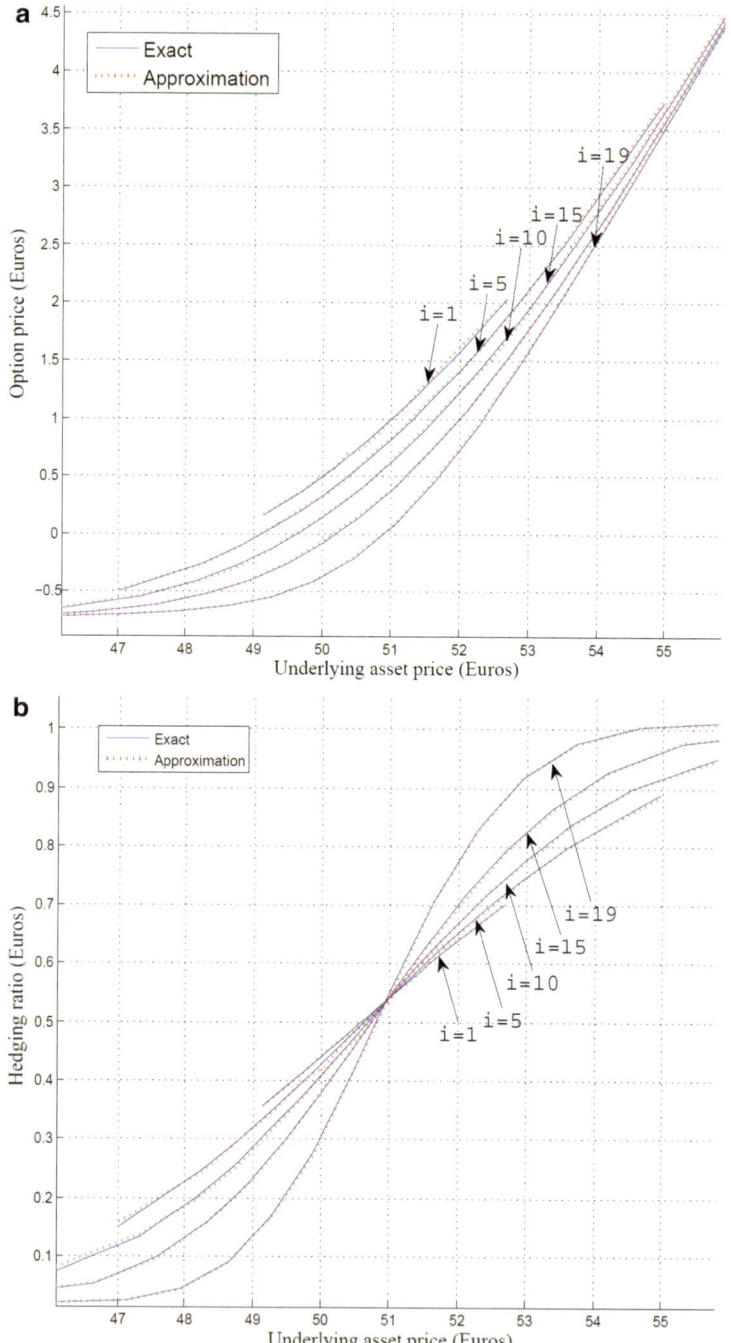

Fig. 2 Comparison between the numerical solution and the explicit formula. (**a**) Value function $x \mapsto u_s(x)$ with $p = 0.1$ Euros and time steps $i =\in \{1, 5, 10, 15, 19\}$. (**b**) Optimal strategy $x \mapsto v(t, x, p)$ for $p = 0.1$ Euros and time steps $i \in \{1, 5, 10, 15, 19\}$

3. The loss function is the partial moment loss function of Sect. 5 with $k = 2$, and the threshold p varies enough to evaluate its impact. In the following comparison, we consider the square root of the obtained error in order to express it in euros. This justifies the terminology *shortfall*, which is a monetary homogeneous quantity.
4. Strategies are rebalanced daily, $T = 128$ days and $T^* = 184$ days, 1 year corresponds to 250 days and $X_0 = 50.89$.

Figure 3 sums up the simulations and compares, for the different values of K, the value function as a function of p. Figure 4 provides a comparison between the two approaches for another criterion: the conditional Value-at-Risk, or expected shortfall. These two figures lead us to the two following conclusions. The first one is that the partial hedging procedure SR allows to hedge the quadratic loss function more efficiently and with less initial amount of money than the BS strategy. The second figure illustrates the fact that this new strategy stays more interesting for another risk criteria than the one used in the specified control problem, providing also possible robustness of the consideration of the shaping factor Λ in our model.

Appendix: Geometric Dynamic Principle and HJB Equation

In what follows, we put ourselves in the Brownian filtration setting of [22] and [7], which encompasses our framework. We omit the presence of Λ by assuming $\mathbb{P}[\Lambda = 1] = 1$ and place ourselves on the interval $[0, T^*]$. We provide one side of the GDP used to derive the supersolution property.

Theorem 1 (Th 3.1, [22]). *Fix* $(t, x, p, y) \in \mathbf{S} \times [-\kappa, \infty)$ *such that* $y > V(t, x, p)$ *and a family of stopping times* $\{\theta^{\nu,\alpha} : (\nu, \alpha) \in \mathcal{U}_{t,y} \times \mathscr{A}_{t,p}\}$. *Then there exists* $(\nu, \alpha) \in \mathcal{U}_{t,y} \times \mathscr{A}_{t,p}$ *such that*

$$Y_{\theta^{\nu,\alpha}}^{t,x,y,\nu} \geq V(\theta^{\nu,\alpha}, X_{\theta^{\nu,\alpha}}^{t,x}, P_{\theta^{\nu,\alpha}}^{t,p,\alpha}) \; \mathbb{P} - a.s.$$

and $Y_{s \wedge \theta^{\nu,\alpha}}^{t,x,y,\nu} \geq -\kappa$ *for all* $s \in [t, T^*] \; \mathbb{P} - a.s.$

Let V_* be defined by $V_*(t, x, p) := \liminf \{V(t', x', p') : B \ni (t', x', p') \to (t, x, p)\}$ where B denotes an open subset of $[0, T^*] \times \mathbb{R}_+^* \times \mathbb{R}_-^*$ with $(t, x, p) \in \mathrm{cl}(B)$. We assume that V is locally bounded on $\mathbf{S}$, so that V_* is finite. In what follows, we introduce only the supersolution property for V_*, deriving from Theorem 1, in the special case given by dynamics (1), (3) and (11).

For $\varepsilon \geq 0$, we introduce the relaxed operator $\Theta \mapsto \bar{F}_\varepsilon(\Theta)$ for the variable $\Theta = (t, x, p, d, d_x, d_p, d_{xx}, d_{pp}, d_{xp}) \in \mathbf{S} \times \mathbb{R}^6$ given by

$$\bar{F}_\varepsilon(\Theta) := \sup_{(u,a) \in \mathcal{N}_\varepsilon(\Theta)} \left\{ (u - d_x) \, \mu(t, x) - \frac{1}{2} \left(\sigma^2(t, x) d_{xx} + a^2 p^2 d_{pp} + 2ap\sigma(t, x) d_{xp} \right) \right\}$$

$$(63)$$

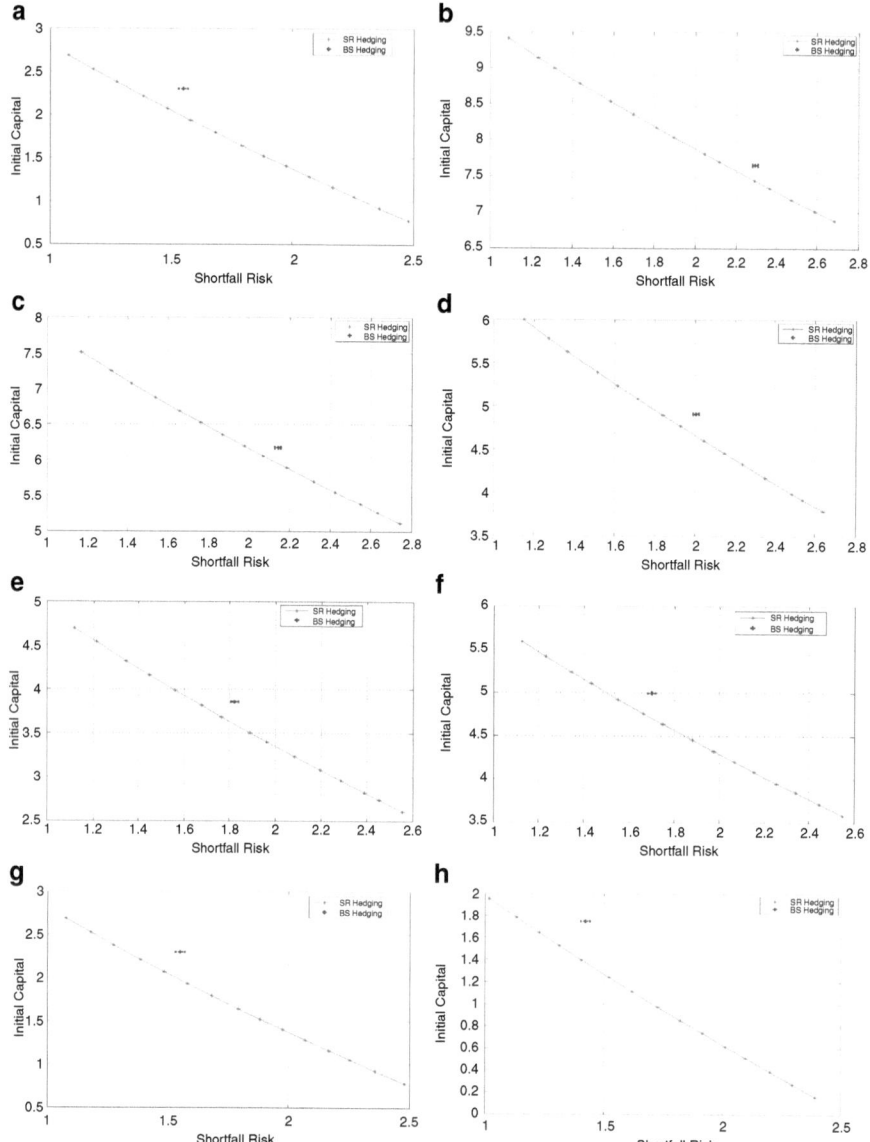

Fig. 3 Initial capital w.r.t. the associated shortfall risk of the Black–Scholes strategy (*blue*) and the shortfall strategy (*red*) with 95 % confidence interval (in *dotted lines*). (**a**) $K = 0,85X(t_0)$. (**b**) $K = 0,90X(t_0)$. (**c**) $K = 0,95X(t_0)$. (**d**) $K = X(t_0)$. (**e**) $K = 1,05X(t_0)$. (**f**) $K = 1,10X(t_0)$. (**g**) $K = 1,15X(t_0)$. (**h**) $K = 1,15X(t_0)$

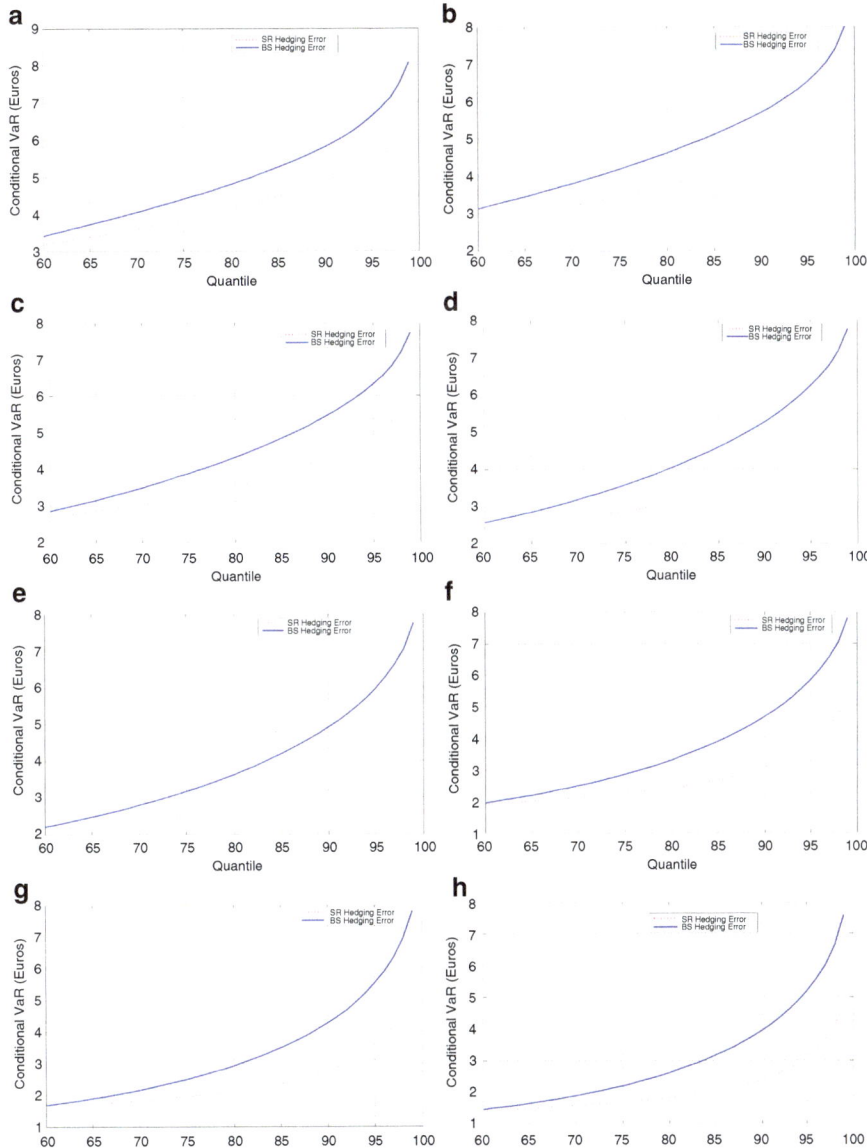

Fig. 4 CVaR value w.r.t. the quantile level of the strategies BS (*blue*) and SR (*red*). (**a**) $K = 0,85X(t_0)$ (**b**) $K = 0,90X(t_0)$. (**c**) $K = 0,95X(t_0)$. (**d**) $K = X(t_0)$. (**e**) $K = 1,05X(t_0)$. (**f**) $K = 1,10X(t_0)$. (**g**) $K = 1,15X(t_0)$. (**h**) $K = 1,15X(t_0)$

with

$$\mathscr{N}_{\varepsilon}(\Theta) := \left\{ (u,a) \in \mathbb{R}^2 \; : \; \left| \sigma(t,x)(u - d_x) - apd_p \right| \leq \varepsilon \right\} . \tag{64}$$

to finally introduce $\bar{F}^*(\Theta) := \limsup \left\{ \bar{F}_{\varepsilon}(\Theta') \; : \; \varepsilon \searrow 0, \Theta' \to \Theta \right\}$. We adopt the convention $\sup \emptyset = -\infty$ and

$$\bar{F}^*\varphi = \bar{F}^*(t,x,p,\varphi(x,p),\varphi_x(t,x),\varphi_p(t,x),\varphi_{xx}(t,x),\varphi_{pp}(t,x),\varphi_{xp}(t,x))$$

for a smooth function φ. We hence formulate the supersolution property of V_*. For definitions and use of viscosity solutions, we refer to [11]. The supersolution property inside the domain is given by Theorem 2.1 and Corollary 3.1 in [7]. The boundary condition at $t = T^*$ is given by Theorem 2.2 in [7]. In our case, by assuming the concavity of Ψ in y, we have the convexity of Ψ^{-1} in the p variable. We also have $\mathscr{N}_0(\Theta) \neq \emptyset$ for any Θ. According to these two properties, the terminal condition takes a much more simple form. Altogether, we obtain the following.

Theorem 2 (Th. 2.1-2.2, [7]). *The function V_* is a viscosity supersolution of*

$$\begin{cases} -\varphi_t(t,x,p) + \bar{F}^*\varphi(t,x,p) = 0 \text{ on } [0,T^*) \times \mathbb{R}_+^* \times \mathbb{R}_-^* \\ V_*(T,x,p) \geq \Psi^{-1}(x,p) \qquad\qquad on \ \mathbb{R}_+^* \times \mathbb{R}_-^* \end{cases} . \tag{65}$$

There is a special Cauchy boundary problem for $p = 0$ we elude here. In our case, since $\Psi(0) = 0$, the stochastic target problem reduces to the superhedging problem. The target must be reached $\mathbb{P}$-almost surely and we obtain directly the HJB equation of [22]. In our complete market framework, it straightly provides the Black–Scholes equation (24) to which $V_*(.,0)$ is also a viscosity solution.

References

1. Ankirchner, S. Imkeller, P., Reis, G. D.: Pricing and hedging of derivatives based on non-tradable underlyings. Math. Financ., **20**(2), 289–312 (2008)
2. Bauwens, L., Hafner, C.M., Pierret, D.: Multivariate volatility modeling of electricity futures. J. Appl. Econ. **28.5**, 743–761 (2013)
3. Becherer, D.: Rational hedging and valuation with utility-based preferences Universitätsbibliothek, Technischen Universität Berlin (2001)
4. Benth, F., Koekebakker, S.: Stochastic modeling of financial electricity contracts. J. Energy Econ. **30**, 1116–1157 (2007)
5. Bouchard, B., Dang, N.: Generalized stochastic target problems for pricing and partial hedging under loss constraints-application in optimal book liquidation. Finance Stochast. **17**(1), 31–72 (2011)
6. Bouchard, B., Dang, N.: Optimal control versus stochastic target problems: an equivalence result. Syst. Control Lett. **61**(2), 343–346 (2012)
7. Bouchard, B., Elie, R., Touzi, N.: Stochastic target problems with controlled loss. SIAM J. Control Optim., **48**, 3123–3150 (2009)

8. Bouchard, B., Moreau, L., Nutz, M.: Stochastic target games with controlled loss. Ann. Appl. Probab., **24**(3), 899–934 (2014)
9. Bouchard, B., Moreau, L., Soner, M.H.: Hedging under an expected loss constraint with small transaction costs. arXiv:1309.4916 (2013, preprint)
10. Broadie, M., Glasserman, P.: Estimating security price derivatives using simulation. Manage. Sci. JSTOR, 269–285 (1996)
11. Crandall, M.G., Ishii, H., Lions, P.L.: User's guide to viscosity solutions of second order partial differential equations. Bull. Am. Math. Soc. **27**(1), 1–67 (1992)
12. Cvitanic, J.: Minimizing expected loss of hedging in incomplete and constrained markets. SIAM J. Control Optim. **38**(4), 1050–1066 (2000)
13. Eichhorn, A., Romisch, W., Wegner, I.: Mean-risk optimization of electricity portfolios using multiperiod polyhedral risk measures. In: Power Tech, 2005 IEEE Russia, pp. 1–7 (2005)
14. Föllmer, H., Leukert, P.: Quantile hedging. Finance Stochast. **3**(3), 251–273 (1999)
15. Föllmer, H., Leukert, P.: Efficient hedging: cost versus shortfall risk. Finance Stochast. **4**, 117–146 (2000)
16. Föllmer, H., Penner, I.: Convex risk measures and the dynamics of their penalty functions. Stat. Decis. **24**, 61–96 (2006) [Oldenbourg Wissenschaftsverlag GmbH]
17. Hildmann, M., Andersson, G., Caro, G., Daly, D., Rossi, S.: What makes a good hourly price forward curve? In: 2013 10th International Conference on the European Energy Market (EEM), pp. 1–7. IEEE, Piscataway (2013)
18. Janson, S., Tysk, J.: Feynman-Kac formulas for Black-Scholes type operators. Bull. Lond. Math. Soc. **38**(2), 262–289 (2006)
19. Lindell, A., Raab, M.: Strips of hourly power options—approximate hedging using average-based forward contracts. Energy Econ. **31**, 348–355 (2009) [Elsevier]
20. Moreau, L.: Stochastic target problems with controlled loss in jump diffusion models. SIAM J. Control Optim. **49**, 2577–2607 (2011)
21. Soner, H., Touzi, N.: Dynamic programming for stochastic target problems and geometric flows. J. Eur. Math. Soc. **4**, 201–236 (2002) [Springer]
22. Soner, H.M., Touzi, N.: Stochastic target problems, dynamic programming, and viscosity solutions. SIAM J. Control Optim. **41**(2), 404–424 (2002)
23. Verschuere, M., Von Grafenstein, L.: Futures hedging in power markets: evidence from the EEX. Working paper (2003)

Calibration of Electricity Price Models

Olivier Féron and Elias Daboussi

Abstract This paper addresses the issue of model calibration to electricity prices. The non-storability of electricity introduces new problems in terms of modeling and calibration, especially when the objective is to represent both spot prices and forward products, the latter showing a particular time interval: the delivery period. The two main approaches to model electricity prices are: (i) models on a fictitious forward curve from what we can deduce spot prices and forward products with any delivery period, and (ii) models on spot prices from what we can deduce any forward products. In this paper we study both approaches and we focus on the calibration issues. The first part of the paper studies different calibration methods for a classic Gaussian factorial model as described in Benth and Koekebakker (2008), Kiesel, Schidlmayr, and Börger (2009) and mostly based on Heath-Jarrow-Morton approach (Heath, Jarrow, and Morton, Econometrica, 1992). In this case different calibration methods can be proposed, based on spot and/or forward prices, but the main objective is to compare or validate these estimation procedures. We compare these procedures on the valuation of specific portfolios and we then stress the high impact of the calibration method. The second part concerns the calibration issues of a structural model proposed in Aïd, Campi, Langrené (2013). In particular we study the reconstruction performances of forward prices and we address the issue of model calibration in terms of determining the parameters to exactly fit the observable forward products. We propose a modification in the structural model to ensure its ability to be calibrated on all the observed forward products and we give some illustrations of calibration performances.

O. Féron (✉)
EDF Research & Development, 1 avenue du Général de Gaulle, 92140, Clamart, France

FiME Lab, Finance of Energy Markets joint Laboratory between University Paris Dauphine, CREST, École polytechnique and EDF R&D
e-mail: olivier-2.feron@edf.fr

E. Daboussi
EDF Research & Development, 1 avenue du Général de Gaulle, 92140, Clamart, France
e-mail: elias.daboussi@gmail.com

© Springer Science+Business Media New York 2015
R. Aïd et al. (eds.), *Commodities, Energy and Environmental Finance*,
Fields Institute Communications 74, DOI 10.1007/978-1-4939-2733-3_7

1   Introduction

Modeling electricity prices is a very exciting challenge, as their behaviour is unique compared to other assets like equities or even other commodities, a fact mostly due to its non-storability. The underlying incompleteness of the recently deregulated electricity markets makes possible a vast range of models to price electricity contracts, and thus scientific literature abounds with models that try to capture the well-known stylized facts of electricity prices.

An electricity producer needs price models for different applications: electricity price prediction (in a short-term horizon), risk-management, hedging, pricing (in a mid-term horizon) and investment decisions (in a long-term horizon). The price models used must be adapted to the application of interest. In this paper we are interested in a mid-term horizon and the financial application of risk-management, pricing and the determination of hedging strategies. We position our study on the case of an electricity producer who has to manage financial risks. His portfolios are composed of physical and financial assets:

- *Production units.* The power generating plants can be represented, in a first approximation, as a basket of European spread options whose underlying assets are the spot prices of power and the fuel used to produce electricity. In the case of thermal power plants the carbon spot price is a third underlying asset. For example, a gas power plant can be represented as a basket of European options of payoff $\left(S_t^e - h^g S_t^g - h^c S_t^c - K\right)^+$, with S_t being the spot prices and the superscripts e, g and c are respectively for electricity, gas and carbon, h^g and h^c are coefficients determining the performances of the plant (the "heat rate"), and K is the fixed production cost. Of course the modelization of power plants may be more complex once we consider dynamic constraints, starting and stopping costs... But the most important point to note is that the underlying assets are the spot prices.
- *Storage assets.* Gas storage and hydraulic dams are the most common storage assets of a power producer. They are classically represented as swing options that let the option holder buy a predetermined quantity of energy at a predetermined price while having some flexibility in the amount purchased and the price paid. The underlying assets are the spot prices of power and gas for the gas storage asset.
- *Electricity supply contracts.* On the other hand, an electricity producer has contracts for supplying electricity. These contracts may have optionalities[1] that allow them to be represented as swing options, moving average options or more exotic derivatives. The underlying assets are spot prices of power and fuels, but also forward prices in the case where the sale price depends on some historical forward prices.

---

[1]For example, a load curve contract allows the owner to buy at a fixed price an undetermined quantity q_t of power in an interval $[q_t^{min} ; q_t^{max}]$ around a specific load curve.

In order to manage the risks of such portfolios, a price model is needed to represent both spot prices and forward products, on several commodities in the Energy market. Therefore we focus on adapted models for this objective, in particular we are not interested in models that exclusively represent spot prices (see [7] for some examples) or, on the other hand, exclusively represent forward products as it is proposed, for example, in [6, 16].

We can classify the electricity price models into two main categories depending on the element considered as the basis of modelization, but all the models aiming to represent both spot prices and forward products need to determine the "forward curve", i.e. a function $F_t(T)$ defining fictitious forward contracts delivering 1MWh of electricity at dates T during one unit of time (1 h or 1 day).

The first class of models is dedicated to directly represent the forward curve and is mostly related to classic interest rate models like in [13]. We refer to [6] and [16] for some examples and [14] to justify using interest rate models for electricity prices. The main advantage of this class of model is the ability to use the broad literature of interest rate models which leads, in general, to a lot of closed-form formulas of pricing and an easy determination of hedging strategies. However, the calibration of these models is a real issue in the case of the power market because the observed quotations are not some points of the forward curve, but a weighted average of the forward curve over different periods (the delivery periods, as detailed in Sect. 2). Therefore the relationship between the model parameters and the observed products is more complex.

The second class of models focuses on the spot price representation. The starting point is then to model power spot prices as finely as possible, using sophisticated processes, as done for example in the popular jump-diffusion model from [10]. This has also lead to a new methodology for forecasting and modeling spot prices, and we refer to [19] for a complete panel of statistical methods that are used with reduced-form models. These models may depend on several hidden factors [5] or other observable factors. In particular, structural models define a relationship between the power spot price, the fuel spot prices and other observable variables like demand and production capacities, temperature... Various structural models exist which we refer to [9] for a complete survey, underlying the fact that they differ depending on the drivers they take into account. For example, some authors decide to link the prices only to the demand, as [4] did in what is often referred to as the first structural model. The most relevant drivers are the capacity, the demand, and the prices of fuels needed to produce electricity, which are directly observable, and thus some have studied the performances of their models by confronting their simulated spot prices against historical data, as it was done in [2]. As for any spot price model, forward prices are deduced using no-arbitrage arguments, leading, for structural models, to a relationship between electricity forward prices and fundamental drivers like forward prices on fuels.

The objective of this paper is to study the calibration issues for each class of models. Firstly we study a factorial model representing the forward curve, with two Gaussian factors, as proposed in [16]. In the case of forward curve models, the calibration on initial forward products is trivial, it is sufficient to determine the appropriate initial forward curve. However the model parameters have to be estimated. In this part the proposed estimation procedures are not especially original, though the calibration on forward volatilities has not been described, to our knowledge, in previous literature. But the main objective is to highlight the practical problem of calibration due to the complex relationship between parameters and observed products, and also due to the real need to represent both spot prices and forward products.

In a second part, we study the structural model proposed in [2] in terms of forward price reconstruction and its ability to be calibrated. To our knowledge, this topic has not been treated in the literature for structural models, hence this will form the main contribution of this paper. The efficient method of calibration we propose allows to widen the application scope of such a structural model. In particular it opens up the possibility of using structural models for pricing applications.

The paper is organized as follows. Section 2 concerns the study of a 2-factor model and proposes a comparison of calibration results in a simple example of pricing application. Section 3 is focused on a structural model and its ability to represent forward prices and to be calibrated. The conclusion and some perspectives are proposed in Sect. 4.

2   Parameters Estimation for a 2-Factor Model

In this section we study the estimation problem for a very classic factorial model used to represent power prices. The exposed procedures are not original but the objective is to stress the estimation issue once the model is used to represent both spot and forward prices. The factorial representation of the power forward curve was already studied, for example in [18] and justified in [14]. The authors in [17] highlight a decomposition in two factors for modeling power prices in the Norwegian market, with a weak correlation between those two factors. And, in [16] an explicit two-factor model is proposed in the risk-neutral probability:

$$\frac{dF_t(T)}{F_t(T)} = \sigma_s(t)e^{-\alpha(T-t)}dW_t^{(s)} + \sigma_l(t)dW_t^{(l)} \tag{1}$$

where $\alpha \in \mathbb{R}_*^+$ and $\sigma_s(t)$ and $\sigma_l(t)$ are positive integrable functions. This model is very close to the well known Gabillon model [3] and exactly the same for a specific form of $\sigma_l(t)$. In all this paper the time is measured in years, the "mean-reverting" parameter α will then be measured in year^{-1}.

Because of the presence of $e^{-\alpha(T-t)}$ in the first factor we call it the "short-term factor", and the second term will be called the "long-term factor". Also the form of the short-term factor allows us to represent the specific behavior of power prices: increasing volatility when the maturity goes to zero. For simplicity we consider no correlation between the two Brownian motions but all that follows can be easily extended with a non-zero correlation.

The main advantage of using Heath-Jarrow-Morton [13] type factorial models for power prices is its ability to be calibrated on observed forward products by specifying an appropriate initial forward curve $F_0(T)$, and, because of the broad literature of this type of modeling (especially for interest rate models), an easy use for pricing applications. We note also that in the case of commodity price modeling, using HJM framework is simpler than for interest rates because no drift condition must be satisfied, except drift equals zero.

In the following we aim to highlight, however, the estimation issue when, as a power producer, both spot prices and forward products must be well represented. In particular we will expose two different estimation methods.

- The first estimation method is based on the observed forward products where the objective is to fit the volatility curve. In this context we stress the difficulty due to the forward product properties (with a delivery period) that makes its process generally non Markovian and we then propose an approximation on the diffusion process to make the parameters estimation feasible.
- The second estimation method is based on both spot prices and forward products. In this context we describe the spot price model in terms of an "observation-state equations" system to use the classic Kalman filter and estimate the short-term factor parameters. Long-term forward products are used to estimate the long-term volatility $\sigma_l(t)$.

With a simple example of pricing application we propose to stress the high impact of estimation procedures to the indicators of interest. The objective is, in this simple example, to give a performance measurement by comparing the different results to a "benchmark" value. However the objective is not to come to a conclusion on any ranking of calibration methods but only to stress their impact in the context of the power market.

2.1   Method 1: Calibration on Forward Volatilities

In this section we develop a calibration method based on forward price observations. The proposed approach aims to fit the forward price volatilities. Although this approach is classic, the main issue is due to the specificities of power forward prices. Indeed the difference between the forward prices represented in model (1) and the observed forward products (with a delivery period) makes the calibration more complex. In this context we propose some approximations to make the parameters estimation feasible. In particular we propose an approximation of forward product diffusion by a Markovian process.

Equation (1) gives the dynamics of a "unitary" forward price, i.e. a forward price of an instantaneous (or unitary) delivery period. The available observed products are defined by $F_t(T, \theta)$ as the price at time t of 1MWh delivered from T to $T + \theta$, θ being called the "delivery period". Let us consider a discretization time step

h (1 h for example). From Eq. (1) and the assumption of absence of arbitrage opportunity we can deduce the relationship between forward products and unitary forward prices:

$$F_t(T, \theta) = \frac{h}{\theta} \sum_{i=1}^{\frac{\theta}{h}-1} F_t(T + ih) \tag{2}$$

and their dynamics:

$$dF_t(T, \theta) = \frac{h}{\theta} \sum_{i=0}^{\frac{\theta}{h}-1} \left[\sigma_s(t) e^{-\alpha(T+i-t)} dW_t^{(s)} + \sigma_l(t) dW_t^{(l)} \right] F_t(T + ih) \tag{3}$$

The presence of $F_t(T + ih)$ illustrates that, in general, the SDE for $F_t(T, \theta)$ is not Markovian, as already shown in [6], which makes the calibration intricate. The approximation we propose is based on the introduction of shaping factors, defined as follows:

$$\lambda_i^{t,T,\theta} = \frac{F_t(T + ih)}{F_t(T, \theta)}, \quad \forall i = 0, \ldots, \frac{\theta}{h} - 1 \tag{4}$$

These shaping factors can be interpreted as weighting factors applied in hour (or day) i of the delivery period $[T \; ; \; T + \theta]$ with respect to the mean value $F_t(T, \theta)$ of forward prices over $[T \; ; \; T+\theta]$. One can note that the shaping factors are normalized by definition because $\sum_{i=0}^{\frac{\theta}{h}-1} \lambda_i^{t,T,\theta} = \frac{\theta}{h}$. One can also note that these shaping factors are random and depend on the quotation date t. With the introduction of shaping factors the SDE on $F_t(T, \theta)$ can be rewritten:

$$\frac{dF_t(T, \theta)}{F_t(T, \theta)} = \sigma_s(t) e^{-\alpha(T-t)} \Psi(t, T, \theta) dW_t^{(s)} + \sigma_l(t) dW_t^{(l)} \tag{5}$$

with $\Psi(t, T, \theta) = \frac{h}{\theta} \sum_{i=0}^{\frac{\theta}{h}-1} \lambda_i^{t,T,\theta} e^{-\alpha ih}$ being the weighted average of the shaping factors over the delivery period. The fact that the dynamics on $F_t(T, \theta)$ is neither Markovian nor Gaussian is now reflected in the fact that $\Psi(t, T, \theta)$ is random and depends on time t.

Let $(t_n)_{n=0,\ldots,N}$ be a time discretization with $t_0 = 0$ and constant time step[2] $\delta t = t_{n+1} - t_n$. We consider the following approximations.

- The functions $\sigma_s(t)$ and $\sigma_l(t)$ are constant over each interval $[t_n \; ; \; t_{n+1}]$.
- The shaping factors are constant with respect to time t: $\lambda_i^{t,T,\theta} = \lambda_i^{T,\theta}$.

---

[2]The time step δt will be related to the observed prices, therefore δt may be different from the discretization step h of the delivery period.

Therefore the function $\Psi(t, T, \theta) = \Psi(T, \theta) = \frac{h}{\theta} \sum_{i=0}^{\frac{\theta}{h}-1} \lambda_i^{T,\theta} e^{-\alpha ih}$ does not depend on time t and the return $R_n(T, \theta)$ of forward products between t_n and t_{n+1} is given by:

$$R_n(T, \theta) = \log \frac{F_{n+1}(T, \theta)}{F_n(T, \theta)}$$

$$= -\frac{1}{2}\sigma_n^2(T, \theta) + \sigma_s(t_n) e^{-\alpha(T-t_n)} \Psi(T, \theta) \sqrt{v(2\alpha, \delta t)} \varepsilon_n^s + \sigma_l(t_n) \sqrt{\delta t} \varepsilon_n^l$$

with ε_n^s and ε_n^l two independent Gaussian random variables of zero-mean unit-variance and

$$v(a, \delta t) = \frac{e^{a\delta t} - 1}{a}$$

$$\sigma_n^2(T, \theta) = \sigma_s^2(t_n) \Psi^2(t_n, T, \theta) e^{-2\alpha(T-t_n)} v(2\alpha, \delta t) + \sigma_l^2(t_n) \delta t$$

Now suppose $N + 1$ observations $(F_n(T, \theta))_{n=0,\dots,N}$ at dates $(t_n)_{n=0,\dots,N}$ and constant volatility functions $\sigma_s(t) = \sigma_s$ and $\sigma_l(t) = \sigma_l$, we can therefore compute the theoretical forward returns depending on the three parameters σ_s, σ_l and α and the corresponding volatility:

$$V_{th}^2(T, \theta, \alpha, \sigma_s, \sigma_l) = \frac{1}{N} \sum_{n=0}^{N-1} Var\left[R_n(T, \theta)\right] \qquad (6)$$

In the particular case where we assume constant shaping factors $\lambda_i^{T,\theta} = 1$ we obtain:

$$V_{th}^2(T, \theta, \alpha, \sigma_s, \sigma_l) = \Phi(\Delta) \Psi^2(\theta) \sigma_s^2 v(2\alpha, \theta) + \sigma_l^2 \delta t \qquad (7)$$

with $\Delta = Nh$ the quotation period, and

$$\Phi(\Delta) = \frac{1}{N} \frac{1 - e^{-2\alpha\Delta}}{1 - e^{-2\alpha h}} \quad \text{and} \quad \Psi(\theta) = \frac{h}{\theta} \frac{1 - e^{-\alpha\theta}}{1 - e^{-\alpha h}} \qquad (8)$$

The calibration consists in estimating three parameters, σ_s, σ_l and α. A first solution would be to estimate them by maximizing the likelihood function, therefore to estimate parameters that fit as well as possible the observed values of forward returns. However empirical studies have shown that the parameter values are very sensitive to the choice of products considered for the estimation. Instead we propose a calibration method consisting in fitting the volatilities of the observed forward products. More precisely from the observed forward returns $R_n^{obs}(T, \theta) = \log \frac{F_{n+1}(T,\theta)}{F_n(T,\theta)}$ we can compute the empirical volatility:

$$V_{emp}^2(T, \theta) = \frac{1}{N-1} \sum_{n=1}^{N} (R_n^{obs}(T, \theta) - \overline{R}(T, \theta))^2, \quad \overline{R}(T, \theta) = \frac{1}{N} \sum_{n=1}^{N} R_n^{obs}(T, \theta)$$

$$(9)$$

Remark. If derivatives on forward products are available, it is possible to compute the implied (Black) volatility instead of the historical volatility.

The calibration procedure then consists in optimizing the distance between theoretical and empirical volatilities for all observed forward products:

$$(\hat{\alpha}, \hat{\sigma}_s, \hat{\sigma}_l) = \arg\min(\alpha, \sigma_s, \sigma_l) \sum_{(T,\theta)} \left(V_{emp}^2(T, \theta) - V_{th}^2(T, \theta, \alpha, \sigma_s, \sigma_l) \right)^2 \qquad (10)$$

2.2   Method 2: Calibration on Spot Prices and Long-Term Forward Products

By integration of (1) and taking the limit $T \to t$ we obtain:

$$\log S_t = \log F_t(t) = \log F_0(t) - \frac{1}{2}\left[\sigma_s^2 \frac{1 - e^{-2\alpha t}}{2\alpha} + \sigma_l^2 t\right]$$
$$+ \int_0^t \sigma_s e^{-\alpha(t-u)} dW_u^{(s)} + \int_0^t \sigma_l dW_u^{(l)}$$

By noting

$$X_t^s = \int_0^t \sigma_s e^{-\alpha(t-u)} dW_u^{(s)} \quad \text{and} \quad X_t^l = \int_0^t \sigma_l dW_u^{(l)} \qquad (11)$$

We can rewrite the spot price dynamics as a (state - observation equations) system:

$$d(\log S_t) = \left(\frac{\partial F_0(t)}{\partial t} + \mu(t)\right) dt + dX_t^s + dX_t^l \qquad (12)$$

$$dX_t^s = -\alpha X_t^s dt + \sigma_s dW_t^{(s)} \qquad (13)$$

$$dX_t^l = \sigma_l dW_t^{(l)} \qquad (14)$$

with $\mu(t) = \frac{1}{2}\left[\sigma_s^2 e^{-2\alpha t} + \sigma_l^2\right]$.

The drift part can be treated as a seasonality component of the spot price. Its estimation can be made by a deseasonalization step. In the following this seasonality will be represented by seven daily parameters, 12 monthly parameters and one parameter per year, which are estimated by a classic linear regression, with additional constraints of normalization for the daily and monthly parameters. After this deseasonalization step, the maximum likelihood estimation of $(\sigma_s, \sigma_l, \alpha)$ can proceed from the residual by using a Kalman filter [12, 15] to compute the likelihood. In order to also use forward products in the calibration we propose to first estimate σ_l from long-term forward products (year-ahead or season-ahead, depending on the market) with the approximation:

$$\frac{dF_t(T,\theta)}{F_t(T,\theta)} \approx \sigma_l dW_t^l, \quad \text{if } T - t >> 0 \tag{15}$$

The maximum likelihood estimation from spot prices then proceed to estimate the short-term parameters σ_s and α, with this pre-estimated σ_l.

2.3 Results

This section shows an illustration on the impact of the calibration methods to a simple pricing application. As already said the objective is not to give a ranking of calibration methods, but only to stress and quantify the difference of results, in terms of value for a simple portfolio, due to the choice of the calibration method.

2.3.1 Data Set

We consider a portfolio composed of a strip of European options on forward products. More precisely, we consider 24 European options on monthly forward products: product "April-2013" to product "March-2015". The date of pricing is $t_0 =$ March 12th, 2013 and all the options are at the money. We consider two different markets: the UK power market and the French power market. The main advantage of considering options on monthly forward products is the possibility to have a "benchmark" value. Indeed, one can consider a model directly on forward monthly products, as proposed in [6], calibrated on observed empirical volatilities. let us denote by $M_t = T - t$, then the benchmark model can be written as follows:

$$\frac{dF_t(t + M_t)}{F_t(t + M_t)} = \sigma(M_t)dW_t \tag{16}$$

where $\sigma(M_t)$ is a piecewise constant function fitting exactly the empirical forward volatilities.

We consider 1 year of historical data for the calibration on forward products. The products used for calibration depend on the market.

- The products used for the calibration in the UK power market are: 1 to 4 Week-ahead, 1 to 4 Month-ahead, 1 to 4 Quarter-ahead and 1 to 6 Season-ahead.
- The products used for the calibration in the French power market are: 1 Week-ahead, 1 to 3 Month-ahead, 1 to 3 Quarter-ahead and 1 to 2 Year-ahead.

The shaping factors $\lambda_i^{T,\theta}$ are all considered equal to 1. This strong approximation is only used for the calibration purpose because it has no significant impact on the reconstructed volatilities.

Concerning the calibration on spot prices, we consider 2 years of historical data to estimate the seasonality part and 1 year of the residual signal to estimate the short term parameters σ_s and α. The long term volatility σ_l is estimated as an averaged volatility of the 1 and 2 year-ahead products for the French power market, and as an averaged volatility of the 1 to 6 Season-ahead products for the UK power market.

2.3.2 Calibration Results

Table 1 shows the estimated parameters on the UK power market with respect to the calibration method and Fig. 1 illustrates the resulting forward volatilities reconstructed by the 2-factor model in comparison to the empirical forward volatilities. We can note that the empirical volatilities do not seem to decrease monotically with the maturity. This effect can be mainly explained by the overlapping delivery period of the forward products. Concerning the reconstructed volatility curves we can observe that the long-term volatility values are similar due to the quasi-similarity of its estimation procedure. The most important point is the difference in value for the

Table 1 Estimation results on UK and French power: estimated parameters with respect to the calibration method

Parameter	Calibration on forward volatilities		Calibration on spot prices	
	UK power market	French power market	UK power market	French power market
σ_s (%)	19.1	45	84.5	302
$\alpha(Y^{-1})$	1.37	8.73	162.65	88.15
σ_l (%)	9.8	11	9.8	11

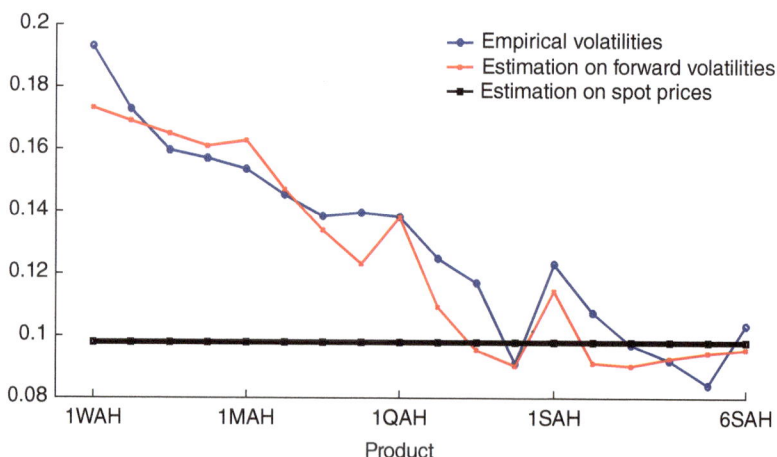

Fig. 1 Estimation results on UK power market: empirical volatility (*blue line with circles*), reconstructed volatility of the 2-factor model calibrated on forward volatilities (*red line*) and on spot prices (*black line with squares*)

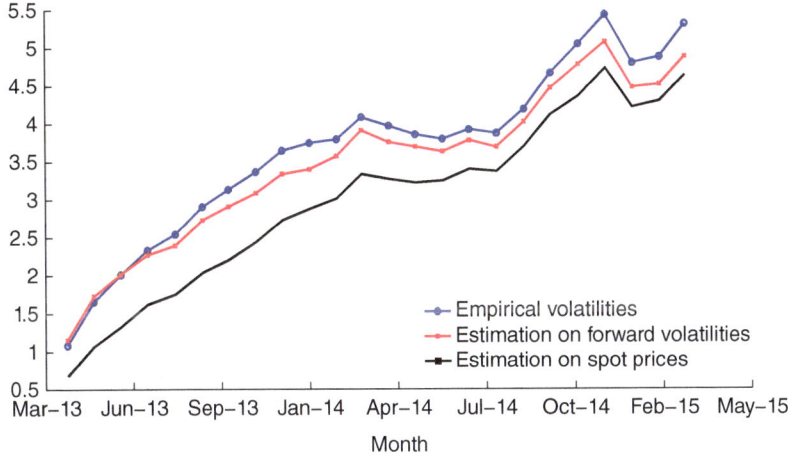

Fig. 2 European option pricing on UK power market: benchmark value (*blue line with circles*), values from the 2-factor model calibrated on forward volatilities (*red line*) and on spot prices (*black line with squares*)

short-term parameters: a factor 4 in σ_s and a factor 100 in α from the estimation on forward volatilities to the estimation on spot prices. In the case of calibration on spot prices, the parameter α can be interpreted as a "mean-reverting" coefficient driving the spot prices. Its estimated value then shows that the spot price presents highly mean-reverting behavior, this being mostly due to the presence of spikes. This high value of α leave a single constant volatility factor (the long-term factor) to fit the whole forward volatility term structure. In this case the estimated forward volatility curve, as shown in Fig. 1, is nearly completely flat and the well known Samuelson effect cannot be captured. On the other hand, because the same parameter α drives the decreasing speed of the forward volatility curve, it becomes obvious that the estimated values are completely different and depend on the calibration method.

In Fig. 2 we illustrate the impact of the estimation methods in the value of the "toy" portfolio. Compared to the benchmark value, we notice a weak error with the value computed from the two-factor model calibrated on forward volatilities. This confirms the results of [17]: two factors can be sufficient to represent forward products. However this does not take into account the need to also represent spot prices. And, as shown in Fig. 2, the resulting value, when the two-factor is used with parameters estimated on spot prices, highly underestimates the benchmark value with an error of around 20 %. Similar remarks can be made in the French power market (Table 1, Figs. 3 and 4) where the error can reach 30 % with the two-factor model calibrated on spot prices.

The main conclusion is not to reject the calibration on spot prices, because the chosen application context (pricing European options on forward products) is completely adapted to a calibration on forward volatilities. This context allowed us to build a benchmark value and then to make an objective comparison of calibration methods. Another consideration, for example with a portfolio exposed on spot

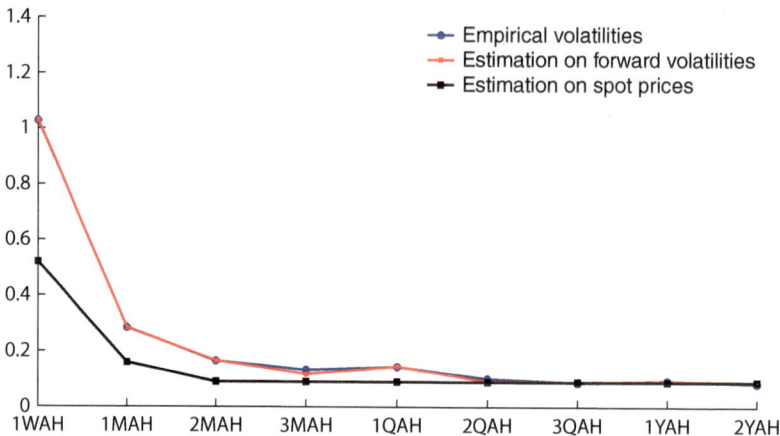

Fig. 3 Estimation results on French power market: empirical volatility (*blue line with circles*), reconstructed volatility of the 2-factor model calibrated on forward volatilities (*red line*) and on spot prices (*black line with squares*)

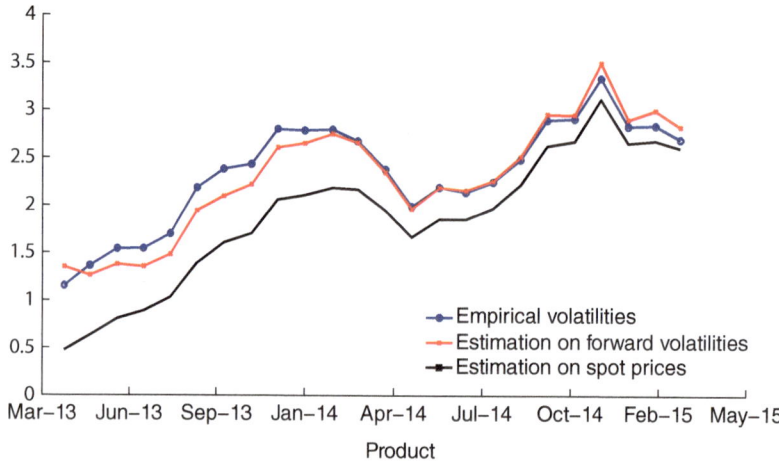

Fig. 4 European option pricing on French power market: benchmark value (*blue line with circles*), vale from the 2-factor model calibrated on forward volatilities (*red line*) and on spot prices (*black line with squares*)

prices, would have shown that the calibration on spot prices is more adapted, but in this case a benchmark value cannot be built easily and is still an open problem. The main conclusion is that, if the objective is to represent both spot forward prices, two factors are not sufficient and, before the calibration methodology, a study of more complex models is necessary. For example, adding at least a decay factor (i.e. a second mean-reverting parameter) in the long-term factor would allow us to capture both mean-reverting behaviour of spot prices and the Samuelson effect on the forward volatility curve. However this change of model will create a decreasing volatility toward 0, which may be contrary to the observed volatility curves (see, for example, Fig. 3 which shows that the forward volatility curve seems to converge toward a positive value). Another solution would be to add a third factor of the same form as the short-term factor, therefore to have two mean-reverting parameters as previously and to keep a non-zero limit value of the forward volatility level. However the issue of calibration will remain and increase with an increasing number of factors.

3   Calibration of a Structural Model

In this section we study the second class of commonly used models for power prices: the class of structural models. This approach is more recent and adapted to the stylized facts of electricity. In particular it allows to represent a strong relationship between the power spot price and the factors that explain it: the fuel spot prices, the demand and the production capacities. This section focuses on a particular structural model, proposed first in [1] and modified in [2]. By definition the model is adapted to represent power spot prices. However, to our knowledge, there is no literature on the performances of this kind of models in terms of forward price representation. This is the first objective of this section and, as a natural sequel, we address the issue of calibration on observed forward prices.

3.1   Reminder of the Model

The structural risk-neutral model we study in this paper is a modified version of the one introduced in [1], and its complete presentation can be found in [2]. In this section, we recall the approach and the main results obtained by the authors.

3.1.1   Approach and Main Results

The model is derived from the aggregation of two essential observations. On the one hand, when the market is not in a period of stress, the price of the marginal fuel of the generation system will be the dominant part of the electricity spot price. On the other hand, at times of market stress, the well-known spikes of the electricity spot prices will occur when the demand reaches the system maximal capacity. In this model, such behaviour is captured by a "scarcity" function, that will explode to form the prices spikes, thus leading to the following form for the spot price:

$$S_t = g\left(C_t^{max} - D_t\right) \sum_{i=1}^{n} h^i S_t^i \mathbf{1}_{\{D_t \in I_t^i\}}, \tag{17}$$

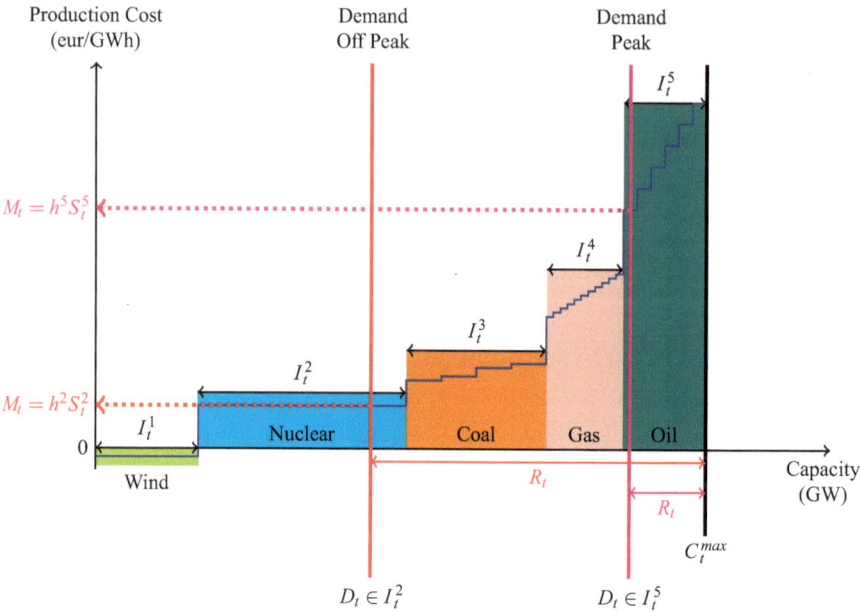

Fig. 5 Illustration scheme of electricity price construction

with D_t the demand, C_t^i, $i = 1, \ldots, n$ the capacities of the fuels used in the production system, h^i the corresponding heat rates, S_t^i the corresponding spot prices of fuels, I_t^i the capacity interval where fuel i is marginal (see Fig. 5), $C_t^{max} = \sum_{k=1}^{n} C_T^k$ the total capacity of the production system, and g the "scarcity" function:

$$g(x) = \min\left(M, \frac{\gamma}{x^v}\right) \mathbf{1}_{x>0} + M \mathbf{1}_{x<0}.$$

Equation (17) that defines the model can be summarized in a simpler way: when the demand is in the marginal interval of the ith fuel, the spot price of electricity is equal to the cost needed to produce 1MWh of electricity from the ith fuel, times the scarcity factor.

Remark. In this model the merit order of fuels is assumed to be fixed. For this study we keep the same assumption for more simplicity and also in order to keep acceptable computation time.

This spot model has been backtested, using historical data for demand, capacities, and fuel spot prices. The parameter M of the scarcity function is estimated so as to roughly match the high cap on electricity spot price, defined by the market as 3,000 €MWh.[3] Its estimated value is 30. The other parameters of the scarcity function

---

[3]http://www.epexspot.com/en/product-info/auction/france.

Table 2 Hourly parameters of the scarcity function g

Hour (h)	γ	ν	Hour (h)	γ	ν
0	0.44634	2.5137	12	0.73737	6.517
1	0.67993	4.0184	13	0.53414	4.1539
2	0.82051	5.2762	14	0.6109	4.6532
3	0.86753	5.0142	15	0.68395	5.0617
4	0.84629	4.7554	16	0.9543	8.9007
5	0.68841	4.2384	17	1.5229	29.843
6	0.7833	6.4935	18	1.8399	57.121
7	0.97349	12.1802	19	1.1405	14.1943
8	0.77457	7.5105	20	1.0153	9.7041
9	0.8497	8.521	21	0.56393	3.8245
10	0.72403	6.5117	22	0.55286	3.973
11	0.63956	5.6207	23	0.6688	4.6805

have been estimated to best fit the model price with the historical spot prices, for each hour of the day. The results can be seen in Table 2 and more performance illustrations can be found in [2].

The main objective is not to pursue the performance evaluation of the spot model. We intend to take a closer look at the underlying structural relationship for the forward prices, and then we start by recalling the pricing methodology adopted for this model. The presence of the demand and capacities implies an incomplete market setting, and thus an infinity of no-arbitrage prices for any derivative or, equivalently, an infinite number of risk-neutral measures. The criterion used to value an electricity derivative for this model is the Local Risk Minimization approach, which allows us to choose a risk-neutral measure $\hat{\mathbb{Q}}$. Technical details concerning the choice of this measure, which uses the Local Risk Minimization principle (introduced in [11]), can be found in the original paper. The main results are detailed in Appendix 1 and lead to a no-arbitrage price for a forward price $F_t^e(T)$ on electricity that takes the form:

$$F_t^e(T) = \mathbb{E}^{\hat{\mathbb{Q}}}[S_T | \mathscr{F}_t],$$

with $\mathscr{F}_t = \mathscr{F}_t^S \vee \mathscr{F}_t^C \vee \mathscr{F}_t^D$ the filtration representing the market information at time t, which is the filtration generated by the randomness of the fuels, the demand and the capacities, and $\hat{\mathbb{Q}}$ obtained using Local Risk Minimization, satisfying $\hat{\mathbb{Q}} = \mathbb{P}$ on $\mathscr{F}_t^C \vee \mathscr{F}_t^D$. Consequently, the forward price takes the simple form:

$$F_t^e(T) = \sum_{i=1}^{n} h^i G^i(t, T, C_t, D_t) F_t^i(T),$$

$$G^i(t, T, C_t, D_t) = \mathbb{E}\left[g(C_T^{max} - D_T)\mathbf{1}_{D_T \in I_T^i} | \mathscr{F}_t^{D,C}\right].$$

If we want to reconstruct the real forward prices that are exchanged on electricity markets, we need to average the forward price with instantaneous delivery over the delivery period $[T, T + \theta]$ leading to the final form:

$$F_t(T, \theta) = \sum_{i=1}^{n} \underbrace{\left(\frac{1}{\theta} \sum_{T'=T}^{T+\theta} G^i(t, T', C_t, D_t) \right)}_{\text{stochastic weights}} h^i F_t^i(T') \qquad (18)$$

This result is the core of this study: starting from a spot model, we necessarily have a result on forward prices involving the forward prices on fuels, the demand process and the capacities processes. In the following we will study the performances of this forward relationship as well as classic studies on spot prices like for example in [1, 8]. Because this relationship shows the historical probability of demand D_t and capacities C_t^i the global model needs a specification of their dynamics.

3.1.2 Demand and Capacities Modeling

We now need to model the behaviour of the electricity demand and capacities that we take into account in the model. We follow the same model introduced in [2], and thus we decide to decompose the demand and capacities processes into two parts: a deterministic part $f_*(t)$ and a stochastic part $Z_*(t)$:

$$D_t = f_D(t) + Z_D(t),$$
$$C_t^i = f_i(t) + Z_i(t), \text{ for } i = 1, \ldots, n$$

The deterministic part will model the seasonal trend of the demand or capacities, while the stochastic part will capture the randomness of these processes. We choose to slightly modify the original model for the seasonality functions. For all the processes, we will take into account the yearly seasonality, and a week trend, that will capture the trend of every hour of the week, leading to a 168 parameters. This can be interpreted as a week scheme, that will be reproduced all year long, following a yearly seasonality drift. We decide to assume:

$$f_D(t) = \mathbf{week}_D(t) + d_1 + d_2 \cos\left(2\pi(t - d_3)\right),$$
$$f_i(t) = \mathbf{week}_i(t) + c_1^i + c_2^i \cos\left(2\pi(t - c_3^i)\right) \text{ for } i = 1, \ldots, n$$

We keep the Ornstein-Uhlenbeck form of the stochastic parts, leading to:

$$dZ_D(t) = -\alpha_D Z_D(t)dt + \sigma_D dW_t^D, \qquad (19)$$
$$dZ_i(t) = -\alpha_i Z_i(t)dt + \sigma_i dW_t^i, \text{ for } i = 1, \ldots, n \qquad (20)$$

3.2   Results on Reconstructed Forward Prices

3.2.1   Description of the Dataset

The whole dataset consists in 4 years of historical data from January 1st, 2009 to December 31st, 2012. Data from Demand and capacities have been retrieved from Réseau de Transport Electrique (RTE,[4]) whereas data of fuel and electricity prices come from Platts.[5] We consider a simple case with three types of production units: nuclear, gas/coal and oil power plants. This allows us to consider a fixed ranking of production cost, the cheapest cost being for nuclear plants and the most expensive being for the oil power plant. Carbon emission taxes are taken into account. Table 3 presents the datasets used in the estimation. It is interesting to note that we used day-ahead demand, because the demand, as it appears in the spot model, is used to model the spot price of electricity, which is a day-ahead price as well. Thus, in this forward framework, we need to stay consistent with the spot model, and use day-ahead demand.

3.2.2   Estimation Results for Demand and Capacities Parameters

To estimate the parameters of the demand or capacities processes, we follow the same framework: first we estimate the parameters of the seasonality function, and then we estimate the Ornstein-Uhlenbeck parameters. The seasonality estimation is done by using classic statistical tools like linear regression methods and the details can be found in Appendix 2. For the stochastic parameters, we used ordinary least squares to estimate the parameters of the Ornstein-Uhlenbeck processes, securing confidence intervals (Table 4).

Table 3 Description of the dataset

Name	Source	Data	Frequence/type	Dates covered/value
Demand	RTE	D_t	Hourly	2009–2012
Nuclear capacity		C_t^1		
Coal+gas capacity		C_t^2		
Oil capacity		C_t^3		
Fuel forwards	Platts	$F_t^i(T)$	Daily	2009–2012
Nuclear heat rate		h_1	Constant	$0.84.10^{-4}$
Coal+gas heat rate		h_2		0.45
Oil heat rate		h_3		1.5

---

[4] www.rte-france.com.

[5] www.platts.com.

200
O. Féron and E. Daboussi

Table 4 Demand and capacities parameters estimated from the whole dataset

Parameter	Estimate	Confidence interval	Parameter	Estimate	Confidence interval
d_1	0.37389	0.3464-0.40137	c_1^2	4.6952	4.6855-4.7049
d_2	1.967	1.9123-2.0218	c_2^2	1.3929	1.3753-1.4105
d_3	0.14137	0.14095-0.14181	c_3^2	0.19436	0.19345-0.19529
α_D	32.7958	27.2804-38.3112	α_2	21.367	16.7558-25.9782
σ_D	16.7316	16.1176-17.4223	σ_2	5.0362	4.8514-5.2441
c_1^1	7.7918	7.785-7.7985	c_1^3	49.6449	49.6182-49.6716
c_2^1	0.91707	0.90447-0.9297	c_2^3	7.0054	6.9545-7.0563
d_3^1	0.18299	0.18218-0.18382	c_3^3	0.17465	0.17428-0.17501
α_1	26.4656	21.4961-31.4351	α_3	7.842	4.8483-10.8358
σ_1	4.8794	4.7003-5.0808	σ_3	11.7357	11.305-12.2201

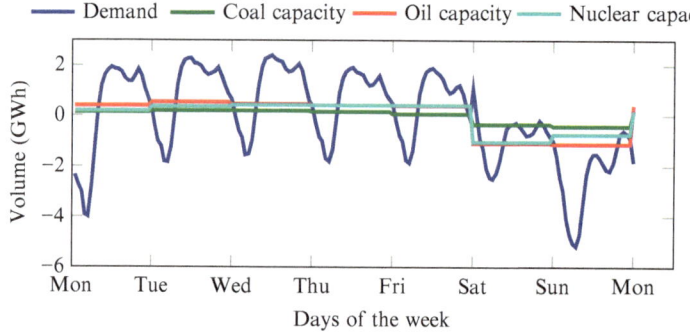

Fig. 6 Estimated **week** parameters

Figure 6 shows the estimated values of the weekly parameter for demand and capacities. It underlines the fact that no weekly seasonality can be seen for the capacities, while it is a major feature of the demand.

3.2.3 Reconstructed Forward Prices

In order to study the performances of forward price reconstruction we implemented the following algorithm: at each date t from January 1st, 2009 to December 31st, 2012,

1. Estimate parameters of demand and capacities processes from 2 years of historical data.
2. Consider the observed forward fuel prices $F_t^i(T, \theta)$ for all observable (T, θ) and assume $F_t^i(T') = F_t^i(T, \theta)$ for all $T' \in [T, ; T + \theta]$
3. Compute Eq. (18) to build the electricity forward price and compare to the observed one.

The algorithm is implemented in Matlab 2010a on a laptop.[6] The computation time is mainly due to the computation of the functions H and $\mathscr{G}$ defined in the section "Computing the Stochastic Weights" in Appendix 1. If we use the series expansion approach proposed in [2], based on the extended incomplete gamma function, the reconstruction of the 1-Year-ahead product for only one date takes 21 s. If, instead, we approximate the functions H and $\mathscr{G}$ by Monte Carlo approach with 400 samples,[7] the reconstruction of the 1-year-ahead product takes 2.3 s.

Figure 7 shows the reconstruction results for 1-month-ahead, 1-quarter-ahead and 1-year-ahead products. We can remark satisfactory reconstruction of the 1-month-ahead product, capturing level and seasonality quite well. The fundamental relationship (18), i.e. the link between electricity forward prices and an expectation of demand and capacities levels, seems dominant in the explanation of the 1-month-ahead. The 1-year-ahead reconstruction is also efficient, but we have to note a level underestimation for the period 2009 to 2010. Although no changes in demand and capacities have been observed at the end of 2010, the reconstruction results are more efficient during the period 2011 to 2012. There is no explanation at this time about this particular change of behavior and it remains an open question. Reconstruction results are less efficient for the 1-quarter-ahead product, where, in particular, the level is not well captured. The relationship (18) is then not the fundamental element that drives the price of this product. We must further investigate the comprehension of the market actors to understand this particular effect.

3.3 Calibration

In the previous section, we have seen the ability of the model when it comes to reconstructing the forward prices. But a new question arises: is it possible, with this model, to reproduce exactly the forward prices that we can observe, at a certain time? This question is extremely important on the markets, as it is very important to be able to fit a model to the real prices in order to avoid any arbitrage possibility. In this section, the aim will be to calibrate the model on the forward contracts that we can observe on the markets, which is completely different from a standard spot model calibration: indeed, if we follow an implied volatility framework, we need to find a parameter of the model that makes the price a strictly monotonous one, and thus we can obtain an implied parameter, at each calibration date, for the price. This is feasible, as the model has a lot of parameters, and that their behavior is flexible, but it does not actually give us the calibration that we are looking for. When it comes

---

[6]Intel(R) Core(TM) i3-2375M CPU @ 1.50 GHz.

[7]The number of samples is empirically chosen so as to obtain a difference between the series expansion and Monte Carlo computations lower than 0.1€ on the reconstructed price.

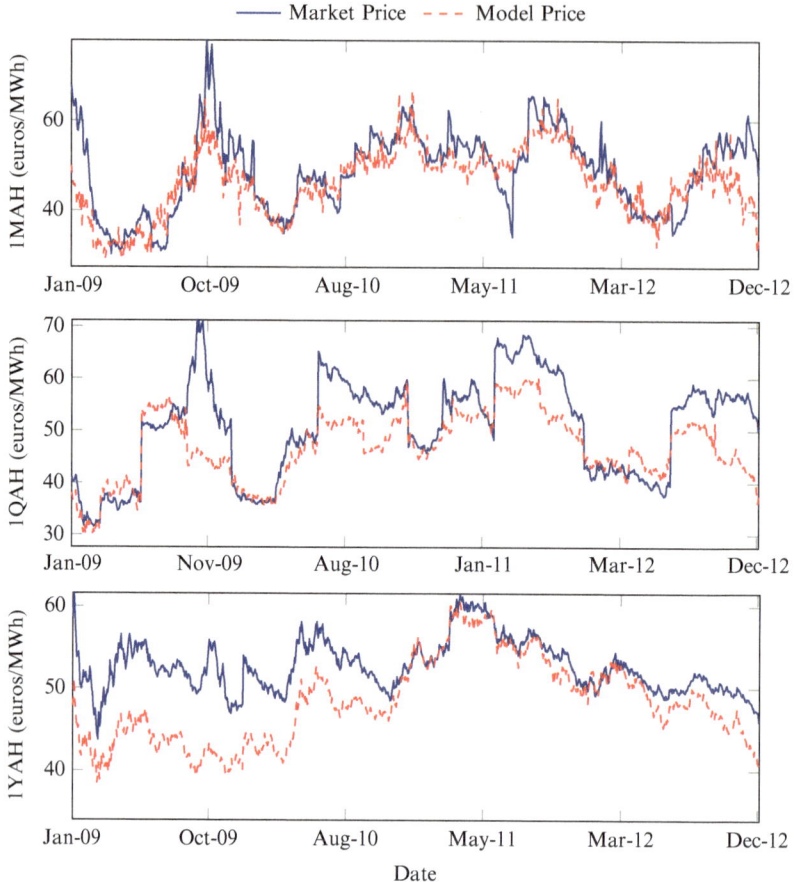

Fig. 7 Results of forward product reconstruction: observed (*solid-blue line*) and reconstructed (*dashed-red line*) 1 month-ahead (*top*), 1-quarter-ahead (*middle*) and 1-year-ahead (*bottom*) prices from January 1st, 2009 to December 31st, 2012

to calibrating on forward contracts, the problem is much more complicated, we have to calibrate the model to fit a given curve: the forward curve, that contains all the available contracts at a certain date.

3.3.1 Adjusting the Demand Parameter

In our model, we looked closely at the influence of the parameters of the demand and capacities processes, and we found the deterministic part of the demand to be a pertinent adjustment parameter, as it was already proposed in [9]. Let us recall the model for the demand process:

$$D_t = f_D(t) + Z_D(t)$$

In this case the expected value of D_T conditional to $\mathcal{F}_t$:

$$\mathbb{E}\left[D_T | \mathcal{F}_t^{D,C}\right] = f_D(T) + e^{-\alpha_D(T-t)}(D_t - f_D(t)) := m_{t,T}^D$$

The introduction of a non-zero long-term mean $\varepsilon \in \mathbb{R}$ in the process of $Z_D(t)$:

$$dZ_D(t) = \alpha_D(\varepsilon - Z_D(t))dt + \sigma_D dW_t^D$$

will only affect the mean of the demand process, conditional to $\mathcal{F}_t$:

$$\mathbb{E}\left[D_T | \mathcal{F}_t^{D,C}\right] = m_{t,T}^D + \varepsilon(1 - e^{-\alpha_D(T-t)}) \tag{21}$$

This modification may be interpreted as a change of probability for the demand process, leading to a difference between the risk-neutral and the historical probabilities. In this case the resulting pricing measure is no longer the one obtained by Local Risk Minimization.

The stochastic weights (the details of calculus are available in Appendix 1) are all affected:

$$G^1(\varepsilon, t, T, D_t, C_t) = H(m_{t,T}^2 + m_{t,T}^3, m_{t,T}^1 - m_{t,T}^D - \varepsilon(1 - e^{-\alpha_D(T-t)}), \sigma_2^3, \sigma_1^{1,D})$$

$$G^2(\varepsilon, t, T, D_t, C_t) = H(m_{t,T}^3, m_{t,T}^1 + m_{t,T}^2 - m_{t,T}^D - \varepsilon(1 - e^{-\alpha_D(T-t)}), \sigma_3^3, \sigma_1^{1,D})$$

$$-H(m_{t,T}^2 + m_{t,T}^3, m_{t,T}^1 - m_{t,T}^D - \varepsilon(1 - e^{-\alpha_D(T-t)}), \sigma_2^3, \sigma_1^{1,D})$$

$$G^3(\varepsilon, t, T, D_t, C_t) = \mathcal{G}(m_{t,T}^1 + m_{t,T}^2 + m_{t,T}^3 - m_{t,T}^D - \varepsilon(1 - e^{-\alpha_D(T-t)}), \sigma_1^{3,D})$$

$$-H(m_{t,T}^3, m_{t,T}^1 + m_{t,T}^2 - m_{t,T}^D - \varepsilon(1 - e^{-\alpha_D(T-t)}), \sigma_3^3, \sigma_1^{1,D})$$

with $m_{t,T}^i = \mathbb{E}\left[C_{T'}^i | \mathcal{F}_t^{D,C}\right]$, $i = 1, 2, 3$ and functions H and $\mathcal{G}$ defined in Appendix 1. We also remark that the function $\varepsilon \longmapsto F_t(T, \theta, \varepsilon)$ is strictly increasing, and that we have the following asymptotical results:

$$\lim_{\varepsilon \to +\infty} F_t(T, \theta, \varepsilon) = M h_n F_t^n(T) \simeq 3000 \text{ eur/MWh}$$

$$\lim_{\varepsilon \to -\infty} F_t(T, \theta, \varepsilon) = 0$$

This shows that, for any contract $F_t^{obs}(T, \theta)$ observed at time t, we can find a unique value of ε able to exactly reproduce it. We can use a dichotomic algorithm to solve the equation $F_t(T, \theta, \varepsilon) = F_t^{obs}(T, \theta)$, as our interest function is strictly monotonous, but the regularity of this function in fact allows us to use the Newton-Raphson algorithm, which gives a quadratic convergence instead of a linear convergence, using the derivatives of $F_t(T, \theta, \varepsilon)$:

$$\begin{cases} \varepsilon_0 = 0 \\ \varepsilon_{n+1} = \varepsilon_n - \dfrac{F_t(T,\theta,\varepsilon) - F_t^{obs}(T,\theta)}{\frac{\partial F_t}{\partial \varepsilon}(T,\theta,\varepsilon)} \end{cases}$$

To compute the derivatives of $F_t(T, \theta, \varepsilon)$, we compute the derivatives of the weights:

$$\frac{\partial G^1}{\partial \varepsilon} = -(1 - e^{-\alpha_D(T-t)}) \left[\frac{\partial H}{\partial x_2} (m^2_{t,T} + m^3_{t,T}, m^1_{t,T} - m^D_{t,T} - \varepsilon(1 - e^{-\alpha_D(T-t)}), \sigma^3_2, \sigma^{1,D}_1) \right]$$

$$\frac{\partial G^2}{\partial \varepsilon} = -(1 - e^{-\alpha_D(T-t)}) \left[\frac{\partial H}{\partial x_2} (m^3_{t,T}, m^1_{t,T} + m^2_{t,T} - m^D_{t,T} - \varepsilon(1 - e^{-\alpha_D(T-t)}), \sigma^3_3, \sigma^{1,D}_1) \right. $$

$$\left. - \frac{\partial H}{\partial x_2} (m^2_{t,T} + m^3_{t,T}, m^1_{t,T} - m^D_{t,T} - \varepsilon(1 - e^{-\alpha_D(T-t)}), \sigma^3_2, \sigma^{1,D}_1) \right]$$

$$\frac{\partial G^3}{\partial \varepsilon} = -(1 - e^{-\alpha_D(T-t)}) \left[\frac{\partial \mathcal{G}}{\partial x_1} (m^1_{t,T} + m^2_{t,T} + m^3_{t,T} - m^D_{t,T} - \varepsilon(1 - e^{-\alpha_D(T-t)}), \sigma^{3,D}_1) \right. $$

$$\left. - \frac{\partial H}{\partial x_2} (m^3_{t,T}, m^1_{t,T} + m^2_{t,T} - m^D_{t,T} - \varepsilon(1 - e^{-\alpha_D(T-t)}), \sigma^3_3, \sigma^{1,D}_1) \right]$$

The derivatives of the functions $\mathcal{G}$ and H can be found in the appendix of the original paper [2]. In practice we define a function $\varepsilon(T)$ added in the deterministic part of the demand process in order to calibrate the model on all observable forward products. This function will be piecewise constant, with constant parts inside the delivery periods.

3.3.2 Calibration Results

The results of the calibration are given in Figs. 8 and 9. Figure 8 shows an example of estimated $\varepsilon(T)$ at date June 28th, 2011 considering the observable baseload forward products in the French power market: 1 to 6 Month-ahead, 1 to 3 Quarter-ahead and 1 Year-ahead. This example of result shows that the bias is more important for small maturities, but is reasonable (-1.5 GW in maximum) compared to the total available capacity in France (between 80 and 130 GW). When it comes to contracts with a longer granularity, the calibration shows that the values of ε needed to fit the model given by the model, with the real prices, are small, especially for the 1YAH contract. In Fig. 9 we repeated the calibration procedure from January 1st, 2011 to December 31st, 2012 (with a weekly frequency) for three values of $\varepsilon(T)$: the ones needed to exactly retrieve the 1 Month-ahead, 1 Quarter-ahead and 1 Year-ahead products. This result confirms the previous remarks on a decreasing level of ε with an increasing maturity. It also confirms that the resulting values are acceptable (less than 3 GW) compared to the total capacity.

4   Conclusion

In this paper we exposed specific calibration issues for electricity price models. In the first part we stressed that interest rate models, currently used in practice for electricity price modeling, present additional difficulties for calibration. This is mostly

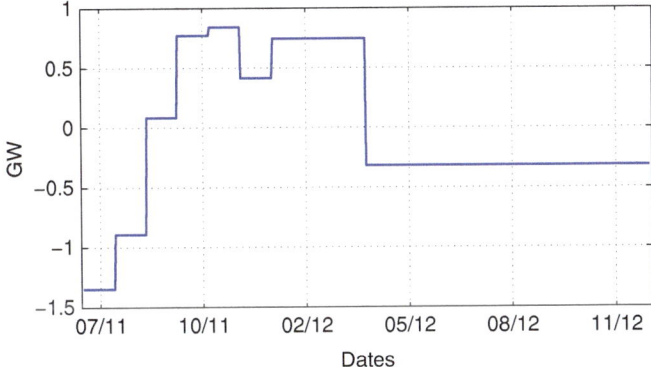

Fig. 8 Calibrated $\varepsilon(T)$ on observed baseload forward prices (1MAH → 6MAH, 1QAH → 3QAH and 1YAH in French power market) at date June 28th, 2011

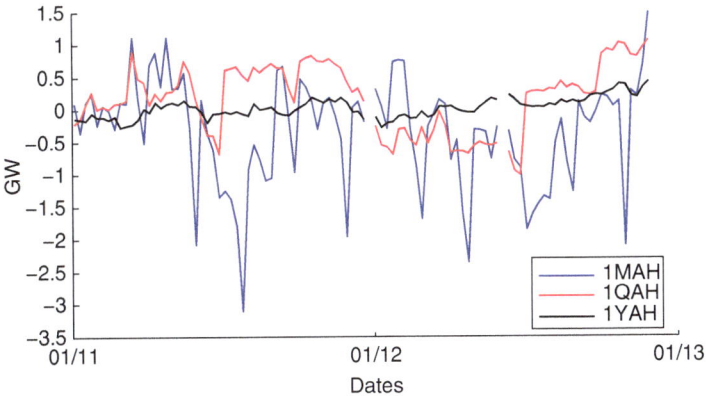

Fig. 9 Evolution of calibrated ε for 1MAH (*blue*), 1QAH (*red*) and 1YAH (*black*) from January 1st, 2011 to December 31st, 2012

due to the non storability of electricity and, hence, the presence of a delivery period in the observed forward products. In this specific context, the calibration procedures must introduce some approximation in the diffusion processes to make the parameter estimation feasible. Also, when the objective is to represent both spot prices and forward products, as an electricity producer aims to, the previous studies [17] about a sufficient number of factors to represent all the products, must be revisited.

In the second part we proposed an original study of how a structural model for electricity prices, initially dedicated to a good representation of spot prices, is able to model forward products. This kind of study is essential for practitioners and can help in the modeling choice. We proposed an easy algorithm, with a modification of the demand model in the risk-neutral probability, to calibrate the model from observed forward products. This idea has already been proposed in [9], for example, and we show how to use it in the structural model described in [2].

In further work the forward price reconstruction will be deepened by introducing a price confidence interval induced by the uncertainties on estimated parameters for demand and capacities processes. This will allow us to measure the impact of these uncertainties to reconstructed forward products. The problem of fixed merit order, as it is assumed in [2], will be addressed as a direct extension of the model. Also, following the calibration results, the future objective will be to price specific power derivatives like, for example, a power plant and study the resulting hedging strategy.

Appendix 1: Structural Model Description

In this appendix, we recall the main results concerning the model, and we detail the computation of the forward prices and stochastic weights. We refer to [1, 2] for more details and proofs.

Forward Pricing

In this section, we detail the computation of a forward contract $F_t(T, \theta)$. We first need to compute the forward price $F_t(T')$, for any $T' \in [T, T + \theta]$, and thus we need a pricing formula, i.e. we need to take a closer look at the EMM that we will use. To do so, we consider the submarket composed by the ith fuel only: assuming this submarket to be complete, there is a unique risk-neutral measure $\mathbb{Q}^i$ that is a risk-neutral measure for the ith fuel. It is shown, in the original paper, that this measure is one of the EMM for the spot price of electricity S_t. Also, the Local Risk Minimization approach, used in this structural model, leads to a Föllmer and Schweizer minimal EMM $\hat{\mathbb{Q}}$ that corresponds to zero risk premiums for the demand and the capacities. In other words, $\hat{\mathbb{Q}}$ is an EMM for the fuels, and coincides with the historical measure $\mathbb{P}$ for the demand and capacities.

These remarks, along with the mutual independence between the demand, fuel prices and capacities, lead to:

$$F_t^e(T') = \hat{\mathbb{E}}[S_{T'}|\mathscr{F}_t] = \hat{\mathbb{E}}\left[g\left(C_{T'}^{max} - D_{T'} \right) \sum_{i=1}^{n} h_i S_{T'}^i \mathbf{1}_{D_{T'} \in I_{T'}^i} \middle| \mathscr{F}_t \right]$$

$$= \sum_{i=1}^{n} h_i \mathbb{E}\left[g\left(C_{T'}^{max} - D_{T'} \right) \mathbf{1}_{D_{T'} \in I_{T'}^i} \middle| \mathscr{F}_t \right] \mathbb{E}^i\left[S_{T'}^i | \mathscr{F}_t \right]$$

$$= \sum_{i=1}^{n} h^i G^i(t, T, C_t, D_t) F_t^i(T) ,$$

for any $T' > t$. To obtain the real forward price with delivery period $[T, T + \theta]$, we only have to take the mean of the instantaneous prices $F_t(T')$, for any $T' \in [T, T+\theta]$. The next section will focus on the computation of the stochastic weights.

Computing the Stochastic Weights

Using the models introduced in Sect. 3.1.2 for the demand and capacities, the random variables $D_{T'}$ et $C_{T'}^i$ are Gaussian, conditional to the filtration $\mathscr{F}_t^{D,C}$, and we have:

$$\begin{aligned}
\mathbb{E}\left[D_{T'}|\mathscr{F}_t^{D,C}\right] &= m_{t,T'}^D = f_D(T') + e^{-\alpha_D(T'-t)}(D_t - f_D(t)) \\
\mathbb{V}\mathrm{ar}\left[D_{T'}|\mathscr{F}_t^{D,C}\right] &= \left(\sigma_{t,T'}^i\right)^2 = \frac{\sigma_D^2}{2\alpha_D}\left[1 - e^{-2\alpha_D(T'-t)}\right] \\
\mathbb{E}\left[C_{T'}^i|\mathscr{F}_t^{D,C}\right] &= m_{t,T'}^i = f_i(T') + e^{-\alpha_i(T'-t)}(C_t^i - f_i(t)) \\
\mathbb{V}\mathrm{ar}\left[C_{T'}^i|\mathscr{F}_t^{D,C}\right] &= \left(\sigma_{t,T'}^i\right)^2 = \frac{\sigma_i^2}{2\alpha_i}\left[1 - e^{-2\alpha_i(T'-t)}\right]
\end{aligned} \tag{22}$$

We are trying to compute the following quantities:

$$G^i(t, T', C_t, D_t) = \mathbb{E}\left[g(C_{T'}^{max} - D_{T'})\mathbf{1}_{D_{T'} \in I_{T'}^i}|\mathscr{F}_t^{D,C}\right]$$

In the original paper, it is shown that:

$$\begin{aligned}
G^1(t, T', C_t, D_t) &= H(m_2^n, m_1^{1,D}, \sigma_2^n, \sigma_1^{1,D}) \\
G^i(t, T', C_t, D_t) &= H(m_{i+1}^n, m_1^{i,D}, \sigma_{i+1}^n, \sigma_1^{i,D}) - H(m_i^n, m_1^{i-1,D}, \sigma_i^n, \sigma_1^{i-1,D}) \\
&\qquad\qquad\qquad\qquad\qquad\qquad\qquad\qquad \forall i = 2 \ldots n - 1 \\
G^n(t, T', C_t, D_t) &= \mathscr{G}(m_1^{n,D}, \sigma_1^{n,D}) - H(m_{i+1}^n, m_1^{i,D}, \sigma_{i+1}^n, \sigma_1^{i,D})
\end{aligned}$$

with:

$$m_i^n = \sum_{k=i}^n m_{t,T'}^k \qquad\qquad \left(\sigma_i^n\right)^2 = \sum_{k=i}^n \left(\sigma_{t,T'}^k\right)^2$$

$$m_1^{i,D} = \sum_{k=1}^i m_{t,T'}^k - m_{t,T'}^D \qquad\qquad \left(\sigma_1^{i,D}\right)^2 = \sum_{k=1}^i \left(\sigma_{t,T'}^k\right)^2 + \left(\sigma_{t,T'}^D\right)^2$$

and:

$$\mathscr{G}(m,\sigma) = \int_{\mathbb{R}} g(x)\phi_N(x,m,\sigma)dx$$

$$H(m_1,m_2,\sigma_1,\sigma_2) = \int_0^\infty \mathscr{G}(x+m_1,\sigma_1)\phi(x,m_2,\sigma_2)dx$$

where $x \longmapsto \phi(x,\mu,\sigma)$ the probability density function of a Gaussian random variable with mean μ and variance σ^2.

In our case, $n = 3$, these computations lead to, using the more convenient notation $m^i = m^i_{t,T'}$ and $m^D = m^D_{t,T'}$:

$$G^1(t,T',C_t,D_t) = H(m^2 + m^3, m^1 - m^D, \sigma_2^3, \sigma_1^{1,D})$$

$$G^2(t,T',C_t,D_t) = H(m^3, m^2 + m^1 - m^D, \sigma_3^3, \sigma_1^{2,D}) - H(m^2 + m^3, m^1 - m^D, \sigma_2^3, \sigma_1^{1,D})$$

$$G^3(t,T',C_t,D_t) = \mathscr{G}(m^2 + m^3 + m^1 - m^D, \sigma_1^{3,D}) - H(m^3, m^2 + m^1 - m^D, \sigma_3^3, \sigma_1^{2,D})$$

Appendix 2: Estimation of Demand and Capacities Parameters: The Deterministic Part

The model for the deterministic part is:

$$f_D(t) = d_1^{(D)} + d_2^{(D)} \cos\left(2\pi(t - d_3^{(D)})\right) + \textbf{week}_D(t)$$

We denote by $(Y_i) = (D_{t_i})$ the demand data, and t_i the dates corresponding with the data. To estimate the deterministic part **minus** the weekly scheme, we start with the following least-square regression:

$$Y_i = p_1 + p_2 \cos(2\pi t_i) + p_3 \sin(2\pi t_i) + \varepsilon_i, \text{ with } \varepsilon_i \sim \mathscr{N}(0,\sigma^2)$$

This equation becomes, using a more convenient matrix notation:

$$Y = Xp + \varepsilon,$$

with:

$$X = \begin{pmatrix} 1 & \cos(2\pi t_1) & \sin(2\pi t_1) \\ \vdots & \vdots & \vdots \\ 1 & \cos(2\pi t_n) & \sin(2\pi t_n) \end{pmatrix}$$

We can then estimate the parameters, by using the least-square estimator:

$$\hat{p} = (X^t X)^{-1} X^t Y$$

We can have access to $(1 - \alpha)$-confidence intervals for the first three deterministic parameters:

$$\left[\hat{p} + t_{n-2}^{1-\frac{\alpha}{2}} S, \hat{p} - t_{n-2}^{1-\frac{\alpha}{2}} S \right]$$

With:

$$S = \sqrt{\frac{\frac{1}{n-2} \sum_{i=1}^{n} \hat{\varepsilon}_i^2}{\sum_{i=1}^{n} (x_i - \bar{x})^2}} = \sqrt{\frac{Var(Y)}{Var(X)}} \frac{1}{\sqrt{n-2}}$$

We can then transform the previous regression into one that fits our model:

$$p_2 \cos(2\pi t_i) + p_3 \sin(2\pi t_i) = \sqrt{p_2^2 + p_3^2} \cos(2\pi t_i + \phi)$$

$$= d_2^{(D)} \cos(2\pi (t - d_3^{(D)}))$$

Thus we change p into $d^{(D)}$, using :

$$d_2^{(D)} = \sqrt{p_2^2 + p_3^2}$$

$$d_3^{(D)} = \frac{1}{2\pi} \arccos \left(\frac{p_3}{d_2^{(D)}} \right)$$

This transformation also allows us to compute confidence bounds for $d_1^{(D)}$, $d_2^{(D)}$ and $d_3^{(D)}$. This first part of the estimation procedure gives us estimated parameters such as:

$$\boxed{f_D(t) = d_1^{(D)} + d_2^{(D)} \cos \left(2\pi (t - d_3^{(D)}) \right)}$$

We now need to estimate the weekly scheme **week**$_D(t)$. To do so, we use a the following method, which is quite classic when it comes to estimating a weekly pattern:

- First, we compute the weekly mean of the data, and store it in a variable called $W \in \mathbb{R}^{n_w}$, if n_w is the number of weeks for which we have data.
- Then we compute the weekly residuals, which are the distance between data and the mean of the corresponding week: $R_{t_i} = D_{t_i} - W_{k_i}$, if t_i is in the weekly number k_i. The residuals are then a centered version of the demand.

- Finally, we compute the mean of these residuals, for every hour of the week (which means that we take 168 means of the corresponding residuals):

$$\mathbf{week}_D(t_i) = \frac{1}{n_w} \sum_{t_k \equiv t_i[24]} R_{t_k}$$

References

1. Aïd, R., Campi, L., Nguyen Huu, A., Touzi, N.: A structural risk-neutral model of electricity. Int. J. Theor. Appl. Financ. **12**(7), 925–947 (2009)
2. Aïd, R., Campi, L., Langrené, N.: A structural risk-neutral model for pricing and hedging power derivatives. Math. Financ. **23**(3), 387–438 (2013)
3. Gabillon, J., Analysing the forward curve. In: Risk Books (1st edn.) Managing Energy Price Risk, London, 1995
4. Barlow, M.T.: A diffusion model for electricity prices. Math. Financ. **12**(4), 287–298 (2002)
5. Barlow, M.T., Gusev, Y., Lai, M.: Calibration of multifactor models in electricity markets. Int. J. Theor. Appl. Financ. **7**(2), 101–120 (2004)
6. Benth, F.E., Koekebakker, S.: Stochastic modeling of financial electricity contracts. Energy Econ. **30**(3), 1116–1157 (2008)
7. Bunn, D.W., Karakatsani, N.: Forecasting electricity prices. London Business School Working Paper, 2003
8. Carmona, R., Coulon, M., Schwartz, D.: Electricity price modeling and asset valuation: a multi-fuel structural approach. Math. Financ. Econ. **7**(2), 167–202 (2013)
9. Carmona, R., Coulon, M.: A survey of commodity markets and structural models for electricity prices. In: Quantitative Energy Finance: Modeling, Pricing, and Hedging in Energy and Commodity Markets. Springer, New York. ISBN 9781461472476 (2013)
10. Cartea, A., Figueroa, M.: Pricing in electricity markets: a mean-reverting jump diffusion model with seasonality. Appl. Math. Financ. **12**(4), 313–335 (2005)
11. Follmer, H., Schweizer, M.: Hedging of contingent claims under incomplete information. In: Applied Stochastic Analysis. Gordon and Breach, New York (1991)
12. Hamilton, J.D.: Time Series Analysis. Princeton University Press, Princeton (1994)
13. Heath, D., Jarrow , R., Morton, A.: Bond pricing and the term structure of interest rates: a new methodology for contingent claims valuation. Econometrica **60**(1), 77–105 (1992)
14. Hinz, J., von Grafenstein, L., Verschuere, M., Wilhelm, M.: Pricing electricity risk by interest rate methods. Quant. Financ. **5**(1), 49–60 (2005)
15. Kalman, R.E.: A new approach to linear filtering and prediction problems. Trans. ASME J. Basic Eng. **82**(Series D), 35–45 (1960)
16. Kiesel, R., Schidlmayr, G., Börger, R.H.: A two-factor model for electricity forward market. Quant. Financ. **9**(3), 279–287 (2009). Available at SSRN: http://ssrn.com/abstract=982784
17. Koekebakker, S., Ollmar, F.: Forward curve dynamics in the nordic electricity market. Manag. Financ. **31**(6), 73–94 (2005)
18. Miltersen, K.R., Schwartz, E.S.: Pricing of options on commodity futures with stochastic term structure of convenience yiels and interest rates. J. Financ. Quant. Anal. **33**(1), 33–59 (1998)
19. Weron, R.: Modeling and Forecasting Electricity Loads and Prices: A Statistical Approach. Wiley Finance, Chichester (2006)

Part III
Real Options

Incorporating Managerial Information into Real Option Valuation

Sebastian Jaimungal and Yuri Lawryshyn

Abstract The adoption of real options analysis (ROA) by practitioners, despite being widely viewed as a superior method for valuing managerial flexibility, remains limited due to varied difficulties in its implementation. In this work, we propose an approach that utilizes cash-flow estimates from managers as key inputs and results in project value cash-flows that exactly match the arbitrarily distributed estimates. We achieve this through the introduction of an observable, but not tradable, market stochastic driver process which drives the project's cash-flow, rather than modeling the project value directly. Our framework can be used to value managerial flexibilities and obtain hedges in an easy to implement manner for a variety of real options such as entry/exit, multistage, abandonment, etc. As well, our approach to ROA provides a co-dependence between cash-flows, is consistent with financial theory, requires minimal subjective input of model parameters, and bridges the gap between theoretical ROA frameworks and practice.

1 Introduction

Real options analysis (ROA) has been recognized as a superior method to quantify the value of real-world investment opportunities where managerial flexibility can influence their worth, as compared to standard net present value (NPV) and discounted cash-flow (DCF) analysis (see, e.g., [6, 9, 21]). Moreover, the literature tends to fall into two streams: practical methods and implementation, and theoretical frameworks. Practical methods aim to develop valuation techniques that can be implemented by practitioners, and, by design, are generally not mathematically complex. However, a number of these methods make simplifying assumptions that may be hard to accept, and some are inconsistent with financial theory. Theoretical frameworks tend to be mathematically rigorous and aim to highlight issues

S. Jaimungal (✉)
Department of Statistical Sciences, University of Toronto, Toronto, ON, Canada
e-mail: sebastian.jaimungal@utoronto.ca

Y. Lawryshyn
Department of Chemical Engineering, University of Toronto, Toronto, ON, Canada
e-mail: yuri.lawryshyn@utoronto.ca

© Springer Science+Business Media New York 2015   213
R. Aïd et al. (eds.), *Commodities, Energy and Environmental Finance*,
Fields Institute Communications 74, DOI 10.1007/978-1-4939-2733-3_8

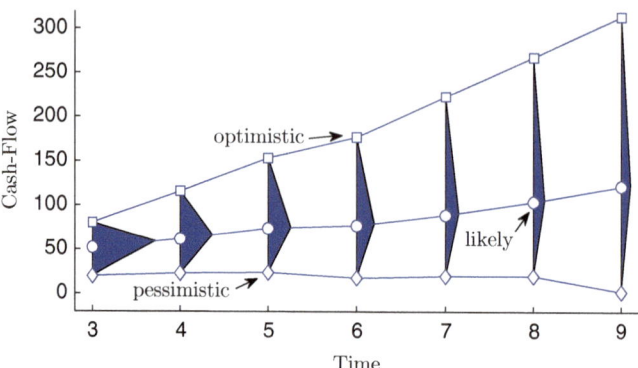

Fig. 1 An example of typical, optimistic and pessimistic cash-flow stream provided by a manager. The *triangles* depict the triangular distribution corresponding to that cash-flow on that date

associated with real option valuation and decision making. Theoretical approaches are difficult for managers to accept, and therefore use in practice, because they rely on assumptions, such as a project's value is modeled as a geometric Brownian motion (GBM), which may not reflect the manager's view of future cash-flows. Indeed, managers tend to have a very simplified and condensed view of what the potential cash-flows might be worth and certainly do not have a stochastic process in mind, even one as simple as a GBM, when coming to these views, nor do they believe that a stochastic process provides an approximation which helps in decision making.

For example, in the most simplistic situation, a manager may have three views on a future cash-flow (as in Fig. 1): (i) a typical scenario, (ii) an optimistic scenario, and (iii) a pessimistic scenario. The manager will then typically provide a sequence of these views representing a sequence of cash-flows from a project. If the manager is particularly sophisticated, she may even provide a sequence of distributions (e.g., normal, triangular, etc...) for the set of the cash-flows. Furthermore, the manager must take decisions based on their cash-flow distributions and any ad-hoc matching of a sequence of cash-flow distributions to a stochastic process will not lead to valuations, and, therefore strategies, that respect their condensed views. It is the aim of this paper to incorporate the manager's cash-flow views into real options analysis. More specifically, we will study two classical examples: (i) irreversible investment (at a fixed future time) into a sequence of cash-flows and (ii) an entry-exit problem where the manager decides to invest at a fixed future date, but may stop the cash-flow (with a cost) once invested. Our approach is, however, very generic and can be applied to essentially all real option problems. Before providing an overview of our approach, we first provide a limited review of the recent literature on ROA.

Borison [3] categorizes the main practical approaches to ROA into five categories: (i) the Classical, (ii) the Subjective, (iii) the Market Asset Disclaimer (MAD), (iv) the Revised Classical and (v) the Integrated. Each method has its strengths and weaknesses and for a thorough review we refer the reader to [3].

Most of the practical approaches assume that the risk embedded in the real option is spanned by a traded asset (i.e., that the market is complete). The error in this assumption is somewhat taken into account by modifying the discount factor using the weighted average cost of capital (WACC). This modification, despite being partly grounded in capital asset pricing model (CAPM) theory, is in the best case, adhoc. Further, this assumption causes the value of the option to irreversibly invest in a project to be strictly increasing in volatility. Such a result contradicts the observations of an empirical study by [18] as well as the results of theoretical frameworks that will be discussed below. Our approach agrees with the theoretical frameworks as well as the empirical study.

Berk et al. [2] develop a real options model of a staged R&D project. The cash flows are modeled as a GBM with partial correlation to a pricing kernel, i.e. a traded asset. Included in the model are uncertainties related to catastrophic failure, technical uncertainty and investment costs. Projects are valued by assuming managers will invest optimally given the state of observed exogenous and endogenous parameters. While the model brings insights regarding the type of information available to managers and its impact on decision making, the author provides no way to tie the model to manager specified cash-flow distributions, making it difficult to implement in practice.

Recently, a number of theoretical formulations for treating managerial flexibility with market incompleteness have been developed. For example, [16] present four incomplete market real options models and show that standard real options approaches, which assume complete markets, can lead to contradictory results. Under certain conditions, contrary to the standard real options approaches (see [9], for example), in the presence of *enough* risk aversion, the option value decreases as idiosyncratic risk increases, as does the optimal timing of investment. This work highlights issues associated with standard real options models, and especially, the complete market assumption.

Henderson [12] presents an incomplete market model for a perpetual (irreversible) option to invest in a lump-sum payment. She assumes that the project value is a GBM and that the Brownian driver is partially correlated to a risky traded asset (itself a GBM). Henderson utilizes the indifference pricing paradigm, whereby the manger solves two optimization problems: one with the real-option payoff and one without, and determines the amount of wealth she is willing to pay, which renders both value functions equal (see, e.g., [5]). She assumes exponential utility without consumption and derives a closed form solution for the perpetual option to her model, and establishes the existence of three distinct parameter regions. Depending on the region, Henderson shows that in incomplete markets, the optimal exercise time may or may not behave in a consistent manner when considering different levels of cash-flow risk, compared to standard real option approaches. We note that [11] presents a finite time horizon version of the [12] model and provides similar conclusions through numerical experiments. Both [12] and [11] highlight issues associated with the standard real options models. Also noteworthy, but not directly applicable, is the work of [14] who study early exercise employee stock options using indifference pricing on an underlier which the employee is not allowed to (or cannot—because it is private equity) trade.

Hugonnier and Morellec [13] examine managerial agency issues associated with investment opportunities. The authors show that a manager's investment decisions are biased by the manager's need to reduce his own idiosyncratic risk associated with the idiosyncratic risk of the project and reward mechanisms associated with project success, which leads to earlier than optimal investment, reducing value for potentially well diversified shareholders. Many other theoretical approaches have been presented in the literature. We highlight, however, that no formulation presented to date is able to, almost universally, meld these theoretical approaches to managerial supplied cash-flow estimates.

We now provide a brief overview of our approach to valuing real-options that take into account the manager's condensed views on the project's value, as well as market incompleteness, and highlight some of our key results. First, we assume that future cash-flow estimates are provided by the manager in the form of a sequence $\{F_k^* : k = 1, \dots, n\}$ of probability distributions at a sequence $\{t_k : k = 1, \dots, n\}$ of payment dates[1]—this can be viewed as the input data into the model and encapsulates the manager's view, however she developed them, on the future cash-flow of the project (once invested). This input "data" can simply be triangular (representing typical, optimistic and pessimistic scenarios), normal, log-normal, or any other continuous density. Second, we assume that there exists a non-tradable underlying stochastic process S_t called the *stochastic driver*,[2] taken in this work to be an Ito process, which will drive the cash-flows. Specifically, we uniquely determine a sequence of mappings $\varphi_k : \mathbb{R} \mapsto \mathbb{R}$ between the stochastic driver and the manager specified cash-flow distributions which renders the transformed stochastic driver $\varphi_k(S_{t_k})$ equal in distribution to the manager specified distribution F_k^*. In this manner, although we do utilize an underlying stochastic process, that process is mapped so that it exactly matches the collection of manager specified distributions. Hence, as far as the manager is concerned, she need only supply her condensed views and be confident in the fact that the real options analysis provides strategies that are entirely consistent with those views. Next, we assume that the stochastic driver is correlated to a traded market index I_t, which we assume to be a GBM. This market index will allow us to partially hedge the real options, but correlation may be zero in which case there is no partial hedge. Finally, we will utilize the notation of indifference pricing, along the lines of [12], but account for the sequence of cash-flows as well as the mapping between the stochastic driver and the cash-flow distributions.

Beyond the new modeling paradigm, this work contains two main results. As in most models that account for risk-aversion, the value of the real option decreases with increase risk-aversion. This is not new. However, our results show that there is a critical level of risk-aversion above which the manager assigns a value of zero to the real-option regardless of the option. This is a new result, and one which corresponds to the manner in which manager's behave.

---

[1] This can be easily generalized to a project which pays out a continuous dividend.

[2] In earlier versions of this work, we referred to this process as the market sector indicator.

2 Replicating Cash-Flow Distributions

When managers think about investing in projects, they typically have in mind a cash-flow associated with three scenarios: (i) the most likely scenario (ii) the optimistic scenario and (iii) the pessimistic scenario. Datar and Mathews [7] propose a methodology based on such scenarios and we formulate our approach in a similar manner. An example of the three cash-flow scenarios is depicted graphically in Fig. 1. The three scenarios may be derived through Monte Carlo simulations representing the technical risk inherent in the project, the corporations potential market share, the market value of the end product and so on. Regardless of how the manager comes to this cash-flow distribution, one of our main goals is to provide a consistent dynamic model which leads to a project cash-flow possessing any distribution which a manager provides. Our analysis is not limited to the triangular distribution shown in the example, although this distribution is, perhaps, the simplest form that managers employ widely.

Here, we introduce an underlying observable, but not tradable, process S_t which drives the project cash-flows. This underlying process can be thought of as a *stochastic driver*. We model the stochastic driver as an Ito process as follows

$$dS_t = v(t, S_t)\,dt + \eta(t, S_t)\,dW_t, \tag{1}$$

where W_t is a standard Brownian motion under the real-world measure $\mathbb{P}$, $v : \mathbb{R}_+ \times \mathbb{R} \mapsto \mathbb{R}$ represents the stochastic driver's drift and $\eta : \mathbb{R}_+ \times \mathbb{R} \mapsto \mathbb{R}$ represents its volatility, assumed to satisfy the usual conditions so that (1) admits a weak solution. Throughout this article we work on the filtered completed probability space $(\Omega, \mathbb{P}, \mathscr{F} = \{(\mathscr{F}_s)_{0 \leq s \leq T_n}\})$ where the filtration $\mathscr{F}$ is the natural filtration generated by the two correlated Brownian motions W_t and B_t (which will be used to drive a tradable market index later on). Here T_n is the time at which the last cash-flow is received or incurred (in the case of a cost) and designates the terminal time of the project. Specific stochastic driver examples of interest are

$$\begin{cases} v(t, S) = v\,S\,, & \eta(t, S) = \eta\,S\,, & \text{Geometric Brownian Motion,} \\ v(t, S) = \kappa\,(\bar{S} - S)\,, & \eta(t, S) = \eta\,, & \text{Ornstein-Uhlenbeck process,} \\ v(t, S) = \kappa\,(\bar{S} - S)\,, & \eta(t, S) = \eta\,\sqrt{S}\,, & \text{Feller process.} \end{cases} \tag{2}$$

Ornstein-Uhlenbeck (OU) processes have been argued as more realistic than GBM as early as [9] and [15], and utilized by many others in various real options contexts. Two recent works include [20], who studies real options with stochastic costs, and [22], who extends Sarkar's work. Carlos and Nunes [4] study the case of constant elasticity of variance model for real option valuation.

GBM and OU processes are often used in the real options literature since they capture the essential source of uncertainty inherent in valuation. However, both processes are typically applied directly to the project value (or its log). Here, we do not model the project value, instead, the stochastic driver is only an underlying source of uncertainty for the cash-flows. Specifically, the cash-flows $\{V_k : k = 1,\ldots,n\}$ received at time $\{T_k : k = 1,\ldots,n\}$ are each modeled as a function $\varphi_k(\cdot)$ of the underlying stochastic driver, so that

$$V_k = \varphi_k(S_{T_k}). \tag{3}$$

As a result, the project value can be viewed as a strip of European contingent claims on the stochastic driver with payoff functions φ_k at time T_k. When the stochastic driver is high, the project value will be high, and when the stochastic driver is low, the project value will be low. Furthermore, since the cash-flows are all driven by the same underlying stochastic driver, and the stochastic driver is dependent on its past, there is a natural correlation between cash-flows induced by the path dependence in the stochastic driver. If the stochastic driver is high at the time of one cash-flow, resulting in a large cash-flow, then the probability of a large cash-flow at the next time-step is also high. This is a very desirable feature which has clear economic grounding. Contrastingly, in a number of practical approaches, the cash-flow distributions are typically assumed to be independent or correlation is introduced in a rather adhoc manner (as in the [7] approach).

Although the stochastic driver S_t is not itself tradable, we do assume it is *correlated* with a market index I_t that is tradable. This market index is, for simplicity, assumed to be a GBM

$$dI_t = \mu I_t dt + \sigma I_t dB_t, \tag{4}$$

where B_t is a standard Brownian motion under the real-world measure. Since the stochastic driver and the market index are likely correlated, we assume that the processes W_t and B_t are correlated with correlation coefficient ρ. The assumption that such a traded market index exists will allow us to partially hedge the cash-flows and the option to invest in the project.

Focusing on a single cash-flow distribution, our task is to determine φ such that at some cash-flow date T, V_T possesses the manager specified distribution $F^*(v)$—we use an asterisk to remind the reader that this distribution is provided by the manager. This requirement can be restated as, find φ such that

$$\mathbb{P}(V_T < v) = \mathbb{P}(\varphi(S_T) < v) = F^*(v), \tag{5}$$

and can be visualized as in Fig. 2 for the case of a triangular managerial distribution and GBM stochastic driver. We will require the marginal distribution function of the stochastic driver, specifically let $F(T,S) := \mathbb{P}(S_T < S)$, which we assume is known either in closed form, or through solving the appropriate forward equation numerically (for the three cases, GBM, OU and Feller, the pdf are known in closed form).

It is not difficult to see that if $\varphi(S)$ is assumed invertible and the cash-flow distribution $F^*(v)$ is invertible, then the solution is unique. However, invertibility is by no means necessary. Nonetheless, we restrict our analysis to this case as it

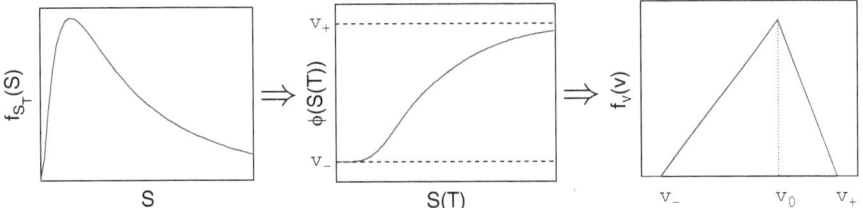

Fig. 2 The underlying pdf $f_{S_T(S)}$ is mapped through the function $\varphi(S)$ to match the triangular distribution

leads to sound economically meaningful results. Note that the probability matching Eq. (5) can be also be interpreted as a quantile matching relationship where φ acts as a probability distortion function. One of our main results is recorded in the following Proposition.

Proposition 1 (The Replicating Payoff). *The payoff function $\varphi_k(S)$ which produces the manager specified distribution $F_k^*(v)$ for the cash-flow at time T_k, when the stochastic driver satisfies* (1), *is given by*

$$\varphi_k(S) = F_k^{*-1}\left(F(T_k, S)\right). \tag{6}$$

For the GBM and OU cases, we have $F(T, S) = \Phi(z(T, S))$ where Φ is the standard normal cdf and

$$z(T, S) = \begin{cases} \dfrac{1}{\eta\sqrt{T}} \ln \dfrac{S}{S_0} - \dfrac{(v - \frac{1}{2}\eta^2)}{\eta}\sqrt{T}, & \text{GBM case} \\[2ex] \dfrac{(S - \bar{S}) - (S_0 - \bar{S})e^{-\kappa T}}{\eta\sqrt{(1 - e^{-2\kappa T})/2\kappa}}, & \text{OU case}. \end{cases} \tag{7}$$

We emphasize that the payoff function φ and the resulting valuations will depend on the process chosen for the stochastic driver. However, numerical examples have shown that the sensitivity of the valuations to the driver process are significantly less than the sensitivity associated with managerial specified cash-flow distributions. In a practical setting, it is likely that managers will have a much stronger opinion on the form of the cash-flow distributions than the driver process. As mentioned earlier, in the real options literature, the GBM and OU processes are generally the processes most often chosen, and most often are used to drive the *value* of the cash-flows, not the cash-flows themselves, as done here.

An important feature of our approach is that the cash-flow distributions inherit a co-dependence structure from the underlying dynamics of the stochastic driver. In Fig. 3, the stochastic driver is shown to evolve through time and at cash-flow dates, transformed via the cash-flow functions φ_k into the actual cash-flow amount. The implied co-dependence can be formalized and for the GBM and OU cases reduce to a Gaussian copula (see, e.g., [1] and [17]). Cash-flows which are far apart from

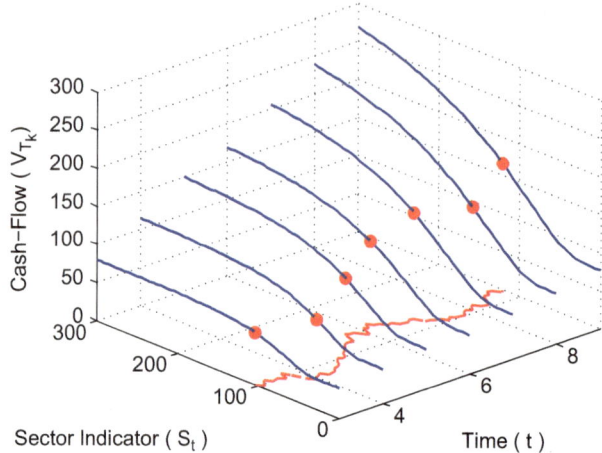

Fig. 3 The cash-flows between dates inherit a Gaussian co-dependence structure since the underlying stochastic driver (shown here on the x-y plane) which drives the cash-flow is a gaussian process

one another are less dependent because the correlation of the stochastic driver's level decays as the time separation increases. This intuitive feature is captured in the copula's correlation matrix which decreases as one moves away from the center band of the matrix.

Proposition 2 (The Implied Cash-Flow Co-Dependence (GBM/OU)). *The co-dependence of the cash-flow stream is governed by a Gaussian copula. Specifically, the joint distribution of the cash-flows* V_k *is given by the expression*

$$\mathbb{P}(V_1 < v_1, \ldots, V_n < v_n) = \Phi_\Omega \left(\Phi^{-1}(F_1(v_1)), \ldots, \Phi^{-1}(F_n(v_n)) \right), \qquad (8)$$

where, the elements of the matrix Ω *are*

$$\Omega_{jk} = \begin{cases} \sqrt{T_{\min(j,k)} / T_{\max(j,k)}}, & \textit{GBM case} \\[2mm] \sqrt{\dfrac{1 - e^{-2\kappa T_{\min(j,k)}}}{1 - e^{-2\kappa T_{\max(j,k)}}}}, & \textit{OU case}. \end{cases} \qquad (9)$$

and $\Phi_\Omega(\cdot)$ *denotes the multi-variate standard normal cdf with correlation matrix* Ω.

The above result shows that as the time between cash-flows increase, their co-dependence decreases. Our resulting co-dependence is consistent with the underlying economics of nearby cashflows and does not require any adhoc assumptions on generating the co-dependence structure. Notice that the OU case allows the manager to tune the co-dependence structure by choosing κ. In principle, if the mean-reversion rate and/or volatility is made time-dependent (e.g., piecewise constant between cash-flows), then quite arbitrary co-dependence structures can be generated, and the expressions can be found in closed form. This can be useful if the manager has a very specific view on the co-dependence structure, in addition to their view on the cash-flow distribution.

The cash-flow matching described here can be embedded into many practical real-option valuation questions ranging from an irreversible investment in a project at a fixed future, to the timing option to invest in a project, or abandon a project, and so on. As well, our cash-flow matching approach can be layered on top of theoretical approaches. In the next section we will demonstrate how to layer our approach on top of indifference valuation for the case of an irreversible investment and in Sect. 5 we address the optimal exit problem.

3 Indifference Valuation

In this section, we address how our framework can be used to incorporate a manager's aversion to risk, and in particular, the implication this risk-aversion has on valuing the idiosyncratic risks embedded in the project valuation. Recall that in our approach the cash-flows are themselves options on the stochastic driver and are chosen to exactly match the manager's specified distribution. To value the cash-flows, as well as the option to invest, we adopt the principle of indifference valuation along the lines of [12]. Henderson studies the perpetual option to invest in a project which provides a single lump-sum payment upon investment. Grasselli [11] studies a finite time horizon version of the Henderson framework and develops a tree approach to valuation. Moreover, [14] study the American version of Henderson in the context of employee stock options with early exercise. Our formulation instead focuses on the matching of manager specified sequence of cash-flow distributions with the investment decision made on a fixed future date. The optimal timing problem, where the investment can be made on any date between now and a fixed future time, is not considered here, but is a relatively straightforward generalization. We will address an optimal exit version of the problem in Sect. 5 to illustrate how timing problems can easily be treated.

3.1 Indifference Valuation Methodology

Indifference valuation requires solving two optimal investment problems: (i) the investment problem in the absence of the option to invest, and (ii) the investment problem in the presence of the option to invest. The indifference price is then defined as the amount of wealth f the manager is willing to give up in scenario (ii) such that her value function equals the value function in scenario (i) without giving up f. A precise definition follows the problem set up and discussions below.

As before, we assume the manager has supplied a series of uncertain cash-flow projections $v_1, v_2, \ldots, v_k$ at a sequence of dates $T_1, T_2, \ldots, T_n$. To value the option, the manager must provide a distributional assumption on the set of future cash-flows—e.g. in the form of the triangular distributions discussed earlier. Let $F_k^*(v)$ denote the distribution of the cash-flow occurring on date T_k for $k = 1, \ldots, n$. Given these cash-flow distributions, Proposition 1 provides the payoff function $\varphi_k(S)$ for which $v_k = \varphi_k(S_{T_k})$ has the manager specified distribution $F_k^*(v)$.

The optimal investment problem without the option to invest is the classical Merton problem. Here we adopt exponential utility, i.e., the manager's utility $u(x) = -e^{-\gamma x}$ when his/her discounted wealth is x and allow the manager to invest in the traded index I_t and the risk-free money-market. The manager's value function without investment in the option is then

$$V^{(0)}(t, x, I) = \sup_{\pi \in \mathscr{A}_t} \mathbb{E}_{t,x}[-e^{-\gamma X_{T_n}^\pi}], \tag{10}$$

where $\mathbb{E}_{t,x,I}[\cdot]$ denotes $\mathbb{P}$-expectation conditional on $X_t = x$ and $I_t = I$, $\mathscr{A}_t$ is the set of self-financing admissible strategies $\mathscr{A}_t = \{(\pi_s)_{t \le s \le T_n} : \mathbb{E}[\int_t^{T_n} \pi_s^2 \, ds] < +\infty\}$ and π_s denotes the dollar amount invested in the index at time s so that $dX_s^\pi = (\mu - r)\pi_s \, ds + \sigma \pi_s \, ds$, $X_t^\pi = x$. As is well known, when managers have exponential utility, the value function is in fact independent of I and is explicitly given by

$$V^{(0)}(t, x, I) = -e^{-\gamma x - \frac{1}{2}\lambda^2(T_n - t)}, \tag{11}$$

where $\lambda = \frac{\mu - r}{\sigma}$ is the market-price-of-risk of the traded index and the optimal investment is constant

$$\pi_t^{*(0)} = \frac{1}{\gamma} \frac{\mu - r}{\sigma^2}. \tag{12}$$

Next, we consider the optimal investment problem when the manager invests in the option. In this case, the manager faces the idiosyncratic risks embedded in the future project cash-flows together with the potential of investing in the hedging asset. Consequently, the manager solves the following optimal control problem

$$V^{(1)}(t, x, S, I) = \sup_{\pi \in \mathscr{A}_t} \mathbb{E}_{t,x,S,I}\left[-e^{-\gamma X_{T_n}^\pi}\right] \tag{13}$$

where $\mathbb{E}_{t,x,S,I}[\cdot]$ denotes $\mathbb{P}$-expectation conditional on $X_t = x$, $S_t = S$, and $I_t = I$. Moreover, the discounted wealth process now satisfies the SDE

$$dX_t^\pi = (\mu - r)\pi_t \, dt + \sigma \pi_t \, dB_t, \qquad \forall t \in (T_{k-1}, T_k), \quad \text{and} \tag{14a}$$

$$X_{T_0}^\pi = X_{T_0^-}^\pi - \tilde{K} \mathbb{I}\{\mathscr{E}\}, \tag{14b}$$

$$X_{T_k}^\pi = X_{T_k^-}^\pi + \tilde{\varphi}_k(S_{T_k}) \mathbb{I}\{\mathscr{E}\}, \tag{14c}$$

for $k = 1, \ldots, N$, $\tilde{K} = e^{-rT_0} K$, $\tilde{\varphi}_k(S) = e^{-rT_k} \varphi(S)$ and the $\mathscr{F}_{T_0}$-measurable set $\mathscr{E} := \{V^{(1)}(T_0, X_{T_0^-}^\pi - \tilde{K}, S_{T_0}, I_{T_0}) > V^{(0)}(T_0, X_{T_0^-}^\pi, I_{T_0})\}$. The manager uses the set $\mathscr{E}$ to decide whether or not to invest, because the manager is assumed to be a utility maximizer, hence, at time T_0, she will pick the option (invest or not) which provides the highest value function at time T_0. The intuition behind the dynamics in (14) is that (i) between cash-flow dates the manager's wealth evolves continuously, according to the self-financing investment policy π; (ii) at T_0 the manager must decide whether to invest in the option—which reduces her discounted wealth by $\tilde{K}$ if investment is made, otherwise there is no change and the value function reduces to the Merton value function; (iii) if it was optimal to invest in the option at time T_0, then on cash-flow dates T_k her wealth jumps by the cash-flow amounts.[3]

The indifference price is then defined as the amount of wealth which the manager is willing to resign to receive the option such that her value function remains unchanged relative to not receiving the option. More precisely, the indifference price f solves

$$V^{(1)}(t, x - f(t, x, S, I), S, I) = V^{(0)}(t, x, I). \tag{15}$$

The Merton value function $V^{(0)}$ is already provided in (11), so our remaining task is to obtain $V^{(1)}$.

The optimization problem (13) for $V^{(1)}$ is similar to the classical Merton problem, except that the manager is faced with the series of uncertain cash-flows $\varphi_k(S_{T_k})$. This sequence of cash-flows is a strip of European options on the stochastic driver S_t, which is not tradable, but is partially spanned by the hedging asset I_t.

To solve the optimal investment problem, it is convenient to introduce a third value function representing the value function of a manager who receives the project cash-flows for free—i.e., with a strike of $K = 0$. This value function is given by

$$V^{(2)}(t, x, S, I) = \sup_{\pi \in \mathscr{A}_t} \mathbb{E}_{t,x,S,I} \left[-e^{-\gamma X_{T_n}^\pi} \right] \tag{16}$$

where the discounted wealth process satisfies the SDE

$$dX_s^\pi = (\mu - r)\pi_s \, ds + \sigma \, \pi_s \, dB_s, \qquad \forall \, s \in (T_{k-1}, T_k), \quad \text{and} \tag{17a}$$

$$X_{T_k}^\pi = X_{T_k^-}^\pi + \tilde{\varphi}_k(S_{T_k}) \, \mathbb{I}\{\mathscr{E}\}. \tag{17b}$$

This differs from the full problem (13) in that the cost to receive the future cash-flows K is not paid at T_0. For all times $t > T_0$ and on the set of events in which

---

[3]In principle, it is possible to consider the cash-flow as a continuous stream of cash-flows, in which case (17a) would be modified to $dX_s = (\mu - r)\pi_s \, ds + \sigma \, \pi_s \, dB_s + \varphi_s(S_s) \, ds$. However, Managers rarely specify a continuous stream of cash-flows, and although operations can be viewed as providing income on a continuous basis, we opt to leave the cash-flows discrete as this is how managers typically estimate cash-flow streams.

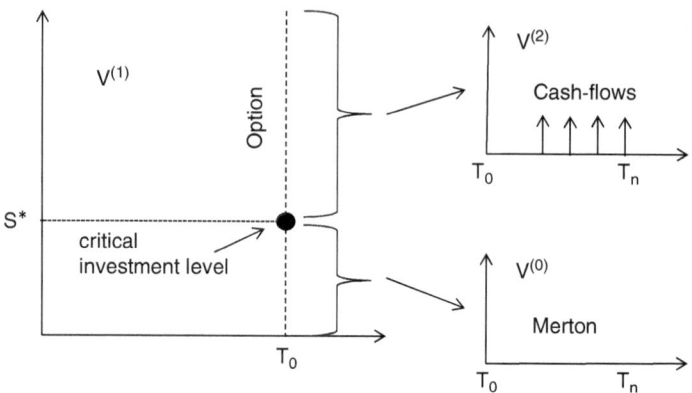

Fig. 4 The value function for the manager's optimization problem (13) when holding the option to invest can be viewed as first solving for (i) the Merton problem (10) and (ii) the cash-flow optimization problem (16) in the region $[T_0, T_n]$; and then pasting them together at the critical level S^*

$\mathbb{I}\{\mathscr{E}\} = 1$ (i.e., on the set of events in which the manager exercises the option to invest), the two value functions are equal, $V^{(1)}(t, x, S, I) = V^{(2)}(t, x, S, I)$. Moreover, for all times $t > T_0$ and on the set of events in which $\mathbb{I}\{\mathscr{E}\} = 0$ (i.e., on the set of events in which the manager does not exercise the option), the full value function equals the Merton solution, $V^{(1)}(t, x, S, I) = V^{(0)}(t, x, I)$. Thus, we can solve the full problem by solving for $V^{(2)}$ and $V^{(0)}$ on the interval $t \geq T_0$, paste them together at T_0, along the boundary of the set $\mathscr{E}$, to obtain $V^{(1)}$ at T_0 and then propagate backwards to the interval $t < T_0$. This decomposition of the problem can be visualized as in Fig. 4. Since the boundary conditions and cash-flows do not depend on the hedging asset (in this case the traded index I_t), the value functions $V^{(0)}$, $V^{(1)}$, $V^{(2)}$ (and therefore S^*) are in fact independent of I (but not S). Consequently, the indicator $\mathbb{I}\{\mathscr{E}\}$ switches from 1 to 0 at the critical point S^* defined by $V^{(2)}(T_0, x - \tilde{K}, S^*, I) = V^{(0)}(T_0, x, I)$. This critical point is unique because the value function is increasing in S since the embedded option is a call on a strip of options which have payoffs $\varphi_k(S)$ increasing in S. Moreover, this asset value S^* can be viewed as the critical stochastic driver level above which the future cash-flow has an indifference value greater than $\tilde{K}$. We will see in Sect. 4 that having a strip of cash-flows, rather than a lump-sum at the investment time, may lead to the real-option being valued at exactly zero.

The dynamic programming principal (DPP) implies that the value functions $V^{(2)}$ and $V^{(1)}$ satisfy the Hamilton-Jacobi-Bellman (HJB) equations

$$(\partial_t + \mathscr{L}_S) V^{(2)}$$
$$+ \sup_{\pi} \left\{ (\mu - r)\pi \, \partial_x V^{(2)} + \tfrac{1}{2}\pi^2\sigma^2 \, \partial_{xx} V^{(2)} + \rho\eta\sigma \, \pi \, S\partial_{xS} V^{(2)} \right\} = 0, \tag{18a}$$

for $t \in (T_{k-1}, T_k)$, $k = 1, \ldots, n$, subject to the sequence of boundary conditions

$$V^{(2)}(T_k^-, x, S) = V^{(2)}(T_k, x, S) \, \exp\{-\gamma \, \tilde{\varphi}_k(S)\}, \quad \text{for} \quad k = 1, \ldots, n-1 \quad \text{and}$$
(18b)

$$V^{(2)}(T_n, x, S) = \exp\{-\gamma \, \tilde{\varphi}_n(S)\},$$
(18c)

and

$$(\partial_t + \mathscr{L}_S) V^{(1)}$$
$$+ \sup_\pi \{(\mu - r)\pi \, \partial_x V^{(1)} + \tfrac{1}{2}\pi^2\sigma^2 \, \partial_{xx} V^{(1)} + \rho\eta\sigma \, \pi \, S \partial_{xS} V^{(1)}\} = 0,$$
(19a)

for $t \in [0, T_0)$, subject to the T_0 boundary condition

$$V^{(1)}(T_0^-, x, S) = V^{(2)}(T_0, x - \tilde{K}, S) \, \mathbb{I}\{S > S^*\} + V^{(0)}(T_0, x, S) \, \mathbb{I}\{S \le S^*\}.$$
(19b)

In the above $\mathscr{L}_S$ denotes the generator of the stochastic driver S_t, i.e.,

$$\mathscr{L}_S = \nu(t, S) \, \partial_S + \tfrac{1}{2}\eta^2(t, S) \, \partial_{SS}.$$
(20)

The T_0 boundary condition in (19) accounts for the option to invest in the cash-flow, and as commented on earlier, occurs at the critical stochastic driver level S^* above which the indifference value of the cash-flow is greater than $\tilde{K}$.

It is possible to reduce the HJB equations into a system of linear PDEs through a sequence of transformations. The result of these computations are recorded in the following Theorem.

Theorem 1 (Value Function and Optimal Investment.). *The solution to the HJB equations* (18) *and* (19) *admit the representation*

$$V^{(a)}(t, x, S, I) = V^{(0)}(t, x, I) \left(H^{(a)}(t, S)\right)^\beta$$

where $\beta = (1-\rho^2)^{-1}$ *for* $a = 1, 2$. *Further, let* $\hat{\mathscr{L}}_S$ *denote the infinitesimal generator of the stochastic driver under the minimal martingale measure* $\hat{\mathbb{Q}}$ *induced by the Radon-Nikodym derivative*

$$\frac{d\hat{\mathbb{Q}}}{d\mathbb{P}} = e^{-\frac{1}{2}\lambda^2 T_n + \lambda W_{T_n}}.$$
(21)

Specifically,

$$\hat{\mathscr{L}}_S = \hat{\nu}(t, S) \, \partial_S + \frac{1}{2}\eta^2(t, S) \, \partial_{SS}, \quad \text{and} \quad \hat{\nu} = \nu(t, S) - \rho \, \eta(t, S) \, \lambda.$$
(22)

Then, $H^{(a)}(t, S)$ satisfy the linear PDEs

$$\partial_t H^{(2)} + \hat{\mathcal{L}}_S H^{(2)} = 0, \quad t \in (T_{k-1}, T_k), \qquad k = 1, \ldots, n \tag{23a}$$

with terminal/pasting conditions

$$H^{(2)}(T_k^-, S) = H^{(2)}(T_k, S) \exp\left\{-\tfrac{\gamma}{\beta}\, \tilde{\varphi}_k(S)\right\}, \quad \text{for} \quad k = 1, \ldots, n-1, \quad \text{and} \tag{23b}$$

$$H^{(2)}(T_n, S) = \exp\left\{-\tfrac{\gamma}{\beta}\, \tilde{\varphi}_n(S)\right\}, \tag{23c}$$

and

$$\partial_t H^{(1)} + \hat{\mathcal{L}}_S H^{(1)} = 0, \quad t \in [0, T_0), \tag{24a}$$

with terminal condition

$$H^{(1)}(T_k^-, S) = 1 - \left(1 - H^{(2)}(T_k, S) e^{\frac{\gamma}{\beta} \tilde{K}}\right) \mathbb{I}\{S > S^*\}, \tag{24b}$$

where S^* is the unique solution to

$$H^{(2)}(T_0, S^*) = e^{-\frac{\gamma}{\beta} \tilde{K}}. \tag{25}$$

Moreover, the optimal investment policy is

$$\pi^{(a)} = \frac{1}{\gamma} \frac{\mu - r}{\sigma^2} + \rho \frac{\eta}{\sigma} \frac{\beta}{\gamma} S \, \partial_S \ln H^{(a)}. \tag{26}$$

The previous theorem shows that manager's value function is obtained by solving the PDE (23) in the region $t \in [T_0, T_n]$ and then (24) in the region $t \in [0, T_0)$. Although analytical solutions are in general not available, standard finite difference techniques (such as implicit-explicit schemes) can be used to approximate the solution. In the region $t < T_0$, the manager's value function is unique, however, for $t > T_0$, the value function is either given by $V^{(2)}$ or $V^{(0)}$ depending on whether the option to invest was exercised at T_0.

Armed with the value functions, the indifference price easily follows and a simple application of Feyman-Kac leads to a stochastic representation for the indifference value. These results are recorded in the Corollary below.

Corollary 1 (Indifference Price Stochastic Representation). *The indifference price $f(t, S)$ (in discounted dollars) of the cash-flow is given by*

$$f(t, S) = \begin{cases} -\frac{\beta}{\gamma} \ln H^{(1)}(t, S), & t \in [0, T_0), \\[2mm] -\frac{\beta}{\gamma} \ln H^{(2)}(t, S)\, \mathbb{I}\{S > S^*\}, & t \in [T_0, T_n], \end{cases} \tag{27}$$

where, $H^{(a)}(t, S)$ solve the PDEs (23a) and (24a). Moreover, $H^{(a)}(t, S)$ admit the expectation representation

$$H^{(2)}(t, S) = \mathbb{E}_{t,S}^{\hat{\mathbb{Q}}}\left[e^{-\frac{\gamma}{\beta}\sum_{m:t<T_m}^{n}e^{-r(T_m-t)}\varphi(S_{T_m})}\right], \qquad t \in [0, T_n], \qquad (28a)$$

$$H^{(1)}(t, S) = \mathbb{E}_{t,S}^{\hat{\mathbb{Q}}}\left[e^{-\frac{\gamma}{\beta}\mathbb{I}\{S_{T_0}>S^*\}\sum_{m=1}^{n}e^{-r(T_m-t)}\varphi(S_{T_m})}\right], \qquad t \in [0, T_0). \qquad (28b)$$

Proof. From the definition of the indifference price in (15), and applying the form of the value functions in Theorem 1, Eq. (27) follows immediately. The stochastic representation follows from an inductive application of Feynman-Kac applied to the collection of linear PDEs in (23a) and (24a). □

When the risk-aversion of the manager tends to zero, it is well known that the indifference price converges to the so-called Davis price ([8])—also known as the marginal price. In particular, the value of the option reduces to an expectation under the minimal martingale measure. Indeed, from Corollary 1, it is easy to see that

$$f(t, S) \xrightarrow[\gamma\downarrow 0]{} \begin{cases} \mathbb{E}_{t,S}^{\hat{\mathbb{Q}}}\left[\sum_{m:t<T_m}^{n} e^{-r(T_m-t)}\varphi(S_{T_m})\right], & t \in [0, T_0), \\ \mathbb{E}_{t,S}^{\hat{\mathbb{Q}}}\left[\sum_{m=1}^{n} e^{-r(T_m-t)}\varphi(S_{T_m})\,\mathbb{I}\{S_{T_0} > S^*\}\right], & t \in [T_0, T_n]. \end{cases} \qquad (29)$$

It is also well known that the valuation in incomplete markets using the minimal martingale measure leads to hedges which are locally risk minimizing—see e.g., [10].

4   An Illustrative Example

In this section, we illustrate how our methodology can be applied to the example of [7], who consider the option to invest in developing an unmanned aerial vehicle (UAV). The project consists of a significant investment of \$200 k required in 2 years, after which, the company will receive the estimated cash-flow stream of Table 1, where management has provided optimistic, likely and pessimistic cash-flow estimates. Clearly, management has the option to not invest at the end of year 2, at which point the future cash-flows would be forfeited. Thus, the project should be treated as a *real* option. The traded index parameters are $\mu = 8\%$ and $\sigma = 20\%$, and the risk-free rate is taken as $r = 5\%$; moreover, for the stochastic driver we choose the OU process with parameters $S_0 = 50$, $\kappa = 0.25$, $\bar{S} = 50$ and $\eta = 10$. Finally, the correlation $\rho = 0.5$. These parameters are used in the remainder of the document unless otherwise stated.

Table 1 An example cash-flow for a UAV project which costs $200 to invest in at year 2

| | End of year | | | | | | |
Scenario	3	4	5	6	7	8	9
Optimistic	80	116	153	177	223	268	314
Most Likely	52	62	74	77	89	104	122
Pessimistic	20	23	24	18	20	20	22

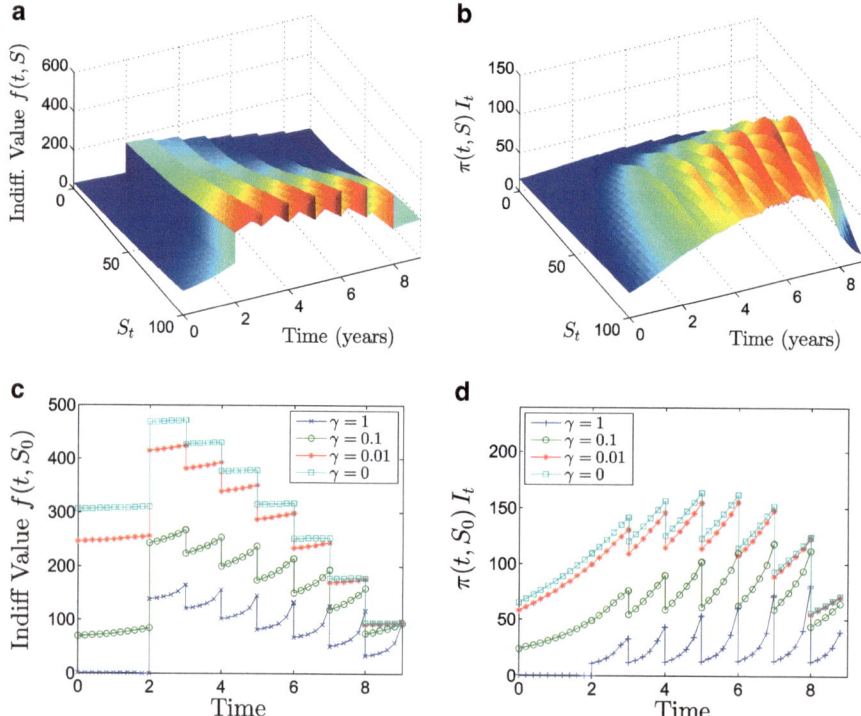

Fig. 5 Real option value and optimal investment strategy for the cash-flow in Table 1 when the stochastic driver is an OU process. (**a**) Value Surface $\gamma = 0.1$. (**b**) Investment Surface $\gamma = 0.1$. (**c**) Value along $S_t = S_0$. (**d**) Investment along $S_t = S_0$

To solve the system of PDEs (23) and (24) we utilize a simple a Crank-Nicholson finite-difference scheme on a finite grid of points in time and driver space. At the upper/lower boundaries ($\overline{S}/\underline{S}$) in S, we apply the boundary conditions that $\partial_{SS}H|_{\overline{S}} = \partial_{SS}H|_{\underline{S}} = 0$. Between each epoch period, we use the finite-difference scheme to recursively solve the PDEs backwards. At the cash-flow dates, the function H is modified according to the appropriate pasting conditions, and the solver continues in the new epoch using the modified H as the terminal boundary condition.

The real option value as a function of the market stochastic driver and time is shown in Fig. 5a for the case of $\gamma = 0.1$. There are two main observations. First, the value of the RO is increasing in the stochastic driver, but flattens out towards small and large values. The reason is that as one moves to large/smaller values the pessimistic/optimistic scenario are more likely to occur, and these scenarios are bounded. Second, the RO value increases as a cash-flow approaches, and then drops immediately after a cash-flow, reflecting the payout. The optimal investment $I\pi(t, S)$ in the traded asset is shown in Fig. 5b. Note that the optimal investment is smooth across the investment date T_0, but jumps at the investment times since there are fewer cash-flows remaining once a cash-flow has been paid out. Secondly, the optimal investment contains a peak around the mean-reverting level of 50 and also has a peak in the time dimension. The peak in S is because the optimal investment depends on the sensitivity of the RO with respect to the stochastic driver and as we can see from Fig. 5a, although the RO increases in S, it is bounded above and below, and therefore flattens out for large and small S. This leads to a peak in the optimal investment in the S direction. The peak in the time seems at first contradictory since there are less cash-flows remaining. However, between every cash-flow, the optimal investment amount increases in time because as a cash-flow approaches, its outcome is more certain and, just as in a standard call option, the investment increases in time. After a cash-flow the optimal investment drops.

The lower panels of Fig. 5 (panels (c) and (d)) show slices (from a fixed S) through time of the value and optimal investment when the stochastic driver is fixed at its mean-reverting point for four different risk aversion levels. Both the value and optimal investment decreases as risk aversion increases. What is interesting, however, is that for the largest risk-aversion level of $\gamma = 1$, while the RO has non-zero value post exercise, prior to exercise it is worthless. The reason is that at the time of investment, for this level of risk-aversion, the RO's indifference value is less than the strike K (no matter what is the outcome for the stochastic driver), hence, the manager will never exercise the option and it is worthless prior to T_0. This is quite different from approaches which consider lump-sum payments upon investment. If a lump-sum payment is provided upon investment, then, since the lump-sum is measurable on the investment day (i.e., $\mathscr{F}_{T_0}$), the indifference price of the lump-sum is the lump-sum amount itself, and as long as there is a positive probability that the lump-sum is greater than the investment amount, then the real-option will have a non-zero value (albeit it may be arbitrarily small).

5   Indifference Valuation for Optimal Exit

In many circumstances, if an investment is made at T_0, then the manager may have the option to permanently shutdown (with shutdown cost of C) the project prior to the last cash-flow date of T_n. It may be optimal to shutdown early if the stochastic driver is so low that the future remaining cash-flows provide the manager with lower utility than not shutting the project down. In this case, the valuation becomes a compound option where the option to shutdown the project acts as an American-styled option which feeds into the European option to invest in the project at time T_0. In terms of indifference pricing, the valuation problem in the region $t > T_0$ requires optimally comparing the value function provided by holding onto the option to keep running and receive the cash-flows versus paying the shutdown costs and then reducing to the Merton problem for the remainder of the time to T_n.

Consequently, the manager solves the following optimal control problem

$$V^{(3)}(t, x, S, I) = \sup_{\pi \in \mathscr{A}_t} \sup_{\tau \in \mathscr{T}} \mathbb{E}_{t,x,S,I}\left[-e^{-\gamma X_{T_n}^{\pi}}\right] \tag{30}$$

where the discounted wealth process now satisfies the SDE

$$dX_t^{\pi} = (\mu - r)\pi_t \, dt + \sigma \, \pi_t \, dB_t, \qquad \forall t \in (T_{k-1}, T_k), \quad \text{and} \tag{31a}$$

$$X_{T_0}^{\pi} = X_{T_0^-}^{\pi} - \tilde{K}\mathbb{I}\{\mathscr{E}\}, \tag{31b}$$

$$X_{T_k}^{\pi} = X_{T_k^-}^{\pi} + (\tilde{\varphi}_k(S_{T_k}) - F)\mathbb{I}\{\mathscr{E} \cap \tau > T_k\}, \tag{31c}$$

$$X_{\tau}^{\pi} = X_{\tau^-}^{\pi} - C. \tag{31d}$$

and, as before, the set $\mathscr{E} := \{V^{(3)}(T_0, X_{T_0^-}^{\pi} - \tilde{K}, S_{T_0}, I_{T_0}) > V^{(0)}(T_0, X_{T_0^-}^{\pi}, I_{T_0})\}$ represents the set on which the option to invest in the project is exercised at T_0. Moreover, $\mathscr{T}$ represents the set of $\mathscr{F}$-adapted stopping times in the interval $(T_0, T_n]$. Under general assumptions,[4] such optimal stopping/control problems can be solved by searching for stopping times which are of the form $\tau = \inf\{t : (t, S_t) \notin \mathscr{D}\}$ for some compact domain $\mathscr{D} \subset (T_0, T_n] \times \mathbb{R}_+$ (see for example [19]). In this case, the interior of the domain $\mathscr{D}$ represents the shutdown region and its boundary is the optimal shutdown boundary.

As before, to aid in the distinction between the pre ($t < T_0$) and post ($t > T_0$) investment regions, we introduce another value function corresponding to a manager who pays nothing to receive the cash-flows (together with the option to shutdown). To this end, let

$$V^{(4)}(t, x, S, I) = \sup_{\pi \in \mathscr{A}_t} \sup_{\tau \in \mathscr{T}} \mathbb{E}_{t,x,S,I}\left[-e^{-\gamma X_{T_n}^{\pi}}\right] \tag{32}$$

where the discounted wealth process now satisfies the SDE

$$dX_t^{\pi} = (\mu - r)\pi_t \, dt + \sigma \, \pi_t \, dB_t, \qquad \forall t \in (T_{k-1}, T_k), \quad \text{and} \tag{33a}$$

$$X_{T_k}^{\pi} = X_{T_k^-}^{\pi} + (\tilde{\varphi}_k(S_{T_k}) - F)\mathbb{I}\{\tau > T_k\}, \tag{33b}$$

$$X_{\tau}^{\pi} = X_{\tau^-}^{\pi} - C. \tag{33c}$$

Then, on the investment date T_0, the value function

---

[4]Since we have diffusion processes driving the relevant dynamics.

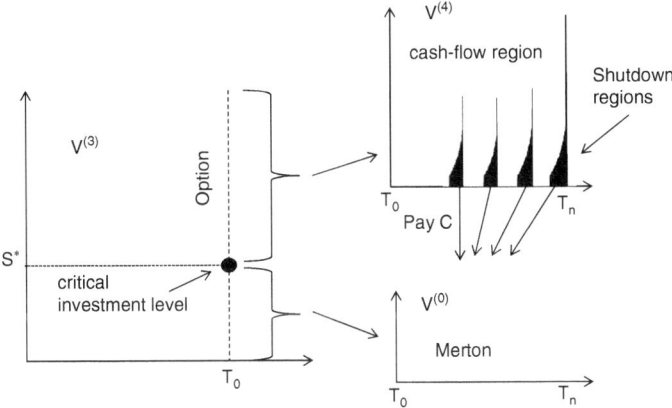

Fig. 6 The value function for the manager's optimization problem (30) when holding the option to invest and shutdown after investment can be viewed as first solving for (i) the Merton problem (10) and (ii) the cash-flow optimization problem (32) with shutdown option in the region $(T_0, T_n]$; and then pasting them together at the critical level S^*

$$V^{(3)}(T_0, x, S, I) = V^{(4)}(T_0, x, S, I) \, \mathbb{I}\{\mathcal{E}\} + V^{(0)}(T_0, x, S, I) \, \mathbb{I}\{\mathcal{E}^c\}.$$

This decomposition of the problem is similar to the previous case without the shutdown option and can be visualized as in Fig. 6. The difference is that now, the option to shutdown after investment allows the manager to pay the shutdown cost of C and revert to the Merton problem. Once again, the optimal investment set $\mathcal{E} = \{S_{T_0} > S^*\}$ for some S^*.

The DPP now implies that value function $V^{(4)}(t, x, S, I)$ satisfies the quasi-variational inequality (QVI)

$$\max \Bigg[(\partial_t + \mathcal{L}_S) \, V^{(4)}$$

$$+ \sup_\pi \left\{ (\mu - r)\pi \, \partial_x V^{(4)} + \tfrac{1}{2}\pi^2\sigma^2 \, \partial_{xx} V^{(4)} + \rho\eta\sigma \, \pi \, S\partial_{xS}V^{(4)} \right\} ; \qquad (34a)$$

$$V^{(0)}(t, x - C, S, I) - V^{(4)}(t, x, S, I) \Bigg] = 0,$$

for $t \in (T_{k-1}, T_k)$, $k = 1, \ldots, n$, subject to the sequence of terminal/pasting conditions

$$V^{(4)}(T_k^-, x, S, I) = V^{(4)}(T_k, x, S, I) \, e^{-\gamma \tilde{\varphi}_k(S)}, \quad \text{for} \quad k = 1, \ldots, n-1 \quad \& \tag{34b}$$

$$V^{(4)}(T_n, x, S, I) = e^{-\gamma \tilde{\varphi}_n(S)}. \tag{34c}$$

Clearly, $V^{(4)}$ is independent of I and we will drop I from its arguments from here onwards. The DPP implies that the value function $V^{(3)}(t, x, S, I)$ satisfies the same HJB equation as $V^{(1)}$, but with a new terminal condition, specifically,

$$(\partial_t + \mathscr{L}_S)\, V^{(3)} + \sup_{\pi} \left\{ (\mu - r)\pi\, \partial_x V^{(3)} + \tfrac{1}{2}\pi^2\sigma^2\, \partial_{xx} V^{(3)} + \rho\eta\,\pi\, S\partial_{xS} V^{(3)} \right\} = 0 \,,$$
(35a)

for $t \in [0, T_0)$, subject to the T_0 terminal condition

$$V^{(3)}(T_0^-, x, S) = V^{(4)}(T_0, x - \tilde{K}, S)\, \mathbb{I}\{S > S^*\} + V^{(0)}(T_0, x, S)\, \mathbb{I}\{S \le S^*\} \,.$$
(35b)

It is once again possible to reduce the dynamic programming equations into a system of quasi-variational inequalities and a linear PDE through a sequence of transformations. The final result is recorded in the following Theorem.

Theorem 2 (Value Function and Optimal Investment for Exit Problem). *The solution to Eqs. (34) and (35) admit the representation*

$$V^{(a)}(t, x, S, I) = V^{(0)}(t, x, I)\left(H^{(a)}(t, S)\right)^{\beta}$$

where $\beta = (1 - \rho^2)^{-1}$ for $a = 3, 4$. Moreover, $H^{(4)}(t, S)$ satisfies the linear QVI

$$\max\left\{ \partial_t H^{(4)} + \hat{\mathscr{L}}_S H^{(4)} \, ; \, e^{\frac{\gamma}{\beta}C} - H^{(4)} \right\} = 0, \quad t \in (T_{k-1}, T_k), \qquad k = 1, \dots, n$$
(36a)

with terminal/pasting conditions

$$H^{(4)}(T_k^-, S) = H^{(4)}(T_k, S)\, \exp\left\{ -\tfrac{\gamma}{\beta}\, \tilde{\varphi}_k(S) \right\}, \quad \textit{for} \quad k = 1, \dots, n-1, \quad \textit{and}$$
(36b)

$$H^{(4)}(T_n, S) = \exp\left\{ -\tfrac{\gamma}{\beta}\, \tilde{\varphi}_n(S) \right\},$$
(36c)

and $H^{(3)}(t, S)$ satisfies the linear PDE

$$\partial_t H^{(3)} + \hat{\mathscr{L}}_S H^{(3)} = 0, \quad t \in [0, T_0),$$
(37a)

with terminal condition

$$H^{(3)}(T_k^-, S) = 1 - \left(1 - H^{(4)}(T_k, S)e^{\frac{\gamma}{\beta}\tilde{K}}\right) \mathbb{I}\{S > S^*\},$$
(37b)

where S^* is the unique solution to

$$H^{(4)}(T_0, S^*) = e^{-\frac{\gamma}{\beta}\tilde{K}}.$$
(38)

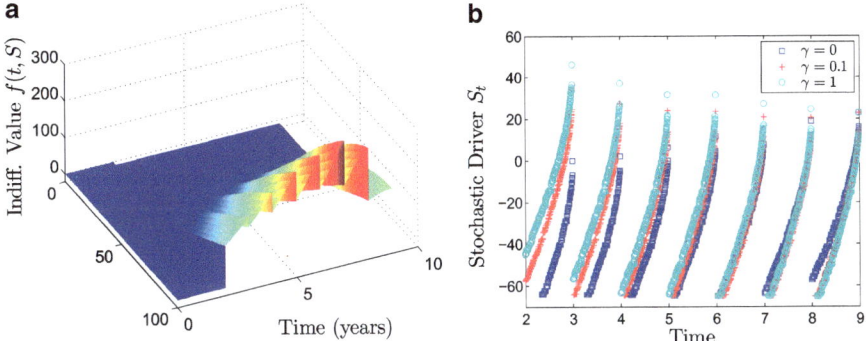

Fig. 7 The indifference valuation surface and optimal exit boundaries for the entry-exit problem with the cash-flow in Table 1 when the stochastic driver is an OU process. The fixed cost $F = 50$, shutdown costs $C = 10$. (**a**) Indifference Value with $\gamma = 0.1$. (**b**) Exit Boundaries

Moreover, the optimal investment policy is

$$\pi^{(a)} = \frac{1}{\gamma} \frac{\mu - r}{\sigma^2} + \rho \frac{\eta}{\sigma} \frac{\beta}{\gamma} S \partial_S \ln H^{(a)}. \tag{39}$$

To solve this system numerically, we once again applied a simple Crank-Nicholson finite-difference scheme to obtain the continuation value at each step and compare with the immediate exercise value between epoch dates. At epoch dates, the pasting conditions are applied and the modified H used as the terminal condition for the next epoch. The additional boundary conditions along $(\overline{S}/\underline{S})$ of $\partial_{SS}H|_{\overline{S}} = \partial_{SS}H|_{\underline{S}} = 0$ were once again applied. Naturally, there are many more efficient and methods that have faster convergence, such as SOR or penalty methods; however, such numerical issues are not the focus here.

We next apply these results to the example RO introduced in Sect. 4 and illustrate the results in Fig. 7. Here, in addition, we include a running cost F, so that all cash-flows are reduced by this fixed amount. Figure 7a shows the indifference valuation surface. Notice that this surface has some negative values when the stochastic driver is significantly small which reflects the fact that in those scenarios the resulting cash-flows will be pushed towards the pessimistic predictions and therefore the manager is more likely to pay the exit cost and shutdown the project. Figure 7b shows the effect that risk aversion has on the optimal exit boundaries. For the early cash-flows, the exit boundaries move upwards as the manager becomes more risk averse and hence the manager shuts down the project earlier to avoid potential losses. Closer to the end of the investment horizon, however, the optimal exit boundaries decrease. This reversal of the manager's behaviour is because we have assumed that the manager also pays the shutdown costs at the end of investment horizon in all scenarios.

6   Conclusions

In this work we present a real options approach that is practical to implement, and utilizes manager's specified cash-flows distributions rather than postulating an adhoc stochastic model for project values. We developed an approach where any cash-flow estimate can be matched exactly and demonstrate how our matching approach can be layered on top of indifference pricing through two classical real-options problems: the option to invest in a project and the optimal exit problem. One key economic finding from our approach is that if the manager's risk aversion is high, the real-option value may drop to zero. This occurs because the indifference value of the future cash-flows, on the date of the investment, may be lower than the strike price of the option if the manage is risk averse enough. Many indifference pricing approaches instead always assign a value to the real option regardless of the level of risk aversion. The main aspect of standard approaches which results in this behaviour is that they typically assume the project value is paid as a lump-sum upon investment, while here we assume the manager receives a cash-flow stream.

Several other directions are still open for future research. For example, investigating other real option problems such as valuing the early entry problem, early entry and exit, the duopoly problem, among others. As well, it would be interesting to assess the resulting hedging strategy and hedging performance using our framework. Finally, incorporating regime-switching and/or jumps in the stochastic driver process could lead to interesting features in the valuation as well as the resulting hedging strategies.

Acknowledgements SJ would like to thank NSERC for partially funding this research.

Appendix: Proof of Results

In this Appendix we provide concise proofs of the main results.

Proof of Proposition 1

We seek $\varphi(.)$ such that $\mathbb{P}(\varphi(S_T) \le v | \mathscr{F}_0) = F^*(v)$. Since,

$$S_T | \mathscr{F}_0 \overset{d}{=} S_0 \exp\left\{(v - \tfrac{1}{2}\eta^2)T + \eta\sqrt{T}Z\right\} \quad \text{where} \quad Z \underset{\mathbb{P}}{\sim} \mathscr{N}(0,1),$$

we have that

$$\mathbb{P}(\varphi(S_T) \le v | \mathscr{F}_0) = \Phi\left(\frac{\ln \frac{\varphi^{-1}(v)}{S_0} - (v - \tfrac{1}{2}\eta^2)T}{\eta\sqrt{T}}\right) \overset{\triangle}{=} F^*(v). \tag{40}$$

Consequently, if $F^*(.)$ is invertible then

$$\varphi(S) = F^{*-1}\left(\Phi\left(\frac{\ln\frac{S}{S_0} - (v - \frac{1}{2}\eta^2)T}{\eta\sqrt{T}}\right)\right) \tag{41}$$

and the proof is complete.   □

Proof of Proposition 2

Here, we prove that the co-dependence structure of the cash-flow distribution is governed by a Gaussian copula. We require the following joint distribution function:

$$\mathbb{P}(V_1 < v_1, \ldots, V_n < v_n)$$
$$= \mathbb{P}(\varphi_1(S_{T_1}) < v_1, \ldots, \varphi_n(S_{T_n}) < v_n)$$
$$= \mathbb{P}(F_1^{*-1}\left(\Phi(z(T_1, S_{T_1}))\right) < v_1, \ldots, F_n^{*-1}(\Phi(z(T_n, S_{T_n}))) < v_n) ,$$

where $z(T, S) = \frac{1}{\eta\sqrt{T}} \ln\frac{S}{S_0} - \frac{v - \frac{1}{2}\eta^2}{\eta}\sqrt{T}$. Clearly,

$$S_{T_k} \overset{d}{=} S_0 \exp\left\{(v - \frac{1}{2}\eta^2)T_k + \eta\sqrt{T_k}Z_k\right\}$$

where $\{Z_1, \ldots, Z_k\}$ are jointly normal with mean zero and covariance matrix $\Omega_{ij} = \sqrt{T_{\min(i,j)}/T_{\max(i,j)}}$.

Since each distribution function F_k^* is assumed invertible, we then have

$$\mathbb{P}(V_1 < v_1, \ldots, V_n < v_n) = \mathbb{P}\left(Z_1 < \Phi^{-1}\left(F_1^*(v_1)\right), \ldots, Z_n < \Phi^{-1}\left(F_n^*(v_n)\right)\right)$$
$$= \Phi_\Omega\left(\Phi^{-1}\left(F_1^*(v_1)\right), \ldots, \Phi^{-1}\left(F_n^*(v_n)\right)\right) .$$

This completes the proof.   □

Proof of Theorem 1

Here we provide a sketch of the proof. The first order condition in the HJB equations (18) and (19) provide the optimal investment policy in feedback control form as

$$\pi^{(a)} = -\frac{(\mu - r)\partial_x V^{(a)} + \rho\eta\sigma S\partial_{xS} V^{(a)}}{\sigma^2\partial_{xx} V^{(a)}}. \tag{42}$$

The HJB equations then reduce to

$$(\partial_t + \mathscr{L}_S)\, V^{(a)} - \frac{1}{2}\frac{\left[(\mu - r)\partial_x V^{(a)} + \rho\eta\sigma S\partial_{xS} V^{(a)}\right]^2}{\sigma^2 \partial_{xx} V^{(a)}} = 0, \qquad (43)$$

subject to the appropriate terminal conditions. Writing

$$V^{(a)}(t, x, S, I) = V^{(0)}(t, x, I)h^{(a)}(t, S),$$

the above PDE reduces to

$$(\partial_t + \mathscr{L}_S)\, h^{(a)}(t, S) + \rho\eta\lambda\, S\partial_S h^{(a)}(t, S) + (\rho\eta)^2\frac{(S\partial_S h^{(a)}(t, S))^2}{h(t, S)} = 0, \qquad (44)$$

Now setting $h^{(a)}(t, S) = \left(H^{(a)}(t, S)\right)^{\beta}$ after some tedious computations, the above non-linear PDEs for $h^{(a)}$ reduces into the linear PDEs (23) and (24) for $H^{(a)}$. Moreover, the boundary conditions for $V^{(a)}$ become the stated boundary conditions for $H^{(a)}$. Since classical solutions exist for the linear PDE system (23) and (24), and the resulting feedback controls are admissible, the usual arguments imply that the solution of DPE is the solution to the original optimal control problem. The uniqueness of S^* follows from the fact the terminal conditions and the subsequent pasting conditions are decreasing functions of S. Hence, the H function inherits this property and, therefore, the solution to (25) is unique.   □

Proof of Theorem 2

Here we provide a sketch of the proof. The first order condition in the HJB equations (46) and (35a) provide the optimal investment policy in feedback control form as $(a = 3, 4)$

$$\pi^{(a)} = -\frac{(\mu - r)\partial_x V^{(a)} + \rho\eta\sigma S\partial_{xS} V^{(a)}}{\sigma^2 \partial_{xx} V^{(a)}}. \qquad (45)$$

The DPEs then reduce to

$$\max\left\{(\partial_t + \mathscr{L}_S)\, V^{(4)} - \frac{1}{2}\frac{\left[(\mu - r)\partial_x V^{(4)} + \rho\eta\sigma S\partial_{xS} V^{(4)}\right]^2}{\sigma^2 \partial_{xx} V^{(4)}}\;;\right.$$

$$\left. V^{(0)}(t, x - C, S, I) - V^{(4)}(t, x, S, I)\right\} = 0,$$

$$(46)$$

and

$$(\partial_t + \mathcal{L}_S) V^{(3)} - \frac{1}{2} \frac{\left[(\mu - r)\partial_x V^{(3)} + \rho\eta\sigma S\partial_{xS}V^{(3)}\right]^2}{\sigma^2\partial_{xx}V^{(3)}} = 0, \qquad (47)$$

subject to the appropriate terminal conditions. Writing

$$V^{(a)}(t, x, S, I) = V^{(0)}(t, x, I)h^{(a)}(t, S),$$

and setting $h^{(a)}(t, S) = \left(H^{(a)}(t, S)\right)^{\beta}$ after some tedious computations, the above non-linear DPEs for $h^{(a)}$ reduce into the linear DPEs (36) and (37) for $H^{(a)}$. Moreover, the boundary conditions for $V^{(a)}$ become the stated boundary conditions for $H^{(a)}$. Standard results imply that the viscosity solution of the linear DPE system (36) and (37) is the solution to the original optimal control problem. The exercise point S^* is unique once again due to the boundary conditions and pasting conditions being decreasing in S. □

References

1. Ané, T., Kharoubi, C.: Dependence structure and risk measure. J. Bus. **76**(3), 411–438 (2003)
2. Berk, J., Green, R., Naik, V.: Valuation and return dynamics of new ventures. Rev. Financ. Stud. **17**(1), 1–35 (2004)
3. Borison, A.: Real options analysis: where are the emperor's clothes? J. Appl. Corp. Finance **17**(2), 17–31 (2005)
4. Carlos, J., Nunes, J.: Pricing real options under the constant elasticity of variance diffusion. J. Futur. Mark. **31**(3), 230–250 (2011)
5. Carmona, R. (ed.): Indifference Pricing: Theory and Applications. Princeton University Press, Princeton (2008)
6. Copeland, T., Antikarov, V.: Real Options: A Practitioner's Guide. W. W. Norton and Company, New York (2001)
7. Datar, V., Mathews, S.: European real options: an intuitive algorithm for the black- scholes formula. J. Appl. Finance **14**(1), 45–51 (2004)
8. Davis, M.: Option pricing in incomplete markets. In:Mathematics of Derivative Securities, pp. 216–226. The Newton Institute, Cambridge University Press, Cambridge (1998)
9. Dixit, A., Pindyck, R.: Investment Under Uncertainty. Princeton University Press, Princeton (1994)
10. Föllmer, H., Schweizer, M.: Hedging of contingent claims under incomplete information. In: Davis, M.H., Elliott, R.J. (eds.) Applied Stochastic Analysis: Stochastics Monographs, pp. 389–414. Gordon and Breach, London (1991)
11. Grasselli, M.: Getting real with real options: a utility-based approach for finite-time investment in incomplete markets. J. Bus. Finance Account. **38**(5&6), 740–764 (2011)
12. Henderson, V.: Valuing the option to invest in an incomplete market. Math. Finan. Econ. **1**, 103–128 (2007)
13. Hugonnier, J., Morellec, E.: Corporate control and real investment in incomplete markets. J. Econ. Dyn. Control. **31**, 1781–1800 (2007)
14. Leung, T., Sircar, R.: Accounting for risk aversion, vesting, job termination risk and multiple exercises in valuation of employee stock options. Math. Financ. **19**(1), 99–128 (2009)

15. Metcalf, G., Hasset, K.: Investment under alternative return assumptions. comparing random walk and mean revertion. J. Econ. Dyn. Control. **19**, 1471–1488 (1995)
16. Miao, J., Wang, N.: Investment, consumption, and hedging under incomplete markets. J. Financ. Econ. **86**(3), 608–42 (2007)
17. Nelsen, R.B.: An Introduction to Copulas. Springer, New York (2006)
18. Oriani, R., Sobrero, M.: Uncertainty and the market valuation of R&D within a real options logic. Strateg. Manag. J. **29**(4), 343–361 (2008)
19. Pham, H.: Continuous-time Stochastic Control and Optimization with Financial Applications. Springer, Berlin (2009)
20. Sarkar, S.: The effect of mean reversion on investment under uncertainty. J. Econ. Dyn. Control. **28**, 377–396 (2003)
21. Trigeorgis, L.: Real Options: Managerial Flexibility and Strategy in Resource Allocation. MIT Press, Cambridge (1996)
22. Tsekrekos, A.E.: The effect of mean reversion on entry and exit decisions under uncertainty. J. Econ. Dyn. Control. **34**(4), 725–742 (2010)

Real Options with Regulatory Policy Uncertainty

Christian Maxwell and Matt Davison

Abstract Energy Finance as a field is particularly bedeviled by regulatory uncertainty. This is notably the case for the real option analysis of long-lived energy infrastructure. How can one decide optimal build times on a 50 year project horizon when regulations regarding pricing and costs change on a much shorter time scale? In this paper we present a quantitative framework for modelling and interpreting regulatory changes for energy real options as a Poisson jump process, in a context where other relevant prices follow diffusion processes. We illustrate this conceptual framework with a case study involving the US corn ethanol market for which subsidy levels have experienced frequent changes. Subsidy levels have an easily quantified impact on operations and profitability, making this a nice arena to introduce ideas which might later be extended to less easily quantified regulatory changes. Numerical techniques are presented to solve the resulting partial integro differential variational inequalities. These solution techniques are deployed to solve instructive numerical examples, and conclusions for public policy are drawn.

1 Introduction

All large energy and natural resource projects are subject to government policy or regulation of some kind. These regulations are intended to achieve public policy goals and their effects should be taken into account by firms planning to enter into energy or resource investments. Energy and resource projects often have long project horizons and operating life spans on the order of decades. Consider the example of a firm deciding to enter into a 50 year energy production investment. Policy in terms of taxation, environmental regulations and other laws may materially affect project cash flows. These policies have been known to change. Some policy

C. Maxwell
Department of Applied Mathematics, University of Western Ontario, London, ON, Canada
e-mail: cmaxwel5@uwo.ca

M. Davison (✉)
Department of Statistical and Actuarial Sciences, University of Western Ontario, London, ON, Canada
e-mail: mdavison@uwo.ca

© Springer Science+Business Media New York 2015

R. Aïd et al. (eds.), *Commodities, Energy and Environmental Finance*,
Fields Institute Communications 74, DOI 10.1007/978-1-4939-2733-3_9

239

amendments are well broadcast and announced while others are not. Although policy changes may appear "predictable" in the short term, forecasting onto a 50 year project horizon renders the policy changes apparently random, and hence requiring models of *policy uncertainty*.

Policy Uncertainty is characterized by changes in taxation, legal and other regulatory policies that affect a business' operations and profitability. The uncertainty derives from the inability to predict policy in the long term; uncertainty about forthcoming policy or announcements of policy changes; or sudden and abrupt changes in policy. Some anecdotal examples of policy uncertainty in energy and resource markets from recent North American news headlines follow:

Ontario Looks Set to Cut Green Energy Subsidies Solar rates expected to be cut substantially. Industry has 6 weeks to provide input. [31]

Ontario Drops Plan for TransCanada Power Plant Ontario cancels planned Trans Canada power plant with province to discuss compensation with TransCanada. Costs may exceed $1 billion CAD and affect off peak pricing. [24, 30]

Ivanhoe 'Surprised' by New Mongolian Windfall Tax Mongolia sets surprise windfall tax on (among possibly others) Ivanhoe's Oyu Tolgoi mine of 68 % when gold hits $500 per ounce. [7]

This does not by any means represent an exhaustive list. Attempts have been made to quantify and measure policy uncertainty (e.g. [3]). In [3] and [17] the authors also note that policy uncertainty can make firms hesitate or delay to enter into long term projects as they wait for more policy certainty before making decisions. This has caught the eye of Canadian and American macroeconomic policy makers noting both that firms appear to accumulate cash and hesitate to make business decisions amidst regulatory uncertainty [16, 35].

In this paper we present a quantitative framework for modelling and interpreting regulatory changes for energy real options as a jump diffusion process, in a context where other relevant prices follow pure diffusion processes. Policy uncertainty is by its nature very difficult to hedge and leads to market incompleteness even if the remaining underlying prices could otherwise be traded.

This real option method of modelling resource project management decisions was introduced by [6] in a seminal paper that considered the problem of optimally starting and stopping production to maximize the profits of a natural resource project. The optimal entry and exit from investment projects was also considered by [11] in another classical real option paper. A collection of illustrative real option papers can be found in [12].

In particular, we consider a firm contemplating the option to invest in an ethanol from corn production plant. We build on the analysis of our past work [23] which intended to quantify the impact (both intended and unintended consequences) of ethanol policy on production. This current work adds the complication of policy uncertainty deriving from a volumetric production tax subsidy which has changed several times over the past 35 years. We aim to understand the effects of ethanol policy uncertainty on production from the producer's perspective. An example of the application of real option analysis to understand the effects of windfall taxes on mining operations can be found in [33]. A complementary and interesting analysis on policy uncertainty and real options can be found in [17]. The authors of [17] use empirical data to determine how regulatory uncertainty in American electricity markets affects start up and shut down decisions for power plants; their evidence supports the anecdotal claims mentioned above that uncertainty leads management to defer decision making. Our real option model sets out to design a framework to quantitatively model this added uncertainty and capture its effects on decision making.

1.1 Corn Ethanol Production and Subsidy Policy

The ethanol market in the US is large, estimated at 13.3 billion gallons produced in 2012 by over 209 plants [32]. Efforts to promote US energy independence and initiatives to obtain fuel from environmentally friendly sources have led to the subsidization of the production of ethanol biofuel from corn. Subsidies have historically been provided to ethanol producers by means of a volumetric ethanol excise tax credit for blenders and a small ethanol producer tax credit. The subsidy amount has changed from \$0.40/gallon at its introduction in 1978 (Energy Tax Act) and been adjusted several times until its final level \$0.45/gallon in the 2008 Farm Bill followed by termination (by non-renewal) at the end of 2012 [13, 15]. Table 1 shows the history of ethanol subsidy policy changes and amendments since its inception.

A year following the lapse of many of the energy subsidies, about one quarter of Nebraska's ethanol plants were in idle status [27]. The loss of the subsidy was a possible contributing factor to the shut downs as [21] note that, without subsidies, ethanol plants may lose their economic viability.

Table 1 Historical ethanol subsidies. Source: [15]

Act	Year	Subsidy (\$/gallon)
Energy tax act	1978	0.40
Surface transportation assistance act of 1982	1983	0.50
Tax reform act	1984	0.60
Omnibus budget reconciliation act	1990	0.54
1998 policy adjustment effective 2001	2001	0.53
1998 policy adjustment effective 2003	2003	0.52
Extension of policy with adjustment	2005	0.51
Farm bill	2008	0.45
Expiration of tax credit	2012	–

1.2 Outline

Our paper uses a crush spread analysis to value a facility which produces ethanol from corn using a real options analysis following our framework in [23]. The outline is as follows: Section 2 specifies the plant characteristics, management decisions, and associated costs and profits. Section 3 derives the stochastic optimal control problem for the optimal plant operating rule. Section 4 illustrates the numerical results. Finally Sect. 5 draws conclusions about policy uncertainty and its effects on ethanol production, closing off with some policy recommendations.

2 The Real Option Model

Management contemplating the decision to invest in an ethanol production plant has the flexibility to enter or defer the project given price conditions and expected future profitability [12]. After initiating and building the ethanol plant, management again has the flexibility to switch production on (1) and off (0) given prevailing economic conditions. The goal of this paper is to examine how ethanol price and policy uncertainty affects a producer's business entry and subsequent operating decisions given price conditions, subsidy policy expectations, and the remaining project life.

Following our analysis [23], throughout this paper all currency is in United States dollars (USD); liquid volume is in gallons; solid volume is in bushels; weight is in tons; and interest is percent per year appropriate to USD deposits continuously compounded.

2.1 Plant Specification and Operating Costs

The following costs are scaled in terms of gallon of production capacity per year and were estimated by [34]. The model is based on our detailed ethanol real option analysis in [23]. This valuation considers the income stream associated with the production of ethanol from corn along with the ethanol-gasoline blender subsidy.

The capitalized construction cost B is estimated at \$1.40/gallon for a "typical" sized facility with nameplate capacity of 40,000,000 gallons/year. The plant salvage value Q is estimated at 10 % of capitalized cost. The switching cost D_{01} to resume production from an idle state is estimated at 10 % of capitalized cost per gallon of annual production capacity. Similarly, the switching cost D_{10} to pause production from an active operating state is estimated at 5 % of capitalized cost per gallon of annual production capacity.

2.2 Running Profits

The plant produces ethanol L_t (priced in USD/gallon) from corn C_t (priced in USD/bushel). The running profit from the corn ethanol crush spread is developed in [23] on a per bushel per year basis assuming the popular dry grind process for producing ethanol [4].

$$\text{corn} \rightarrow \text{ethanol} + \text{by-products} \tag{1}$$

The profit function while operating, f_1, is given by

$$f_1(L_t, C_t, Z_t) = \kappa(L_t + Z_t - K_1) - C_t \tag{2}$$

where Z_t is the government volumetric subsidy (USD/gallon). The conversion factor $\kappa = 2.8$ is the yield in terms of gallons of ethanol produced per bushel of corn [4] and is consistent with the CME Group's references on trading ethanol crush spreads [8].

The net running cost while on can be decomposed in terms of the fixed running cost p of \$0.68/gallon, less the average by-product distillers dried grains G (USD/ton) produced per bushel of corn [21, 23, 34]

$$K_1 = p - \frac{\omega}{\kappa} G. \tag{3}$$

The process produces 17 lbs of by-product per bushel and hence the yield factor $\omega = 17/2000$ [4].[1]

While production is idle, [34] estimated that fixed running costs K_0 are roughly 1 % of capitalized construction costs per gallon of production capacity or 20 % of fixed running cost while in production (note that, while idle, no ethanol is produced and consequently no subsidy is applied). The profit function while off, f_0, is

$$f_0(L_t, C_t, Z_t) = -\kappa K_0 \tag{4}$$

where the midpoint between the two estimates is used [23]

$$K_0 = \frac{0.01B + 0.20p}{2}. \tag{5}$$

Finally, the interest rate r is taken to be a target return of 8 % per annum continuously compounded to capture the risk associated with the ethanol project cash flows [23, 34]. Our analysis uses only the physical measure for the stochastic assets. We note however that the price risk associated with corn and ethanol can

---

[1] There are 2000 lbs in a ton.

be hedged using futures and the arbitrage free return can be determined by noting that the jumps are not correlated with the market following an argument popularized in [25].

2.3 Stochastic Price Models

Following our analysis in [23], ethanol L_t and corn C_t are modelled by a joint geometric Brownian motion (GBM) diffusion

$$dL_t = \mu L_t dt + \sigma L_t dW_{1t} \tag{6}$$

$$dC_t = aC_t dt + bC_t dW_{2t} \tag{7}$$

$$\text{Corr}[W_{1t}, W_{2t}] = \rho \tag{8}$$

where (W_{1t}, W_{2t}) is a two-dimensional Brownian motion defined on a filtered probability space $(\Omega, \mathscr{F}_t, P)$ which satisfies the usual conditions [28].

The econometric model parameters are estimated by ordinary least-squares regression of the log time series $\ln \frac{X_t}{X_{t-1}}$ using the 10 year monthly historical price series from Dec/02–Jan/11 capitalizing on earlier work in [23]. Prices for no. 2 yellow corn Omaha, NE underlying the CME corn futures contract were obtained from [36]. Average rack prices freight on board for ethanol were obtained from [26]. The correlation estimate ρ was obtained via the sample correlation of the residuals. Parameter estimation results are in Table 2. Note that both drifts were found to be statistically zero at the 95 % confidence interval. The estimate for the average distillers dried grains price $\hat{G}$ was estimated by regressing the time series against a constant.

The stochastic subsidy Z_t is modelled as a pure Poisson arrival time jump process with arrival rate λ and jumps of size J.

$$dZ_t = (J - Z_t)dN_t \tag{9}$$

Table 2 Regression estimation results

Parameter estimate	Value	t-test
$\hat{\mu}$	0	$P\left(\frac{\hat{\mu}-\mu}{s.e.} > t \middle| \mu = 0\right) = 0.409$
$\hat{\sigma}$	0.156	–
$\hat{a}$	0	$P\left(\frac{\hat{a}-a}{s.e.} > t \middle| \mu = 0\right) = 0.202$
$\hat{b}$	0.123	–
$\hat{\rho}$	0.105	–
$\hat{G}$	\$115.6	$G \in [108.4, 122.8]^a$

[a] based on 95 % confidence interval Student-t with 119 degrees of freedom

Table 3 Maximum likelihood estimation results

Parameter estimate	Estimator	Value	95 % confidence interval
$\hat{\lambda}$	$\left(\frac{1}{n}\sum_{i=1}^{n} t_i - t_{i-1}\right)^{-1}$	0.24	$[0.10, 0.42]$
$\hat{\alpha}$	$\frac{1}{n}\sum_{i=1}^{n}\ln x_i$	-0.69	$[-0.79, -0.58]$
$\hat{\beta}^2$	$\frac{1}{n-1}\sum_{i=1}^{n}(\ln x_i - \hat{\alpha})^2$	0.015^{a}	$[0.0066, 0.062]$

[a] Corrected unbiased estimator

where dN_t, defined on the probability space, is a continuous-time counting process $\{N_t, t \geq 0\}$ that counts the number of jumps over time dt and

$$dN_t = \begin{cases} 1 & \text{with probability } \lambda dt \\ 0 & \text{otherwise.} \end{cases} \qquad (10)$$

The times between jumps $t_i - t_{i-1}$ are seen to be quite well modelled by independently exponentially distributed Poisson arrivals. The jumps J are assumed to be drawn from a lognormal distribution with parameters $LogN(\alpha, \beta^2)$. The parameters are estimated via maximum likelihood using the data in Table 1. The estimation results are summarized in Table 3.

The sample set for the subsidy policy is small (8 observations) and requires a test of the goodness of fit. By our model choice, the time between arrivals Δt of subsidy changes is exponentially distributed with parameter λ $(Exp(\lambda))$ and the series $\frac{\ln Z_t - \hat{\alpha}}{\hat{\beta}}$ has a Student's t-distribution since $\ln Z_t \sim N(\alpha, \beta^2)$. The plots of the estimated theoretical cumulative distribution functions (CDFs) versus the empirical distributions are included in Fig. 1 along with the QQ plots. By visual inspection, both data appear to be well suited to the proposed subsidy model.

Lilliefors tests (a nonparametric variant of the Kolmogorov-Smirnoff test) were applied to test for normality in the log subsidy series and exponentiality in the subsidy arrival times using MATLAB's `lilliefors.m` function. Both samples accepted the null hypothesis of normality and exponentiality at the 5 % significance level. This statistical evidence further supports our proposed model.

2.4 Policy Uncertainty "at its Worst"

Since the policy uncertainty cannot be hedged and is presumably not correlated with any market assets, there is cause for concern in terms of how to price this ethanol real option. Not only is there risk in the randomness of the process, but there is an

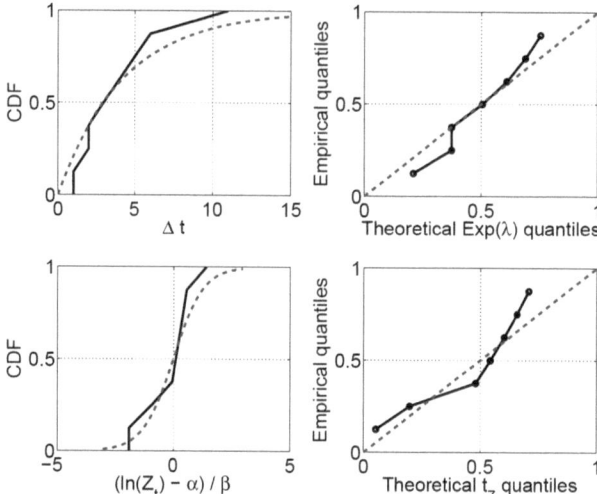

Fig. 1 The empirical CDF (*solid black*) vs the theoretical CDF (*grey dashed*) of the time between arrivals $\Delta t \sim Exp(\hat{\lambda})$ (*upper left*). The QQ plot of the time between arrivals (*upper right*). The empirical CDF (*solid black*) vs the theoretical CDF (*grey dashed*) of the normalized subsidy series $\frac{\ln Z_t - \hat{\alpha}}{\hat{\beta}} \sim t_7$ (*lower left*). The QQ plot of the subsidy series (*lower right*)

added complexity of uncertainty risk in the choice of model so-called "Knightian" uncertainty. To account for this model risk, uncertainty around the jump process parameters is included.

There are several possible ways to deal with model uncertainty and market incompleteness including: (1) cautiously deploying assumptions to simplify the problem; (2) utility indifference pricing with model uncertainty [19, 22]; and (3) best/worst case pricing (similar to the idea of good deal bounds and super-replication) [2]. Our analysis follows alternative (3) due to its financial intuition, transparency, and lack of subjectivity around economic aversion parameters or choice utility functions associated with utility-based pricing (which produce a subjective "personal price"). There is a connection between (2) and (3) however, in that as the risk aversion parameter tends to infinity, the utility indifference price tends to the worst-case price. Management buying into an ethanol project can be considered "long" the real option. The worst case price is what a strongly risk averse buyer may consider when purchasing an option.

Management contemplating investment in an ethanol project may ask the question: *Given the uncertainty around subsidy policy over the past 35 years, what is the expected case and worst case project value?* To answer this question, the reference policy uncertainty distribution is adjusted within the following heuristically determined parameter bounds to form best and worst case bounds for the project value.

2.4.1 Bounds on α

Suppose management assumes VaR_{05} style bounds on α.[2] In order to choose a lower bound for α, management chooses a parameter α_{min} such that the probability of observing a subsidy level J lower than the lowest historical subsidy $Z_{min} = 0.40$ is 95 %, i.e. $P(J < Z_{min}) = 0.95$. For a lognormal distribution with variance $\beta^2 = 0.015$, $\alpha_{min} = -1.118$. An upper bound can be chosen as α_{max} such that the probability of observing a lower subsidy J than the historical maximum $Z_{max} = 0.60$ is also less than 5 %, i.e. $P(J < Z_{max}) = 0.05$. In this case, the upper bound is $\alpha_{max} = -0.309$.[3]

2.4.2 Bounds on λ

Similarly, the average arrival time of subsidy changes is bounded by infinity (i.e. no changes at all) where $\lambda_{min} = 0$. Reasoning that the US Farm Bill is the primary means by which ethanol subsidy policies are amended and that a new omnibus bill is passed every 5 years or so, λ_{max} can be chosen such that the probability of observing at least one jump in a 5 year cycle is at least 95 %. Thus management seeks λ_{max} such that $P(k = 0; \lambda_{max}, t = 5) \leq 0.05$ (i.e. the probability of observing zero jumps is at most 5 %) where the probability of exactly k jumps occurring over t is $P(k; \lambda, t) = \frac{(\lambda t)^n}{n!} e^{-\lambda t}$. This is given by $e^{-\lambda_{max} 5} \leq 0.05 \Rightarrow \lambda_{max} = \frac{\ln(0.05)}{5}$ or $\lambda_{max} = 0.60$.

2.4.3 The Best and Worst Case Bounds

The best and worst case bounds can be summarized by the following:

$$\alpha \in [\alpha_{min}, \alpha_{max}] = [-1.118, -0.309] \tag{11}$$

$$\lambda \in [\lambda_{min}, \lambda_{max}] = [0, 0.60]. \tag{12}$$

---

[2] We note that management could use another technique to choose bounds such as the 95 % confidence intervals on the mean estimate for example in Table 3.

[3] We note that these bounds were chosen heuristically based on ethanol policy history and with reference to political precedent of the subsidy level. They do not represent a rigorous mathematical treatment of the small sample population time series.

3   The Stochastic Control Problem

In this section, we develop the jump diffusion counterpart of our model in [23] which leads to a system of interconnected obstacle problems, i.e. partial integro differential (PID) variational inequalities.

The total expected earnings V_i over the life of the project is given by the sum of its profits, plus the sum of any switching costs incurred over its operating life

$$V_i(l, c, z, t) = \sup_{\tau, u} E\left[\int_t^T e^{-r(s-t)} f_{I_s}(L_s, C_s, Z_s) ds \right.$$

$$\left. + \sum_{k=1}^n e^{-r(\tau_k - t)} D_{u_{k-1}, u_k} \,\middle|\, (L_t, C_t, Z_t, u_0) = (l, c, z, i)\right] \quad (13)$$

The pair (τ, u) is the control that the manager has over the facility in his ability to toggle production on and off. It consists of a set of switching times τ_k and states to be switched into u_k with $I_t = u_k, t \in [\tau_k, \tau_{k+1})$. Thus τ_k is an increasing set of switching times with $\tau_k \in [t, T]$ and $\tau_k < \tau_{k+1}$ given the initial operating state $u_0 = i$.

If management assumes a worst case pricing scenario for the policy parameters (λ, α), then

$$V_i(l, c, z, t) = \sup_{\tau, u} \inf_{\lambda, \alpha} E\left[\int_t^T e^{-r(s-t)} f_{I_s}(L_s, C_s, Z_s) ds \right.$$

$$\left. + \sum_{k=1}^n e^{-r(\tau_k - t)} D_{u_{k-1}, u_k} \,\middle|\, (L_t, C_t, Z_t, u_0) = (l, c, z, i)\right] \quad (14)$$

where $\lambda \in [\lambda_{min}, \lambda_{max}]$ and $\alpha \in [\alpha_{min}, \alpha_{max}]$. The limits on λ and α prevent the optimization argument from growing unbounded and becoming singular [29]. The controls $(u, \tau, \alpha, \lambda)$ come from an admissible set of non-anticipating controls (i.e. $\mathcal{F}_t$-measurable and Markovian).

3.1   An Intuition Building One-Dimensional Simplified Model

To make the full model exposition easier and to develop intuition consider, for the time being, a simplified one-dimensional approximation of the spread less fixed running costs

$$X_t = \kappa L_t - C_t - K \quad (15)$$

where X_t follows a simple Brownian stochastic differential equation

$$dX_t = a dt + b dW_t \quad (16)$$

where a and b are naively chosen to fit the model. To further simplify the process, assume now that Z_t has normally distributed jumps such that

$$dZ_t = JdN_t \tag{17}$$

where $J \sim N(\alpha, \beta^2)$. The two $(X_t + Z_t)$ can be combined into a jump diffusion process Y_t

$$dY_t = adt + bdW_t + JdN_t \tag{18}$$

with solution

$$Y_t = Y_0 + at + bW_t + \sum_{k=1}^{n} J_k \tag{19}$$

where $\sum_{k=1}^{n} J_k \sim N(n\alpha, n\beta^2)$.

The expected income of the facility over its lifespan is

$$V_i(y, t) = \sup_{\tau, u} \inf_{\lambda, \alpha} E\left[\int_t^T e^{-r(s-t)} f_{I_s}(Y_s)ds + \sum_{k=1}^{n} e^{-r(\tau_k - t)} D_{u_{k-1}, u_k} \middle| (Y_t, u_0) = (y, i) \right] \tag{20}$$

By application of dynamic programming (see [5] or [29]) for optimal switching problems, the value function can be written as

$$V_i(y, t) = \sup_{\tau} \inf_{\lambda, \alpha} E\left[\int_t^\tau e^{-r(s-t)} f_i(Y_s)ds + e^{-r(\tau - t)} \left\{ V_j(Y_\tau, \tau) - D_{ij} \right\} \right] \tag{21}$$

where $i, j \in \{0, 1\}$ and τ is the first time it is optimal to switch production regimes. Now the problem consists of finding the optimal sets of prices and times to either

- hold production in its current state i, denoting this continuation or (hold) set as H_i, or
- switch production into the other state j, denoting this switching set as S_{ij}.

By another application of dynamic programming and Ito's lemma for jump diffusions, this equation leads to a coupled system of free boundary PID equations (PIDEs). The free boundary problem can be written in complementary form by noting that either it is optimal to switch and $V_i = V_j - D_{ij}$ or it is optimal to hold and V_i satisfies a PIDE subject to $V_i \geq V_j - D_{ij}$. Thus the equation extends on the whole space easing the need to track the switching boundary as a PID variational inequality (see [28] for an excellent reference on controlled jump diffusions). Thus the system of equations may be expressed as

$$\max\left[\underbrace{\frac{\partial V_i}{\partial t} + \mathcal{L}[V_i] + \inf_{\lambda, \alpha} \mathcal{I}[V_i] + f_i - rV_i}_{H_i}, \underbrace{(V_j - D_{ij}) - V_i}_{S_{ij}} \right] = 0. \tag{22}$$

where the spatial differential part of the generator is

$$\mathcal{L}[V] = a\frac{\partial V}{\partial y} + \frac{1}{2}b^2\frac{\partial^2 V}{\partial y^2} \qquad (23)$$

and the integro part is

$$\mathcal{I}[V] = \lambda(E[V(y+J)] - V(y)). \qquad (24)$$

The expectation E is taken with respect to a normal $N(\alpha, \beta^2)$ kernel g_N

$$E[V(y+J)] = \int_{-\infty}^{\infty} V(y+J)g_N(J)dJ. \qquad (25)$$

Theorem 1 (Worst Case Price). *The minimal optimal control is given by*

$$\alpha = \alpha_{min}, \quad \lambda = \begin{cases} \lambda_{min} & \text{if } E[V(y+J)] - V(y) \geq 0, \\ \lambda_{max} & \text{if } E[V(y+J)] - V(y) < 0 \end{cases} \qquad (26)$$

Theorem 2 (Best Case Price). *The maximal optimal control is given by*

$$\alpha = \alpha_{max}, \quad \lambda = \begin{cases} \lambda_{max} & \text{if } E[V(y+J)] - V(y) \geq 0, \\ \lambda_{min} & \text{if } E[V(y+J)] - V(y) < 0 \end{cases} \qquad (27)$$

Theorem 3 (Worst and Best Case Price if $\alpha = 0$). *The minimal optimal control is given by*

$$\lambda = \lambda_{min}, \qquad (28)$$

and the maximal optimal control is given by

$$\lambda = \lambda_{max}, \qquad (29)$$

if $\alpha = 0$ for all y.

See Appendix 2 for proofs of the above.

An interpretation of the maximal (respectively minimal) optimal control is as follows: (1) If the expected value post-jump $E[V(y + J)]$ is better than its current value $V(y)$, assume that the jump arrives as (in)frequently as possible $1/\lambda_{max}$ $(1/\lambda_{min})$. (2) Assume that the jumps are in general as (un)favourable as possible α_{max} (α_{min}).

3.1.1   Lessons from Merton

In the simplification where (1) the policy parameters (λ, α) are constant and (2) switching costs D_{ij} are zero, the problem reduces to a PIDE which yields the option price

$$\frac{\partial V}{\partial t} + a\frac{\partial V}{\partial y} + \frac{1}{2}b^2\frac{\partial^2 V}{\partial y^2} + \lambda(E[V(y+J)] - V(y)) - rV + f(y) = 0 \qquad (30)$$

where $f(y) = y^+ = \max(y, 0)$.

Using the Feynman-Kac Formula [28] and following Merton's classical paper on jump diffusions [25], the solution to the PIDE is

$$V(y, t) = E\left[\int_t^T e^{-r(s-t)}f(Y_s)ds \,\middle|\, Y_t = y\right]. \qquad (31)$$

Theorem 4 (Constant Coefficient Option Price). *The option price* $V(y, t)$ *satisfies*

$$V(y, t) = \sum_{n=0}^{\infty}\int_t^T e^{-\lambda(s-t)}\frac{\lambda^n(s-t)^n}{n!}e^{-r(s-t)}\left(A_{s,n}\Phi(d) + \frac{B_{s,n}}{\sqrt{2\pi}}e^{-\frac{d^2}{2}}\right)ds \qquad (32)$$

where $A_{s,n} = y + a(s-t) + n\alpha$, $B_{s,n}^2 = b^2(s-t) + n\beta^2$, $d = A_{s,n}/B_{s,n}$ *and* $\Phi(x)$ *is the standard normal cumulative distribution function.*

See Appendix 2 for the derivation of the governing PIDE and option price.

3.2   The Complete Problem

Return now to the stochastic control problem for the real option

$$V_i(l, c, z, t) = \sup_{\tau, u}\inf_{\lambda, \alpha} E\left[\int_t^T e^{-r(s-t)}f_{l_s}(L_s, C_s, Z_s)ds\right.$$

$$\left. + \sum_{k=1}^n e^{-r(\tau_k - t)}D_{u_{k-1}, u_k}\,\middle|\,(L_t, C_t, Z_t, u_0) = (l, c, z, i)\right] \qquad (33)$$

where $\lambda \in [\lambda_{min}, \lambda_{max}]$ and $\alpha \in [\alpha_{min}, \alpha_{max}]$. We follow a similar argument as before using dynamic programming to reduce the switching problem to a single decision τ

$$V_i(l, c, z, t)$$

$$= \sup_\tau\inf_{\lambda, \alpha} E\left[\int_t^\tau e^{-r(s-t)}f_i(L_s, C_s, Z_s)ds + e^{-r(\tau-t)}\{V_j(L_\tau, C_\tau, Z_\tau, \tau) - D_{ij}\}\right]. \qquad (34)$$

Using Ito's lemma for jump diffusions and noting as in [5, 29, 37] that the problem can be written in complementary form as a variational inequality

$$
\max \left[\underbrace{\frac{\partial V_i}{\partial t} + \mathscr{L}[V_i] + \inf_{\lambda, \alpha} \mathscr{I}[V_i] + f_i - rV_i}_{H_i}, \ \underbrace{(V_j - D_{ij}) - V_i}_{S_{ij}} \right] = 0. \tag{35}
$$

where the spatial differential part of the generator is

$$
\mathscr{L}[V] = \mu l \frac{\partial V}{\partial l} + ac \frac{\partial V}{\partial c} + \frac{1}{2} \sigma^2 l^2 \frac{\partial^2 V}{\partial l^2} + \rho \sigma lbc \frac{\partial^2 V}{\partial l \partial c} + \frac{1}{2} b^2 c^2 \frac{\partial^2 V}{\partial c^2} \tag{36}
$$

and the integro part is

$$
\mathscr{I}[V] = \lambda (E[V(l, c, J)] - V(l, c, z)). \tag{37}
$$

Theorem 5 (Worst Case Price). *The minimal optimal control is given by*

$$
\alpha = \alpha_{min}, \quad \lambda = \begin{cases} \lambda_{min} & \text{if } E[V(l, c, J)] - V(l, c, z) \geq 0, \\ \lambda_{max} & \text{if } E[V(l, c, J)] - V(l, c, z) < 0 \end{cases} \tag{38}
$$

Theorem 6 (Best Case Price). *The maximal optimal control is given by*

$$
\alpha = \alpha_{max}, \quad \lambda = \begin{cases} \lambda_{max} & \text{if } E[V(l, c, J)] - V(l, c, z) \geq 0, \\ \lambda_{min} & \text{if } E[V(l, c, J)] - V(l, c, z) < 0 \end{cases} \tag{39}
$$

See Appendix 2 for proofs of the above.

3.3 The Decision to Enter

Management's optimal decision time to enter into the business τ maximizes the expected value

$$
V(l, c, z, t)
$$
$$
= \sup_{\tau} \inf_{\lambda, \alpha} E \left[e^{-r(\tau - t)} \max\{V_1, V_0\}(L_\tau, C_\tau, Z_\tau, \tau) - B \middle| (L_t, C_t, Z_t) = (l, c, z) \right] \tag{40}
$$

and is a classical "American" style exercise call option. By dynamic programming, the optimal stopping problem satisfies the PID variational inequality

$$\max \left[\underbrace{\frac{\partial V}{\partial t} + \mathcal{L}[V] + \inf_{\lambda,\alpha} \mathcal{I}[V] - rV}_{H}, \quad \underbrace{(\max(V_1, V_0) - B) - V}_{S} \right] = 0. \qquad (41)$$

This completes the jump diffusion analogue of [23] and represents the optimal entry strategy for investment into a corn-ethanol biofuel production plant.

4 Numerical Results

This section begins with a numerical investigation of the behaviour of the constant coefficient analytical model. The section then proceeds with an investigation of the effects of policy uncertainty on the one-dimensional model including (i) the loss in value and (ii) the effects on switching decisions (which is also a proxy investigation of the effects on the entry decision). Finally, the section concludes with an investigation of the change in value between the full model with both policy uncertainty and model certainty or uncertainty.

4.1 The Constant Coefficient Model

Consider $V(y,t)$ in Eq. 32. Its behaviour is monotone increasing in y. Figure 2 shows that the function is increasing in α. This is as expected since if the jumps tend to be more positive ($\alpha > 0$), the spread tends to jump non-locally to a higher value of y (recall the option is monotone increasing in y), and vice versa if α tends to be more negative.

Figure 3 indicates V is an increasing function of λ (although it is generally insensitive to λ). This makes sense intuitively since as the frequency of jumps increases, more volatility is added to the option in terms of $B_{s,n}$, and Black-Scholes style options are increasing functions in volatility.

Figure 4 shows that V is sensitive to λ when there is an expected direction with the jumps (i.e. $\alpha \neq 0$).

4.1.1 Impact on Value

The parameters λ and α can be interpreted as measures of how infrequently policy changes occur and where management expects the subsidy to level move to, respectively. If the subsidy is expected to move up in value $\alpha > 0$, the jumps make the project more favourable. The opposite occurs if $\alpha < 0$: The future policy outlook is negative, and the project/option loses value.

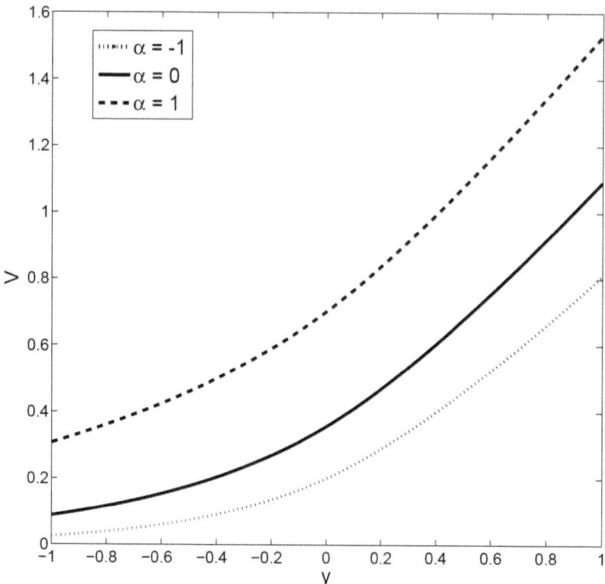

Fig. 2 The option value $V(y, t)$ at various levels of α (expected jump level) given standard parameters of $\lambda = 1$ (Poisson arrival rate of jumps), $\beta = 1$ (volatility of jump distribution), $a = 0$ and $b = 1$ (drift and volatility of diffusion), $r = 0.01$ (discount rate), and $T - t = 1$ (remaining option tenor)

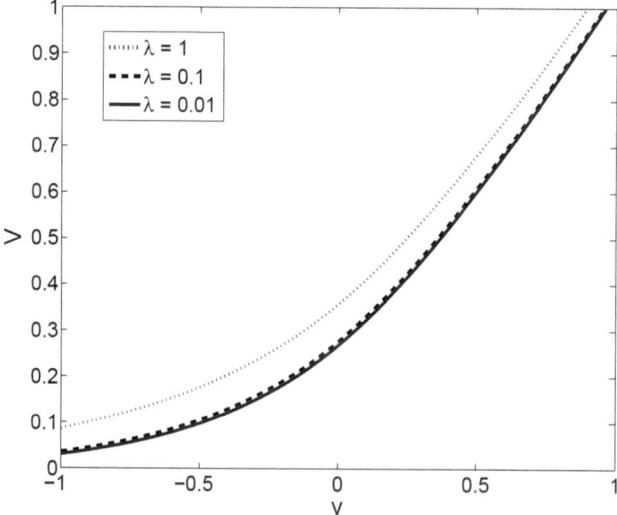

Fig. 3 The option value $V(y, t)$ at various levels of λ (Poisson arrival rate of jumps) given standard parameters of $\alpha = 0$ (expected jump level), $\beta = 1$ (volatility of jump distribution), $a = 0$ and $b = 1$ (drift and volatility of diffusion), $r = 0.01$ (discount rate), and $T - t = 1$ (remaining option tenor)

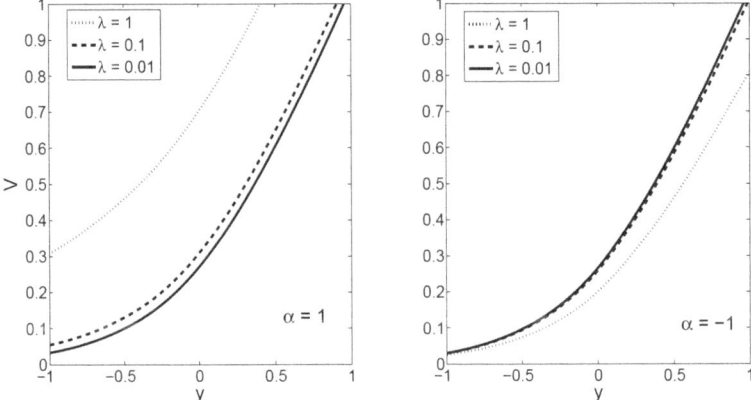

Fig. 4 The option value $V(y,t)$ at various levels of λ (Poisson arrival rate of jumps) and α (expected jump level) given standard parameters of $\beta = 1$ (volatility of jump distribution), $a = 0$ and $b = 1$ (drift and volatility of diffusion), $r = 0.01$ (discount rate), and $T - t = 1$ (remaining option tenor). *On the left,* $\alpha = 1$ *and on the right* $\alpha = -1$

As λ increases, policy changes occur more frequently which adds project/option value by means of the increased volatility associated with each jump. As the option to switch production off mitigates downside jumps on value V, the upside value of the jump volatility disproportionately increases the option's value. Figure 3 also reveals that the option is very insensitive to λ when there is no expected "directionality" in the jumps, i.e. when $\alpha = 0$.

4.2 The One-Dimensional Model

We now turn to an investigation of the effects of model uncertainty for a risk averse investor into the real option ethanol project. In this analysis, $f_1(y) = y$ and $f_0(y) = 0$ while $D_{01} = 0.2$ and $D_{10} = 0.1$.

Figure 5 shows the project valuation results for the expected price with policy uncertainty, best and worst case prices given policy uncertainty where $\alpha = 0$ is fixed and $\lambda \in [0, 1]$. The underlay shows the switching boundaries S_{ij} in y. Figure 6 shows the same information as Fig. 5 but in this case there is model uncertainty $\alpha \in [-0.2, 0.2]$ with expected parameter $\alpha = 0$.

4.2.1 Impact on Value

The gap between the best and worst case prices can be significantly large if α is allowed to vary indicated in Fig. 6; otherwise the difference is small (Fig. 5) as expected from our results with the constant coefficient model. Since this function is convex, the integral operator is single-signed and the parameter λ assumes either

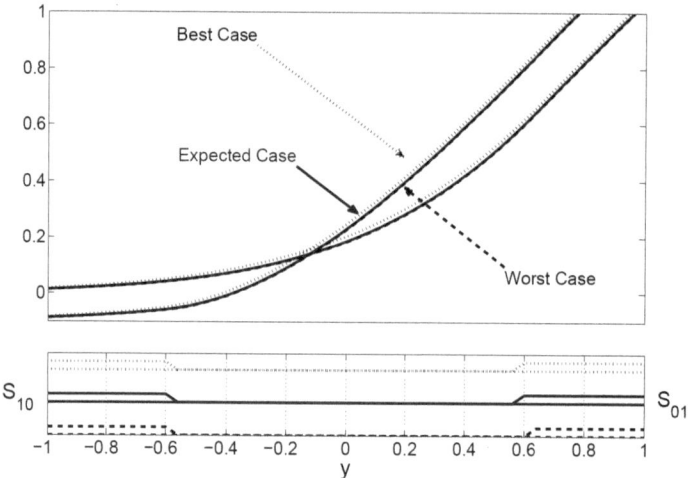

Fig. 5 The option value $V(y, t)$ at an "expected case" of $\lambda = 0.1$ (Poisson arrival rate of jumps) and $(\lambda_{min}, \lambda_{max}) = (0, 1)$ (parameter boundaries), $\alpha = 0$ and $\beta^2 = 0.1$ (mean and variance of jump distribution), $a = 0$ and $b = 1$ (drift and volatility of diffusion), $r = 0.01$ (discount rate), and $T - t = 1$ (option tenor). Switching costs are $D_{01} = 0.2$ and $D_{10} = 0.1$

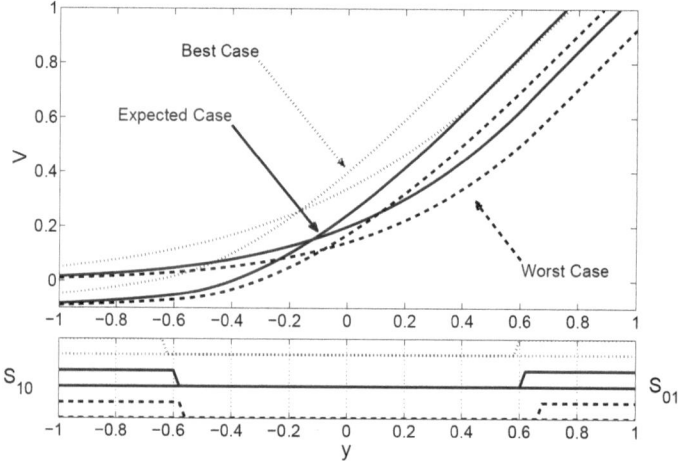

Fig. 6 The option value $V(y, t)$ at an "expected case" of $\lambda = 0.1$ (Poisson arrival rate of jumps) and $\alpha = 0$ (expected mean jump size), but where $\lambda \in [0, 1]$ and $\alpha \in [-0.2, 0.2]$ (parameter boundaries). The remaining parameters are $\beta^2 = 0.1$ (variance of jump distribution), $a = 0$ and $b = 1$ (drift and volatility of diffusion), $r = 0.01$ (discount rate), and $T - t = 1$ (option tenor). Switching costs are $D_{01} = 0.2$ and $D_{10} = 0.1$

λ_{min} in the worst case or λ_{max} in the best case when $\alpha = 0$ in the example in Fig. 5 by Jensen's inequality. The constant coefficient expected case model will always be bounded by the best and worst case project prices. In these examples, the expected case is nearer to the worst case since $\lambda = 0.1$ is closer to $\lambda_{min} = 0$ than $\lambda_{max} = 1$.

4.2.2 Impact on Switching Decision

Although the effects are not very pronounced on the 1 year time horizon, model uncertainty has an impact on switching decisions. The lower charts in Figs. 5 and 6 represent the switching boundaries

- $S_{01} = \{y : V_0(y, 0) = V_1(y, 0) - D_{01}\}$, the set of prices where the operating status is optimally switched on from idle, and
- $S_{10} = \{y : V_1(y, 0) = V_0(y, 0) - D_{10}\}$, the set of prices where the operating status is optimally switched off from running.

It can be seen that in the...

...worst case scenario: The operator switches production on later than in the expected case (i.e. at $y > y^*$ if y^* is where the operator would switch production on in the expected case). Similarly, the operator switches production off earlier compared to the expected case (i.e. at $y < y^*$ if y^* is where the operator would switch production off in the expected case).

...best case scenario: The operator switches production on earlier and switches production off later compared to the expected case.

In the example where $\alpha = 0$ is fixed, the differences in switching boundaries between the best, worst and expected cases are almost negligible (Fig. 5). However in the other example where $-0.2 \leq \alpha \leq 0.2$ can vary, the differences in switching boundaries between the best, worst and expected cases can deviate a great deal. Thus it is not so much when management thinks a change in policy might occur (i.e. λ-driven) but rather how management expects that policy to change with respect to its current policy conditions—that is, α-driven.

4.3 The Complete Model

This section concludes with a numerical investigation of the ethanol plant value in the presence or absence of policy uncertainty and model uncertainty. The ethanol plant is assumed to have a 10 year investment horizon, $T - t = 10$.

4.3.1 With and Without Policy Uncertainty

We compare the real option project valuation of the ethanol plant in two cases where:

- Management ignores the uncertainty in the ethanol subsidy policy and assumes $Z_t = Z$ (constant) to take its Jan/2011 value (Table 1),
  - in this case, $f_1(L_s, C_s, Z) = \kappa(L_s - K_1 + Z) - C_s$ where $Z = \$0.45/\text{gallon}$ is constant (also $\lambda = 0$); and

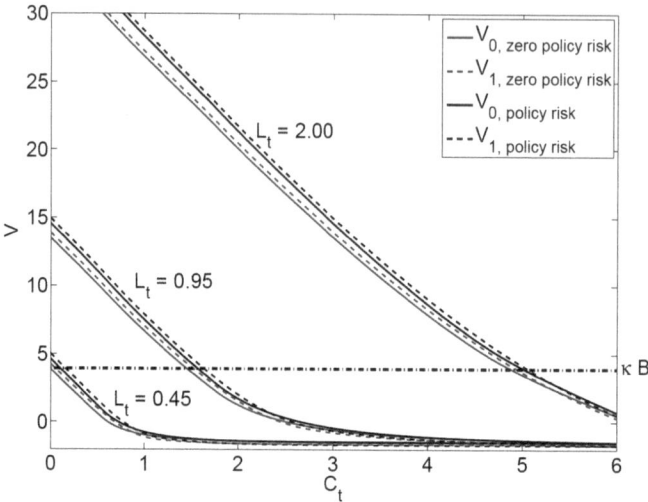

Fig. 7 $V(L_t, C_t, Z, t)$ without policy uncertainty vs $V(L_t, C_t, Z_t, t)$ with policy uncertainty. Parameters (from Tables 2 and 3) are $\mu = 0$ and $\sigma = 0.156$ (drift and volatility of ethanol), $a = 0$ and $b = 0.123$ (drift and volatility of corn), $Z = 0.45$ without policy uncertainty and $Z_t = 0.45$, $\lambda = 0.24$, $\alpha = -0.69$ and $\beta^2 = 0.015$ (arrival rate, mean and variance of jumps) with policy uncertainty

- Management considers the uncertainty in the ethanol subsidy policy with known parameters (model certainty) and assumes the model parameters in Table 3 subject to the initial subsidy level being its Jan/2011 value as above,
  - in this case, $f_1(L_s, C_s, Z_s) = \kappa(L_s - K_1 + Z_s) - C_s$ where $Z_t = \$0.45/\text{gallon}$.

Figure 7 shows the value functions at various levels of C_t in the presence and absence of policy uncertainty. Figure 8 shows the switching boundaries in both cases.

Impact of Policy Uncertainty on Value As inferred from our one-dimensional analysis in Sect. 3.1, policy uncertainty adds more value to the real option due to two distinct factors: (1) Given $Z_t = 0.45 < 0.51 = e^{\alpha + \frac{1}{2}\beta^2} = E[J]$, it is likely that the subsidy policy will jump to a higher level giving the option more value in the presence of policy uncertainty. (2) The extra volatility provided by the jump process adds volatility value to the option. The downside of policy switches on an ethanol plant can be mitigated by switching production off, while the upside value is maintained by keeping (or switching) production on when prices favourably allow for it. The capitalized cost of construction on a per bushel basis κB is also included in Fig. 8.

Impact of Policy Uncertainty on Switching Decisions The boundary at which production is switched on from an idle state is ∂S_{01} and the boundary at which production is turned off from a running state is ∂S_{10}. In this case, the initial subsidy

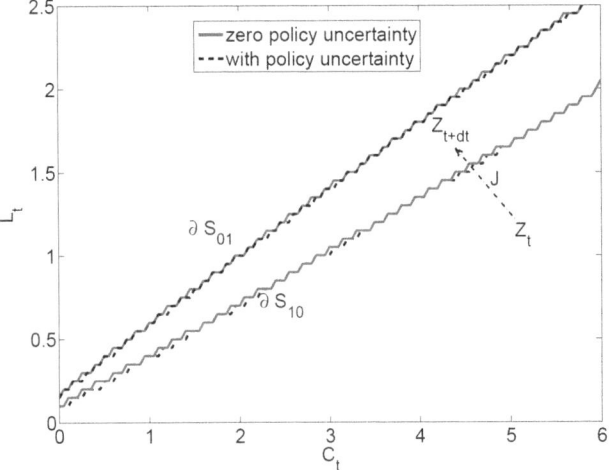

Fig. 8 The switching boundaries ∂S_{01} and ∂S_{10} in the presence and absence of policy uncertainty. Parameters (from Tables 2 and 3) are $\mu = 0$ and $\sigma = 0.156$ (drift and volatility of ethanol), $a = 0$ and $b = 0.123$ (drift and volatility of corn), $Z = 0.45$ without policy uncertainty and $Z_t = 0.45$, $\lambda = 0.24$, $\alpha = -0.69$ and $\beta^2 = 0.015$ (arrival rate, mean and variance of jumps) with policy uncertainty

level Z_t is less than the long run average $E[J] = e^{\alpha + \frac{1}{2}\beta^2}$, $Z_t = 0.45 < 0.51 = e^{-0.69 + \frac{1}{2}0.015}$. Thus, the operator generally waits longer before turning production off, due to a positive outlook that the subsidy might jump up to its long term average. Similarly, the operator generally turns production on sooner in hope that the subsidy might again jump to its (higher) long run average. More precisely, given a point (c, l) on ∂S_{01} in the absence of policy uncertainty, if (c, l^*) is on ∂S_{01}^* in the presence of policy uncertainty, then $l^* < l$ (respectively $l^* > l$) when production is shut down earlier (later).

Changes in z shift value and switching decisions up or down non-locally as Z_t jumps. The general direction of the jumps is illustrated in Fig. 8 by the arrow $Z_t \xrightarrow{J} Z_{t+dt}$.

It should be noted that if management were expecting the subsidy to jump to a lower level, the opposite situation as described above would occur. Management would switch production off earlier and turn production on later for fear that the subsidy might fall.

4.3.2 Policy Uncertainty with Model Uncertainty

In the likely event that the distribution and parameters of the regulatory uncertainty process are unknown, management may choose a worst case valuation for the ethanol plant project value. The assumed boundaries for policy change arrival rate are $\lambda \in [0, 0.60]$ and expected mean subsidy policy $\alpha \in [-1.118, -0.309]$.

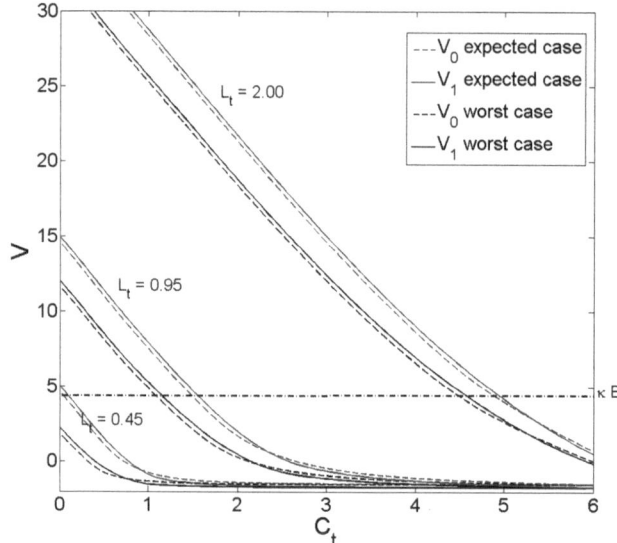

Fig. 9 $V(L_t, C_t, Z_t, t)$ vs $\inf_{\lambda,\alpha} V(L_t, C_t, Z_t, t)$ with policy (and model) uncertainty. Constant parameters (from Tables 2 and 3) are $\mu = 0$ and $\sigma = 0.156$ (drift and volatility of ethanol), $a = 0$ and $b = 0.123$ (drift and volatility of corn), and $Z_t = 0.45$, $\lambda = 0.24$, $\alpha = -0.69$ and $\beta^2 = 0.015$ (arrival rate, mean and variance of jumps). Non-constant parameters for model uncertainty are $\alpha \in [-1.118, -0.309]$ and $\lambda \in [0, 0.60]$

Figure 9 illustrates the worst case value compared to the expected case given by the model parameters in Tables 2 and 3. The switching boundaries are illustrated in Fig. 10 comparing the worst case operating decisions to the expected case.

For completeness, Fig. 11 shows the envelope of best case, worst case and expected project values in the presence of policy and model uncertainty. The bounds can be quite large between the best and worst project values even for "seemingly small" parameter boundaries. The switching boundaries are illustrated in Fig. 12 comparing the best case operating decisions to the expected case.

Impact of Worst Case Model Uncertainty on Value The worst case real option ethanol plant value represents a lower bound in project value. Figure 9 also includes the capitalized cost of construction on a per bushel of capacity basis κB. As expected, fewer projects are net present value positive in the worst case project value compared to the expected case. That is, given the two sets of prices at a time t the set of prices that are "Net Present Value (NPV) positive" for entering into the project are

$$NPV = \{(l, c) : \max(V_1, V_0) - B > 0\} \text{ and } NPV^* = \{(l, c) : \inf_{\lambda,\alpha} \max(V_1, V_0) - B > 0\},$$

$$\tag{42}$$

then

$$NPV^* \subseteq NPV \tag{43}$$

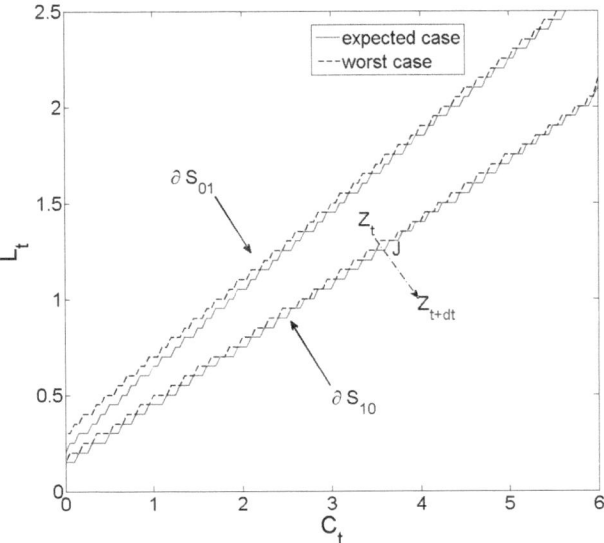

Fig. 10 The switching boundaries ∂S_{01} and ∂S_{10} in the presence of policy uncertainty and model uncertainty in the worst case. Constant parameters (from Tables 2 and 3) are $\mu = 0$ and $\sigma = 0.156$ (drift and volatility of ethanol), $a = 0$ and $b = 0.123$ (drift and volatility of corn), and $Z_t = 0.45$, $\lambda = 0.24$, $\alpha = -0.69$ and $\beta^2 = 0.015$ (arrival rate, mean and variance of jumps). Non-constant parameters for model uncertainty are $\alpha \in [-1.118, -0.309]$ and $\lambda \in [0, 0.60]$

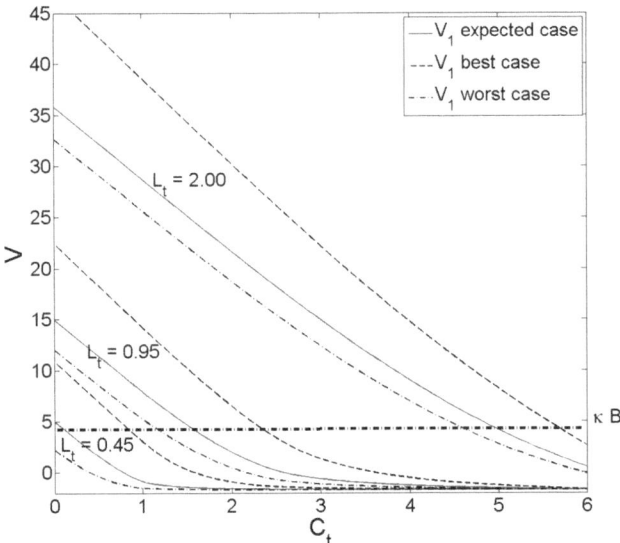

Fig. 11 $V_1(L_t, C_t, Z_t, t)$ vs $\inf_{\lambda,\alpha} V_1(L_t, C_t, Z_t, t)$ vs $\sup_{\lambda,\alpha} V_1(L_t, C_t, Z_t, t)$ with policy uncertainty. Constant parameters (from Tables 2 and 3) are $\mu = 0$ and $\sigma = 0.156$ (drift and volatility of ethanol), $a = 0$ and $b = 0.123$ (drift and volatility of corn), and $Z_t = 0.45$, $\lambda = 0.24$, $\alpha = -0.69$ and $\beta^2 = 0.015$ (arrival rate, mean and variance of jumps). Non-constant parameters for model uncertainty are $\alpha \in [-1.118, -0.309]$ and $\lambda \in [0, 0.60]$

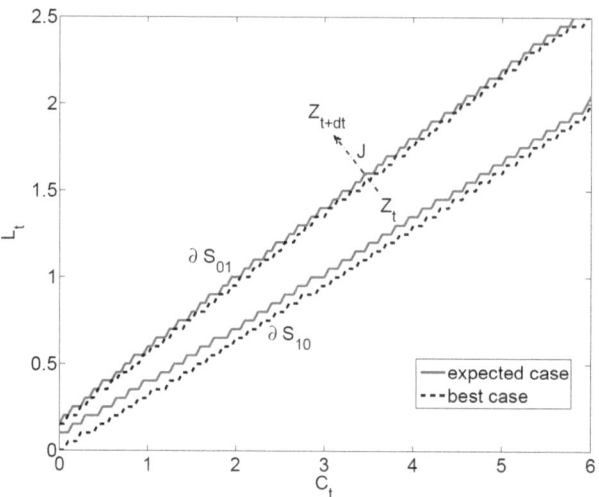

Fig. 12 The switching boundaries ∂S_{01} and ∂S_{10} in the presence of policy uncertainty and model uncertainty in the best case. Constant parameters (from Tables 2 and 3) are $\mu = 0$ and $\sigma = 0.156$ (drift and volatility of ethanol), $a = 0$ and $b = 0.123$ (drift and volatility of corn), and $Z_t = 0.45$, $\lambda = 0.24$, $\alpha = -0.69$ and $\beta^2 = 0.015$ (arrival rate, mean and variance of jumps). Non-constant parameters for model uncertainty are $\alpha \in [-1.118, -0.309]$ and $\lambda \in [0, 0.60]$

This means that fewer investments are entered into during times of high policy uncertainty if management is risk averse.

In certain cases, the integral operator may be $\mathcal{I}[V] = E[V(l, c, J)] - V(l, c, 0.45) > 0$ and accordingly $\lambda = \lambda_{min} = 0$ in the minimization. This is similar to the case with zero policy uncertainty. Thus, the worst case option value may at times approach the option value in the absence of policy uncertainty.

Impact of Worst Case Model Uncertainty on Operating Decisions The possible subsidy outcomes in the worst case scenario have a much more negative outlook than the expected case. Thus in the worst case scenario, the optimal strategy tends to be conservative when making switching decisions (Fig. 10). The net result is that management switches production on much later and switches production off much earlier compared to the expected case operating strategy.

Comments on the Best Case Model Figure 11 shows that the gap between the best and worst case prices can be quite large. This is an artifact of the stochastic optimization problem that leads to very large arbitrage free price good deal bounds in practice with financial derivatives. Similar to before, management switches production on earlier and switches production off later compared to the expected case operating strategy (Fig. 12).

5   Conclusions

The goal of our paper is to develop a quantitative model for managing and pricing regulatory risk. The accomplishments and overall theme of our paper are summarized in what follows.

5.1   Summary

Our paper laid out several research goals to contribute to the existing real options literature and the less developed body of research in policy uncertainty.

We presented a real option model to attempt to quantitatively model policy uncertainty using a jump diffusion process. This model allows for the valuation of long term energy projects in the presence of policy uncertainty. For a corn-ethanol case study (following [23]), we presented a real option model involving both standard price uncertainty modelled using a simplified one dimensional jump diffusion process for the relevant price spread and stochastic subsidy. We followed this with a more sophisticated multivariate model which independently modeled both the input and the output price. In addition, this model included the impact of policy uncertainty using a randomly fluctuating subsidy level. This fluctuating subsidy was quantified using a pure jump process. Given that there may be model uncertainty for the subsidy policy process, our proposed model includes a "worst case" (modelled using a *VaR* level) policy uncertainty scenario which allows the project investor to quantify and manage his worst case regulatory downside risk. This work allowed us to draw some general conclusions with policy level implications, as summarized and described in the next section.

5.2   Policy Conclusions

We outline the policy effects and numerical conclusions from our analysis in Sect. 4.

5.2.1   Policy Uncertainty

In the case of policy certainty versus uncertainty, for the convex (or "long vol") real options considered here, the effects of policy uncertainty always increase the value of the option when there is no directionality in the subsidy jumps.

More generally, the effects of policy uncertainty may be positive or negative for the project valuation. For example, if the subsidy is currently low and the future subsidy level is expected to be higher, the possibility of a jump in policy increases the overall value of the option. The opposite holds when the subsidy is high and the future subsidy is expected to be lower than today.

5.2.2 Model Uncertainty

Typically, the effect of ambiguity in policy uncertainty models on project valuation is negative: A strongly risk averse manager taking a long position in the option should price the project using the worst case of possible parameters.

The optimal operating strategy in terms of the sets of prices, times, and subsidy levels to switch production vary based on the scenario. The strategy however generally obeys the following rules: (1) If the scenario is a worst case (respectively best case), then production is switched off earlier (later) compared to the constant parameter expected case, and production is switched on later (earlier) compared to the expected case. This represents an pessimistic (optimistic) outlook on regulatory policy changes. (2) If the scenario is a constant parameter case with policy uncertainty, then production is switched on earlier (later) if the current policy regime is lower (higher) than the expected long run trend. Similarly production is switched off later (earlier) if the current policy regime is lower (higher) than the expected long run trend.

The anecdotal evidence that suggests businesses delay investment longer in periods of high policy uncertainty is seen to be consistent with our model, supporting those claims [3, 17, 35]. In particular, given the tendency is generally to delay during periods of policy uncertainty suggests that investors use pessimistic model outlooks when making investment decisions. Given that fewer projects were net present value positive in the model uncertainty case versus the policy uncertainty with known parameters case, our model supports the claim that fewer investments are entered into during periods of high policy uncertainty.

5.3 Possible Extensions

The lognormal distribution for the policy subsidy jump process was chosen for several reasons: (1) subsidies cannot become negative; (2) model familiarity, since geometric Brownian motion itself leads to a lognormal distribution and Merton's seminal jump diffusion paper [25]; (3) analytical tractability; and (4) its second moments exist. The distribution however has large positive skew with a fat tail. This choice of distribution can lead to results which are relatively indifferent toward downside risk in the subsidy process, as the probability of observing very low subsidies is much smaller than the probability of observing very high subsidies. For reference, plots of the expected, worst and best case subsidy jump probability distribution functions are shown in Fig. 13 along with a reference case to better illustrate the positive skew and fat tail.

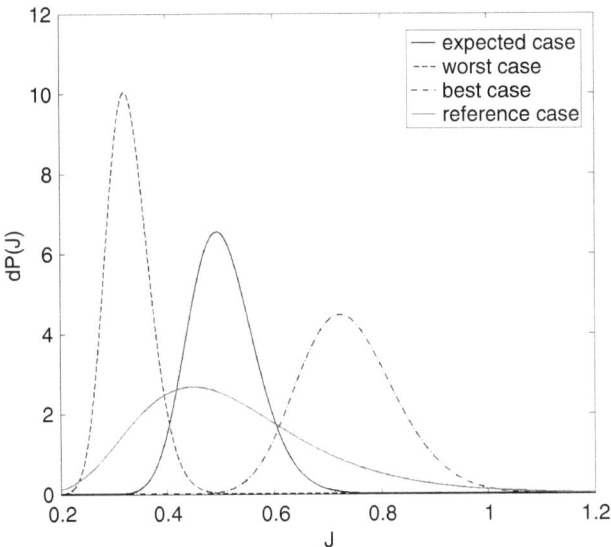

Fig. 13 The probability distribution functions $dP(J)$ of the jumps J of the expected case $LogN(-0.69, 0.015)$, worst case $LogN(-1.118, 0 : 015)$, best case $LogN(-0.309, 0.015)$, and a reference case $LogN(-0.7, 0.1)$ highlight the skew

To improve the model, more classes of jump distributions or non-constant Poisson arrival rates could be considered for future work. Another possible improvement to the expected subsidy jump model would be to incorporate management's views on the probability of possible policy outcomes or cases, each with an associated probability determined by management (an idea motivated by [20] but here simplified). This is both easier to justify to industry practitioners and greatly simplifies the analysis as it effectively removes the continuous variable J and replaces it with a discrete variable J_i. This reduces the dimensionality of the PID variational inequality system, which greatly reduces the computational time by reducing the problem to solving discrete weighted probabilities for each outcome J_i. For completeness, the integro operator would be replaced with $\mathscr{I}[V] = \lambda(\sum_i V_i P_i - V)$ and a PID variational inequality solved for each outcome i with associated value function V_i and management probability estimate P_i.

Appendix 1: Numerical Method

A brief exposition of the numerical method used to solve this PID variational inequality system is presented below. We refer the reader to [9, 14, 18, 28] for a more detailed analysis of the finite difference solutions to stochastic control problems and PIDEs.

The general PID variational inequality is of the form

$$\max \left[\frac{\partial V}{\partial t} + \mathscr{L}[V] + \mathscr{I}[V] + f - rV, \ h - V \right] = 0. \tag{44}$$

where the differential operator is (occasionally suppressing any l, c, z dependence of μ, σ, a, b)

$$\mathscr{L}[V] = \mu \frac{\partial V}{\partial l} + a \frac{\partial V}{\partial c} + \frac{1}{2}\sigma^2 \frac{\partial^2 V}{\partial l^2} + \rho \sigma b \frac{\partial^2 V}{\partial l \partial c} + \frac{1}{2} b^2 \frac{\partial^2 V}{\partial c^2} \tag{45}$$

and the integro operator is

$$\mathscr{I}[V] = \lambda (E[V(l, c, J)] - V(l, c, z)) \tag{46}$$

and the constraint is

$$h = V_u - D_u \tag{47}$$

The numerical solution is obtained via finite differences at grid points $V(l_i, c_j, z_p, t_k) = V_{i,j,p}^k$ usually using second order centred differences except possibly at the boundary conditions. The grid points are

$$t_k = t_0 + k\Delta t \tag{48}$$

$$l_i = l_0 + i\Delta l \tag{49}$$

$$c_j = c_0 + j\Delta c \tag{50}$$

$$z_p = z_0 + p\Delta z \tag{51}$$

where the increments Δ need not necessarily be uniform. Divided differences are used to approximate the derivatives. Two are shown below for reference

$$\frac{\partial V}{\partial l} \approx \frac{V_{i+1,j,p}^k - V_{i-1,j,p}^k}{2\Delta l} \tag{52}$$

$$\frac{\partial V}{\partial t} \approx \frac{V_{i,j,p}^{k+1} - V_{i,j,p}^k}{\Delta t} \tag{53}$$

The integral $E[V(l, c, J)]$ is simply truncated and approximated along a grid as well

$$E[V(l, c, J)] \approx \int_0^{J_{max}} V(l, c, J)P(J)dJ \approx \sum_{p=0}^{P} V_{i,j,p}^k g(z_p)\Delta z \tag{54}$$

where the expectation is truncated by a point $J_{max} = z_P$ at which the error in the approximation is small. Note any kind of quadrature rule can be used along with non-uniform grid spacing besides the rule shown above.

A fitted scheme is used to write out a system of equations for $V^k_{i,j,p}$ at the grid points

$$\frac{V^{k+1} - V^k}{\Delta t} + \theta L V^{k+1} + (1-\theta)LV^k + \phi I V^{k+1} + (1-\phi)IV^k + f \leq 0 \qquad (55)$$

where L is the differentiation matrix associated with the partial differential operator $\mathscr{L}$ including the source term $-rV$ and I is the integration matrix associated with the integro operator $\mathscr{I}$. The parameters θ and ϕ blend averages of the discretized PIDE at time steps k and $k+1$ (e.g. $\theta = 0$ is fully implicit and $\theta = \frac{1}{2}$ yields a Crank-Nicholson scheme). A small abuse of notation V^k refers to the entire collection of grid points i,j,p at time step k. The running profit function at all grid points is simply f. The differentiation matrix L tends to be stiff whereas the integration matrix I tends to be non-stiff allowing for the use of IMEX style time marching schemes.[4]

For reference, L can be considered a tensor that operates on a square $V_{i,j}$ at all p. In tensor notation, at the interior points L is for example

$$L_{i,j,i,j} = -\frac{2}{\Delta l^2}\frac{1}{2}\sigma^2_{i,j} - \frac{2}{\Delta c^2}\frac{1}{2}b^2_{i,j} - r \qquad (56)$$

$$L_{i,j,i,j-1} = -\frac{1}{2\Delta c}a_{i,j} + \frac{1}{\Delta c^2}\frac{1}{2}b^2_{i,j} \qquad (57)$$

$$L_{i,j,i-1,j-1} = \frac{1}{4\Delta l \Delta c}\rho\sigma_{i,j}b_{i,j} \qquad (58)$$

where $L_{i,j,i+i^*,j+j^*} = 0$ if $|i^*|, |j^*| \geq 2$. Conditions must be applied along the boundary (e.g. linearity at far field). The integration matrix I is applied to a column $V_{i,j,p}$ across all p at a point (i,j), like a matrix in p constant across all i,j. For example,

$$I_{p,p} = \lambda \left[\frac{1}{2}g(z_p)(z_{p+1} - z_p) - 1\right] \qquad (59)$$

$$I_{p,q} = \lambda \frac{1}{2}g(z_q)(z_{q+1} - z_{q-1}). \qquad (60)$$

using a trapezoidal quadrature rule.

The system is solved subject to a known final condition $V(l, c, z, T) = Q(l, c, z)$ (being a backward Kolmogorov type equation). If there is no salvage value at the

---

[4] We note that using a Crank-Nicholson scheme in both L and I appeared to deliver good results.

end of the facility life $V_{i,j,p}^K = Q_{i,j,p} = 0$ (where $T = t_0 + K\Delta t$) but in general the salvage value should satisfy some inequalities around the switching costs D_{ij}.

This is a complementary problem

$$MV^k - b \le 0, \quad h \le V^k, \quad \left(MV^k - b\right)^T (V^k - h) = 0 \qquad (61)$$

where superscript T denotes the matrix transpose. The matrix M is an aggregation of the integration and differentiation matrix pre-multipliers of V^k while b is a vector of collected knowns at time k (from $k + 1$). This matrix system is then solved using an iterative fixed point method similar to projected successive over-relaxation. Several iterative schemes for non-linear control problems are described in [1, 9, 10, 14, 18, 28, 37].

Appendix 2: Optimal Control

The intuition behind the proofs of the theorems in Sect. 3 are presented in this appendix.

Regarding the One-Dimensional Model Optimal Stochastic Control 3.1

Proof (Theorems 1 and 2). Consider the optimization with respect to λ

$$\inf_{\lambda_{min} \le \lambda \le \lambda_{max}} \mathscr{I}[V]. \qquad (62)$$

Due to the boundedness of λ, this problem is nonsingular. Since $\mathscr{I}[V] = \lambda(E[V(y + J)] - V(y))$ is linear in λ, it achieves its critical values at the endpoints $[\lambda_{min}, \lambda_{max}]$ and the optimal λ satisfies

$$\lambda = \begin{cases} \lambda_{min} & \text{if } E[V(y + J)] - V(y) \ge 0, \\ \lambda_{max} & \text{if } E[V(y + J)] - V(y) < 0. \end{cases} \qquad (63)$$

Turning now to the optimization with respect to α,

$$\inf_{\alpha_{min} \le \alpha \le \alpha_{max}} \lambda(E[V(y + J)] - V(y)) \Rightarrow \inf_{\alpha} E[V(y + J)] \qquad (64)$$

where we drop the α bounds for notational brevity. The expectation can be written as

$$\inf_{\alpha} E[V(y+J)] = \inf_{\alpha} \int_{-\infty}^{\infty} V(y+J)g_N(J)dJ, \quad g_N \text{ is the normal kernel } N(\alpha, \beta^2) \quad (65)$$

$$= \int_{-\infty}^{\infty} \inf_{\alpha} \{V(y+\alpha+z)\}g_z^*(z)dz, \quad g_N^* \text{ is the kernel } N(0, \beta^2) \quad (66)$$

$$= \int_{-\infty}^{\infty} V(y+\alpha_{min}+z)g_z^*(z)dz \quad (67)$$

if $V(y)$ is monotone increasing in y which is true of the class of profit functions $f(y)$ considered in this analysis. (This result follows from the Feynman-Kac or Green's formula for $V(y)$ given $f(y)$ is monotone increasing.)

A similar argument applies for deriving the maximal optimal control (Theorem 2) but applied in the opposite direction.

Summarizing, the *worst case* project value is given by the minimal optimal control and the *best case* is given by the maximal optimal control subject to certain regularity conditions on V and f (namely monotonicity). □

Proof (Theorem 3). Note that the integro operator $\mathscr{I}$ is single-signed almost everywhere if f is such that $V(y)$ is convex and $\alpha = 0$. The justification follows from Jensen's inequality $V(E[y+J]) \leq E[V(y+J)]$ and that $E[y+J] = y+\alpha = y$. Thus

$$E[V(y+J)] - V(E[y+J]) = \quad (68)$$

$$E[V(y+J)] - V(y) = \frac{1}{\lambda}\mathscr{I}[V] \geq 0 \quad (69)$$

and accordingly $\lambda = \lambda_{min}$ for all y (and vice versa for the maximal control). □

Regarding the Constant Coefficient Option Price 3.1.1

Proof (Theorem 4). For a function $u(Y_t = y, t)$, applying Ito's lemma for jump diffusions results in

$$u(Y_T, T) - u(y, t) = \int_t^T b\frac{\partial u}{\partial y}dW_s + \int_t^T \left(\frac{\partial u}{\partial t} + a\frac{\partial u}{\partial y} + \frac{1}{2}b^2\frac{\partial^2 u}{\partial y^2}\right)ds$$

$$+ \int_t^T [u(Y_s + J, s) - u(Y_s, s)]dN_t. \quad (70)$$

Taking the expectation causes the Ito integral to become zero (since $E[\int_t^T u\,dW_s|\mathscr{F}_t] = 0$ for smooth functions u). The expectation of the jump term becomes

$$E\left[\int_t^T [u(Y_s + J, s) - u(Y_s, s)]dN_t\right] = \int_t^T E_J[u(Y_s + J, s) - u(Y_s, s)]\,\lambda\,ds \tag{71}$$

since the Poisson arrivals dN_t and Brownian motion dW_t are independent, and $dN_t = 1$ with probability λds or 0 otherwise. Here E_J denotes an expectation with respect to J only (recall $J \perp W_t$).

When $u(\cdot, T) = 0$, and the jumps J and Brownian motion are independent, the expectation is

$$E[u(Y_T, T) - u(y, t)]$$

$$= -u(y, t)$$

$$= E\left[\int_t^T \left(\frac{\partial u}{\partial t} + a\frac{\partial u}{\partial y} + \frac{1}{2}b^2\frac{\partial^2 u}{\partial y^2} + \lambda(E_J[u(Y_s + J, s)] - u(Y_s, s))\right)ds\right] \tag{72}$$

If $u(y, t)$ satisfies the nonhomogeneous PIDE

$$\frac{\partial u}{\partial t} + a\frac{\partial u}{\partial y} + \frac{1}{2}b^2\frac{\partial^2 u}{\partial y^2} + \lambda(E_J[u(y + J, t)] - u(y, t)) = -f(y), \tag{73}$$

the solution has the probabilistic (Feynman-Kac) representation

$$u(y, t) = E\left[\int_t^T f(Y_s)ds \,\Big|\, Y_t = y\right] \tag{74}$$

The discounted value function $V(Y_s, s) = e^{-r(s-t)}u(Y_s, s)$ satisfies the PIDE of Theorem 4 and has probabilistic representation

$$V(y, t) = E\left[\int_t^T e^{-r(s-t)}f(Y_s)ds \,\Big|\, Y_t = y\right]. \tag{75}$$

The key to solving this expectation is to condition Y on n, the number of jumps so far, denoted $Y_{s,n}|n$. Note that the probability of observing n Poisson jumps over a time period $s - t$ is $P(n, s - t) = e^{-\lambda(s-t)}\frac{\lambda^n(s-t)^n}{n!}$. Thus

$$V = E\left(E\left[\int_t^T f(Y_{s,n})ds \,\Big|\, n\right]\right) \tag{76}$$

$$= \sum_{n=0}^{\infty} \int_{-\infty}^{\infty} \int_t^T e^{-\lambda(s-t)}\frac{\lambda^n(s-t)^n}{n!} e^{-r(s-t)}y_{s,n}^+ \frac{1}{\sqrt{2\pi B_{s,n}^2}} e^{-\frac{(y_{s,n}-A_{s,n})^2}{2B_{s,n}^2}} ds\,dy_{s,n} \tag{77}$$

where $A_{s,n} = y + a(s - t) + n\alpha$ and $B_{s,n}^2 = b^2(s - t) + n\beta^2$.

$$V = \sum_{n=0}^{\infty} \int_{-\infty}^{\infty} \int_{t}^{T} e^{-\lambda(s-t)} \frac{\lambda^n (s - t)^n}{n!} e^{-r(s-t)} (A_{s,n} + B_{s,n}z)^+ \frac{1}{\sqrt{2\pi}} e^{-\frac{z^2}{2}} \, ds \, dz \quad (78)$$

$$= \sum_{n=0}^{\infty} \int_{t}^{T} \int_{-d}^{\infty} e^{-\lambda(s-t)} \frac{\lambda^n (s - t)^n}{n!} e^{-r(s-t)} (A_{s,n} + B_{s,n}z) \frac{1}{\sqrt{2\pi}} e^{-\frac{z^2}{2}} \, dz \, ds \quad (79)$$

where $d = A_{s,n}/B_{s,n}$. Changing variables $x = -z$ and flipping the limits of integration yields

$$V = \sum_{n=0}^{\infty} \int_{t}^{T} e^{-\lambda(s-t)} \frac{\lambda^n (s - t)^n}{n!} e^{-r(s-t)}$$

$$\times \left(\int_{-\infty}^{d} A_{s,n} \frac{1}{\sqrt{2\pi}} e^{-\frac{x^2}{2}} \, dx - \int_{-\infty}^{d} B_{s,n}x \frac{1}{\sqrt{2\pi}} e^{-\frac{x^2}{2}} \, dx \right) ds \quad (80)$$

$$V(y, t) = \sum_{n=0}^{\infty} \int_{t}^{T} e^{-\lambda(s-t)} \frac{\lambda^n (s - t)^n}{n!} e^{-r(s-t)} \left(A_{s,n}\Phi(d) + \frac{B_{s,n}}{\sqrt{2\pi}} e^{-\frac{d^2}{2}} \right) ds \quad (81)$$

where $\Phi(x)$ is the standard normal cumulative distribution function.   □

Regarding the Complete Stochastic Control Problem 3.2

Proof (Theorems 5 and 6). The argument for obtaining the optimal λ is identical to the one-dimensional case. Determining the optimal α is similar to the previous case, but slightly more delicate. Again, it rests on the monotonicity of f. Recall

$$f_1(l, c, z) = \kappa(l + z - K_1) - c, \quad f_0(l, c, z) = -\kappa K_0 \quad (82)$$

and thus f_1 is monotone increasing in z and f_0 is unaffected by z. By the Feynman-Kac representation for V_1 in Eq. 34, V_1 is monotone increasing in z. Similarly V_0, via the free boundary condition $V_0 = V_1 - D_{01}$ in Eq. 35, is monotone increasing in z by virtue of the boundary condition and regularity results along the free boundary [28, 29]. Now it remains to show that the expectation has a minimum

$$\inf_{\alpha} E[V(l, c, J)]$$

$$= \inf_{\alpha} \int_{0}^{\infty} V(l, c, J) g_{LN}(J) dJ, \quad g_{LN} \text{ is the lognormal kernel } LogN(\alpha, \beta^2) \quad (83)$$

$$= \int_0^\infty \inf_\alpha \{V(l, c, xe^\alpha)\} g_{LN}^*(x)dx, \ g_{LN} \text{ is the kernel } LogN(0, \beta^2) \qquad (84)$$

$$= \int_0^\infty V(l, c, xe^{\alpha_{min}}) g_{LN}^*(x)dx \qquad (85)$$

Summarizing, the PID variational inequality yields the *worst case* project value (minimal optimal control) when

$$\alpha = \alpha_{min}, \quad \lambda = \begin{cases} \lambda_{min} & \text{if } E[V(l, c, J)] - V(l, c, z) \geq 0, \\ \lambda_{max} & \text{if } E[V(l, c, J)] - V(l, c, z) < 0 \end{cases} \qquad (86)$$

and following a similar argument as above yields the *best case* value (maximal optimal control) when

$$\alpha = \alpha_{max}, \quad \lambda = \begin{cases} \lambda_{max} & \text{if } E[V(l, c, J)] - V(l, c, z) \geq 0, \\ \lambda_{min} & \text{if } E[V(l, c, J)] - V(l, c, z) < 0 \end{cases} \qquad (87)$$

□

References

1. Arnarson, T., Djehiche, B., Poghosyan, M., Shahgholian, H.: A PDE approach to regularity of solutions to finite horizon optimal switching problems. Nonlinear Anal. **79** (2009)
2. Avellaneda, M., Levy, A., Parás, A.: Pricing and hedging derivative securities in markets with uncertain volatilities. Appl. Math. Finance **27**, 73–88 (1995)
3. Baker, S.R., Bloom, N., Davis, S.J.: Measuring economic policy uncertainty. www.policyuncertainty.com (2013)
4. Bothast, R., Schlicher, M.: Biotechnological processes for conversion of corn into ethanol. Appl. Microbiol. Biotechnol.**67**, 19–25 (2005)
5. Brekke, K.A., Øksendal, B.: Optimal switching in an economic activity under uncertainty. SIAM J. Control. Optim. **32**, 1021–1036 (1994)
6. Brennan, M.J., Schwartz, E.S.: Evaluating natural resource projects. J. Bus. **58**, 135–157 (1985)
7. CBC: Ivanhoe 'surprised' by new Mongolian windfall tax. CBC News (2006). http://www.cbc.ca/news/business/ivanhoe-surprised-by-new-mongolian-windfall-tax-1.622422. Accessed 6 Jan 2014
8. CME: Chicago Mercantile Exchange commodity products: Trading the corn for ethanol crush
9. Cont, R., Voltchkova, E.: A finite difference scheme for option pricing in jump diffusion and exponential levy models. SIAM J. Numer. Anal. **3**(4), 1596–1626 (2005)
10. Cryer, C.W.: The solution of a quadratic programming problem using systematic overrelaxation. SIAM J. Control **9**, (1994)
11. Dixit, A.: Entry and exit decisions under uncertainty. Eur. J. Polit. Econ. **97**, 620–638 (1989)
12. Dixit, A., Pindyck, R.: Investment Under Uncertainty. Princeton University Press, Princeton (1994)

13. DOE: Department of Energy alternative fuels data center: laws and incentives "small ethanol producer tax credit" (2013). http://www.afdc.energy.gov/laws/law/US/352. Accessed 1 June 2013
14. Duffy, D.J.: Finite Difference Methdos in Financial Engineering: A Partial Differential Equation Approach. Wiley, West Sussex (2006)
15. EIA: US Energy Information Administration: Energy timelines ethanol (2012). http://www.eia.gov/kids/energy.cfm?page=tl_ethanol. Accessed 15 June 2012
16. Fisher, R.W.: Business decision-making "well nigh impossible" amid US policy uncertainty: Dallas fed chief. Financial Post (2012). http://opinion.financialpost.com/2013/06/04/u-s-economy-needs-clear-road-map/. Accessed 6 Jan 2014
17. Fleten, S.E., Haugom, E., Ullrich, C.J.: Keeping the lights on until the regulator makes up his mind! Midwest Finance Association 2012 Annual Meetings Paper (2012)
18. Forsyth, P.A., Vetzel, K.R.: Numerical methods for nonlinear PDEs in finance. In: Duan, J.C., Gentle, J., Hardle, W. (eds.) Handbook of Computational Finance, pp. 503–528. Springer, Berlin (2012)
19. Jaimungal, S.: Irreversible investments and ambiguity aversion. SSRN (2011). http://ssrn.com/abstract=1961786. Available at SSRN
20. Jaimungal, S., Lawryshyn, Y.: Incorporating managerial information into real option valuation (2011)
21. Kirby, N., Davison, M.: Using a spark-spread valuation to investigate the impact of corn-gasoline correlation on ethanol plant valuation. Energy Econ. **32**, 1221–1227 (2010)
22. Maenhout, P.J.: Robust portfolio rules and asset pricing. Rev. Financ. Stud. **17**, 951–983 (2004)
23. Maxwell, C., Davison, M.: Using real option analysis to quantify ethanol policy impact on the firm's entry into and optimal operation of corn ethanol facilities. Energy Econ. **42C**, 140–151 (2014)
24. McCarter, J.: Missauga Power Plant Cancellation Costs Special Report April 2013. Office of the Auditor General of Ontario (2013)
25. Merton, R.C.: Option pricing when underlying stock returns are discontinuous. J. Financ. Econ. **3**, 125–144 (1976)
26. NEO: Nebraska Energy Office: Nebraska energy statistics ethanol rack prices
27. NEO: Nebraska Energy Office: Ethanol production capacity by plant archive (2013). http://www.neo.ne.gov/statshtml/122_archive.htm. Accessed 1 June 2013
28. Øksendal, B., Sulem, A.: Applied Stochastic Control of Jump Diffusions. Springer, Berlin (2007)
29. Pham, H.: Continuous-Time Stochastic Control and Optimization with Financial Applications. Springer, Berlin (2009)
30. Reuters: Ontario drops plan for TransCanada power plant. Thomson Reuters (2010). http://www.reuters.com/article/2010/10/07/transcanada-ontario-idUSN0716951920101007. Accessed 6 Jan 2014
31. Reuters: Ontario looks set to cut green energy subsidies. Thomson Reuters (2011). http://uk.reuters.com/article/2011/11/02/canada-energy-renewable-idUKN1E7A10GV20111102. Accessed 6 Jan 2014
32. RFA: Renewable Fuels Association: Historic US fuel ethanol production (2014). http://www.ethanolrfa.org/pages/statistics. Accessed 15 Jan 2014
33. Samis, M., Davis, G., Laughton, D.: Using stochastic discounted cash flow and real option monte carlo simulation to analyse the impacts of contingent taxes on mining projects (2007)
34. Schmit, T., Luo, J., Tauer, L.: Ethanol plant investment using net present value and real options analyses. Biomass Bioenergy **33**, 1442–1451 (2009)
35. Shmuel, J.: Stop sitting on your cash piles, Carney tells corporate Canada. Financial Post (2012). http://business.financialpost.com/2012/08/22/dont-blame-the-loonie-for-weak-exports-carney-tells-caw/. Accessed 6 Jan 2014
36. USDA: US Department of Agriculture: Feed grains database
37. Wilmott, P., Dewynne, J., Howison, S.: Option Pricing: Mathematical Models and Computation. Oxford Financial Press, Oxford (1994)

A Hedged Monte Carlo Approach to Real Option Pricing

Edgardo Brigatti, Felipe Macías, Max O. Souza, and Jorge P. Zubelli

Abstract In this work we are concerned with valuing optionalities associated to invest or to delay investment in a project when the available information provided to the manager comes from simulated data of cash flows under historical (or subjective) measure in a possibly incomplete market. Our approach is suitable also to incorporating subjective views from management or market experts and to stochastic investment costs.

It is based on the Hedged Monte Carlo strategy proposed by Potters, Bouchaud, Sestovic (Phys. A Stat. Mech. Appl. 289(3–4):517–525, 2001) where options are priced simultaneously with the determination of the corresponding hedging. The approach is particularly well-suited to the evaluation of commodity related projects whereby the availability of pricing formulae is very rare, the scenario simulations are usually available only in the historical measure, and the cash flows can be highly nonlinear functions of the prices.

1 Introduction

The use of quantitative finance techniques to evaluate projects while trying to capture the value of active management and flexibility is known by the name of *Real Option Analysis* (ROA). The importance of capturing such "non-passive" value of projects can be a decisive factor when trying to decide upon investment within a portfolio of projects. Most of the classical applications of ROA involves vanilla

E. Brigatti
Instituto de Física, Universidade Federal do Rio de Janeiro, Av. A. da Silveira Ramos, 149, Cidade Universitaria, Rio de Janeiro, RJ 21941-972, Brazil
e-mail: edgardo@if.ufrj.br

F. Macías • J.P. Zubelli (✉)
IMPA, Estrada Dona Castorina 110, Rio de Janeiro, RJ 22460-320, Brazil
e-mail: fmacias@impa.br; zubelli@impa.br

M.O. Souza
Departamento de Matemática Aplicada, Universidade Federal Fluminense
R. Mario Santos Braga, s/n, Niteroi, RJ 24020-140, Brazil
e-mail: msouza@mat.uff.br

© Springer Science+Business Media New York 2015
R. Aïd et al. (eds.), *Commodities, Energy and Environmental Finance*,
Fields Institute Communications 74, DOI 10.1007/978-1-4939-2733-3_10

American options as the case of the option to postpone a project, or to abandon it. However, when considering projects related to capacity planning, chemical or petrochemical plants, oil refining, or indeed any commodities-based project, a significant increase in complexity arises. Under these conditions, recurring problems that are encountered in real options, as dealing with market incompleteness, become particularly acute.

In many cases, the company has access to financial instruments that strongly correlate with the projects, and sometimes, as in the case of commodities companies, even with their final product. Thus, the company can hedge some of its exposition yielded by a project, but usually not all of it, by an appropriate hedging portfolio. This suggests that a hedging approach based on Monte Carlo simulations can be a plausible alternative for pricing such real options. Indeed, on one hand quadratic hedging has been used to price financial options in incomplete markets, and it is based on the local minimization of a proxy to variance, that is readily recognized as a risk measure by managers. On the other hand, Monte Carlo approach has been often used when dealing both with options involving many assets—as baskets, rainbow, etc.—or when asset price models are not readily available.

The aim of this work is to propose the use of the so-called Hedged Monte Carlo Method—Monte Carlo pricing through quadratic hedging—to price such complex options.

The plan for this article goes as follows: We close this introductory section with a description of the project evaluation problem we are considering, a short methodological review of the different approaches to real options, and its analysis by means of hedging with financial instruments. In Sect. 2 we present an approach to evaluating real options based on the Hedged Monte Carlo (HMC) method of [39]. It has a number of desirable features: it uses the dynamics under the historical/subjective measure; it allows for an easy determination of the optimal exercise boundary, it has low variance, and allows for an assessment of the nonhedgeable risk. Furthermore, the oracle approach easily allows to incorporate managerial views in many different levels: it can either accommodate views of different managers of related projects, or more global corporative views and scenarios. The method developed is explored in Sect. 3 with some examples and a few case studies. We conclude in Sect. 4 with some final comments and suggestions for further developments.

1.1 Real Options Analysis

The use of mathematical finance techniques has been continuously growing in recent times as a tool to capture the value of flexibility in projects. A classical account can be found in the books of [9] and [48]. The subject blossomed under different names but is generally known *Real Options*. See also [3, 8, 22, 28, 32, 35, 38, 46, 47, 49].

The original framework identifies the Net Present Value of the project as a stochastic process correlated with a tradable risky asset. The risky asset is termed the *twin* or *spanning* asset whereas the project value is sometimes referred as the *surrogate* asset. Subsequent approaches take this identification very far. Indeed, one cannot expect to have a traded asset with a perfect correlation with the project, since this would mean that project risk is totally diversifiable, and hence replicable via financial markets. An alternative view, is to look for an asset, typically an index, that yields a high correlation with the project returns. This is known as the *modern approach*. Other approaches exist. See [2] for a classification, the discussion in [23], and the remarks in Sect. 1.4.

A very strong critique of the real option approach was presented by [21]. There they show, by means of a simple example, that the use of no-arbitrage techniques to nontradable surrogate assets can lead to arbitrary (very high or very low) no-arbitrage option prices. This in turn shows that the economic use of real options in the context of incomplete markets is highly questionable. In the same work, they also show that a variance minimization of the hedging error could be a way out of the economical impasse caused by the lack of completeness of the market.

1.2 Complex Structured Real Options

We are concerned with the practical problem of quantitatively evaluating projects under uncertainty from different scenarios taking into account flexibility of the projects and the possibility of partial hedging with financial instruments. We assume that we have available a fairly large number of scenarios organized in a time series and that connected to the different scenarios we have an *oracle* that produces the cash flows associated to each scenario. The scenarios in turn are linked to traded assets or financial instruments which may be used for hedging the project. Figure 1 describes the situation.

This framework can arise when planning chemical plants or oil refineries. See for example [30, 34, 36, 42, 45]. It also naturally appears when using real option techniques for capacity planning. See [4, 29, 33]. In most of these problems, the markets are overall incomplete, unless under very simplifying assumptions. In addition, such incompleteness will also imply that data will be only available under the historical measure.

We shall now consider different ingredients in such complex options. The first one, stems from the fact that many corporations predict in a fairly precise way their cash flows using a black box (oracle) whose stochasticity only comes through the

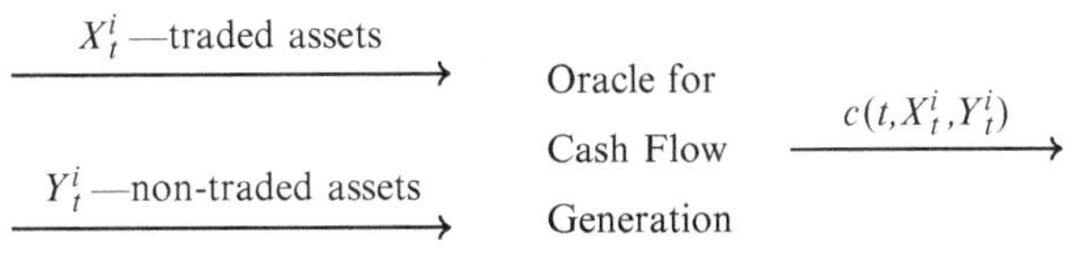

Fig. 1 Description of the oracle producing the cash flows at time t and scenario i

X_t^i —traded assets $\longrightarrow$

Y_t^i —non-traded assets $\longrightarrow$

Oracle for Cash Flow Generation $\xrightarrow{\quad c(t, X_t^i, Y_t^i) \quad}$

inputs from the different assets, supply/demand curves, and production curves. Yet, such oracle depends on the prices of many (stochastic) assets as well as on non-tradable quantities. This is depicted in Fig. 1.

More generally, the cash flows may be produced by simplified models that incorporate algorithms or analytical procedures.

Among the challenges that are present in the evaluation of projects under uncertainty, especially those linked to commodity enterprises, we single out the following:

- *Historical measure:* The simulations are usually presented in the historical measure. Furthermore, the scenarios are provided by management and are loaded with views from advisors or sector specialists. In fact, some corporations delegate the scenario generation to part of the board of directors or an independent division.
- *Managerial views:* It is crucial to incorporate managerial views in the cash flows, as well as automated decisions. An example would be a commodity trading company that has a limited amount of storage capacity for different products. According to the relative prices and profits it may automatically determine how much of each product it would store.
- *Market incompleteness:* The hedging is performed in incomplete financial markets. In fact, sometimes the firm does not have access to the liquidity provided by the financial markets. In other cases, regulations might preclude the management to hold some speculative positions to fully hedge against market variations.
- *Unhedgeable risks:* Investment decisions on commodity related projects have to take into account not only the hedgeable risks, but also the unhedgeable ones. For instance, the decision of exploring an oil field is highly dependent on its production curve and also on ecological risks associated to the operation.
- *Multiple assets:* Investment decisions may depend on the relative value of several traded underlyings. Such assets might have general correlation structures ranging from low to high cross-correlation. Thus, the hedging might have to be very diversified.

1.3 Real Option Analysis Through Hedging

The approach suggested here to attack the general problem mentioned above can be loosely described as a risk minimization one where the project valuation is performed by constructing a portfolio that includes the project delay optionality and the possible hedging of such project by tradable assets. By a methodology introduced by Potters et al. [39] (see also [17]) one can compute different financial options (including American and Bermudian ones) by a recursive risk minimization of *historical-measure* simulated paths. The importance of using historical simulations in the solution of this problem is that managers consider their decisions by looking at observed prices of different commodities and assets. We shall refer to the methodology developed in [39] by the *Hedged Monte Carlo method* (HMC).

Another motivation for the methodology presented here is the critique to the traditional no-arbitrage arguments of real option theory present in the work of [21]. In the latter, the idea of minimizing the variance is considered as an alternative to the shortcomings caused by market incompleteness. A number of different approaches have been developed to deal with incomplete markets. To cite a few: indifference pricing, minimal martingale measure, and minimal entropy measure.

The idea of using HMC or Monte Carlo algorithms to compute option prices in incomplete markets is not new. See, for example, the work of [40] and the references therein. It can also be traced to the preprint of [17]. The novelty of the approach suggested herein is the idea of incorporating the different cash flows in the evaluation, producing the different statistics that may be helpful for the manager and allow for the possibility of incorporating managerial views in the simulations. As it turns out, the HMC methodology corresponds to choosing the minimal martingale measure of Schweizer and Föllmer [44]. See [25] and [12] and references therein for details on such connection.

1.4   Remarks on Alternative Approaches

We shall now briefly review the various methodologies available to price real options.

1.4.1   Hedging Public and Private Risks

As observed in the works of Borison [2] and of Jaimungal and Lawryshyn [23], one of the main issues in evaluating different types of projects is whether the source of risk is public or private. For projects with returns that are highly correlated to the market, risk mitigation should be almost completely achievable by hedging it with traded assets. In most approaches, the project is assumed to be perfectly correlated to a single asset, and hence replicable. Notice that for projects which have a diverse range of products, it might be necessary to use a basket of hedging assets.

On the other hand, projects with mainly private risks, such as for instance R&D, are unlikely to be hedged with the use of traded assets. Moreover, in some cases the valuation of the project can be highly dependent on management estimates. Thus, one can think of such estimates as a non-traded asset that contributes to the value of the project.

From the point of view of utility theory, this can be more precisely measured by specifying the firm's preferences through a utility function, and thus one can think of using indifference pricing. This approach was pursued in a number of works, in particularly in the work of Henderson and Hobson [20], of Grasselli and Hurd [17], and of Grasselli and Hurd [19].

1.4.2   The Classical Method

As mentioned in the introduction, the classical methodology of pricing real options assumes that there is a spanning asset that is highly correlated to the net present value (NPV) of the project. One example of such methodology is the so-called *Marketed Asset Disclaimer (MAD) Approach* is based on the idea of taking the NPV distribution both as the value of the project and as the underlying (tradable) asset. Then, model the asset with a stochastic dynamics and perform Risk-Neutral pricing, perhaps accounting for non-traded issues. See for example [6] and [7]. Among the advantages we mention that it mimics the standard mathematical finance approach, the theory is fairly simple and many out-of-the-box numerical methods are available. As for the disadvantages, besides the general criticism mentioned before in reference to the work of [21], we should also note that often very few data is available for calibration. This makes the choice of the underlying dynamics somewhat arbitrary. Furthermore, for each project, a calibration/choice of underlying dynamics is necessary. This ambiguity is typically tackled by a simplifying assumption on the dynamics, which will hopefully be consistent with the market scenarios.

1.4.3   Monte Carlo Based Approaches

In many situations the project or the firm has a simulator that we shall refer from now on as an *oracle*. Such oracle produces information about the cash flows associated to different projects or optionalities for different scenarios which in turn are generated from inputs of tradable assets. The idea is then to take the oracle output as the payoff distribution, and use the method of Longstaff and Schwartz [26] to compute the corresponding conditional expected values subject to the traded asset prices. This requires the underlying(s) to be simulated in the risk-neutral measure or taking into account the market price of risk in the final result.

Among the pros of such approach, we should mention that it uses fully the oracle information towards the option evaluation, it is easily integrated and automated with the oracle thus leading to a project independent pricing mechanism. Furthermore, it has a good managerial appeal. As for the cons, we have that since the simulation is performed on the oracle data, the realizations are restricted to the ones generated by the oracle. This can impair the quality of the results obtained. Furthermore, the risk-neutral calibration of the scenario generation that will provide inputs to the oracle might be very cumbersome and requires extra work.

1.4.4   Datar-Mathews (DM) Method

In the method proposed by [27] one assumes that it is given the NPV distributions (usually by management). Then, one performs a Monte Carlo simulation to replicate the distribution at the given times and to produce a simulated process for the underlying asset.

Among the advantages, we can mention that it is easily implemented and has great managerial appeal. Yet, there is lack of theory to justify such approach.

1.4.5 Jaimungal-Lawryshyn (JL)

The work of Jaimungal and Lawryshyn [23] includes an extension of DM method as follows: They take the NPV distributions and choose an observable sector index (not-traded on their paper) that is highly correlated with cash flows. They choose a dynamics for this index and based on the dynamics, find the payoff functions that yield the NPV distribution as a function of this market index. Then, they identify the value of the project as expected values of these payoffs (very much line in DM's method). Finally, they choose a correlated (if possible) traded asset or index and perform a risk-neutral valuation using a Minimal Martingale Measure.

Among the advantages of this method, we can cite that as in the DM method, it integrates the managerial view with the Real Option Analysis. Thus it has a good managerial appeal. Furthermore, the theory is more sound. Yet, the market index might not be easily available and one still needs to calibrate the model to the index. This step might be hard if the data is not abundant.

2 The Hedged Monte Carlo Approach and Minimal Martingale Measures

Since the typical data that will be used for the method comes from simulations, it will be naturally discrete in time. Thus, it is natural to adopt a discrete time approach for the algorithm. In this vein, we begin by reviewing the theory for quadratic hedging in discrete time and how it can be used to price contingent claims. This will follow closely the exposition of Föllmer and Schied [11]. Then, we proceed on to discussing the algorithm itself, and present a brief remark about its relation to a continuous version of the problem.

2.1 Hedging in Discrete Time Within an Incomplete Market: A Review

In an incomplete market setting, from its very definition, a self-financing replicating strategy is not usually available. In this scenario, one might give up the replicating property, and look for self-financing hedging strategies that control the down side risk—evaluated by means of a risk measure. See for example the work of Föllmer and Schied [11]. Alternatively, one enforces a replicating strategy and looks for the cheapest strategy with this property. In this latter case, a very popular strategy among practitioners is the minimization of the quadratic tracking error [43]. This choice leads to strategies that are self-financing in the mean under very mild assumptions, that we now briefly review.

As usual, we assume to be in a filtered probability space $(\Omega, \mathscr{F}_T, \mathbb{P})$ and write $L^2(\mathbb{P}) = L^2(\Omega, \mathscr{F}_T, \mathbb{P})$, where $\mathbb{P}$ denotes the historical measure. In what follows, ξ^N denotes the investment (short or long) in the numéraire asset, and ξ denotes the position on d risk assets, with prices given by a d-dimensional stochastic process X. Furthermore, X and V denote *discounted prices* with respect to a risk-free process.

Definition 1. A trading strategy is a pair of stochastic processes (ξ^N, ξ), where ξ^N_t is an adapted process and ξ is a d-dimensional predictable process. The discounted value of the portfolio is

$$V_t := \xi^N_t + \xi_t \cdot X_t$$

The gain process is

$$G_t := \sum_{s=1}^{t} \xi_s \cdot (X_s - X_{s-1}).$$

The cost process is defined as

$$C_t := V_t - G_t.$$

Let H denote a random claim, and assume that

1. $H \in L^2(\mathbb{P})$;
2. $X_t \in L^2(\Omega, \mathscr{F}_T, \mathbb{P}; \mathbb{R}^d)$, for all t.

Definition 2. An admissible L^2-strategy for H is a trading strategy such that it is replicating, i.e.,

$$V_T = H \quad \mathbb{P} \text{ a.s.,}$$

and such that both the value process and the gain process are square-integrable, i.e.,

$$V_t, G_t \in L^2(\mathbb{P}), \ \forall t \in [0, T].$$

We can now introduce a suitable risk process

Definition 3. Let (ξ^N, ξ) be an L^2-admissible strategy. The corresponding local risk process is given by

$$R_t^{\text{loc}}(\xi^N, \xi) = \mathbb{E}[(C_{t+1} - C_t)^2 | \mathscr{F}_t].$$

Let $(\hat{\xi}^N, \hat{\xi})$ be an L^2-admissible strategy with value process $\hat{V}_t$. This strategy is said to be a locally risk-minimizing strategy if, for each t, we have that

$$R_t^{\text{loc}}(\hat{\xi}^N, \hat{\xi}) \leq R_t^{\text{loc}}(\xi^N, \xi), \quad \mathbb{P} \text{ a.s.}$$

for each L^2-admissible strategy whose value process V_t satisfies $V_{t+1} = \hat{V}_{t+1}$.

Definition 4. A trading strategy is a mean self-financing strategy, if its corresponding cost process is a martingale, i.e.:

$$\mathbb{E}[C_{t+1} - C_t | \mathscr{F}_t] = 0.$$

Definition 5. We say that two adapted processes U and V are strongly orthogonal if

$$\text{cov}(U_{t+1} - U_t, V_{t+1} - V_t | \mathscr{F}_t) = 0,$$

where cov denotes the conditional covariance, i.e., $\text{cov}(A, B | \mathscr{F}_t) = \mathbb{E}[AB | \mathscr{F}_t] - \mathbb{E}[A | \mathscr{F}_t]\mathbb{E}[B | \mathscr{F}_t]$.

The following result (see [11]) guarantees the existence of the corresponding hedge:

Theorem 1.

1. *An L^2-admissible strategy is locally risk minimizing if, and only if, it is mean self-financing, and its cost process is strongly orthogonal do X.*
2. *There exists a locally risk minimizing strategy if, and only if, H admits the so-called Follmer-Schweiser decomposition:*

$$H = c + \sum_{t=1}^{T} \xi_t \cdot (X_t - X_{t-1}) + L_T, \quad \mathbb{P}\text{-}a.s.,$$

where c is a constant, ξ is a d-dimensional predictable process, such that $\xi_t \cdot (X_t - X_{t-1}) \in L^2(\mathbb{P})$ for each t, and L is a square-integrable martingale that is strongly orthogonal to X, and satisfies $L_0 = 0$.
   In this case, the locally risk-minimizing strategy $(\hat{\xi}^N, \hat{\xi})$ is given by :

$$\hat{\xi} = \xi$$

$$\hat{\xi}_t^N = c + \sum_{s=1}^{t} \xi_s \cdot (X_s - X_{s-1}) + L_t - \xi_t \cdot X_t.$$

Notice that the associated cost process is $C_t = c + L_t$.

2.2 Pricing by Risk Minimization

The proof of Theorem 1 is actually constructive and yields the following algorithm:

Algorithm 1.

1. Set $\hat{V}_T := H$;
2. For $t = T - 1$ down to $t = 0$ do

   a. Set

$$(\hat{V}_t, \hat{\xi}_{t+1}) := \underset{(V_t, \xi_{t+1})}{\mathrm{argmin}} \, \mathbb{E}\left[\left(\hat{V}_{t+1} - (V_t + \xi_{t+1} \cdot (X_{t+1} - X_t))\right)^2 \Big| \mathscr{F}_t\right]; \quad (1)$$

3. Set $\hat{C}_t := \hat{V}_t - \sum_{s=1}^{t} \hat{\xi}_s \cdot (X_s - X_{s-1})$, $t = 0, \cdots, T$;
4. Set $\hat{c} := \hat{C}_0$;
5. Set $\hat{L}_t := \hat{C}_t - \hat{c}$, $t = 0, \cdots, T$;
6. Set $\hat{\xi}_t^N := \hat{c} + \sum_{s=1}^{t} \hat{\xi}_s \cdot (X_s - X_{s-1}) + \hat{L}_t - \hat{\xi}_t \cdot X_t$, $t = 0, \cdots, T$.

Notice that if $\mathbb{P}$ is a risk-neutral measure, then X_t is a square-integrable martingale. In this case, the Galtchouk-Kunita-Watanabe decomposition ([5]) yields

$$\mathbb{E}[H|\mathscr{F}_t] = \hat{V}_0 + \sum_{s=1}^{t} \hat{\xi}_s \cdot (X_s - X_{s-1}) + L_t$$

and hence we have

$$\mathbb{E}[H|\mathscr{F}_t] = \hat{V}_t.$$

This allows for a consistent interpretation of the value of a local risk minimizing strategy as an arbitrage-free price of H. However, in general, X will not be a martingale under $\mathbb{P}$, and in the incomplete setting there will be many martingale measures that are equivalent to $\mathbb{P}$. It turns out that one of these measures is particularly relevant for hedging under local risk minimization.

Definition 6. Let $\mathscr{P}$ denote the set of martingale measures that are equivalent to $\mathbb{P}$. We say that $\hat{\mathbb{P}} \in \mathscr{P}$ is a minimal martingale measure if

$$\mathbb{E}\left[\left(\frac{d\hat{\mathbb{P}}}{d\mathbb{P}}\right)^2\right] < \infty,$$

and if every square-integrable martingale under $\mathbb{P}$, which is strongly orthogonal to X is also a martingale under $\hat{\mathbb{P}}$.

Theorem 2. *If there exists a minimal martingale measure $\hat{\mathbb{P}}$, and denoting by $\hat{V}$ the value process of the local risk minimizing strategy, then we have that*

$$\hat{V}_t = \hat{\mathbb{E}}\left[H|\mathscr{F}_t\right].$$

We close this section with some practical remarks. The first one is that the crucial part of Algorithm 1, as far as valuation of the contingent claim is concerned, is composed of steps 1 and 2. The second one is that for real options and numerical simulations it is more convenient to work with undiscounted prices of the assets and the contract. Thus, from now on we shall revert to actual prices and use a discounting factor of $\rho = \exp(r\Delta t)$ where r is the risk-free rate.

If we are given a payment stream of cashflows, c_t for $t = T_0, \cdots T_F \leq \infty$, under the minimal martingale measure $\hat{\mathbb{P}}$ and discounting by the constant interest rate, the expected value $\mathscr{V}_t$ is given by

$$\mathscr{V}_t = \hat{\mathbb{E}}\left[\sum_{s=t}^{T_F} c_s/\rho^{s-t}\Big|\mathscr{F}_t\right].$$

In this case, the generalization of Algorithm 1 is straightforward. Under the assumption that we are working in a Markovian setting such value becomes

$$\mathscr{V}_t = \hat{\mathbb{E}}\left[\sum_{s=t}^{T_F} c_s/\rho^{s-t}\Big|X_t = x\right]. \tag{2}$$

We shall now address the question of computing such conditional expectation from historical simulations. If we have a large number N of simulations to the process $\{X_t\}_{t=0,1,\cdots}$, we can approximate the term on the R.H.S. of the local risk term R_t^{loc} by

$$R_t^{\text{loc}} \approx \frac{1}{N}\sum_{i=1}^{N} \left(\rho^{-1}V_{t+1}(X_{t+1}^i) - V_t(X_t^i) - \xi_{t+1}(X_t^i)\left[\rho^{-1}X_{t+1}^i - X_t^i\right]\right)^2.$$

The next step is to make the problem numerically tractable. But this, following the ideas of Longstaff and Schwartz [26] and Potters et al. [39], can be accomplished by introducing a function basis for the unknown function $\xi_{t+1}(x)$ (respec. $V_t(x)$) and considering a finite element expansion. More precisely, let us write

$$V_t(x) = \sum_{a=1}^{b} \gamma_t^a K_a(x)$$

and

$$\xi_{t+1}(x) = \sum_{a=1}^{b} \psi_{t+1}^a H_a(x),$$

286 E. Brigatti et al.

where H_a (respec. K_a) forms a basis for the space of functions ξ_{t+1} (respec. $V_t(x)$). Then, one can substitute the minimization problem in Eq. (1) by the minimization:

$$\underset{\{\gamma_t^j,\psi_{t+1}^j\}_{j=1}^b}{\operatorname{argmin}} \sum_{i=1}^N \left[\rho^{-1}V_{t+1}(X_{t+1}^i)-\sum_{a=1}^b \gamma_t^a K_a(X_t^i)-\sum_{a=1}^b \psi_{t+1}^a H_a(X_t^i)\cdot(\rho^{-1}X_{t+1}^i-X_t^i)\right]^2$$

(3)

In other words, the expected value is computed by expanding the function in $L^2(\Omega, \mathscr{F}_t, d\hat{\mathbb{P}})$ in a suitable basis and truncating at an appropriate level. Needless to say, there are a number of relevant issues, ranging from conditions on the processes to approximation spaces. A more detailed analysis of the *non-Markovian* case and of such approximation spaces would take us too far afield. See for example Section 1.3 of the work of Lipp [25].

2.3 The HMC Algorithm for Real Options

We shall now present the proposed algorithm for the evaluation of the delay option of a project that could be started at any time between say the time $T_0 \geq 0$ and T. In financial terms, this consists of a Bermudian option that could be exercised at any time between T_0 and T. Obviously, it reduces to an American option if $T_0 = 0$ is the present time. In mathematical terms this corresponds to a discrete version of a free boundary problem. We assume further that our cash flows could come at any time till T_F. The main building block of our algorithm is the regression described in Eq. (3).

We assume we are given the following inputs:

- A vector time series of traded assets x_t^i, for a period of times $t = T_0, \cdots, T$, and for the scenarios $i = 1, \cdots, N$.
- The corresponding cash flows associated to the different scenarios c_t^i for $t = T_0, \cdots, T_F$, and $i = 1, \cdots, N$. Such cash flows would be produced by an oracle which takes into account the different traded asset values and the non-traded ones.[1]
- A long term behavior for the project value or the cash flows (possibly under the different scenarios).
- The exercise period of the optionality $T_0, \cdots, T$, where $0 \leq T_0 < T \leq T_F$.

We now perform the following algorithm:

---

[1]In principle, it could be also time dependent and even scenario dependent. Furthermore, it can incorporate managerial views by emphasizing specific regions of the probability space.

Algorithm 2. *[HMC for Real Options]*

1. *Initialize the project value $\mathcal{V}_T(X_T^i)$ for the different scenarios $i = 1, \cdots, N$ by using Eq. (2) for $t = T_0 \cdots T$.*
2. *Initialize for $t = T$ the payoff $\hat{V}_T(X_T^i) = (\mathcal{V}_T(X_T^i) - K)^+$ for the different scenarios $i = 1, \cdots, N$.*
3. *For $t = T - 1, \cdots, T_0$ do:*

   a. *Define the functions:*
   $$V_t(x) := \sum_{a=1}^{b} \gamma_t^a K_a(x) \text{ and } \xi_{t+1}(x) := \sum_{a=1}^{b} \psi_{t+1}^a H_a(x)$$
   b. *Solve the quadratic minimization problem in terms of the coefficients γ_t^a, ψ_{t+1}^a:*

   $$\underset{\{\gamma_t^a, \psi_{t+1}^a\}_{a=1}^{b}}{\text{argmin}} \sum_{i=1}^{N} \left[\rho^{-1}\hat{V}_{t+1}(X_{t+1}^i) - \sum_{a=1}^{b} \gamma_t^a K_a(X_t^i) - \sum_{a=1}^{b} \psi_{t+1}^a H_a(X_t^i) \cdot (\rho^{-1}X_{t+1}^i - X_t^i) \right]^2$$

   c. *Define $\hat{V}_t(X_t^i) := \max\{(\mathcal{V}_t^i - K)^+, \hat{V}_t(X_t^i)\}$.*

4. *Output: The values of $\hat{V}_{T_0}(x)$ for $x \in \{X_0^i\}_{i=1}^{N}$ and the points in the exercise region.*

It $T_0 = 0$ we could continue the downward loop without the comparison and the computed values in V_0 would give an approximation for the option value and the different scenarios[2] at the initial time $t = 0$.

If we were working with the risk neutral simulations in a complete market, this algorithm reduces to a variant of the celebrated algorithm of Longstaff and Schwartz [26].

Remark 1. In the actual implementation, the user may be interested in having access to the exercise region as well as to more information about the suitability of investment by using different statistics. Thus, it may be interesting to refine the Item 3.c. of the algorithm as follows:

3.c.   Define $\hat{V}_t(X_t^i) := \max\{(\mathcal{V}_t(X_t^i) - K)^+, \hat{V}_t(X_t^i)\}$ and store:

   i. $I_t := \{i \in \{1, \cdots, N\}/\hat{V}_t(X_t^i) \le (\mathcal{V}_t(X_t^i) - K)^+\}$
   ii. $v_t := \min\{(\mathcal{V}_t(X_t^i) - K)/i \in I_t\}$
   iii. $Pr_t := P\left((\mathcal{V}_t(X_t) - K)^+ \le v_t\right) \approx \#\{i \in \{1, \cdots, N\}/(\mathcal{V}_t(X_t^i) - K)^+ \le v_t\} \cdot N^{-1}$

The stored values of the points $(t, \hat{V}_t(X_t^i))$ for $i \in I_t$ correspond to an approximate description of the exercise region.

The quantity $\mathcal{V}_t(X_t^i) - K$ will be called *intrinsic value of the investment option* in the sequel. It refers to the best estimate of the stream of cash flows under the minimal martingale measure given the scenario i minus the investment K.

---

[2]Such different scenarios may reduce to a single point in case the initial scenario is known.

The managerial usage of these statistics springs from the fact that, in many cases, the stochastic generated cash flows inherit a corporate view of the market scenarios. As such, these statistics provide a subjective view on the investment scenarios that is appreciated by managers.

2.3.1 Implementation Notes

The attentive reader will notice that the main bottle-neck of the whole procedure is precisely in the minimization of 3.(b). A fast and stable algorithm here would make the difference in practical applications. This minimization can be performed very efficiently by using the QR algorithm to solve an overdetermined system of linear equations. See the text of Golub and Van Loan [16] for the numerical analysis background. The methodology can then be implemented (as we did) in a matlab-like environment with the standard Linpack packages. It can be easily ported to other popular programming languages such as R and Java.

The choice of the basis function is the subject of research by many authors even in the case of the classical LSM algorithm of Longstaff and Schwartz [26]. We follow the suggestion in the work of Potters et al. [39] for the one-dimensional case of taking the elements of the basis for hedge to be derivative of the ones for the option. We also take into account the suggested basis in [13]. In the multidimensional case we consider tensor products of the elements in the different dimensions.

2.4   Remark on the Continuous Limits

In the case of data simulated or estimated from a continuous model, we might consider realizations with arbitrarily small time intervals and refined asset price grids. Then, a very natural question is whether the discrete algorithm has any form of limit as $\Delta t \searrow 0$. This problem then can be divided into two parts. First, the continuous limit of discrete time model. Secondly, the numerical method to solve the limit case, its accuracy and efficiency.

Concerning the first issue, in the case of European options it is well established that the minimal martingale measure of Fölmer and Schweizer is associated to Backward Stochastic Differential Equations (BSDEs). See for example [10] for an early account. In the work of Pham [37] the main results of the theory of quadratic hedging in a general incomplete model of continuous trading with semi-martingale price process are reviewed. In particular, two types of criteria are studied: the mean-variance approach and the (local) risk-minimization, which is connected to the continuous limit of the approach considered here. In the work of Bobrovnytska and Schweizer [1] the mean-variance hedging problem is treated as a linear-quadratic stochastic control problem. They show for continuous semi-martingales in a general filtration that the adjoint equations leads to BSDEs for the three coefficients of the quadratic value process.

Concerning the second issue, the use of regression-like Monte Carlo methods has received a lot of attention recently. See [14, 15, 24] In particular, under appropriate conditions, the convergence of the HMC method can be proved and the error analysis has been performed in [14]. Furthermore, in [25] the HMC method has been implemented to some exotic options and its numerical aspects have been studied. In [12] the HMC method was implemented for actuarial problems.

3 Examples and Case Studies

We shall now exemplify the methodology proposed in the previous sections. The first set of examples will be purely illustrative ones aiming to exemplify the efficacy of the algorithm for option evaluation. They serve as validation and accuracy check for the codes. The second set comes from a large number of real data and practical evaluations. The examples take into account a large number of hedging energy commodities in the evaluation of a potential project in the energy sector. Finally, we present an exploration on a fictitious example involving gas data (Henry Hub index) and a technology stock (Google). The project cash flows would be associated to the difference of (rescaled) values of such underlyings added to an uncorrelated and nonhedgeable noise component.

3.1 Illustrative Theoretical Examples

The first example concerns the running of the algorithm in the classical Black-Scholes market with simulated prices taken in the historical measure. More precisely, we consider a European option expiring in 3 months with strike $K = 100$, current asset price varying around the at-the-money value $X(0) = 100$, volatility $\sigma = 0.3$, and interest rate $r = 0.05$. The number of basis elements (monomials 1, x and x^2) was $b = 3$ and a total of $N = 5,000$ simulations in an arbitrary (fixed) probability measure.

Although this is a very simple text-book example, Fig. 2 conveys the fact that the results are pretty accurate even for such a small number of simulations and small number of basis elements.

In the second example we check the algorithm performance of the difference of two hedgeable assets X_1 and X_2. More precisely we consider a 65 days exchange option with payoff $(X_{1,T_F} - X_{2,T_F})^+$. The variables X_1 and X_2 satisfy geometrical Brownian motion dynamics with $\sigma_1 = 0.3$, $\sigma_2 = 0.2$, and $r = 0.05$. The analytical results are obtained using the Margrabe's formula. In our setting this formula states that the fair price for the option is: $X_{1,0}N(d_1) - X_{2,0}N(d_2)$, where N denotes the cumulative distribution function for a normal distribution and $d_{1,2} = \left(\ln[X_{1,0}/X_{2,0}] \pm \sigma^2 T_F/2\right)/\sigma\sqrt{T_F}$, with $\sigma = \sqrt{0.3^2 + 0.2^2}$. See [31]. Here, we used two monomials and $N = 10,000$ simulations. The results are displayed in Fig. 3.

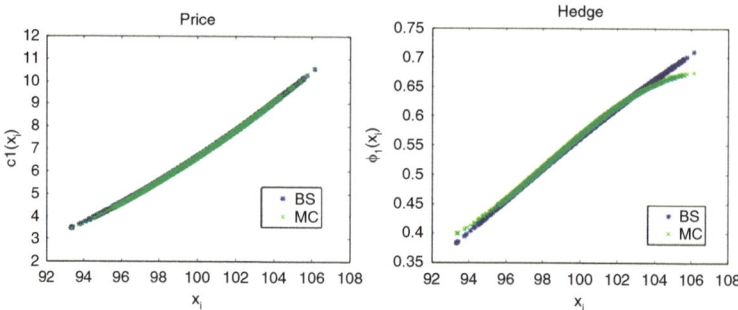

Fig. 2 The results of a comparison of the actual Black-Scholes formula price and the Hedged Monte Carlo algorithm result. On the *left* we display the prices and on the *right* we display the hedge value

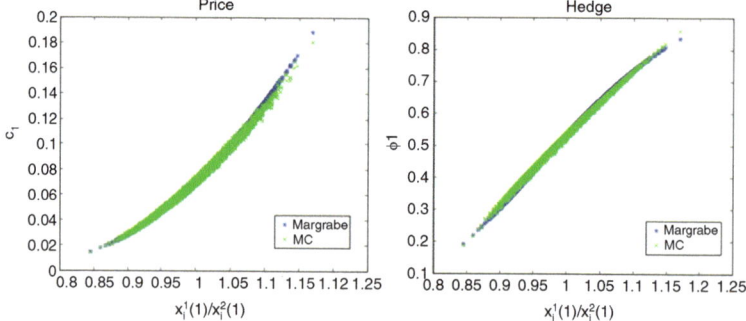

Fig. 3 Results of the comparison between the HMC algorithm and the Margrabe formula

3.2 Practical Examples

3.2.1 First Example

An energy company considers the optionality of starting a new project that would last for 11 years. The project value V_t is dependent on 12 different underlyings. The option is exercizable every year during the first 5 years. The company also has a trading desk that could be used for financial investment in some or all of such different assets.

The optionality was evaluated using several different sets of hedging assets. We now report on the results obtained with one hedging variable (in this example the Brent price) and considering 2,000 paths along 11 years with a (continuously compounded annualized) interest rate $r = 0.08$. We also computed examples with more hedging variables.

In Fig. 4 we present the option evaluation using one hedging variable. In this example the project works as a hedge towards low prices of the Brent. The fact that the intrinsic value of the project is smaller than the optionality indicates that the company should wait to start the project.

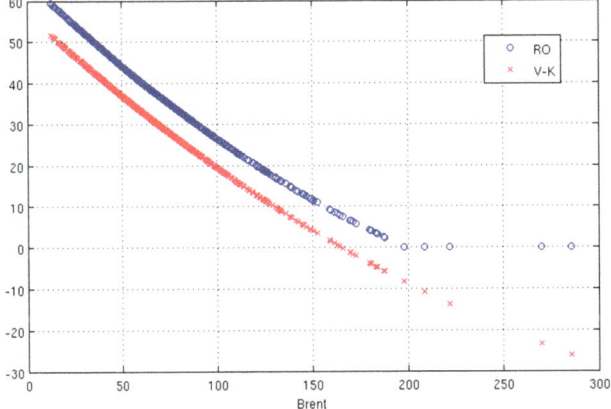

Fig. 4 Option evaluation using one hedging variable as a function of the Brent value. The difference between the project and the investment ($\mathcal{V}_{t=0}(X_1) - K$) is plotted in (*red*) crosses while the optionality $V_{t=0}(X_1)$ is plotted with (*blue*) *circles*. Here, the investment (strike) is $K = 10.89$ and the risk free interest rate $r = 0.08$

Fig. 5 A description of the cash flow under the different scenarios. The *lower line* corresponds to the 5 % quantile and *top* one to the 95 %. The *marked region* indicates the 90 % frequency region

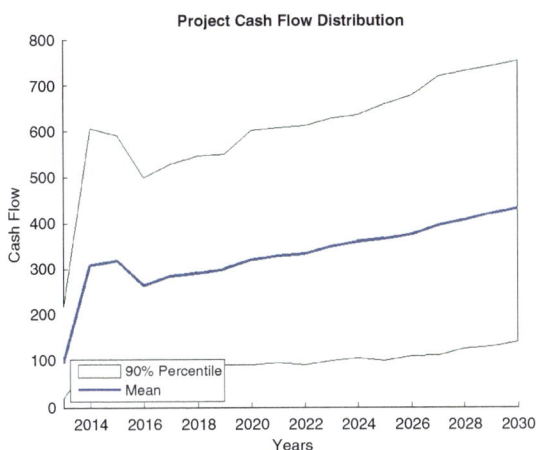

3.2.2 Second Example

In this example we consider a project that would run for 15 years, an investment of 1,500 monetary units and a yearly free interest rate of 8.00 %. The cash flows for this period are the results of an oracle that depends on a number of traded and non-tradable variables and in turn are produced by means of running different scenarios. Some of their descriptive statistics is presented in Fig. 5.

The intrinsic values of the optionality for the different times, including the 5 %, and 95 % quantiles for the project value are presented in Figs 6 and 7. By applying the Hedged Monte Carlo method we compute the value of the delay optionality considering three hedging variables. The project should be exercised if at a certain

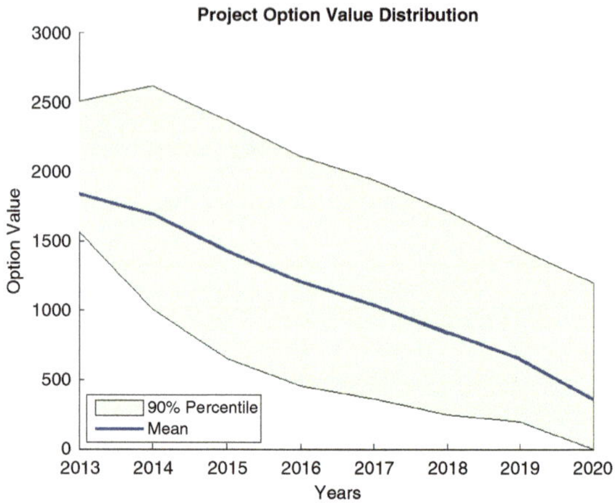

Fig. 6 Value of the project optionality. The *lower line* corresponds to the 5 % quantile and *top* one to the 95 %. The *marked region* indicates the 90 % frequency region

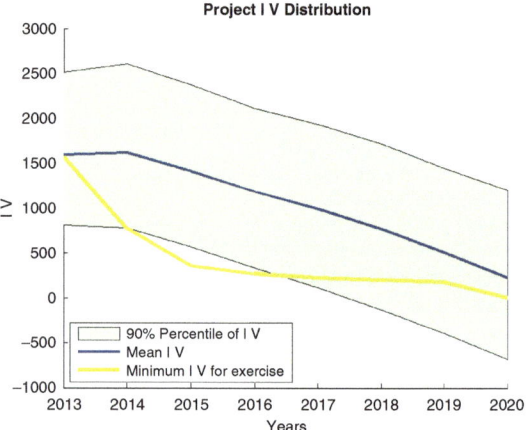

Fig. 7 A description of the project Intrinsic Value (IV) statistics under the different scenarios and the minimum value for exercise. The *lower line* corresponds to the 5 % quantile and *top* one to the 95 %. The *marked region* indicates the 90 % frequency region

time and corresponding scenario the intrinsic project value is more than the delay optionality. This leads to a trigger curve that tells us for each scenario whether to invest or not (Fig. 7).

3.2.3 Third Example

Differently from the previous examples whereby the actual cash flows came from complex (black-box type) oracles, our present example concerns a fictitious project where the cash flows would come from a (fairly) simple mathematical function. It concerns an artificial potential investment on a gas propelled vehicle that could be used by an information technology company to gather geographical data and to use in their web-based advertisements. For simplicity we take the cash flow highly correlated to Google stock through the equation

$$c_t(X, \epsilon) = \mathsf{H}\left(aX_{1,t} - bX_{2,t} - I + \epsilon_t\right), \tag{4}$$

where X_1 is the price of a Google stock, X_2 is Henry Hub (HH) gas index, I is a fixed running cost, ϵ_t is a nonhedgeable noise. The function H in our example is defined as

$$\mathsf{H}(x) = \begin{cases} 0, & x \le 0, \\ x, & x \in (0, 1), \\ 1, & x \ge 0. \end{cases}$$

The rationale behind H is to simulate the saturation given by very large values of the stock and to clip the values below zero.

We performed the data collection using publicly available data downloaded by using public domain R software.[3] The historical results between August 19th, 2004 and November 24th, 2013 are displayed in Fig. 8. We calibrated the historical log-returns of the data with a GARCH(1,1) model, and then performed a principal component analysis of the bi-dimensional innovation time series. From that we generated the simulations of future scenarios (Figs. 8 and 9).

In this example we consider a project that would run for a maximum period of say 3 years and the decisions could be performed monthly. The cash flows for this period are the results of the oracle described in Eq. (4) that depends on a value of Google and HH Gas. Finally we choose an investment of $INV = 3.5$ and a risk-free interest rate of 8.00 %.

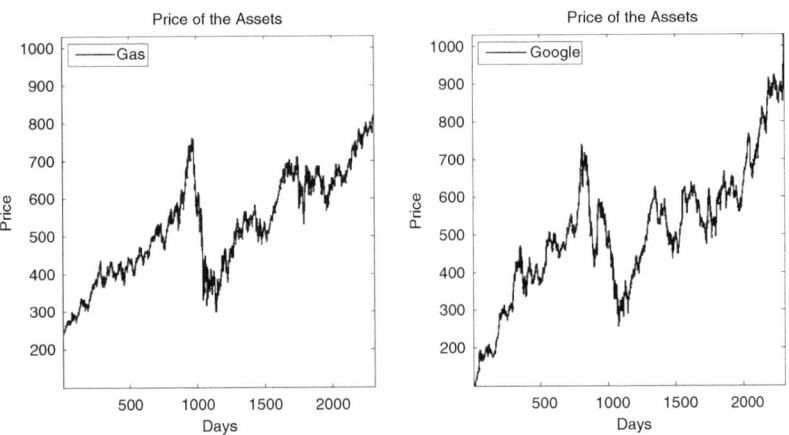

Fig. 8 Time series for the assets between August 19th, 2004 and November 24th, 2013

---

[3]See for example [41].

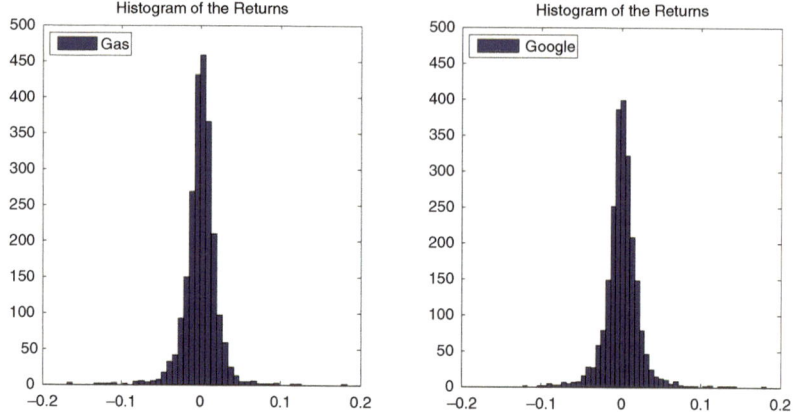

Fig. 9 Histogram of the log returns for the assets between August 19th, 2004 and November 24th, 2013

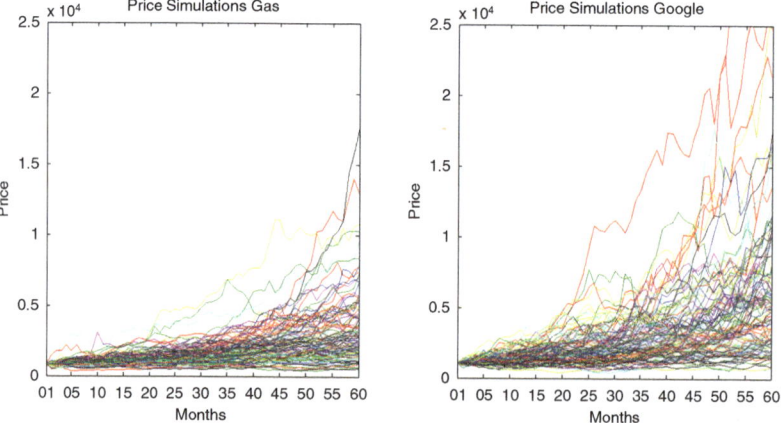

Fig. 10 Asset simulations

In Fig. 10 we present some simulations of the assets, and in Fig. 11 a description of the simulations of the cash flows by showing their mean, their quantiles.

The results in Fig. 13 show how the statistics of the values for the Intrinsic Value (defined as $\mathcal{V} - I$) relates to the curve of minimum value of the Intrinsic Value for exercise (v_t) that was calculated in the refined algorithm leading to Eq. (1). As the time varies between $t = 1$ and $t = 12$, the exercise curve crosses the average of the Intrinsic Values for the different scenarios. The case of v_t being smaller than the Intrinsic Value mean implies a small $Pr_t := P(NPV < v_t)$. These small values of

Fig. 11 Cash flow simulations for the fictitious oracle described by Eq. (4). Using the parameters value $a = 1.2895 \times 10^{-4}$, $b = -5.3191 \times 10^{-5}$, $I = 0.05, \varepsilon_t \sim \mathcal{N}(0, 0.005)$

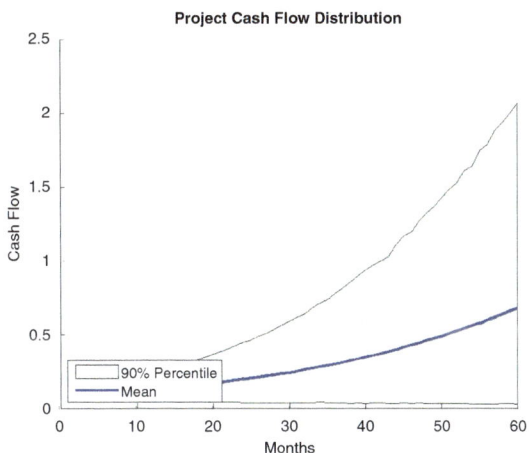

Fig. 12 A description of the option value statistics under the different scenarios

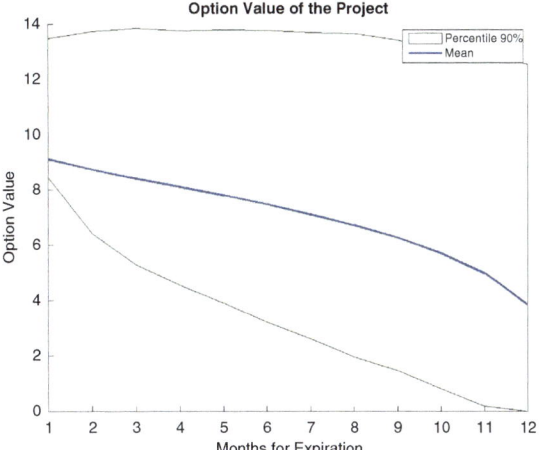

Pr_t give a good suggestion of when to invest. But the decision to invest also has to involve the option value described in Fig. 12 and the expected Intrinsic Value value of Fig. 13.

4 Discussion and Conclusions

In this work we addressed the problem of pricing real options on projects that have their cash flow estimates based on an oracle prediction. Such oracle is typically a combination of asset prices either used for production or obtained as a result of the working project and non-traded specific variables. They can also forecast prices or demand, and they can include managerial views or other non-tradable

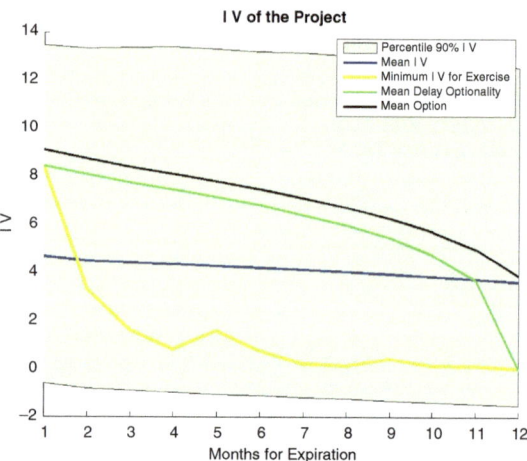

Fig. 13 A description of the project Intrinsic Value statistics under the different scenarios and the minimum value for exercise

information that impacts the project value. These prices and variables may further be processed by an optimization procedure, and this leads to the project cash flows. As discussed in the Introduction, this appears naturally in many situations, in particular for chemical or oil industries.

For such problems, we proposed a method that is based on minimizing the tracking error variance of the hedge. This can be interpreted as assuming that we are in an incomplete market and that the investor is naturally risk averse. In this context, this variance is a natural risk measure for the investor. Under this framework, we show how to price real options using the method of Potters et al. [39]. This lead to a set of consistent prices that reduces to that of the Black-Scholes theory when the market is complete. The obtained price will depend on the set of assets chosen for the hedge. This is natural since companies with access to different markets and vulnerable to different scenarios can have very different values for the same project. Theoretically, one could include all hedging assets on a maximal set, but this is unfeasible from a practical point of view.

Once more, we reinforce the idea that our simulations are all done in the historical measure where the calibration of the models take place. We could also have incorporated managerial views by emphasizing scenarios that would be more likely due to management selective information. On the other extreme, even if the decision maker and the business at hand had access to a completely correlated asset that could be used to hedge the project value, among the advantages of the present approach over a risk-neutral Monte Carlo evaluation we can mention: The reduction of variance of price estimation (for the same precision the number of paths can be up to 100 times smaller). This was already documented in the original work of Potters et al. [39]. The estimation of the hedging strategy, residual risk (in the form of the local variance), and possibly other risk measures (such as VaR and CVaR) at each time step.

As explained in the conclusion of the work of Grasselli [18], it is the time flexibility itself, more than the possibility of replication, that bears the extra value of an investment opportunity. Thus, the fact that we cannot replicate the project value should not be the reason for not trying to quantify such extra value. The work of Grasselli [18] takes the point of view of utility functions and indifference pricing. In contradistinction, here we took the point of view of minimizing risk as measured by the variance. A very natural follow up of the present work would be to compare the different approaches in the case of real world examples, such as the ones presented here. An exploration of the numerical issues related to the choice of the projection basis would also be very welcome.

Acknowledgements E.B. developed this work while visiting IMPA under the Cooperation Agreement between IMPA and Petrobras. MOS was partially supported by CNPq grant 308113/2012-8 and FAPERJ. JPZ was supported by CNPq grants 302161/2003-1 and 474085/2003-1 and by FAPERJ through the programs *Cientistas do Nosso Estado* and *Pensa Rio*. All authors acknowledge the IMPA-PETROBRAS cooperation agreement.

The authors would like to acknowledge and thank a number of discussions with Fernando Aiube (PUC-RJ and Petrobras). We also thank Milene Mondek for the implementation of a number of preliminary examples of the HMC algorithm and Luca P. Mertens for help with the R software and the calibration procedure in the examples.

References

1. Bobrovnytska, O., Schweizer, M.: Mean-variance hedging and stochastic control: beyond the Brownian setting. IEEE Trans. Automat. Control **49**(3), 396–408 (2004)
2. Borison, A.: Real options analysis: Where are the emperor's clothes? J. Appl. Corpor. Finan. **17**(2), 17–31 (2005)
3. Brennan, M.J., Schwartz, E.S.: Evaluating natural resource investments. J. Bus. **58**(2), 135–157 (1985)
4. Chen, C.C., Chiang, Y.S., Chien, C.F.: Real option analysis for capacity investment planning for semiconductor manufacturing. In: International Symposium on Semiconductor Manufacturing, 2007. ISSM 2007, pp. 1–3 (2007)
5. Choulli, T., Stricker, C.: Deux applications de la décomposition de Galtchouk-Kunita-Watanabe. In: Séminaire de Probabilités, XXX. Lecture Notes in Mathematics, vol. 1626, pp. 12–23. Springer, Berlin (1996)
6. Copeland, T., Antikarov, V.: Real Options: A Practitioner's Guide. W. W. Norton and Company, New York (2001)
7. Copeland, T., Tufano, P.: A real-world way to manage real options. Harv. Bus. Rev. **82**(3), 90–99 (2004)
8. Dixit, A.: Entry and exit decisions under uncertainty. J. Polit. Econ. **97**(3), 620–638 (1989)
9. Dixit, A., Pindyck, R: Investment Under Uncertainty. Princeton University Press, Princeton (1994)
10. El Karoui, N., Peng, S., Quenez, M.C.: Backward stochastic differential equations in finance. Math. Finan. **7**(1), 1–71 (1997)
11. Föllmer, H., Schied, A.: Stochastic finance. de Gruyter Studies in Mathematics, vol. 27. Walter de Gruyter, Berlin (2004)
12. Gastel, L.V.: Risk beyond the hedge: options and guarantees embedded in life insurance products in incomplete markets. Master's Thesis, University of Amsterdam, 2013

13. Glasserman, P., Yu, B.: Number of paths versus number of basis functions in American option pricing. Ann. Appl. Probab. **14**(4), 2090–2119 (2004)
14. Gobet, E., Turkedjiev, P.: Linear regression MDP scheme for discrete backward stochastic differential equations under general conditions. Forthcoming in Mathematics of Computation (2015)
15. Gobet, E., Lemor, J.P., Warin, X.: A regression-based Monte Carlo method to solve backward stochastic differential equations. Ann. Appl. Probab. **15**(3), 2172–2202 (2005)
16. Golub, G.H., Van Loan, C.F.: Matrix computations, 4th edn. Johns Hopkins Studies in the Mathematical Sciences. Johns Hopkins University Press, Baltimore, MD (2013)
17. Grasselli, M., Hurd, T.: A Monte Carlo method for exponential hedging of contingent claims (2004). Working paper
18. Grasselli, M.R.: Getting real with real options: a utility–based approach for finite–time investment in incomplete markets. J. Bus. Finan. Account. **38**(5–6), 740–764 (2011)
19. Grasselli, M.R., Hurd, T.R.: Indifference pricing and hedging for volatility derivatives. Appl. Math. Finan. **14**(4), 303–317 (2007)
20. Henderson, V., Hobson, D.G.: Real options with constant relative risk aversion. J. Econ. Dyn. Control **27**(2), 329–355 (2002)
21. Hubalek, F., Schachermayer, W.: The limitations of no-arbitrage arguments for real options. Int. J. Theor. Appl. Finan. **2**(4), 361–373 (2001)
22. Ingersoll, J.E., Ross, S.A.: Waiting to invest - investment And uncertainty. J. Bus. **65**(1), 1–29 (1992)
23. Jaimungal, S., Lawryshyn, Y.: Incorporating managerial information into real option valuation. SSRN eLibrary (2011)
24. Lemor, J.P., Gobet, E., Warin, X.: Rate of convergence of an empirical regression method for solving generalized backward stochastic differential equations. Bernoulli **12**(5), 889–916 (2006)
25. Lipp, T.: Numerical methods for optimization in finance: optimized hedges for options and optmized options for hedging. PhD Thesis, Augsburg University and Pierre and Marie Curie University (2012)
26. Longstaff, F.A., Schwartz, E.S.: Valuing American options by simulation: A simple least-squares approach. Rev. Finan. Stud. **14**(1), 113–147 (2001)
27. Mathews, S., Datar, V., Johnson, B.: A practical method for valuing real options: The Boeing approach. J. Appl. Corp. Finan. **19**(2), 95–104 (2007)
28. McDonald, R., Siegel, D.: The value of waiting to invest. Q. J. Econ. **101**(4), 707–727 (1986)
29. Mittal, G.: Real options approach to capacity planning under uncertainty. Master's Thesis, Massachusetts Institute of Technology (2004)
30. Moro, L., Zanin, A., Pinto, J.: A planning model for refinery diesel production. Comput. Chem. Eng. **22**(1), 1039–1042 (1998)
31. Musiela, M., Rutkowski, M.: Martingale Methods in Financial Modeling. Springer, New York (1997)
32. Myers, S.C.: Determinants of corporate borrowing. J. Finan. Econ. **5**(2), 147–175 (1977)
33. Novaes, A.G.N., Souza, J.C.: A real options approach to a classical capacity expansion problem. Pesquisa Operacional **25**(2), 159–181 (2005)
34. Oldenburg, J., Schlegel, M., Ulrich, J., Hong, T.L., Krepinsky, B., Grossmann. G., Polt, A., Terhorst, H., Snoeck, J.W.: A method for quick evaluation of stepwise plant expansion scenarios in the chemical industry. In: Plesu, V., Agachi, P. (eds.) 17th European Symposium on Computer Aided Process Engineering, p. 2. Elsevier, Amsterdam (2007)
35. Paddock, J.L., Siegel, D.R., Smith, J.L.: Option valuation of claims on real assets - the case of offshore petroleum leases. Q. J. Econ. **103**(3), 479–508 (1988)
36. Papageorgiou, L.G.: Supply chain optimization for the process industries: Advances and opportunties. Comput. Chem. Eng. **33**(1), 931–938 (2009)
37. Pham, H.: On quadratic hedging in continuous time. Math. Meth. Oper. Res. **51**(2), 15–339 (2000)

38. Pindyck. R.S.: Irreversibility, uncertainty, and investment. J. Econ. Lit. **29**(3), 1110–1148 (1991)
39. Potters, M., Bouchaud, J., Sestovic, D.: Hedged Monte-Carlo: low variance derivative pricing with objective probabilities. Phys. A Stat. Mech. Appl. **289**(3–4), 517–525 (2001)
40. Primbs, J.A., Yamada, Y.: A new computational tool for analysing dynamic hedging under transaction costs. Quant. Finan. **8**(4), 405–413 (2008)
41. R Core Team R: A language and environment for statistical computing. R Foundation for Statistical Computing, Vienna, Austria (2013)
42. Sahinidis, N., Grossmann, I., Fornari, R., Chathrathi, M.: Optimization model for long range planning in the chemical industry. Comput. Chem. Eng. **13**, 1049–1063 (1989)
43. Schweizer, M.: Local risk-minimization for multidimensional assets and payment streams. Banach Cent. Publ. **83**, 213–229 (2008)
44. Schweizer, M., Föllmer, H.: Hedging by sequential regression: an introduction to the mathematics of option trading. ASTIN Bull. **18**(2), 147–160 (1988)
45. Shapiro, J.F.: Challenges of strategic supply chain planning and modeling. Comput. Chem. Eng. **28**, 855–861 (2009)
46. Titman, S.: Urban land prices under uncertainty. Am. Econ. Rev. **75**(3), 505–514 (1985)
47. Tourinho, O.: The Option Value of Reserves of Natural Resources. Working Paper 94, University of California, Berkeley (1979)
48. Trigeorgis, L.: Real Options: Managerial Flexibility and Strategy in Resource Allocation. MIT Press, Cambridge (1999)
49. Trigeorgis, L., Mason, S.P.: Valuing managerial flexibility. Midl. Corp. Finan. J. **5**(1), 14–21 (1987)

Transition to Electric Mobility: An Optimal Price Subsidy Rule

René Aïd and Imen Ben Tahar

Abstract Many public policies declare electric mobility as a key lever for insuring the carbon emission target and attaining the objectives of oil-dependence reduction. However, the cost of an electric vehicle (ev) is still way too expensive compared to the conventional fuel-powered vehicle (fv) and constitutes a serious barrier against its diffusion. In this note we formulate a tractable model to analyse the dynamics of the adoption of ev's. The dynamic is driven by increasing marginal production returns and consumer's willingness to pay. We define the social benefit of replacing an fv by an ev as the fuel-economy it allows to realize, and solve for the optimal subsidy rule. We show that in a context of expensive fuel price, a voluntary policy of subsidy can transform the present fuel-powered fleet into an electric one.

1 Introduction

As highlighted in [7, 8], the last decade is marked by new socio-technical developments which have the potential to trigger the emergence of a viable trajectory for electric mobility. These new developments are mainly led by: (i) *Progress in battery technology*: where significant achievements in terms of performance and range have already been realized making ev a more viable product. (ii) *Public policies*: the last decade witnessed greater concerns about climate change. Many governments are committed to binding green-house-gas (GHG) emission reduction targets and number of public policies support electric mobility, declaring it as a key lever for insuring the sustainability of the transport sector while attaining the objectives of GHG limitation and oil-dependence reduction. For instance, the European Commission states a set of objectives among which halving the use of

R. Aïd (✉)
EDF R&D, Finance for Energy Market Research Centre, Clamart, France
e-mail: rene.aid@edf.fr

I. Ben Tahar
Université Paris-Dauphine, CEREMADE, France and Finance for Energy Market Research
Centre, Clamart, France
e-mail: imen@ceremade.dauphine.fr

© Springer Science+Business Media New York 2015
R. Aïd et al. (eds.), *Commodities, Energy and Environmental Finance*,
Fields Institute Communications 74, DOI 10.1007/978-1-4939-2733-3_11

'conventionally-fuelled' cars in urban transport by 2030 and phasing them out in cities by 2050 [9].

Alongside these favorable elements, there are however important obstacles hindering the deployment of evs. A major one is the high cost of the battery. Surveys about *Consumers Willingness to pay for* ev reveal that, in spite of the high premium some consumers are willing to pay, 'battery cost need to drop considerably if ev are to be competitive without subsidy at current gasoline prices' [10]. A critical question is to understand how subsidy policies combine with potential battery-cost reduction via technological learning (*learning-by-doing* and *increasing returns to scale* effects) so that the ev becomes economic.

We formulate a tractable model allowing to quantify the effect of purchase subsidy on the dynamics of ev's adoption. Here, we assimilate the benefit of substituting an fv by an ev to the realized fuel economy over the lifetime of the vehicle. Indeed, while the individual consumer has a short-term view, the public authority has a long-term policy, which gives, from a social perspective, advantage to ev's future fuel economy over present battery expenses. In this note, we stick to a simple deterministic setting where the fv's purchase price and the energy costs for both vehicles are assumed to be constant over time. It constitutes a reference case for a more involved stochastic model which is the object of an upcoming paper.

In Sect. 2 we present the basic hypothesis of our model and introduce the dynamics of the ev adoption. This dynamics is inspired from Brian Arthur [1] seminal paper analyzing competing technologies with increasing returns. It captures the fact that the cost of ev is experiencing a learning curve where the speed at which learning occurs, is spurred by the number of new ev adopters. It implies that the post-subsidy purchase price spread between ev and fv , denoted by $x(t)$, evolves according to

$$\dot{x}(t) = -\alpha \mu x(t) \Phi(x(t) - s(t))$$

where s is the subsidy rule, and Φ is the complementary cumulative distribution function for *the consumers' willingness to pay for* evs. In Sect. 3, we justify the social benefit of subsidizing ev purchases, then we formulate and solve the problem of optimal purchase subsidy rule allowing to maximize the social benefit. We show that the optimal subsidy rule consists in guaranteeing a relatively low net purchase price P^* which is *constant* (up to the interest rate). The value of the subsidy vanishes as the pre-subsidy price of the ev tends to $P^\star$. We show that using this optimal subsidy rule in a context of hugh fuel price, the fuel-powered vehicle fleet can be transform into an electric one in a few decades. Finally, Sect. 4 is dedicated to concluding remarks.

2   Modeling the Dynamics of Electric Vehicle Adoption

Analyzing the potential demand for ev is crucial to model or forecast how manufactures strategies and public policy incentives may influence the deployment of electric mobility. Several studies addressed the demand side, in particular consumers willingness to pay for an electric car compared the reference gasoline powered vehicle, see for example [5, 10]. For our model, we retain two important observations which are often reported:

 (i) significant preference (willingness to pay) heterogeneity across the population,
(ii) required substantial battery price reduction if evs are to meet target volume.

   These facts are illustrated on Fig. 1 with a sample of data extracted from [5] on willingness-to-pay of European consumers. We will use these data for the numerical application in Sect. 3.3.

   As for battery technology, there is an on-going intense R&D activity where close collaborations between automakers and manufacturers are observed [8]. As shown in Fig. 2, the industry projects better performances as well as cost reductions to follow through *learning-by-doing* and *increasing returns to scale* effects: production costs for the not yet mature battery technology shall decline as production cumulates.

2.1   Basic Hypothesis and Notations

For the sake of tractability, we make some simplifying assumptions on the market of new personal vehicles:

Potential Market We assume a constant annual rate of new car purchases, denoted by μ. We suppose that consumers have the choice between two types of standard vehicles, either an electric vehicle (ev) or a fossil-fuel powered car (fv). The hypothesis for standard vehicle we consider are given in Table 1 and correspond to genuine data, except for the gasoline price for which a high price scenario is considered.

Fig. 1 Willingness-to-pay
for ev for European
consumers according to [5]

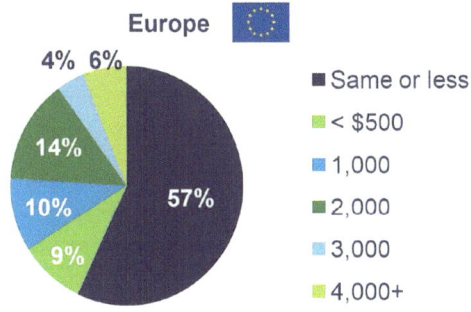

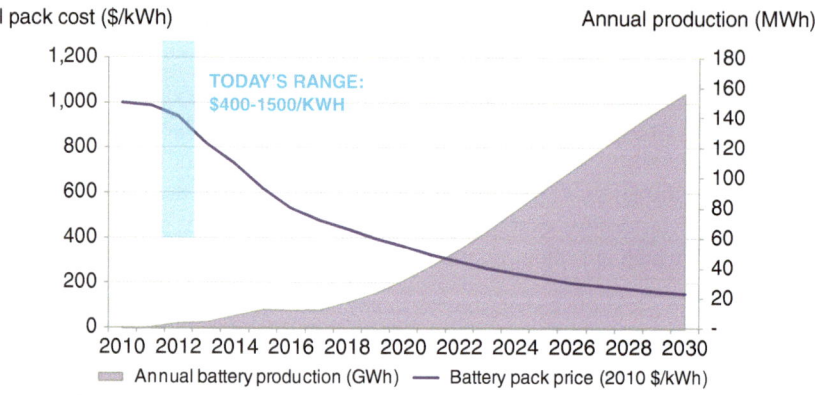

Total pack cost ($/kWh)   Annual production (MWh)

Fig. 2 Battery learning curve: LI-ION battery pack cost and production 2010–2030—*Source*:
Bloomberg new energy finance

Purchase Price and the Battery Learning Curve The purchase price of the conventional fv, p, is supposed to be constant over time. Whereas, accounting for potential technological learning, the purchase price of the ev, p_t^e, is supposed to vary (possibly decline) in the future. We denote by x_t the purchase price spread:

$$x_t := p_t^e - p \ . \tag{1}$$

We assume that this spread is essentially explained by the battery cost. Indeed, relevant literature report that the cost of batteries is the critical factor within the investment cost for electric vehicle [4, 6, 11, 12]. As it is explained in the French Green Book on non-emissive vehicles [12, p. 42] fv and ev share most of their costs (body work, passenger space, communication to drive wheels. . .) and the only differences come for the battery. Following [2], we assume that the battery costs, and consequently the spread x_t, decreases at a *learning rate* proportional to the number of new ev adopters.

Energy Cost We denote, respectively, by f and e the annual energy cost of an fv and an ev. They are assumed to be constant over time. Hence, in our model we do not take into account possible future energy cost reduction. We also do not take into account uncertainty regarding oil prices. Indeed, we intend to isolate the effect of battery technological learning.

2.2 Dynamics of Electric Vehicle Adoption

To model the dynamics of ev deployment, we adapt the framework of Brian Arthur [1] who proposed a simple and insightful model of explore the dynamics of allocation between competing technologies with increasing returns.

Sequentially arriving consumers, indexed by i, choose between the competing ev and fv technologies each agent i is characterized by his willingness-to-pay for the ev, denoted by ω^i. Let t_i denote the date at which consumer i makes his purchase decision. The consumer i chooses an ev if and only if

$$x_{t_i} \leq \omega^i .$$

where x_{t_i} is the purchase price spread defined in (1). The sequence $(\omega^i)_{i \geq 1}$ is assumed to be a sequence of i.i.d. random variables. We denote by Φ the complementary cumulative distribution function of ω^i:

$$\Phi(x) := \mathbb{P}(\omega^i \geq x) . \tag{2}$$

As explained in the previous subsection, we aim to capture the fact that the spread x_t decreases because the cost of ev is experiencing a learning curve where the speed at which learning occurs, is spurred by the number of new ev adopters. For now on, we fix a time step δt, and assume the following dynamics:

$$x_{t+\delta t} = x_t - \alpha\, x_t\, \mathbf{n}_t^{ev} . \tag{3}$$

Here, $\mathbf{n}_t^{ev}$ is number of new ev adopters between the dates t and $t + \delta t$, and α represents the learning rate.

A Limiting o.d.e. It can be shown that the stochastic system (3) can be approximated by the solution of ordinary differential equation (o.d.e.)

$$\dot{\xi}_t = -\alpha\mu\xi_t\Phi(\xi_t) , \quad \xi_0 = x_0 , \tag{4}$$

where μ is the annual rate of new vehicle purchases.

Proposition 2.1. *Define the piecewise continuous linear interpolation $\bar{x}$ by*

$$\bar{x}(t_k) = x_{t_k} \text{ and } \bar{x}(t) = x_{t_k} + (x_{t_{k+1}} - x_{t_k})\frac{(t - t_k)}{(t_{k+1} - t_k)}, \ t \in [t_k, t_{k+1}] , \ t_k := k\delta t .$$

Then, for any $T > 0$

$$e\left[\sup_{t \in [0,T]} |\bar{x}_t - \xi_t|^2\right] = O(\delta t) .$$

Proof. This result follows from a direct application of Lemma 9.2.1 in [3]. $\square$

3   Social Benefit and Optimal Subsidy Rule

3.1   Social Benefit of the Electric Vehicle

In our analysis, we identify the social benefit of the electric mobility with the fuel
economy realized by substituting ev to fv. When estimating the lifecycle cost to
energy, a key issue is the discount rate at which future consumption is valued today.
The individual consumers has a short-term view, reflected by a relative high discount
rate. On the other hand, the public authority has a long-term policy reflected by a
relatively low discount rate. Considering our standard vehicles data, Table 1, this
difference in the discount rate is sufficient to justify from the social perspective the
energy-economy benefits resulting from substituting an ev to an fv ; a collective
benefit which is not perceived at the individual level.

In order to formalize this discussion, we introduce the social cost, P, of a single
fv

$$P = p + \int_0^\infty e^{-\rho t} f dt \ = \ p + f/\rho \,, \tag{5}$$

and the social cost, P_t^e, of a single ev

$$P_t^e = p_t^e + \int_0^\infty e^{-\rho t} e dt \ = \ p_t^e + e/\rho \,, \tag{6}$$

Here, ρ, is the social discount rate supposed to be constant over time. We consider
the lifetime of the vehicle to be sufficiently long to make the approximation of an
infinite time horizon. There is a social benefit to the ev if $P_0^e \leq P$. Consider the cost
difference:

$$P - P_t^e = p - p_t^e + (f - e)/\rho \ = \ b - x_t$$

Table 1   Cost and fuel economy for the electric vehicle

	fv		ev
Energy consumption	5 l/100 km		20 kWh/100 km
Fuel price	1,5 €/l		0,9 €/kWh
Km/year	15.000		15.000
Energy cost/year (€)	1.125		270
Vehicle price (€)	15.000		30.000
Individual discount rate	16 %		16 %
Individual cost	22.000		31.700
Individual benefit of the ev(€)		−9.700	
Social discount rate	4 %		4 %
Social cost (€)	43.125		36.750
Social benefit of the ev(€)		6.375	

where $b := (f - e)/\rho$ is, from the social perspective, the fuel economy resulting from substituting an ev to an fv.

Table 1 summarizes the various costs for the standard vehicles. Energy consumption, fuel price, km per year and vehicle price are taken from the French Green Book on non-emissive vehicle [12, p. 42] . These data corresponds to a urban use of the vehicle. The electricity price is an off-peak price, considering that the vehicles will charge during the night or on non-peaking hours. The hypothesis for the fuel price is high. It corresponds to a situation where the oil would be around 200 USD per baril. The individual discount rate (16 %) is sensibly higher then the social rate (4 %). It reflects the individual's 'impatience' when arbitrating between immediate costs and future benefits. From the social perspective, substituting an fv by an ev results in a benefit of €6,000: a fuel economy of $b = $ €21,000, minus the initial battery cost $x_0 = $ €15,000.

3.2 Optimal Purchase Subsidy

A purchase subsidy is used as a public policy to stimulate the number of ev adopters. We denote by s_t the value of the purchase subsidy at time t, and by x^s the price spread resulting from applying the subsidy rule $s = \{s_t,\ t \geq 0\}$. Here we shall work directly with the approximating dynamics (4). Then, the rate of new ev adopters, when applying the subsidy rule s, is approximated by

$$\mu \, \Phi(x_t^s - s_t) \, , \tag{7}$$

the resulting price-spread dynamics is given by

$$\dot{x}_t^s = -\alpha \mu \Phi(x_t^s - s_t) \, x_t^s \quad \text{with } x_0^s = x_0 \, . \tag{8}$$

Hence, from the social perspective, the subsidy rule $s = \{s_t, t \geq 0\}$ leads to an energy-economy evaluated by

$$\int_0^\infty \mu \Phi(x_t^s - s_t) \left(P - P_t^e\right) dt = \int_0^\infty \mu \Phi(x_t^s - s_t) \left(b - x_t^s\right) dt \, .$$

Here the objective of the public authority is to maximise over a fixed time horizon T the *social surplus* defined as the social energy-economy minus the subvention amount:

$$\max_s \int_0^T \mu \, \Phi(x_t^s - s_t) \left(b - x_t^s - s_t\right) dt. \tag{9}$$

The question of financing this subsidy policy is left aside here.

It turns out that is possible to characterize explicitly the optimal subsidy rule. We shall assume the complementary distribution function Φ satisfies:

$$\Phi \text{ has a bounded support } [\mathbf{x_{min}}, \mathbf{x}_{max}] \tag{10}$$

$$\Phi' > 0 \text{ on }]\mathbf{x}_{min}, \mathbf{x}_{max}] \tag{11}$$

$$h : z \mapsto z + \Phi(z)/\Phi'(z) \text{ is non-decreasing on } [\mathbf{x}_{min}, \mathbf{x}_{max}] \tag{12}$$

and consider as a canonical example the truncated Pareto distribution function:

$$\Phi_p(z) := \left(\frac{\mathbf{x}_{max} - z}{\mathbf{x}_{max} - \mathbf{x}_{min}} \right)^p \mathbf{1}_{[\mathbf{x}_{min}, \mathbf{x}_{max}]}(z) \text{ with } p > 0 . \tag{13}$$

Proposition 3.1. *Assume* (10)–(12) *hold. Let* $x_0 < b$ *be the initial spread value. If* $s^\star$ *is an optimal subsidy strategy, then* $s^\star$ *consists in having the consumer pay for the* ev *a post subsidy price which is constant equal to* $p + x_t^{s^\star} - s_t^\star = p + z^\star$ *where:*

$$\begin{cases} z^\star = \mathbf{x}_{min} & \text{if } 2x_0 e^{-\alpha\mu T} - b \leq h(\mathbf{x}_{min}), \\ z^\star = \mathbf{x}_{max} & \text{if } 2x_0 - b \geq h(\mathbf{x}_{max}), \\ \text{and } h(z^\star) = 2x_0 e^{-\alpha\mu\Phi(z^\star)T} - b \text{ otherwise} . \end{cases} \tag{14}$$

Proof. Let H be the Hamiltonian function for the control problem (9)–(8):

$$H : (x, p) \mapsto \sup_{s \geq 0} \mu\Phi(x - s) \{b + x - s - x(2 + \alpha p)\}$$

Assume that $s^\star := \{s_t^\star, \ t \in [0, T]\}$ is an optimal subsidy strategy. Denote by $x^\star := x^{s^\star}$ the associated price spread and let $z^\star := x^\star - s^\star$. Then, by Pontryagin Maximum principle, there exists an absolutely continuous map $p^\star : [0, T] \to \mathbb{R}$ such that $(x^\star, p^\star)$ satisfies the Hamiltonian system

$$\begin{cases} \dot{x}^\star(t) = \nabla_p H(x^\star(t), p^\star(t)) = -\alpha x^\star \mu\Phi(z^\star(t)), & x(0) = x_0 \\ \dot{p}^\star(t) = -\nabla_x H(x^\star(t), p^\star(t)) = (2 + \alpha p^\star(t))\mu\Phi(z^\star(t)), & p(T) = 0 . \end{cases} \tag{15}$$

and the condition

$$H(x^\star, p^\star) = \mu\Phi(z^\star) \{b + z^\star - x^\star(2 + \alpha p^\star)\} . \tag{16}$$

From (15) we get

$$\frac{d}{dt}(x^\star(2 + \alpha p^\star)) = 0 .$$

Then, by (16), $z^\star$ is constant on $[0, T]$ with: $z^\star = \text{argmax } \Phi(z^\star) \{b + z^\star - x^\star(2 + \alpha p^\star)\}$, and from conditions (10)–(12) we get (14). □

In the case where the willingness to pay follows a truncated Pareto distribution function $\Phi = \Phi_p$, then (14) fully characterizes the subsidy policy:

Proposition 3.2. *Let* $\Phi = \Phi_p$, *and assume that*

$$\text{either (i): } p \leq 1 \quad \text{or (ii): } p > 1 \text{ and } (\mathbf{x}_{max} - \mathbf{x}_{min}) < \frac{p}{1+p}(\mathbf{x}_{max} + b)$$

Let $x_0 < b$ *be the initial spread value. An optimal subsidy strategy* s^* *consists in having the consumer pay for the* ev *a post subsidy price which is constant equal to* $p + z^*(x_0)$ *where* $z^*(x_0)$ *is defined by*

(i) $z^*(x_0) = \mathbf{x}_{min}$ *if* $2x_0 e^{-\alpha\mu T} - b \leq h(\mathbf{x}_{min})$,

(ii) $z^*(x_0)$ *is the unique solution in* $[\mathbf{x}_{min}, \mathbf{x}_{max}]$ *to* :

$$h(z) = 2x_0 e^{-\alpha\mu\Phi(z)T} - b \quad \text{otherwise.}$$

Proof. Let $\Psi : (\tau, x, z) \mapsto h(z) + b - 2xe^{-\alpha\mu\tau\Phi(z)}$, and denote by U the set:

$$U := \{(\tau, x) : x < b \text{ and } \Psi(\tau, x, \mathbf{x}_{min}) < 0\} .$$

Notice that for all $(\tau, x) \in U$, $\Psi(\tau, x, \mathbf{x}_{max}) = \mathbf{x}_{max} + b - 2x_0 > 0$. If either (i) or (ii) is satisfied, then a straightforward, but rather lengthy, analysis of the variations of the function $z \to \Psi(\tau, x, z)$ shows that for all $(\tau, x) \in U$ there exists a unique $z^*(\tau, x) \in]\mathbf{x}_{min}, \mathbf{x}_{max}[$ such that

$$\Psi(\tau, x, z^*(\tau, x)) = 0 \text{ with } \frac{\partial\Psi}{\partial z}(\tau, x, z^*(\tau, x)) > 0.$$

Then $(\tau, x) \in U \mapsto z^*(\tau, x)$ is C^1 on $U \times]\mathbf{x}_{min}, \mathbf{x}_{max}[$ with

$$\frac{\partial z^*}{\partial \tau}(\tau, x) = -\frac{\partial_\tau \Psi(\tau, x, z^*(\tau, x))}{\partial_z \Psi(\tau, x, z^*(\tau, x))} \text{ and } \frac{\partial z^*}{\partial x}(\tau, x) = -\frac{\partial_x \Psi(\tau, x, z^*(\tau, x))}{\partial_z \Psi(\tau, x, z^*(\tau, x))}$$

For $(\tau, x) \notin U$ we set $z^*(\tau, x) = \mathbf{x}_{min}$.

To verify that s^* is indeed the optimal strategy rule, we consider the function:

$$w(t, x) = \int_t^T \mu\Phi(z^*(T - t, x)) \left\{b + z^*(T - t, x) - 2xe^{-\alpha\mu\Phi(z^*(T-t,x))(u-t)}\right\} du$$

$$= \mu\Phi(z^*(T - t, x))(b + z^*(T - t, x))(T - t) - \frac{2x}{\alpha}(1 - e^{-\alpha\mu\Phi(z^*(T-t,x))(T-t)})$$

A direct calculation shows that w solves the Hamilton-Jacobi-Bellman-Equation associated to our problem:

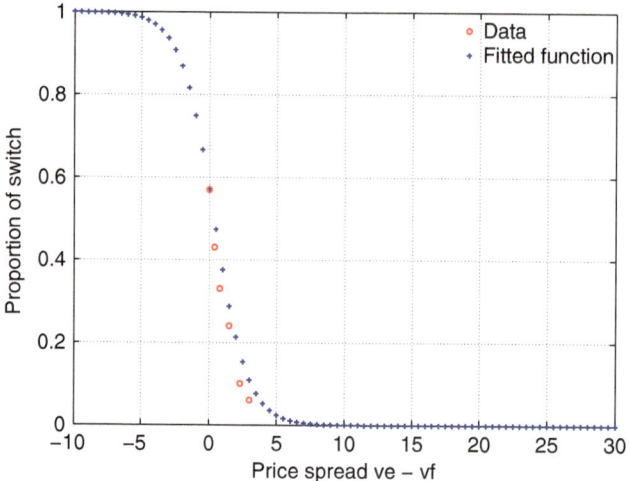

Fig. 3 Complementary cumulative distribution function Φ fitted to the panel data of European consumers from [5]

$$\frac{\partial w}{\partial t_0} + H(x_0, \frac{\partial w}{\partial x_0}) = 0, \ w(T, \cdot) = 0.$$

and we conclude by standard verification argument that $s^\star$ is optimal.   □

3.3 Numerical Experiments

First, we fitted the willing-to-pay function Φ with the data provided by [5] and presented in Sect. 2 on European consumers. Although the sample is very sparse, Fig. 3 shows that the approximation captures the threshold effect around a null value of the spread.

Now we illustrate and compare the evolution of ev adoption over a time horizon of 50 years for three policies:

- the zero purchase subsidy case,
- with the optimal subsidy rule solving (9),
- and with a subsidy capped at €7,000 as it is the case for the current French policy.

The evolution of the price spread x and of the number of ev annual purchases are reported in Fig. 4. We see that the optimal subsidy rule leads to a relatively rapid decrease in the price spread. The optimal policy itself, illustrated in Fig. 5, decreases at the same rate as the price of the ev: it ensures a constant 'post-subsidy' price for the ev which, approximatively exceeds the price of the fv by €1,000.

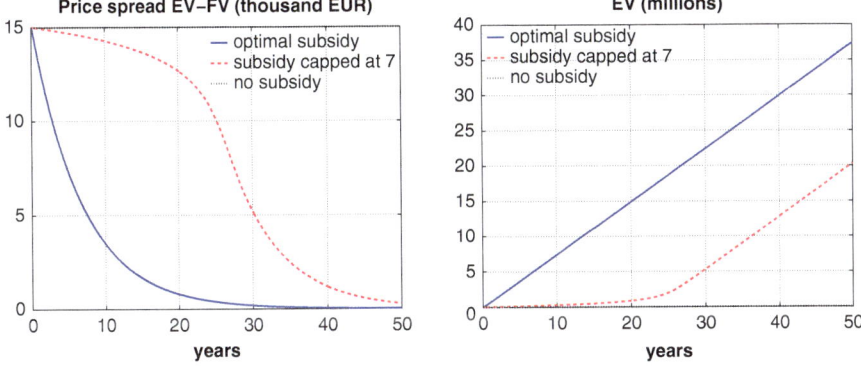

Fig. 4 Evolution of the price spread and of the ev adoption for the three rules

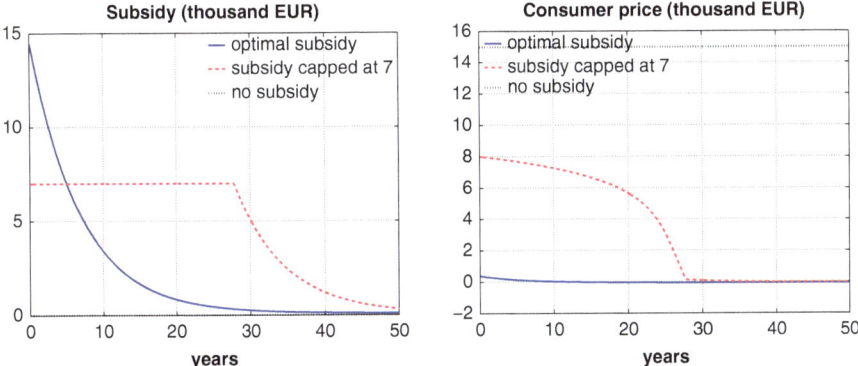

Fig. 5 Optimal subsidy policy and of the 'post-subsidy' purchase price

When applying the optimal subsidy policy, the evolution of ev adoption is immediate and corresponds to a constant annual number of 800,000 ev purchases. Indeed, from the social perspective, there is an immediate social benefit to replacing fvs by evs, and it is optimal to ensure a very rapid transition. Notice that the amount of the optimal subsidy at the initial date is significantly larger than the €7,000 of the current French subsidy policy.

Observe, as it is illustrated in Fig. 4, that the effect current subsidy policy of €7,000 is not significantly different from the no-subsidy case during the first 25 years. It appears that this current subsidy amount is not sufficient—with regard to consumers willingness-to-pay—to insure a rapid transition to the electric mobility. This relatively negligible effect during decades may quickly discourage the public decision-maker to persevere in this policy.

The effect of the optimal subsidy policy is illustrated in Fig. 6. It appears that the optimal policy induces at first losses due to substantial subsidies, before insuring large gains from future fuel-economy. In the first 5 years the cumulated amount of

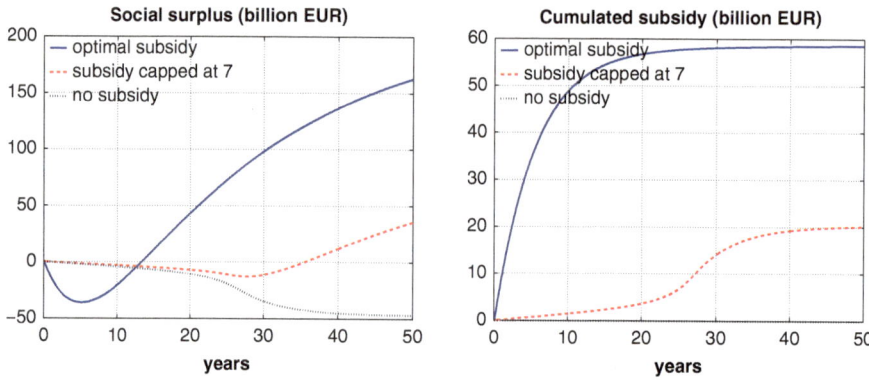

Fig. 6 Evolution of the social surplus and of the cumulated amount of subsidy

subsidies is of € 35 billion, whereas the social surplus is negative. Yet, after 10 years, le social surplus is about € 25 billion for a total amount of subsidy equal to € 50 billion. These short term losses may explain the reluctance of current public policies to put in place the important subsidies needed for massive ev adoption.

4   Conclusion

Motivated by the tractability of the qualitative analysis of the optimal subsidy for ev market, we adopted a quite simple and stylized model for the transition to electric mobility. Hence, the numerical experiments presented here are intended to be illustrative and do not pretend to be accurate. Nevertheless, this model allows to have important insights about the nature of the optimal subsidy rule which may allow a rapid adoption of evs.The optimal subsidy rule derived here, consists in insuring a constant 'post-subsidy' purchase price for the consumer. This result does not depend on the form of the cumulative distribution function of consumers willingness to pay. However, a precise knowledge of the willingness-to-pay distribution is crucial in determining the amount of the subsidy and reveals to be more important that the battery learning rate. The findings of our model allow, also, to question the efficiency of the current subsidy policy, and the modalities with which subsidy amounts are decided.

Perspectives However, in this state of development, this model can not escape certain criticisms. In particular, the fact that the gasoline price is constant and high makes the transition to electric mobility quite natural whereas one main problem is the uncertainty on the oil price. Thanks to the fact that the present model is simple, we have good confidence in our capacity to deal with the introduction of various forms of oil price uncertainty in the same kind of dynamic.

References

1. Arthur, W.B.: Competing technologies, increasing returns and lock-in by historical events. Econ. J. **99**, 116–131 (1989)
2. Bonnel, P., Grootveld, G.V., Junginger, M., Patel, M., Perujo, A., Weiss, M.: On the electrification of road transport - Learning rates and price forecasts for hybrid-electric and battery-electric vehicles. Energy Pol. **48**, 374–393 (2012)
3. Borkar, V.S.: Stochastic Approximation A dynamical Systems Viewpoint. Cambridge University Press, New Delhi (2008)
4. Chalk, S.G., Miller, J.F.: Key challenges and recent progress in batteries, fuel cells, and hydrogen storage for clean energy systems. J. Power Sourc. **159**, 73–80 (2006)
5. Deloitte: Unplugged: Electric Vehicle Realities Versus Consumers Expectations. Deloitte Touche Tohmatsu, London (2011)
6. Delucchi, M., Lipman, T.E.: An analysis of the retail and lifecycle cost of battery-powered electric vehicles. Transp. Res. D **6**, 371–404 (2001)
7. Dijk, M., Orsato, R.J., Kemp, R.: The emergence of an electric mobility trajectory. Energy Pol. **52**, 135–145 (2013)
8. Dijk, M., Yarime, M.: The emergence of hybrid-electric cars: Innovation and path creation through co-evolution of supply and demand. Tech. Forcasting Soc. Chang. **77**, 1371–1390 (2010)
9. European Commission: White paper: roadmap to a single European transport area—towards a competitive and resource efficient transport system, COM(2011) 144 final (2011)
10. Hidrue, M., Parsons, G., Kempton, W., Gardner, M.: Willingness to pay for electric vehicles and their attributes. Resour. Energy Econ. **33**, 686–705 (2011)
11. Matheys, J., Van Autenboer, W., Van den Bossche, P., Van Mierlo, J., Vergels, F.: SUB AT: an assessment of sustainable battery technology. J. Power Sources **162**, 913–919 (2006)
12. Nègre, L.: Livre Vert sur les infrastructures de recharge ouvertes au public pour les véhicules décarbonés, (Green book on charging infrastructure open to public for no-emissive vehicule). Report for the French Ministry of Industry (2011)

Part IV
Dynamic Games in Commodity Markets

Game Theoretic Models for Energy Production

Michael Ludkovski and Ronnie Sircar

Abstract We give a selective survey of oligopoly models for energy production which capture to varying degrees issues such as exhaustibility of fossil fuels, development of renewable sources, exploration and new technologies, and changing costs of production. Our main focus is on dynamic Cournot competition with exhaustible resources. We trace the resulting theory of competitive equilibria and discuss some of the major emerging strands, including competition between renewable and exhaustible producers, endogenous market phase transitions, stochastic differential games with controlled jumps, and mean field games.

1 Introduction

The recent decline in oil prices, from around \$100 per barrel in June 2014 to less than \$50 in January 2015 is a dramatic illustration of the evolution of energy production as a result of competition between different sources. Indeed, the price drop was prompted in large part by OPEC's strategic decision not to decrease its oil output in the face of increased production of *shale oil* in the US, itself arising from new technologies that were spurred by investment in exploration and research in times of higher oil prices. These complex interactions are in addition to long-running concerns about dwindling fossil fuel reserves ('peak oil'), as well as climate change and the transition to sustainable energy sources.

We survey a (necessarily) selective line of work that builds models successively incorporating various of these features starting from a competitive oligopolistic view of an idealized global energy market, in which *game theory* describes the outcome of competition. In particular, the oligopoly will be taken to be in a *Cournot framework*, in which players choose *quantities* of production, and then prices are determined

M. Ludkovski (✉)
Department of Statistics and Applied Probability, University of California Santa Barbara, South Hall, Santa Barbara, CA 93106-3110, USA
e-mail: ludkovski@pstat.ucsb.edu

R. Sircar
ORFE Department, Princeton University, Sherrerd Hall, Princeton, NJ 08544, USA
e-mail: sircar@princeton.edu

© Springer Science+Business Media New York 2015
R. Aïd et al. (eds.), *Commodities, Energy and Environmental Finance*,
Fields Institute Communications 74, DOI 10.1007/978-1-4939-2733-3_12

317

by aggregate supply. This seems reasonable for energy production in which major players determine their output relative to their production costs, as in the expected scenario that OPEC will cut production in order to increase the market price of oil. The complementary framework of Bertrand markets, in which players set *prices*, is more typically suitable for consumer goods markets, among other examples.

We begin with static, or one-period games, as an introduction to some of the effects that can arise, for instance the non-competitiveness of producing a relatively expensive renewable source, such as wind, against a cheap fossil fuel in plentiful supply. However, the very nature of the complexities calls for a dynamic model in which there are (to use a much over-employed cliché) *game changers* over time. Changes in the competitive environment may come from:

- dwindling reserves of oil or coal, ramping up their scarcity value;
- discoveries of new oil reserves (there were over 30 major finds in 2009, for instance);
- technological innovation such as *fracking*, which has led to extraction of shale oil and gas;
- government subsidies for renewable energy sources such as solar and wind power;
- varying costs of energy production, for instance cheaper solar power due to falling silicon prices and improved solar cell efficiency.

Many if not all of these phenomena are unpredictable and dramatic, and motivate the development of *stochastic models*, particularly with potentially significant 'jumps' (for instance in costs or reserves). Moreover, dealing with stochastic dynamic (nonzero-sum) games involving many influential interacting energy producers creates computational challenges, and some approximation methods include numerical discretization, parameter asymptotics, and continuum (mean field) games.

2   Static Cournot Games

The classic work of [10] gives perhaps the earliest example of a Nash equilibrium for describing the outcome of a game. Cournot was concerned with competition between producers of an inexhaustible resource (mineral water): their effect on sales was such that the more each bottled and brought to the market, the lower the price for mineral water that they would receive.

A Cournot market is described by $N \geq 1$ profit-maximizing producers (or players) that compete in a non-cooperative way. In a market with homogenous goods, the players compete based on production quantity (producing identical goods). The market is specified by an inverse demand curve $P(\cdot)$, which maps aggregate production to market price, and as such is a *decreasing* function. The players choose their production levels q_i. Given total production level $Q = q_1 + \ldots + q_N$, the market clearing price is $P(Q)$. A simple illustrative example is linear inverse demand, $P(Q) = \eta - Q$, where η is the saturation level beyond which prices collapse to zero (and may become negative, meaning a producer would have to pay to have his good taken away). Such linear demand can be derived from the behavior of a representative consumer with a quadratic utility function (see, for instance, [35]), and allows to present explicit equilibrium calculations.

The players produce at per-unit (constant) cost of production $s_i \geq 0$, which in general will be different, reflecting the costs of producing from heterogeneous energy sources. The profit of player i is the quantity he produces multiplied by price minus cost:

$$\pi(q_i, Q_{-i}, s_i) = \begin{cases} q_i \left(P(Q_{-i} + q_i) - s_i \right) & \text{if } q_i > 0, \\ 0 & \text{if } q_i = 0, \end{cases} \qquad (1)$$

where $Q_{-i} = \sum_{j \neq i} q_j$ is total production by the players other than i. The last line of (1) allows for the possibility that $P(0^+) = +\infty$, but if a player does not produce anything, then he makes zero profit.

2.1 Nash Equilibrium

Definition 1. A Nash equilibrium is a vector $\boldsymbol{q}^* = (q_1^*, q_2^*, \ldots, q_N^*) \in [0, \infty)^N$ such that, for all i,

$$\pi(q_i^*, Q_{-i}^*, s_i) = \max_{q_i \in [0, \infty)} \pi(q_i, Q_{-i}^*, s_i), \qquad (2)$$

where $Q_{-i}^* = \sum_{j \neq i} q_j^*$. That is, each player's equilibrium production q_i^* maximizes his own profit $\pi(\cdot, Q_{-i}^*, s_i)$ when the other $N - 1$ players produce their equilibrium quantities. If, in addition, $q_i^* > 0$ for all i, then $\boldsymbol{q}^*$ is an *interior* Nash equilibrium.

Under suitable conditions on the price function P and the cost vector $\boldsymbol{s} = (s_1, s_2, \cdots, s_N)$, a Nash equilibrium exists and is unique. An important issue arising from this is that it may be too costly for some players to participate. We refer to [35, Chapter 4] for a discussion and references on general existence results for static Cournot games.

Assumption 1. *The price function P is twice continuously differentiable, with $P' < 0$ everywhere on $(0, \infty)$; and there exists $\eta \in (0, \infty)$ such that $P(\eta) = 0$.*

We order the firms by their costs and assume the latter are strictly less than the *choke price* $P(0^+)$:

$$0 \leq s_1 \leq s_2 \leq \ldots \leq s_N < P(0^+). \qquad (3)$$

When some firms have equal costs, the ordering is arbitrary and does not affect the result that follows. The behavior of P is best characterized in terms of the *relative prudence* of P, namely

$$\rho(Q) = -\frac{QP''(Q)}{P'(Q)}. \tag{4}$$

We also define

$$\bar{\rho} = \sup_{Q \in (0,\infty)} \rho(Q). \tag{5}$$

The following is taken from [20].

Theorem 1. *Suppose that $\bar{\rho} < 2$. Then there is a unique Nash equilibrium which can be constructed as follows. Let $\bar{Q}^* = \max\{Q_n^* \mid 1 \leq n \leq N\}$, where Q_n^* is the unique non-negative solution to the scalar equation*

$$QP'(Q) + nP(Q) = \sum_{j=1}^{n} s_j.$$

The unique Nash equilibrium production quantities are given by

$$q_i^*(s) = \max\left\{\frac{P(\bar{Q}^*) - s_i}{-P'(\bar{Q}^*)}, 0\right\}, \quad 1 \leq i \leq N,$$

and the corresponding profits are

$$G_i(s) = q_i^*(s)(P(\bar{Q}^*) - s_i), \quad 1 \leq i \leq N.$$

In particular, q_i^* and G_i are Lipschitz continuous, and the number of active players (that is, players with $q_i^* > 0$), in the unique equilibrium is $m = \min\{n \mid Q_n^* = \bar{Q}^*\}$. Moreover, in the case of a price function with a constant prudence, $\rho(Q) \equiv \rho$

$$P(Q) = \begin{cases} \dfrac{\eta}{1-\rho}\left(1 - \left(\dfrac{Q}{\eta}\right)^{1-\rho}\right) & \rho \neq 1 \\ \eta(\log\eta - \log Q) & \rho = 1, \end{cases} \tag{6}$$

one can weaken the requirement $\bar{\rho} < 2$ to simply $\rho < N + 1$.

The cost profile s is the main parameter of a Cournot game and the respective sensitivity analysis is crucial, especially in dynamic models. Intuitively, we expect that if player-i costs s_i decrease, her production and profits will rise, and the production and profits of the other players will fall. However, the precise impact depends on the properties of the price function $Q \mapsto P(Q)$; for example there are well known examples where higher costs *increase* production for all players [35]. Analysis of the precise dependence of equilibria on s, including explicit formulas for the sensitivity of q_i^* to s under constant prudence price functions of (6), is given in [20].

2.2 Blockading

The non-negativity constraint on production endogenizes the market structure in terms of the cost profile s. Oligopolies with symmetric production costs generate a trivial market structure, namely either all firms active or all firms inactive. In contrast, in models where firms are asymmetric, some firms may be inactive in equilibrium. Moreover, in dynamic models asymmetric costs induce different entry times into the market. This aspect is especially pertinent to energy markets, where producers using different fuels and technologies have widely different costs of production: for example, oil and coal sources are much cheaper than renewables, such as solar or wind.

To illustrate this effect, consider a very simple case of Theorem 1, namely a duopoly $N = 2$ with linear demand $P(Q) = 1 - Q$ (i.e., $\eta = 1, \rho = 0$). When there is one player with marginal cost of production $s_1 \in [0, 1)$, he chooses his optimal quantity $q_1 \geq 0$ to maximize his monopoly profit function

$$\Pi_1 = q_1(1 - q_1) - s_1 q_1.$$

The optimal quantity and profit are given by

$$q_1^*(s_1) = \frac{1}{2}(1 - s_1), \qquad G_1(s_1) = \frac{1}{4}(1 - s_1)^2.$$

When there are two players with costs $(s_1, s_2) \in [0, 1]^2$ and non-negative production quantities (q_1, q_2), the aggregate quantity is $Q = q_1 + q_2$ and each player's profit function is

$$\Pi_i = q_i(1 - q_i - q_j) - s_i q_i, \qquad i = 1, 2; j \neq i.$$

In a Nash equilibrium $(q_1^*, q_2^*) \in [0, 1]^2$ for the duopoly, each player maximizes profit as a best response to the other player's equilibrium strategy:

$$G_i(s_1, s_2) = \max_{q_i \geq 0} q_i(1 - q_i - q_j^*) - s_i q_i, \qquad i = 1, 2; j \neq i.$$

For costs $s_1, s_2 < \frac{1}{2}$, it is easy to see that both players have positive equilibrium productions

$$q_i^*(s_1, s_2) = \frac{1}{3}(1 - 2s_i + s_j), \qquad G_i(s_1, s_2) = \frac{1}{9}(1 - 2s_i + s_j)^2, \tag{7}$$

where $i = 1, 2; j \neq i$. However, if player j's cost is too high relative to player i's, specifically $s_j > \frac{1}{2}(1 + s_i)$, then he is *blockaded* from production, meaning his equilibrium quantity is zero. In this case, player i has a monopoly and the Nash equilibrium is given by

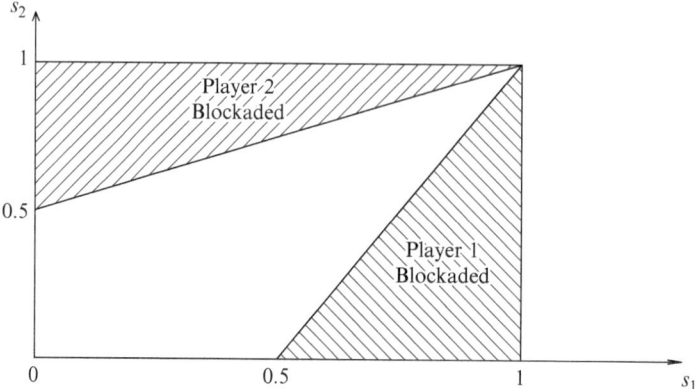

Fig. 1 Type of game equilibrium in a Cournot duopoly with linear demand $P(Q) = 1 - Q$

$$q_i^* = \frac{1}{2}(1 - s_i), \quad q_j^* = 0, \qquad G_i = \frac{1}{4}(1 - s_1)^2, \quad G_j^* = 0.$$

See Fig. 1 for the resulting monopoly wedges.

A current example may be OPEC holding back on cuts in production to drive shale oil producers out of the market and into bankruptcy, which an Op-Ed in The New York Times on 27 January, 2015 described thus: "the plunge in oil prices offers a sobering reminder of the power of markets over policy".

A full characterization of the static N-player game for a wide class of general inverse-demand functions is given in [20, Section 2], and for Bertrand games in [29, Section 2]; a comparison between Cournot and Bertrand in terms of the number of blockaded players is in [30].

3 Exhaustible Resources and Dynamic Games

When a resource, for instance a fossil fuel, is in finite supply, the energy oligopolies are necessarily changing over time due to the increasing scarcity value of the *exhaustible resource*.

3.1 Monopoly and Hotelling's Rule

The seminal work of [21] introduced a mathematical model for management of an exhaustible resource stock. Hotelling considered a single producer (a monopolist), and set up a continuous-time calculus of variations problem for maximizing total discounted value of the resource between now and exhaustion point. A crucial insight of Hotelling is the fact that along the optimum path the marginal value of reserves must grow at the risk-free rate, precisely offsetting the time value of money. This spawned a large body of economic literature based on optimizing social planning in the context of resource management or on Ramsey-type growth models that aim to optimize investment across several economic sectors. The link to exhaustible resources has become especially relevant in the past decade in connection with sustainable production in the face of climate change. For example development of clean energy backstops to guard against exhaustibility of conventional fossil fuels is addressed in [17, 26, 34] among others.

Consider a single oil producer who has reserves $x(t)$ at time t, with the dynamics

$$\frac{dx}{dt} = -q(x(t))\,\mathbf{I}_{\{x(t)>0\}},\tag{8}$$

where $q(x(t))$ is his production (or extraction) rate. When his reserves run out, he no longer participates in the market. The producer extracts to maximize lifetime discounted profit, and his value function v is defined by

$$v(x) = \sup_q \int_0^{\tau_x} e^{-rt}q(x(t))P(q(x(t)))\,dt.$$

Here the maximization is over control strategies $q \geq 0$, $r > 0$ is the discount rate, P is the inverse demand (or price) function satisfying Assumption 1, and τ_x is the exhaustion time

$$\tau_x = \inf\{t > 0 \mid x(t) = 0\}.$$

Moreover x here stands for the initial resource level: $x(0) = x$, and we have assumed zero extraction costs.

By standard dynamic programming arguments, $v(x)$ solves the Hamilton-Jacobi (ordinary) differential equation

$$rv = \sup_{q\geq0} qP(q) - qv', \qquad x > 0,\tag{9}$$

where $v' = dv/dx$, and the boundary condition describing exhaustibility is $v(0) = 0$. Denote by $G(s)$ the solution to the static monopoly problem with cost s:

$$G(s) = \max_{q\geq0} q\,(P(q) - s),$$

and the corresponding optimizer $q^*(s)$, where the parameter again refers to the production costs. Then the ODE (9) is simply $rv = G(v')$, so that v' plays the role of a shadow cost, or scarcity value in the dynamic exhaustible resources monopoly problem, and the optimal extraction policy is $q^*(v'(x(t)))$.

The first-order condition defining q^* is

$$P(q^*) + q^* P'(q^*) = s, \quad \Rightarrow \quad G(s) = -(q^*)^2 P'(q^*). \tag{10}$$

Then, differentiating the ODE (9) with respect to x, we have

$$rv' = -q^{*\prime}(v') \cdot v'' \left(2q^*(v') \cdot P'\left(q^*(v')\right) + (q^*(v'))^2 \cdot P''\left(q^*(v')\right)\right),$$

and differentiating the Eq. (10) for $q^*(s)$ with respect to s gives

$$q^{*\prime} \left(2P'(q^*) + q^* P''(q^*)\right) = 1.$$

Therefore, we have $rv' = -q^*(v')v''$, which evaluated along the optimal trajectory defined by $\frac{dx}{dt} = -q^*(v'(x(t)))$ (up until the exhaustion time τ_x) leads to

$$rv'(x(t)) = -q^* (v'(x(t))) v''(x(t)) = \frac{dx}{dt} v''(x(t)) = \frac{d}{dt} v'(x(t)).$$

This is known as Hotelling's rule (for a monopolist with exhaustible resources):

$$\frac{d}{dt} v'(x(t)) = rv'(x(t)), \tag{11}$$

or $v'(x(t)) = v'(x(0))e^{rt}$.

3.2   Multiple Players with Exhaustible Resources

Incorporating other players into a genuine game framework elevates the single player control/ODE problem to the setting of nonzero-sum differential games and partial differential equations (PDEs). Here existence and regularity theory is scarce (outside of the case of linear-quadratic (LQ) games). General dynamic programming equations are studied in [1], and some applications are presented in [15]. The solution approach generally goes through the feedback strategy representation, which allows to express optimal policies in terms of local properties of the game-value functions. For stationary models, one can utilize Euler-Lagrange methods, while in non-stationary or stochastic contexts, more involved analysis is necessary using Hamilton-Jacobi-Bellman-Isaacs tools. For a nonzero-sum dynamic game between N players, each with their own resources, the computation of a solution generally requires dealing with coupled systems of N fully nonlinear PDEs, with one value function per player. This quickly becomes very challenging as N grows and explains the focus on the duopoly $N = 2$ case.

To illustrate the complexity in the duopoly case, we let $x_i(t)$ be the reserves of each player at time t, which are depleted at their extraction rates q_i:

$$\frac{dx_i}{dt} = -q_i(x(t)), \quad i = 1,2 \quad \text{where} \quad x(t) = (x_1(t), x_2(t)).$$

With assumed zero extraction costs and the same discount rate $r > 0$, each player maximizes lifetime discounted profit in (best) response to the extraction policy of the other. A Nash (or Markov perfect) equilibrium (q_1^*, q_2^*), if it exists, describes the value functions

$$v_i(x) = \sup_{q_i \geq 0} \int_0^{\tau_{x_i}} e^{-rt} q_i(x(t)) P\left(q_i(x(t)) + q_j^*(x(t))\right) dt, \quad i = 1,2; j \neq i, \quad (12)$$

where $\tau_{x_i} = \inf\{t > 0 \mid x_i(t) = 0\}$ are the exhaustion times starting at $x_i = x_i(0)$. The state-space approach in (12) restricts attention to policies specified in closed-loop feedback form $q_i(x)$, linking to the single-agent control frameworks and removing technical challenges related to equilibrium existence. Moreover, it naturally generalizes to the stochastic extensions discussed below. Dynamic programming arguments lead to the following equations for the value functions in $x_1, x_2 > 0$:

$$rv_i = \sup_{q_i \geq 0} \left\{ q_i P(q_i + q_j^*) - q_i \frac{\partial v_i}{\partial x_i} \right\} - q_j^* \frac{\partial v_i}{\partial x_j}, \quad j \neq i,$$

which, using the notation introduced in (1), we can write as

$$rv_i = \sup_{q_i \geq 0} \pi\left(q_i, Q_{-i}^*, \frac{\partial v_i}{\partial x_i} \right) - q_j^* \frac{\partial v_i}{\partial x_j}.$$

This identifies the infinitesimal problem in the dynamic programming equation as the static Nash equilibrium problem with scarcity costs $s_i = \frac{\partial v_i}{\partial x_i}$. The interaction of exhaustibility and blockading allows to endogenize market structure. As players deplete their reserves, their marginal costs may rise sufficiently to make further production uneconomical and causing them to drop out of competition.

Using the notation of Theorem 1 for the solution of the static game, we can write the dynamic game PDEs as

$$rv_i = G_i(\mathscr{D}v) - q_j^*(\mathscr{D}v)\frac{\partial v_i}{\partial x_j}, \quad i = 1,2; j \neq i, \qquad \mathscr{D}v = \left(\frac{\partial v_i}{\partial x_i}, \frac{\partial v_j}{\partial x_j} \right).$$

For the linear pricing function example, the functions G_i and q_i^* were given in (7). In a model of *only* exhaustible resources, when player i runs out, player j has a monopoly until he also exhausts, which lead to boundary conditions on $x_i = 0$ and at $(0,0)$ respectively. A more nuanced (and optimistic) view of future energy production allows that when an oil producer exits, he is replaced by an inexhaustible producer (such as from solar) with infinite (or sustainable) supply. This type of model is analyzed by asymptotic and numerical methods in [20].

Alternatively, one can consider models with a single exhaustible producer, and hence a single state variable, along with $N-1$ renewable producers that do not need to worry about reserves. This maintains game effects but minimizes mathematical complexity (see [20, 30]), and allows to study the effect of blockading: how low must oil reserves go before it becomes profitable to start producing from more expensive but sustainable sources? As (levelized) costs of setting up and maintaining energy production from different sources are different, the entry points for solar and wind, for instance, may likewise be very different. This generates phase transitions in the dynamic game characterized by the (endogenously determined) number of active producers $n(t)$. The respective blockading points can be computed explicitly for the linear price function model, see [30] where it is also shown that a modified, piecewise, version of Hotelling's rule holds in the presence of competition. Namely, there exist blockading times $\tau_1^b \leq \tau_2^b \leq \dots$, such that for $t \in [\tau_{n-1}^b, \tau_n^b)$ there are n energy producers (one exhaustible oil producer and $n-1$ active renewables), and the marginal value function for the oil producer along the equilibrium extraction path grows according to

$$\frac{d}{dt} v'(x(t)) = \left(\frac{1}{2} + \frac{1}{2n}\right) r v'(x(t)), \quad t \in [\tau_{n-1}^b, \tau_n^b). \tag{13}$$

The above relationship recovers the classical Hotelling rule when $n = 1$ (oil monopoly) and blunts the sharp price increases associated with 'peak oil' (the growth rate of v' declines as oil runs out and renewables enter).

4   Renewability and Exploration

Exhaustibility is manifested through consideration of the resource reserves which must remain non-negative. Exhaustible resource stocks can be divided into three types: non-renewable, renewable, and replenishable. Non-renewable resources are only available one time, and once used-up are gone forever. Thus, the level of remaining reserves $x(t)$ is non-increasing. The extreme case of $x(t) = 0$ represents complete exhaustion and requires a boundary condition to specify the resulting utility for the producers. A common example are fossil fuels in a global physical context. Once fossil fuels are removed, one can imagine that production is permanently suspended $v_i(0) = 0$; alternatively one could switch to a more expensive (green) backstop, as described above.

Renewable resources, like fisheries or forests, can grow back on their own if left unexploited. A common setup is a logistic growth model with a finite capacity, such that $\frac{dx}{dt} = F(x(t))$, an ordinary differential equation (see e.g., [5]). With renewable resources the main concern is over-exploitation: if production is too high, stocks can be damaged in the long-term (or completely exhausted). However, sustainable extraction is possible all on its own, and only requires enforceable discipline. Mathematically, sustainability/renewability leads to a stationary model where a local-in-time equilibrium between production and extraction generates a global solution (as a long-term steady state); in contrast non-renewable game equilibria are inherently time-dependent and in particular strongly affected by the "terminal condition" of running out of reserves. While steady-state models are not suited for most energy sources, they are common for describing pollution stock dynamics.

Replenishable resources capture the middle ground—reserves can grow, but this requires separate effort/costs. This is meant to represent costly search for, say, new mines or oil fields and matches the industrial exploration-and-production (E&P) cycle. Indeed, with economic incentives the reserve base is not fixed and can be increased. For example, while oil is exhaustible, it is also replenishable since there is a difference between total abstract reserves on Earth, and what is actually commercially "proven" and drives production decisions. Under replenishable reserves, both the upward and downward dynamics in $x(t)$ are controlled. Mathematically, exploration is modeled via a separate control a_t (which may be coupled to production level q_t). Under exploration, the boundary case $x(t) = 0$ requires separate consideration in terms of whether players can "resurrect" themselves. Assuming that future discoveries can finance present exploration [31, 33] leads to an implicit boundary condition for $v_i(0)$.

In a pure resource model, each player has their own, independent, reserves $x_i(t)$, which she has complete control over. In such models, players interact solely through the price mechanism; reserves then add a separate individual *marginal cost of production*, introducing a new source of asymmetry between the players. Alternatively, especially for renewable resources, one may add further game effects by tying together reserves. This can be done by postulating a single, common reserve stock $x(t)$ (akin to the classical tragedy of the commons [3]), or by considering individual reserves levels that have coupled dynamics. For example [9] used

$$\frac{dx_i}{dt} = \delta x_i(t) 1_{\{X(t) < \bar{X}/2\}} + \delta(\bar{X}_i - x_i(t)) 1_{\{X(t) > \bar{X}/2\}},$$

where $X(t) = \sum_i x_i(t)$ is the total resource stock, and δ is the resource growth rate. Thus, up to the aggregate sustainable level $\bar{X}/2$, resource stocks grow exponentially; beyond $\bar{X}/2$ growth rates lessen linearly, turning negative above the individual carrying capacity $\bar{X}_i$.

Turning our attention to the shocks affecting reserves, the classical formulation provides the deterministic dynamics

$$\frac{dx_i}{dt} = -q_i(t) + F(X_i(t), a_i(t)),$$

where the first term represents lower reserves due to extraction, and the second term represents reserves growth (thanks to either exogenous or endogenous factors).

Thus, the future level of reserves is completely determined by the players' strategies and can be extrapolated to any future date t. While mathematically convenient, this is not very realistic, since practical forecasts of future stocks clearly involve a lot of uncertainty (consider for example forecasting of fishery stocks in 2020, or the fossil fuel exhaustion point somewhere in the next few hundred years). This uncertainty permeates even central planner growth models, so is not solely a feature of uncertainty about future equilibria.

4.1 Shocks to Reserves

Taking a stochastic tack, some models have therefore incorporated stochastic dynamics for reserves, now denoted by a *stochastic process* $X_i(t)$:

$$dX_i(t) = -q_i(t)\,dt + \sigma_i dW_t^i, \tag{14}$$

arguing that reserve levels are uncertain and subject to ongoing up/down revisions described by the Brownian motions W^i. Within a Bertrand competition, [29] justified similar Brownian shocks through small fluctuations in the respective demand levels.

Stochastic shocks become especially pertinent for replenishable stocks, where the attendant exploration efforts yield intrinsically stochastic outcomes. Thus, starting with the seminal work of [25], there has been a long literature on stochastic *exploration*. In particular, discrete upward jumps in reserves, modeled as a Poisson process, have been advocated, leading to $dX_i(t) = -q_i(t)\,dt + \delta dN_t^i$ where δ are (random) increments and N^i is a controlled point process. A common setup is to specify controlled intensity of N^i, i.e., $\lambda_t^i = G(a_i(t))$ where λ^i is the hazard rate of arrivals of N^i. This leads to HJB-I system of equations for the game value functions, see [31].

4.2 More Stochasticity

Beyond the aforementioned stochastic shocks to reserves, one can imagine other factors that generate random environment for the Cournot producers. This is especially so over the medium- and long-run contexts that are often used to motivate the models. Clearly on a longer scale essentially every aspect of the market, including demand, costs, reserves, etc., is subject to unpredictable changes.

To capture macroeconomic cycles, [32] considered stochastic demand, so that $P_t = P(D_t, \boldsymbol{q}_t)$ has exogenous shocks from the stochastic factor D_t. For example, taking D_t to be a 2-state independent Markov chain allows to maintain tractability, while representing the low- and high-demand regimes that can be associated with commodity booms and recessions. In combination with non-renewable resources, stochastic demand generates the phenomenon of *strategic mothballing*, whereby

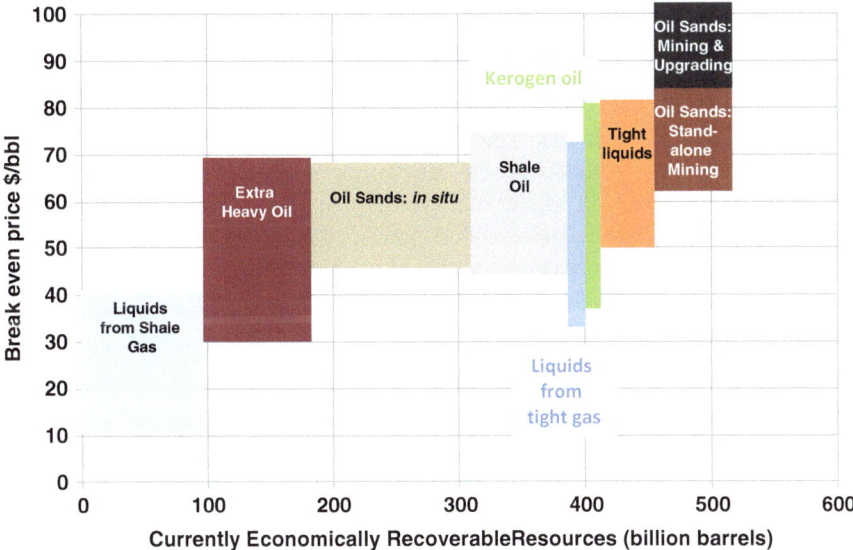

Fig. 2 Estimated oil extraction costs from varying sources. Source: [24]

producers may temporarily shut-down production during low demand periods. Another regime-switching model with exhaustibility but a single agent is in [12].

From a different angle, [11] considered non-constant production costs to mimic the non-stationary economics of extracting more and more difficult to access reserves. Indeed, as well-documented empirically, extraction costs of say crude oil steadily rise as conventional, cheap sources are depleted and replaced by non-conventional oil sands, deep off-shore and shale fields: see Fig. 2.

Accordingly, [11] take costs $s_i(x)$ to depend on reserves, such that $s_i(x)$ increases as x decreases for exhaustible players, and decreases as x decreases for renewable players (due to government subsidies as conventional energy sources are depleted). The resulting dynamic game can force the exhaustible player to leave early, i.e. X_t never reaches zero.

4.3 Other Types of Strategic Interactions

Beyond production/exploration, the literature has also considered other player controls. One major idea from industrial organization (IO) concerns Research and Development (R&D) which generates additional benefits (such as lower production costs or first-mover advantages) to the innovator. In the context of R&D, efforts to innovate may lead to spillovers [7, 13], introducing a different source of coupling between players. Spillovers are well-documented empirically and tend to lower R&D investments and therefore reduce productivity growth. Game-theoretically, spillovers can be viewed as either raising the innovation rate of competitors in a static set-up, or removing first-mover advantages after innovation success. A notable reference is [16] who consider an oligopoly where each of the N firms maximizes R&D effort. The random first innovator is determined stochastically and temporarily collects extra profits. [7] studied a deterministic R&D game with spillovers.

Another link is to the theory of real options, by modeling the strategic opportunities available to producers as one-shot events to be optimally timed. For example, a classical setting concerns producers competing to initiate a new project (such as development of a renewable energy backstop to an exhaustible resource, see [23]) that carries first-mover advantage and leads to a so-called preemption game. Such timing games partition the global model into distinct phases, providing a different mechanism to endogenize market structure. They can be seen as intermediate ground between a static and differential game.

5   Mean Field Games

In dynamic oligopoly problems with a finite number of players, the HJB system of PDEs does not admit an explicit solution, except possibly in the monopoly case. As a result, one needs numerical means for computing the value functions, as well as the equilibrium strategies, which of course quickly becomes infeasible as the number of players goes beyond three. Moreover, even in the two-player case, these equations are hard to handle. To overcome this problem, one may study the market dynamics when the number of firms tends to infinity by using the concept of a mean field game (MFG). MFGs were proposed by Lasry et al. [27, 28] and independently by Huang et al. [22] to handle certain types of competition in the continuum limit of an infinity of small players.

The interaction is modeled by assuming that each player only sees and reacts to the statistical distribution of the states of other players. Optimization against the distribution of other players leads to a backward (in time) Hamilton-Jacobi-Bellman (HJB) equation; and in turn their actions determine the evolution of the state distribution, encoded by a forward Kolmogorov equation. We refer to the survey article by [19] and the recent monograph of [6] for further background. In our context, the mean-field interaction is captured by making the market price be a function of equilibrium global supply, which in turn is affected by the distribution of reserves $m_t(\cdot)$, which is a measure on $\mathbb{R}_+$. Thus, inverse demand curve translates into a functional relationship between reserve distribution $m_t(\cdot)$ and resulting price P_t, according to $D(t, P_t) = \int_{\mathbb{R}_+} q_t^*(x) m_t(dx)$. Numerical resolution of MFG equations is an active area of research; simultaneously dealing with the forward-backward system of PDEs typically requires a fixed-point iteration scheme. At the same time, removing the awkward dependence on the number of players simplifies the equations and provides a unified theory in terms of measure-valued processes.

The analysis of the relationship between Markov perfect equilibria in finite-N Cournot games and the MFG limit was carried in [8] considering the general situation of differentiated goods and asymmetric players. [8] also show that in the continuum MFG limit, the linear Bertrand and Cournot models are equivalent, removing the usual distinction observed with a finite number of players.

In a related work, [18] analyzed a MFG formulation of oil oligopoly for both the deterministic case (reserves endogenously determined by production rates) and the stochastic case (reserves include Brownian noise, generating a parabolic forward Kolmogorov equation for m_t). [18] proposed an iterative numerical scheme and presented some numerical examples for both linear and constant elasticity of substitution (CES) demand curves. One particular focus was on the marginal cost of exhaustibility (Hotelling rent) and also on substitution effects in a two-energy model. See also [2] who treated a *robust* version of above which adds another first-order quadratic term to the MFG equations.

6   Summary of Game Models for Exhaustible Resources

In the table below g refers to a green producer, so that $1 + g$ is a duopoly with one exhaustible and one renewable player, see Sect. 3.2. Infinity of players corresponds to mean-field models. Demand refers to the shape of the price function $P(Q)$.

	# Players	Type	Demand	Randomness	Replenish
Hot31 [21]	1	–	Linear	Determ.	No
DS81 [12]	N	Cournot	Constant	Single-shock	No
DS83 [14]	1	–	Regimes	Poisson	Yes
Ben08 [4]	N	Cournot	Linear	Determ.	Yes
BHW09 [5]	N	Cournot	Linear	Determ.	Yes
HHS [20]	1+g	Cournot	Linear	Brownian	No
LS11 [31]	1+g	Cournot	Regimes	Poisson	Yes
LS12 [30]	1+N	Bertrand	Linear	Determ.	No
LY14 [32]	1+g	Cournot	Linear	Poisson	Yes
CL13 [9]	N	Cournot	Linear	Determ.	Yes
DS14 [11]	1+N	Cournot	Linear	Poisson	Yes
GLL11 [18]	∞	Cournot	CES	Determ.	No
CS14 [8]	∞	Bertrand	Linear	Brownian	No

References

1. Başar, T., Olsder, G.J.: Dynamic Noncooperative Game Theory. SIAM, Philadelphia (1999)
2. Bauso, D., Tembine, H., Başar, T.: Robust mean field games with application to production of an exhaustible resource. In: Proceedings of 7th IFAC Symposium on Robust Control Design, pp 454–459. Aalborg, Denmark (2012)

3. Benchekroun, H.: Unilateral production restrictions in a dynamic duopoly. J. Econ. Theory **111**(2), 214–239 (2003)
4. Benchekroun, H.: Comparative dynamics in a productive asset oligopoly. J. Econ. Theory **138**(1), 237–261 (2008)
5. Benchekroun, H., Halsema, A., Withagen, C.: On nonrenewable resource oligopolies: the asymmetric case. J. Econ. Dyn. Control **33**(11), 1867–1879 (2009)
6. Bensoussan, A., Frehse, J., Yam, P.: Mean Field Games and Mean Field Type Control Theory. Springer, Berlin (2013)
7. Cellini, R., Lambertini, L.: Dynamic R&D with spillovers: competition vs cooperation. J. Econ. Dyn. Control **33**(3), 568–582 (2009)
8. Chan, P., Sircar, R.: Bertrand and Cournot mean field games. Appl. Math. Optim. (to Appear). Doi:10.1007/s00245-014-9269-x (2014)
9. Colombo, L., Labrecciosa, P.: Oligopoly exploitation of a private property productive asset. J. Econ. Dyn. Control **37**(4), 838–853 (2013)
10. Cournot, A.: Recherches sur les Principes Mathématiques de la Théorie des Richesses. Hachette, Paris, English translation by N. T. Bacon published in Economic Classics, Macmillan, 1897, and reprinted in 1960 by Augustus M. Kelly (1838)
11. Dasarathy, A., Sircar, R.: Variable costs in dynamic Cournot energy markets. In: Aid, R., Ludkovski, M., Sircar, R. (eds.) Commodities, Energy and Environmental Finance. Springer, New York (2014)
12. Dasgupta, P., Stiglitz, J.: Resource depletion under technological uncertainty. Econometrica J. Econ. Soc. **49**, 85–104 (1981)
13. Dawid, H., Kopel, M., Kort, P.: R&D competition versus R&D cooperation in oligopolistic markets with evolving structure. Int. J. Ind. Organ. **31**(5), 527–537 (2013)
14. Deshmukh, S.D., Pliska, S.R.: Optimal consumption of a nonrenewable resource with stochastic discoveries and a random environment. Rev. Econ. Stud. **50**(3), 543–554 (1983)
15. Dockner, E.J., Jørgensen, S., Long, N.V., Sorger, G.: Differential Games in Economics and Management Science. Cambridge University Press, Cambridge (2000)
16. Fölster, S., Trofimov, G.: Industry evolution and R&D externalities. J. Econ. Dyn. Control **21**(10), 1727–1746 (1997)
17. Grimaud, A., Lafforgue, G., Magné, B.: Climate change mitigation options and directed technical change: A decentralized equilibrium analysis. Resour. Energy Econ. **33**(4), 938–962 (2011)
18. Guéant, O., Lasry, J.M., Lions, P.L.: Mean field games and oil production. Tech. rep., College de France (2010). http://mfglabs.com/wp-content/uploads/2012/12/cfe.pdf
19. Guéant, O., Lasry, J.M., Lions, P.L.: Mean field games and applications. In: Paris-Princeton Lectures on Mathematical Finance 2010, pp. 205–266. Springer, Berlin (2011)
20. Harris, C., Howison, S., Sircar, R.: Games with exhaustible resources. SIAM J. Appl. Math. **70**, 2556–2581 (2010)
21. Hotelling, H.: The economics of exhaustible resources. J. Polit. Econ. **39**(2), 137–175 (1931)
22. Huang, M., Malhamé, R.P., Caines, P.E. et al.: Large population stochastic dynamic games: closed-loop McKean-Vlasov systems and the Nash certainty equivalence principle. Commun. Inf. Syst. **6**(3), 221–252 (2006)
23. Hung, N., Quyen, N.: On R&D timing under uncertainty: the case of exhaustible resource substitution. J. Econ. Dyn. Control **17**(5), 971–991 (1993)
24. IEA: Medium-term oil and gas markets outlook. Tech. rep., International Energy Agency, Paris, France (2012)
25. Kamien, M.I., Schwartz, N.L.: Optimal exhaustible resource depletion with endogenous technical change. Rev. Econ. Stud. **45**(1), (1978)
26. Lafforgue, G.: Stochastic technical change, non-renewable resource and optimal sustainable growth. Resour. Energy Econ. **30**(4), 540–554 (2008)
27. Lasry, J.M., Lions, P.L.: Jeux à champ moyen. i–le cas stationnaire. Comp. Rend. Math. **343**(9), 619–625 (2006)
28. Lasry, J.M., Lions, P.L.: Mean field games. Jpn. J. Math. **2**(1), 229–260 (2007)

29. Ledvina, A., Sircar, R.: Dynamic Bertrand oligopoly. Appl. Math. Optim. **63**(1), 11–44 (2011)
30. Ledvina, A., Sircar, R.: Oligopoly games under asymmetric costs and an application to energy production. Math. Financ. Econ. **6**(4), 261–293 (2012)
31. Ludkovski, M., Sircar, R.: Exploration and exhaustibility in dynamic Cournot games. Eur. J. Appl. Math. **23**(3), 343–372 (2011)
32. Ludkovski, M., Yang, X.: Dynamic Cournot models for production of exhaustible commodities under stochastic demand. In: Aid, R., Ludkovski, M., Sircar, R. (eds.) Commodities, Energy and Environmental Finance. Springer, New York (2014)
33. Pindyck, R.: The optimal exploration and production of nonrenewable resources. J. Polit. Econ. **86**, 841–862 (1978)
34. Tsur, Y., Zemel, A.: Optimal transition to backstop substitutes for nonrenewable resources. J. Econ. Dyn. Control **27**(4), 551–572 (2003)
35. Vives, X.: Oligopoly pricing: old ideas and new tools. MIT press, Cambridge (2001)

Game Theory Analysis for Carbon Auction Market Through Electricity Market Coupling

Mireille Bossy, Nadia Maïzi, and Odile Pourtallier

Abstract In this paper, we analyze Nash equilibria between electricity producers selling their production on an electricity market and buying CO_2 emission allowances on an auction carbon market. The producers' strategies integrate the coupling of the two markets via the cost functions of the electricity production. We set out a clear Nash equilibrium on the power market that can be used to compute equilibrium prices on both markets as well as the related electricity produced and CO_2 emissions released.

1 Introduction

The aim of this paper is to develop analytic tools in order to design a relevant mechanism for carbon markets, where relevant refers to emissions reduction. For this purpose, we focus on electricity producers in a power market linked to a carbon market. The link between markets is established through a market microstructure approach. In this context, where the number of agents is limited, standard game theory applies. The producers are considered as players behaving on the two financial markets represented here by carbon and electricity. We establish a Nash equilibrium for this non-cooperative J-player game through a coupling mechanism between the two markets.

The original idea comes from the French electricity sector, where the spot electricity market is often used to satisfy peak demand. Producers' behavior is demand driven and linked to the maximum level of electricity production. Each producer strives to maximize its market share. In the meantime, it has to manage the environmental burden associated with its electricity production through a mechanism inspired by the EU ETS (European Emission Trading System) framework:

M. Bossy (✉) • O. Pourtallier
Inria, 06902 Valbonne, France
e-mail: mireille.bossy@inria.fr; odile.pourtallier@inria.fr

N. Maïzi
MINES ParisTech, Centre for Applied Mathematics, CS 10207 rue Claude Daunesse,
06904 Sophia Antipolis Cedex, France
e-mail: nadia.maizi@mines-paristech.fr

© Springer Science+Business Media New York 2015   335
R. Aïd et al. (eds.), *Commodities, Energy and Environmental Finance*,
Fields Institute Communications 74, DOI 10.1007/978-1-4939-2733-3_13

each producer unit of emissions must be counterbalanced by a permit or through the payment of a penalty. Emission permit allocations are simulated through a carbon market that allows the producers to buy allowances at an auction. Our focus on the electricity sector is motivated by its prevalence in the emission share (45 % of the whole emission level worldwide), and the introduction in phase III of the EU ETS of an auction-based allowance allocation mechanism. In the present paper, the design assumptions made on the carbon market aim to foster emissions reduction in the entire electricity sector.

Our approach proposes an original framework for the coupling of bidding strategies on two markets.

Given a static elastic demand curve on the electricity market (referring to the time stages in an organized electricity market, mainly day-ahead and intra-day), we solve the local problem (just a single time period of the same length for both markets) of establishing a non-cooperative Nash equilibrium for the two coupled markets. This simplification is justified here, as we aim to raise the condition under which a carbon market would be a real efficient instrument for carbon mitigation policies.

This analysis is conducted for non-continuous and non-strictly monotone supply functions and bidding strategies on both markets in the complete information framework.

While literature on applied game theory to strategic bidding on power markets mainly addresses profit maximization (see eg [5] with complete information, [6] with private information, [7] with incomplete information), our objective function is share maximization.

The equilibria of the coupled markets are based on the full characterization of the equilibrium electricity price (on the electricity market alone). We prove the uniqueness of the price and shares, for share maximization whereas, to our knowledge this property is not established (under our hypotheses) for profit maximization (see eg [2]).

Moreover, share maximization approach deals with profit by making specific assumptions, i.e. no-loss sales, and a tradeoff between the purchase of allowances and the carbon footprint of the electricity generated. Hence, this work is the first attempt on power and carbon markets coupling through game theory approach. Other coupling approaches use, for instance, models that produce dynamics for both electricity and carbon prices jointly, as in [3, 4].

In Sect. 2, we formalize the market (carbon and electricity) rules and the associated admissible set of players' coupled strategies.

We start by studying (in Sect. 3.1) the set of Nash equilibria on the electricity market alone (see Proposition 1). This set constitutes an equivalence class (same prices and market shares) from which we exhibit a dominant strategy.

Section 3.2 is devoted to the analysis of coupled markets equilibria: given a specific carbon market design (in terms of penalty level and allowances), we compute the bounds of the interval where carbon prices (derived from the previous dominant strategy) evolve. We specify the properties of the associated equilibria.

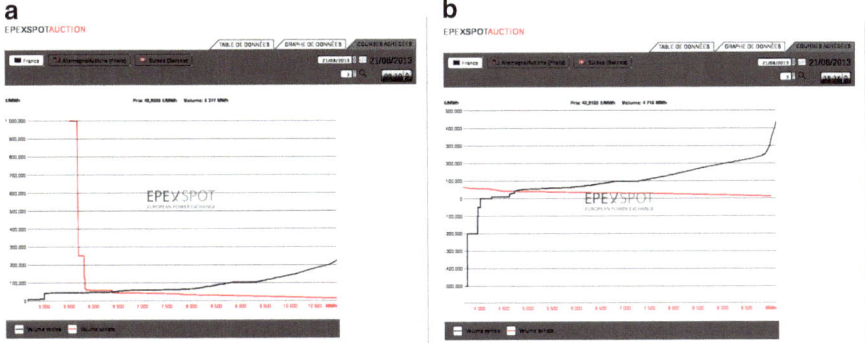

Fig. 1 The *orange curve* is the function $q \mapsto D^{-1}(q)$ on the EPEX market. The evolution of the spot market confirms the relevance of Assumption 1 on the demand function $p \mapsto D(p)$. (**a**) Delivery 9–10 am. (**b**) Delivery 3–4 pm

2 Coupling Markets Mechanism

2.1 Electricity Market

In the electricity market, demand is aggregated and summarized by a function $p \mapsto D(p)$, where $D(p)$ is the quantity of electricity that buyers are ready to obtain at maximal unit price p (see Fig. 1). We assume the following:

Assumption 1. *The demand function $D(\cdot) : \mathbb{R}^+ \rightarrow \mathbb{R}^+$ is non-increasing, left continuous, and such that $D(0) > 0$.*

Each producer $j \in \{1, \dots, J\}$ is characterized by a finite production capacity κ_j and a bounded and non-decreasing function $c_j : [0, \kappa_j] \longrightarrow \mathbb{R}^+$ that associates a marginal production cost to any quantity q of electricity. These marginal production costs depend on several exogenous parameters reflecting the technical costs associated with electricity production e.g. energy prices, O&M costs, taxes, carbon penalties, *etc*. This parameter dependency makes it possible to build different market coupling mechanisms. In the following we use it to link the carbon and electricity markets.

The merit order ranking features marginal cost functions sorted according to their production costs. These are therefore non-decreasing step functions whereby each step refers to the marginal production cost of a specific unit owned by the producer.

The producers trade their electricity on a dedicated market. For a given producer j, the strategy consists of a function that makes it possible to establish an asking price on the electricity market, defined as

$$s_j : \mathscr{C}_j \times \mathbb{R}^+ \longrightarrow \mathbb{R}^+$$

$$(c_j(\cdot), q) \longrightarrow s_j(c_j(\cdot), q),$$

where $\mathcal{C}_j$ the set of marginal production cost functions are explicitly given in the following (see (14)); $s_j(c_j(\cdot), q)$ is the unit price at which the producer is ready to sell quantity q of electricity. An admissible strategy carries out the following sell at no loss constraint

$$s_j(c_j(\cdot), q) \geq c_j(q), \quad \forall q \in \text{Dom}(c_j). \tag{1}$$

A possible example of such strategy is $s_j(c_j(\cdot), q) = c_j(q)$ or $s_j(c_j(\cdot), q) = c_j(q) + \lambda(q)$, where $\lambda(q)$ stands for any additional profit.

The constraint (1) guarantees profitable trade and incorporates an aspect of profit maximization (ie, loss avoidance) into the market share approach. In what follows, we include this profit constraint in the considered class of admissible strategies.

We define the class of admissible strategy profiles on electricity market $\mathbb{S}$ as:

$$\mathbb{S} = \left\{ \begin{array}{c} \mathbf{s} = (s_1, \ldots, s_J); \; s_j : \mathcal{C}_j \times \mathbb{R}^+ \longrightarrow \mathbb{R}^+ \\ (c_j(\cdot), q) \longrightarrow s_j(c_j(\cdot), q) \\ \text{such that } s_j(c_j(\cdot), q) \geq c_j(q), \quad \forall q \in \text{Dom}(c_j) \end{array} \right\}. \tag{2}$$

As a function of q, $s_j(c_j(\cdot), q)$ is bounded on $\text{Dom}(c_j)$. For the sake of clarity, we define for each $q \notin \text{Dom}(c_j)$, $s_j(c_j(\cdot), q) = p_{\text{lolc}}$, where p_{lolc} is the loss of load cost, chosen as any overestimation of the maximal production costs.

For producer j's strategy s_j, we define the associated asking size at price p as

$$\mathcal{O}(c_j(\cdot), s_j; p) := \sup\{q, \; s_j(c_j(\cdot), q) < p\} \tag{3}$$

with $\sup \emptyset = 0$. Hence $\mathcal{O}(c_j(\cdot), s_j; p)$ is the maximum quantity of electricity at unit price p supplied by producer j on the market. We also call $p \mapsto \mathcal{O}(c_j(\cdot), s_j; p)$ the offer function of producer j.

Remark 1.

(i) The asking size function $p \mapsto \mathcal{O}(c_j(\cdot), s_j; p)$ is, with respect to p, an non-decreasing surjection from $[0, +\infty)$ to $[0, \kappa_j]$, right continuous and such that $\mathcal{O}(c_j(\cdot), s_j; 0) = 0$. For a non-decreasing strategy s_j, $\mathcal{O}(c_j(\cdot), s_j; .)$ is its generalized inverse function with respect to q.

(ii) Given two strategies $q \mapsto s_j(c_j(\cdot), q)$ and $q \mapsto s_j'(c_j(\cdot), q)$ such that $s_j(c_j(\cdot), q) \leq s_j'(c_j(\cdot), q)$, for all $q \in \text{Dom}(c_j)$ we have for any positive p

$$\mathcal{O}(c_j(\cdot), s_j; p) \geq \mathcal{O}(c_j(\cdot), s_j'; p).$$

Indeed, if $p_1 \geq p_2$ then $\{q, \; s_j(c_j(\cdot), q) \leq p_2\} \subset \{q, \; s_j(c_j(\cdot), q) \leq p_1\}$ from which we deduce that $\mathcal{O}(c_j(\cdot), s_j; \cdot)$ is non-decreasing. Next, if $s_j(c_j(\cdot), \cdot) \leq s_j'(c_j(\cdot), \cdot)$, for any fixed p, we have $\{q, \; s_j'(c_j(\cdot), q) \leq p\} \subset \{q, \; s_j(c_j(\cdot), q) \leq p\}$ from which the reverse order follows for the requests.

We shall now describe the electricity market clearing. Note that from a market view point, the dependency of the supply with respect to the marginal cost does not need to be explicit. For the sake of clarity, we write $s_j(q)$ and $\mathcal{O}(s_j; p)$ instead of $s_j(c_j(\cdot), q)$ and $\mathcal{O}(c_j(\cdot), s_j; p)$. The dependency will be expressed explicitly whenever needed.

By aggregating the J asking size functions, we can define the overall asking function $p \mapsto \mathcal{O}(\mathbf{s}; p)$ a producer strategy profile $\mathbf{s} = (s_1, \ldots, s_J)$ as:

$$\mathcal{O}(\mathbf{s}; p) = \sum_{j=1}^{J} \mathcal{O}(s_j; p). \tag{4}$$

Hence, for any producer strategy profile $\mathbf{s}$, $\mathcal{O}(\mathbf{s}; p)$ is the quantity of electricity that can be sold on the market at unit price p.

The overall supply function $p \mapsto \mathcal{O}(\mathbf{s}; p)$ is a non-decreasing surjection defined from $[0, +\infty)$ to $[0, \sum_{j=1}^{J} \kappa_j]$, such that $\mathcal{O}(\mathbf{s}; 0) = 0$.

2.1.1 Electricity Market Clearing

Taking producer strategy profile $\mathbf{s} = (s_1(\cdot), \ldots, s_J(\cdot))$ the market sets the electricity market price $p^{\text{elec}}(\mathbf{s})$ together with the quantities $(\varphi_1(\mathbf{s}), \ldots, \varphi_J(\mathbf{s}))$ of electricity sold by each producer.

The market clearing price $p^{\text{elec}}(\mathbf{s})$ is the unit price paid to each producer for the quantities $\varphi_j(\mathbf{s})$ of electricity. The price $p(\mathbf{s})$ may be defined as a price whereby supply satisfies demand. As we are working with a general non-increasing demand curve (possibly locally inelastic), the price that satisfies the demand is not necessarily unique. We thus define the clearing price generically with the following definition.

Definition 1 (The Clearing Electricity Price). Let us define

$$\underline{p}(\mathbf{s}) = \inf \{p > 0; \ \mathcal{O}(\mathbf{s}; p) > D(p)\}$$

$$\text{and} \quad \bar{p}(\mathbf{s}) = \sup \{p \in [\underline{p}(\mathbf{s}), p_{\text{lolc}}]; D(p) = D(\underline{p}(\mathbf{s}))\} \tag{5}$$

with the convention that $\inf \emptyset = p_{\text{lolc}}$. The clearing price may then be established as any $p^{\text{elec}}(\mathbf{s}) \in [\underline{p}(\mathbf{s}), \bar{p}(\mathbf{s})]$ as an output of a specific market clearing rule. To keep the price consistency, the market rule must be such that for any two strategy profiles $\mathbf{s}$ and $\mathbf{s}'$,

$$\text{if } \underline{p}(\mathbf{s}) < \underline{p}(\mathbf{s}') \text{ then } p^{\text{elec}}(\mathbf{s}) < p^{\text{elec}}(\mathbf{s}'),$$

$$\text{if } \underline{p}(\mathbf{s}) = \underline{p}(\mathbf{s}') \text{ then } p^{\text{elec}}(\mathbf{s}) = p^{\text{elec}}(\mathbf{s}'). \tag{6}$$

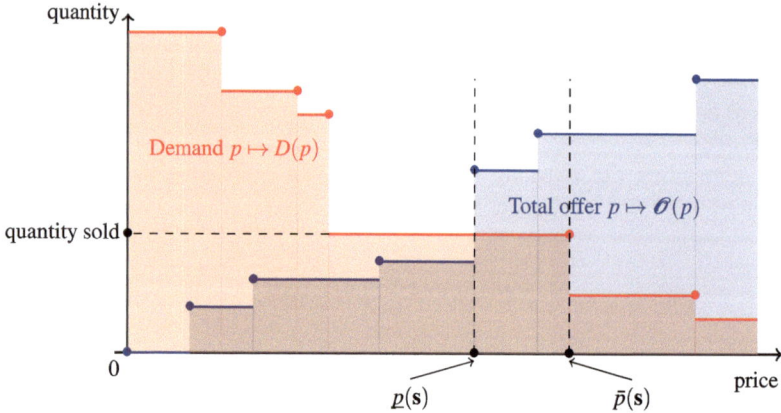

Fig. 2 Electricity price $\underline{p}(\mathbf{s})$ and $\bar{p}(\mathbf{s})$

Note that $\underline{p}(\mathbf{s}) \neq \bar{p}(\mathbf{s})$ only if the demand curve $p \mapsto D(p)$ is constant on some intervals $[\underline{p}(\mathbf{s}), \underline{p}(\mathbf{s}) + \epsilon]$ (see Fig. 2). In that case, $\underline{p}(\mathbf{s})$ corresponds to the best ask price, while $\bar{p}(\mathbf{s})$ is the best bid price. The demand/offer curves that result from the buyer/seller aggregation in a given time period implies some market fixing rules that allocate buyer surplus and seller surplus. In that sense $p^{\mathrm{elec}}(\mathbf{s})$ is a fixing price.[1] Note that $p^{\mathrm{elec}}(\mathbf{s}) = \underline{p}(\mathbf{s})$ maximizes buyer surplus while $p^{\mathrm{elec}}(\mathbf{s}) = \bar{p}(\mathbf{s})$ maximizes seller surplus. Note also that price $\underline{p}(\mathbf{s})$ is well defined in the case where demand does not strictly decrease. This includes the case where demand is constant. In such case, $\underline{p}(\mathbf{s}) = p_{\mathrm{lolc}}$ only if the demand curve never crosses the supply.

Next, we define the quantity of electricity sold at price $p^{\mathrm{elec}}(\mathbf{s})$. When $p^{\mathrm{elec}}(\mathbf{s})$ is such that $\mathscr{O}(\mathbf{s}; p^{\mathrm{elec}}(\mathbf{s})) \leq D(p^{\mathrm{elec}}(\mathbf{s}))$, each producer sells $\mathscr{O}(\mathbf{s}_j; p^{\mathrm{elec}}(\mathbf{s}))$, but cases where $\mathscr{O}(\mathbf{s}; p^{\mathrm{elec}}(\mathbf{s})) > D(p^{\mathrm{elec}}(\mathbf{s}))$ may occur, requiring the introduction of an auxiliary rule to share $D(p^{\mathrm{elec}}(\mathbf{s}))$ among the producers that propose $\mathscr{O}(\mathbf{s}; p^{\mathrm{elec}}(\mathbf{s}))$. Note that in this last case, due to the clearing property (6) on $p^{\mathrm{elec}}(\cdot)$, we have $\mathscr{O}(\mathbf{s}; \underline{p}(\mathbf{s})) > D(p^{\mathrm{elec}}(\mathbf{s})) = D(\underline{p}(\mathbf{s}))$. Hence the $D(p^{\mathrm{elec}}(\mathbf{s}))$ is totally provided by producers with non null offer at price $\underline{p}(\mathbf{s})$. The rule of the market is to share $D(p^{\mathrm{elec}}(\mathbf{s}))$ among these producers only. This gives an explicit priority to the best offer prices $\underline{p}(\mathbf{s})$.

Let us break down supply as follows:

$$\mathscr{O}(\mathbf{s}; \underline{p}(\mathbf{s})) = \sum_{j=1}^{J} \mathscr{O}(s_j; \underline{p}(\mathbf{s})^-) + \sum_{j=1}^{J} \Delta^- \mathscr{O}(s_j; \underline{p}(\mathbf{s})),$$

where $\Delta^- \mathscr{O}(s_j; \underline{p}(\mathbf{s})) := \mathscr{O}(s_j; \underline{p}(\mathbf{s})) - \mathscr{O}(s_j; \underline{p}(\mathbf{s})^-)$ and $f(x^-)$ denotes the left value at x of a function f.

---

[1]One can imagine that Power market participants have access to the detailed fixing rules, but information proves hard to be found.

The market's choice is to fully accept the asking size of producers with continuous asking size curve at point $\underline{p}(\mathbf{s})$. For producers with discontinuous asking size curve at $\underline{p}(\mathbf{s})$, a market rule based on proportionality that favors abundance is used to share the remaining part of the supply: any extra supply available at the clearing price $\mathscr{O}(\mathbf{s};\underline{p}(\mathbf{s})) - D(\underline{p}(\mathbf{s}))$ is split among all generators offering at that price such that they each get the same percentage of their offered quantity allocated to production.

We summarize the market rule on quantities as follows.

Definition 2 (Clearing Electricity Quantities). The quantity $\varphi_j(\mathbf{s})$ of electricity sold by Producer j on the electricity market is

$$
\varphi_j(\mathbf{s}) = \begin{cases} \mathscr{O}(s_j; p^{\text{elec}}(\mathbf{s})) = \mathscr{O}(s_j; \underline{p}(\mathbf{s})), & \text{if } D(p^{\text{elec}}(\mathbf{s})) \geq \mathscr{O}(\mathbf{s}; p^{\text{elec}}(\mathbf{s})), \\ \mathscr{O}(s_j; \underline{p}(\mathbf{s})^-) + \Delta^- \mathscr{O}(s_j; \underline{p}(\mathbf{s})) \dfrac{D(\underline{p}(\mathbf{s})) - \mathscr{O}(\mathbf{s}; \underline{p}(\mathbf{s})^-)}{\Delta^- \mathscr{O}(\mathbf{s}; \underline{p}(\mathbf{s}))}, \\ \qquad \text{if } D(\underline{p}(\mathbf{s})) < \mathscr{O}(\mathbf{s}; \underline{p}(\mathbf{s})), \end{cases}
$$

(7)

where $\Delta^- \mathscr{O}(\mathbf{s}; \underline{p}(\mathbf{s})) := \displaystyle\sum_{j=1}^{J} \Delta^- \mathscr{O}(s_j; \underline{p}(\mathbf{s})) > 0$.

Note that, when $D(\underline{p}(\mathbf{s})) < \mathscr{O}(\mathbf{s}; \underline{p}(\mathbf{s}))$, we have $\Delta^- \mathscr{O}(\mathbf{s}; \underline{p}(\mathbf{s})) > 0$. Note also that

$$
\sum_{i=1}^{J} \varphi_j(\mathbf{s}) = D(p^{\text{elec}}(\mathbf{s})) \wedge \mathscr{O}(\mathbf{s}; p^{\text{elec}}(\mathbf{s})) = D(\underline{p}(\mathbf{s})) \wedge \mathscr{O}(\mathbf{s}; \underline{p}(\mathbf{s})),
$$

(8)

$$
\text{and for all } j, \quad \mathscr{O}(s_j; \underline{p}(\mathbf{s})^-) \leq \varphi_j(\mathbf{s}) \leq \mathscr{O}(s_j; \underline{p}(\mathbf{s})).
$$

(9)

2.2 Carbon Market

Let us recall the CO_2 regulation principle on which we base our analysis. Producers are penalized according to their emission level if they do not own allowances. Hence, in parallel to their position on the electricity market, producers buy CO_2 emission allowances on a separate CO_2 auction market. In the following, we formalize producer strategy on the CO_2 market only.

If they are allowed to, producers buying permits on the CO_2 market will use them either to cap their own power production emissions, either to prevent other players from buying permits. The following assumption introduces some market design rules that control players behavior on that market.

Assumption 2 (Capped Carbon Market).

(i) *The carbon market is capped and has a finite known quantity Ω of CO_2 emission allowances available.*

(ii) *Each producer j can buy a capped number of allowances $\mathscr{E}_j$, related to its own CO_2 emission capacity.*

(iii) *Emissions that are not covered by allowances are penalized at a unit rate $\mathfrak{p}$.*

Note that if one chose $\mathscr{E}_j \geq \Omega$ for all producers then item *(ii)* is void. Other choice for the $\mathscr{E}_j$ can be seen as strengthen regulation tool.

On this market, producers adopt a strategy that consists of an offer function $\tau \mapsto A_j(\tau)$ defined from $[0, \mathfrak{p}]$ to $[0, \mathscr{E}_j]$. Quantity $A_j(\tau)$ is the quantity of allowances that producer j is ready to buy at price τ. This offer may not be a monotonic function. We denote $\mathbb{A}$ the strategy profile set on the CO_2 market,

$$\mathbb{A} := \{\mathbf{A} = (A_1, \ldots, A_J); \text{ s.t. } A_j : [0, \mathfrak{p}] \to [0, \mathscr{E}_j]\}.$$

The CO_2 market reacts by aggregating the J offers by

$$\mathscr{A}(\tau) := \sum_{j=1}^{J} A_j(\tau),$$

and the clearing market price is established following a *second item auction*[2] as:

$$p^{CO_2}(\mathbf{A}) := \sup\{\tau; \mathscr{A}(\tau) > \Omega\}, \quad \text{with the convention } \sup \emptyset = 0. \tag{10}$$

Note that $p^{CO_2}(\mathbf{A}) = 0$ indicates that there are too many allowances to sell. It is worth a reminder here that the aim of allowances is to decrease emissions. In Sect. 3.2, we discuss a design hypothesis (Assumption 5) that guarantees an equilibrium price $p^{CO_2}(\mathbf{A}) > 0$. Therefore, in the following, we assume that the overall quantity Ω of allowances, is such that $p^{CO_2}(\mathbf{A}) > 0$.

Next, we define the amount of allowances bought at price $p^{CO_2}(\mathbf{A})$ by the producers. By Definition (10), we have $\mathscr{A}(p^{CO_2}(\mathbf{A})) \geq \Omega$ and $\mathscr{A}(p^{CO_2}(\mathbf{A})^+) \leq \Omega$. When $\mathscr{A}(p^{CO_2}(\mathbf{A})) > \Omega$, the CO_2 market must decide between the producers with an additional rule. We define

$$\Delta(A_i) := A_i(p^{CO_2}(\mathbf{A})^+) - A_i(p^{CO_2}(\mathbf{A})).$$

For a producer i, $\Delta(A_i) \geq 0$ means that its CO_2 demand does not decrease if the price increases. It is therefore ready to pay more to obtain the quantity of

---

[2] Also called *Dutch auction market* with several units to sell, in a *second item auction* market, the seller begins with a very high price and reduces it. The price is lowered until a bidder accepts the current price.

allowances it is asking for at price $p^{CO_2}(\mathbf{A})$. The CO_2 market gives priority to this kind of producer, which will be fully served. The producers such that $\Delta(A_j) < 0$ share the remaining allowances. This can be written as follows.

Each producers with $A_j(p^{CO_2}(\mathbf{A})) > 0$ obtains the following quantity $\delta_j(\mathbf{A})$ of allowances

$$
\delta_j(\mathbf{A}) := \begin{cases} A_j(p^{CO_2}(\mathbf{A})), & \text{if } \Delta(A_j) \geq 0, \\[2em] A_j(p^{CO_2}(\mathbf{A})^+) + \dfrac{(-\Delta(A_j))^+}{\sum\limits_{i=1}^{J}(-\Delta(A_i))^+} \left(\Omega - \sum\limits_{i=1}^{J} A_i(p^{CO_2}(\mathbf{A})) \mathbb{1}_{\{\Delta(A_i)\geq 0\}} \right), \\[2em] \text{otherwise.} \end{cases}
$$

(11)

2.3 Carbon and Electricity Market Coupling

In the following, we formalize the coordination of a producer's strategy on the CO_2 and electricity markets. This could be seen as if both markets were synchronized during a single time period with the same length (eg, 1 h).

As mentioned earlier, for each producer, the marginal cost function is parametrized by the positions $\mathbf{A}$ of the producers on the carbon market. Indeed, producer j can obtain CO_2 emission allowances on the market to avoid penalization for (some of) its emissions. Those emissions that are not covered by allowances are penalized at a unit rate $\mathfrak{p}$.

A profile of an offer to buy from the producers $\mathbf{A} = (A_1, \ldots, A_J)$, through the CO_2 market clearing, corresponds to a unit price of $p^{CO_2}(\mathbf{A})$ of the allowance and quantities $\delta_j(\mathbf{A})$ of allowances bought by each producer (defined by the market rules (10),(11)).

This yields to the following modified marginal production cost function[3] $c_j^{\mathbf{A}}(\cdot)$, parametrized by the emission regulations:

$$
q \mapsto c_j^{\mathbf{A}}(q) = \begin{cases} c_j(q) + e_j(q)p^{CO_2}(\mathbf{A}), & \text{for } q \in [0, \kappa_j^{CO_2}(\mathbf{A}) \wedge \kappa_j] \\ c_j(q) + e_j(q)\mathfrak{p}, & \text{for } q \in [\kappa_j^{CO_2}(\mathbf{A}) \wedge \kappa_j, \kappa_j] \end{cases}
$$

(12)

where for all producers $\{j = 1, \ldots, J\}$,

- $q \mapsto e_j(q)$ is the emission rate (originally in CO_2 t/Mwh),
- $\kappa_j^{CO_2}(\mathbf{A})$ is the electricity capacity covered by the bought allowances $\delta_j(\mathbf{A}) \leq \mathcal{E}_j$:

---

[3]Note that this representation might also include the allowances possibly stored by the producers in the previous periods.

$$\kappa_j^{\mathrm{CO_2}}(\mathbf{A}) = \mathrm{argmax}\{k; \int_0^k e_j(z)dz \le \delta_j(\mathbf{A})\}.$$

In this coupled market setting, the strategy of producer j thus makes a pair (A_j, s_j). The set of admissible strategy profile is defined as

$$\sigma = \{(\mathbf{A}, \mathbf{s}); \ \mathbf{A} \in \mathbb{A}, \ \mathbf{s} \in \mathbb{S}\}, \tag{13}$$

where in the definition of $\mathbb{S}$ in (2), we use

$$\mathscr{C}_j = \{c_j^{\mathbf{A}}; \ \mathbf{A} \in \mathbb{A}\}. \tag{14}$$

Prices for allowances and electricity, $p^{\mathrm{CO_2}}((\mathbf{A}, \mathbf{s}))$ and $p^{\mathrm{elec}}((\mathbf{A}, \mathbf{s}))$, quantities of allowances bought by each producer, $\delta_j((\mathbf{A}, \mathbf{s}))$ and market shares on electricity market $\varphi_j((\mathbf{A}, \mathbf{s}))$ of each producer corresponds to any strategy profile $(\mathbf{A}, \mathbf{s}) \in \Sigma$, through the market mechanisms described.

3 Nash Equilibrium Analysis

We suppose that the J producers behave non-cooperatively, aiming at maximizing their individual market share on the electricity market. For a strategy profile $(\mathbf{A}, \mathbf{s}) \in \Sigma$, the market share of a producer j depends upon its strategy $(A_j, s_j(\cdot))$ but also on the strategies $(\mathbf{A}_{-j}, \mathbf{s}_{-j})$ of the other producers.[4] In this set-up the natural solution is the Nash equilibrium (see e.g. [1]). More precisely we are looking for a strategy profile

$$(\mathbf{A}^*, \mathbf{s}^*) = ((A_1^*, s_1^*), \cdots, (A_J^*, s_J^*)) \in \Sigma$$

that satisfies Nash equilibrium conditions: none of the producers would strictly benefit, that is, would strictly increase its market share from a unilateral deviation. Namely, for any producer j strategy $(\mathbf{A}_j, \mathbf{s}_j)$ such that $((\mathbf{A}_{-j}^*, \mathbf{s}_{-j}^*), (A_j, s_j)) \in \Sigma$, we have

$$\varphi_j((\mathbf{A}^*, \mathbf{s}^*)) \ge \varphi_j((\mathbf{A}_{-j}^*, \mathbf{s}_{-j}^*), (A_j, s_j)), \tag{15}$$

where φ_j is the quantity of electricity sold. Note that the dependency in terms of $\mathbf{A}$ through the marginal cost $c_j^{\mathbf{A}}$ is now made explicit in φ_j.

Condition (15) has to be satisfied for any unilateral deviation of any producer j. In particular (15) has to be satisfied for a producer j admissible deviation (A_j^*, s_j) such that $((\mathbf{A}_{-j}^*, \mathbf{s}_{-j}^*), (A_j^*, s_j)) \in \Sigma$ where producer j would only change its behavior on the electricity market.

---

[4] Here we use the generic notation $\mathbf{b}_{-j}$ that stands for the profile set $(b_1, \cdots, b_{j-1}, b_{j+1}, \cdots, b_J)$.

Remark 2. The electricity strategy component s^* of the Nash equilibrium $(\mathbf{A}^*, \mathbf{s}^*)$ is also a Nash equilibrium for the restricted electricity game, where producers only behave on the electricity market with marginal electricity production costs $c_j^{\mathbf{A}^*}(\cdot)$, $j = 1, \cdots J$.

The next section focuses on determining a Nash equilibrium on the game restricted to the electricity market.

3.1 Equilibrium on the Power Market

In this restricted set-up, we consider that the marginal costs $\{c_j, j = 1 \ldots, J\}$ are known data, possibly fixed through the position $\mathbf{A}$ on the CO_2 market. In this section, we refer to $\mathbb{S}$ as the set of admissible strategy profiles, in the particular case where $\mathscr{C}_j = \{c_j\}$ for each $j = 1, \ldots, J$.

The Nash equilibrium problem is as follows: find $\mathbf{s}^* = (s_1^*, \ldots, s_J^*) \in \mathbb{S}$ such that

$$\forall j, \forall \ s_j \neq s_j^*, \quad \varphi_j(\mathbf{s}^*) \geq \varphi_j(\mathbf{s}_{-j}^*, s_j). \tag{16}$$

The following proposition exhibits a Nash equilibrium, whereby each producer must choose the strategy denoted by C_j, and referred to as *marginal production cost strategy*. It is defined by

$$C_j(q) = \begin{cases} c_j(q), & \text{for } q \in \text{Dom}(c_j) \\ p_{\text{lolc}}, & \text{for } q \notin \text{Dom}(c_j). \end{cases} \tag{17}$$

Proposition 1.

(i) *For any strategy profile* $\mathbf{s} = (s_1, \ldots, s_J)$, *no producer* $j \in \{1, \ldots, J\}$ *can be penalized by deviating from strategy* s_j *to its marginal production cost strategy* C_j, *namely,*

$$\varphi_j(\mathbf{s}) \leq \varphi_j(\mathbf{s}_{-j}, C_j). \tag{18}$$

In other words, for any producer j, C_j *is a dominant strategy.*

(ii) *The strategy profile* $\mathbf{C} = (C_1, \ldots C_J)$ *is a Nash equilibrium.*

(iii) *If the strategy profile* $\mathbf{s} \in \mathbb{S}$ *is a Nash equilibrium, then we have* $\bar{p}(\mathbf{s}) = \bar{p}(\mathbf{C})$, *and for any producer* j, $\varphi_j(\mathbf{s}) = \varphi_j(\mathbf{C})$.

Point *(ii)* exhibits a Nash equilibrium strategy profile as a direct consequence of the dominance property *(i)*. Clearly the Nash equilibrium is non-unique, since we can easily show that a producer's given supply can follow from countless different strategies. Nevertheless point *(iii)* shows that there is a unique associated quantities of electricity sold by producers. The market coupling mechanism that we propose in the following section is based on this uniqueness property which allows the computation of the equilibrium shares on electricity and carbon markets. Moreover, any Nash equilibrium price evolves in the interval $p^{\text{elec}}(\mathbf{s}) \in [\underline{p}(\mathbf{C}), \bar{p}(\mathbf{C})]$, which reduces to the point $\{\bar{p}(\mathbf{C})\}$ in various situations, in particular when $D(\cdot)$ strictly decreases at $\underline{p}(\mathbf{C})$, or when p^{elec} is chosen equal to $\bar{p}$.

Proofs of *(i)* and *(iii)*, which are rather tedious due to non-strictly monotony and possible discontinuity of supply and offers, are postponed to "Proof of Proposition 1" in Appendix.

3.2 Coupled Market Design Using the Nash Equilibrium

From this point we restrict our attention to a particular market design. In the following, the scope of the analysis applies to a special class of producers, a specific electricity market price clearing (satisfying Definition 1) and a range of quantities Ω of allowances available on the CO_2 market. Although not necessary, the following restriction simplifies the development.

Assumption 3 (On the Producers). *Each producer j operates a single production unit, for which*

(i) *the initial marginal cost contribution (that does not depend on the producer positions $\mathbf{A}$ in the CO_2 market) is constant, $q \mapsto c_j(q) = c_j$. The related emission rate $q \mapsto e_j(q) = e_j$ is also assumed to be a positive constant,*
(ii) *the producers are different pairwise: $\forall i, j \in \{1, \cdots J\}, (c_i, e_i) \neq (c_j, e_j)$.*

In what follows (according to Assumption 2), in order to limit the number of parameters involved in the discussion, the maximal cap of allowances that each producer j may buy is set to $\mathscr{E}_j = e_j \kappa_j$. This arbitrary but natural choice does not penalize producers capacity level, and does not bring any restriction to the following equilibrium analysis.

As a consequence of Assumption 3, the marginal production cost in (12) can simply be written as

$$q \mapsto c_j^{\mathbf{A}}(q) = \begin{cases} c_j + e_j p^{\text{CO}_2}(\mathbf{A}), & \text{for } q \in [0, \dfrac{\delta_j(\mathbf{A})}{e_j} \wedge \kappa_j] \\ c_j + e_j \mathfrak{p}, & \text{for } q \in [\dfrac{\delta_j(\mathbf{A})}{e_j} \wedge \kappa_j, \kappa_j]. \end{cases} \tag{19}$$

For a given strategy profile on the electricity market, Definition 1 gives a range of possible determinations for the electricity price. Previously, the analysis of the Nash Equilibrium restricted to the electricity market did not require a precise clearing price determination. Nevertheless to extend our analysis to the coupling we need to make explicit this determination and assume the following:

Assumption 4 (On the Electricity Market). *For a given strategy profile* **s** *of the producers, the clearing price of electricity is* $p^{\mathrm{elec}}(\mathbf{s})$. *The market rule fixes* $p^{\mathrm{elec}}(\cdot) = \bar{p}(\cdot)$ *or* $p^{\mathrm{elec}}(\cdot) = \underline{p}(\cdot)$ *as defined in* (5).

We will illustrate below that this choice of clearing price ensures the increasing behavior of $p^{\mathrm{elec}}(\cdot)$ and right continuity in terms of the carbon price (see Lemma 1).

The quantity Ω of CO_2 allowances available plays a crucial role in the market design. If this quantity is too high, the allowance's market price will drop to zero, leaving the market incapable of fulfilling its role of decreasing CO_2 emissions. Therefore we clearly need to make an assumption that restricts the number of allowances available. Appropriately capping the maximum quantity of allowances available requires information on which producers are willing to obtain allowances. This is the objective of the following paragraph where we define a *willing to buy* function that plays a central role in the analysis of Nash equilibria.

3.2.1 Willing to Buy Functions

In this paragraph, we aim at guessing a Nash equilibrium candidate. We base our reasoning on the dominant strategies on the electricity market alone (see Proposition 1). Remark 2 allows us to fix the electricity market strategy as a *marginal production cost strategy*, given the marginal cost functions $\mathbf{C^A} = \{c_j^A, j = 1, \ldots J\}$ imposed by the output of the CO_2 clearing, as in (19).

In particular, when $\mathbf{A} \in \mathbb{A}$, we observe that the strategies $(\mathbf{A}, \{c_j^A, j = 1, \ldots J\})$ are in the set of admissible strategies defined in (13).

From now on, all the strategy profiles that we consider on the carbon market are assumed to be admissible.

In the following, as the discussion will mainly focus on the impact of strategies $\mathbf{A}$ through the carbon market, we denote the electricity market output as:

$$
\begin{aligned}
p^{\mathrm{elec}}(\mathbf{A}) &\quad \text{instead of} \quad p^{\mathrm{elec}}(\mathbf{C^A}) \\
(\varphi_1(\mathbf{A}), \ldots, \varphi_J(\mathbf{A})) &\quad \text{instead of} \quad (\varphi_1(\mathbf{C^A}), \ldots, \varphi_J(\mathbf{C^A})).
\end{aligned}
\tag{20}
$$

To begin with, we consider an exogenous CO_2 cost τ similar to a CO_2 tax rate: the producers' marginal cost becomes for any $\tau \in [0, \mathfrak{p}]$, $c_j^{\tau}(\cdot)$,

$$
c_j^{\tau}(q) = c_j + \tau e_j, \text{ for } q \in [0, \kappa_j], \ j = 1, \ldots, J.
$$

In this *tax* framework, the dominant strategy on the electricity market is also parametrized by τ as $\mathbf{C^{\tau}} = \{c_j^{\tau}, j = 1, \ldots J\}$ defined in (17). The clearing electricity price and quantities follow as

$$
\begin{aligned}
p^{\mathrm{elec}}(\tau) &= p^{\mathrm{elec}}(\mathbf{C^{\tau}}), \\
(\varphi_1(\tau), \ldots, \varphi_J(\tau)) &= (\varphi_1(\mathbf{C^{\tau}}), \ldots, \varphi_J(\mathbf{C^{\tau}})).
\end{aligned}
\tag{21}
$$

Price $p^{elec}(\tau)$ will be referred to as the *taxed* electricity price, by contrast with price $p^{elec}(\mathbf{A})$ issued from the *marginal production cost strategy* that results from the position $\mathbf{A}$ on the carbon market.

Remark 3. Considering a carbon tax τ and a carbon market strategy $\mathbf{A}$ such that $\tau = p^{CO_2}(\mathbf{A})$, we emphasize the fact that the corresponding electricity prices are not equivalent, but we always have the following inequality

$$p^{elec}(\tau) \leq p^{elec}(\mathbf{A}).$$

This follows from the fact that for all i, $c_i^\tau(\cdot) \leq c_i^A(\cdot)$ and hence $\mathscr{O}(c_i^{\mathbf{A}}; \cdot) \leq \mathscr{O}(c_i^\tau; \cdot)$. The gap between $\mathbf{C}^\tau(\cdot)$ and $\mathbf{C}^A(\cdot)$ comes both from the width (Ω effect) and the height (penalty effect) of their steps.

We start with the following:

Lemma 1. *Under Assumption 4, the map* $\tau \mapsto p^{elec}(\tau)$ *is non-decreasing and right continuous.*

We determine the *willing-to-buy-allowances functions* $\mathscr{W}_j(\cdot)$ and $\mathscr{W}(\cdot)$, as follows:

$$\mathscr{W}_j(\tau) = e_j \varphi_j(\tau) \quad \text{and} \quad \mathscr{W}(\tau) = \sum_{j=1}^{J} \mathscr{W}_j(\tau). \tag{22}$$

For producer j, $\mathscr{W}_j$ is the quantity of emissions it would produce under the penalization τ, and consequently the quantity of allowances it would be ready to buy at price τ. Given the CO_2 value τ, the total amount $\mathscr{W}(\tau)$ represents the allowances needed to cover the global emissions generated by the players who have won electricity market shares. We also define the functions

$$\overline{\mathscr{W}}_j(\tau) = e_j \kappa_j \mathbb{1}_{\{\varphi_j(\tau) > 0\}}, \quad \text{and} \quad \overline{\mathscr{W}}(\tau) = \sum_{j=1}^{J} \overline{\mathscr{W}}_j(\tau). \tag{23}$$

Given that the CO_2 value τ, $\overline{\mathscr{W}}(\tau)$ is the amount of allowances needed by the producers who have won electricity market shares and want to cover their overall production capacity κ_j. Obviously we have

$$\mathscr{W}(\tau) \leq \overline{\mathscr{W}}(\tau), \ \forall \tau \in [0, \mathfrak{p}].$$

Moreover,

Lemma 2. *The function* $\tau \mapsto \mathscr{W}(\tau)$ *is non-increasing:*

$$\mathscr{W}(t') \leq \mathscr{W}(t), \ \forall \ 0 \leq t < t' \leq \mathfrak{p}.$$

The proofs of both Lemma 1 and Lemma 2 can be found in Proofs of Lemmas 1 and 2 Appendix.

3.2.2 Towards an Equilibrium Strategy

The main result of the section is the computation of the bounds of the interval in which the coupled carbon market Nash equilibria prices evolve: we demonstrate that there is no possible deviation enabling a Nash equilibrium carbon price outside this interval. The price bounds are elaborated as specific carbon prices associated to two explicit strategies, build from the *willing-to-buy-allowances* functions: the *Lower price strategy*, and the *Higher price strategy*.

In order to characterize further Nash equilibria candidates, evolving in this price interval, we analyze a third set of strategies that are *intermediate strategies*.

Those strategies rely on our last design assumption which prevents the carbon market from market failure:

$$\mathscr{W}(0) \leq \Omega : \text{ no auction}, \qquad \overline{\mathscr{W}}(\mathfrak{p}) \geq \Omega : \text{ allowances shortage}.$$

Assumption 5 (On the Carbon Market Design). *The available allowances Ω satisfy*

$$\overline{\mathscr{W}}(\mathfrak{p}) < \Omega < \mathscr{W}(0).$$

Moreover, $\mathfrak{p}$ is chosen such that no producer is sidelined from the game: for all j, $\tau \mapsto \overline{\mathscr{W}}_j(\tau)$ is not identically zero on $[0, \mathfrak{p}]$.

Assumption 5 allows to define two prices of particular interest for the game analysis:

$$\tau^{\text{lower}} = \sup\{\tau \in [0, \mathfrak{p}] \text{ s.t. } \mathscr{W}(\tau) > \Omega\}, \tag{24}$$

$$\text{and} \quad \tau^{\text{higher}} = \sup\{\tau \in [0, \mathfrak{p}] \text{ s.t. } \overline{\mathscr{W}}(\tau) > \Omega\}. \tag{25}$$

Observe that we always have $\tau^{\text{lower}} \leq \tau^{\text{higher}}$.

3.2.3 Lower Price Through Lower Price Strategy

Lemma 3. *Consider any strategy $\mathbf{A}^{\mathscr{W}} = (A_1^{\mathscr{W}}, \ldots, A_J^{\mathscr{W}})$ such that*

$$A_j^{\mathscr{W}}(\tau) = \begin{cases} \mathscr{W}_j(\tau^{\text{lower}}), & \text{for } 0 \leq \tau \leq \tau^{\text{lower}}, \\ \text{anything admissible}, & \text{for } \tau > \tau^{\text{lower}}. \end{cases} \tag{26}$$

(i) $p^{CO_2}(\mathbf{A}^{\mathscr{W}}) \geq \tau^{\text{lower}}$.
(ii) *In the case where $p^{CO_2}(\mathbf{A}^{\mathscr{W}}) = \tau^{\text{lower}}$, there is no unilateral favorable deviation that clears the market at a CO_2 price lower than τ^{lower}.*

We call the *Lower price strategy* $(\mathscr{W}_1, \ldots, \mathscr{W}_J)$, as $p^{CO_2}((\mathscr{W}_1, \ldots, \mathscr{W}_J)) = \tau^{\text{lower}}$ by price definitions (10) and (24).

Proof. Point *(i)* is a consequence of the definition of $\tau^{\text{lower}} = \sup\{\tau \in [0,\mathfrak{p}],$ s.t. $\mathscr{W}(\tau) > \Omega\}$. Since $A_j^{\mathscr{W}}(\tau) = \mathscr{W}_j(\tau)$ for $\tau \le \tau^{\text{lower}}$, it follows that $p^{\text{CO}_2}(\mathbf{A}^{\mathscr{W}}) = \sup\{\tau \in [0,\mathfrak{p}], \text{ s.t. } \sum_j A_j^{\mathscr{W}}(\tau) > \Omega\} \ge \tau^{\text{lower}}$.

To prove *(ii)*, first note that, since we assume $p^{\text{CO}_2}(\mathbf{A}^{\mathscr{W}}) = \tau^{\text{lower}}$, we have $\varphi_j(\mathbf{A}^{\mathscr{W}}) \le \varphi_j(\tau^{\text{lower}}) = \frac{1}{e_j}\mathscr{W}_j(\tau^{\text{lower}})$. Indeed, the carbon market clearing can decrease the global function $\mathscr{O}(\mathbf{C}^{\tau^{\text{lower}}}; \cdot)$ to $\mathscr{O}(\mathbf{A}^{\mathscr{W}}; \cdot)$, but the demand function stay unchanged. So, we still have $\varphi_j(\mathbf{A}^{\mathscr{W}}) = \frac{1}{e_j}\delta_j(\mathbf{A}^{\mathscr{W}})$.

Suppose one producer, say Producer 1, deviates and chooses $\tilde{A}_1(\cdot)$ instead of $A_1^{\mathscr{W}}(\cdot)$. Suppose the new carbon price $\tilde{\tau} := p^{\text{CO}_2}(\mathbf{A}_{-1}^{\mathscr{W}}, \tilde{A}_1) < \tau^{\text{lower}}$. Since $A_j^{\mathscr{W}}(\tilde{\tau}^+) = A_j^{\mathscr{W}}(\tilde{\tau})$ for $j \ne 1$, necessarily we have $\tilde{A}_1(\tilde{\tau}^+) \le \tilde{A}_1(\tilde{\tau})$, by definition of $\tilde{\tau}$. Then $\Delta(\tilde{A}_1) \ge 0$ and it follows that $\delta_1(\mathbf{A}_{-1}^{\mathscr{W}}, \tilde{A}_1) \le \delta_1(\mathbf{A}^{\mathscr{W}})$, but $\delta_j(\mathbf{A}_{-1}^{\mathscr{W}}, \tilde{A}_1) \ge \delta_j(\mathbf{A}^{\mathscr{W}})$ for the others $j \ne 1$.

If $p^{\text{elec}}(\mathbf{A}_{-1}^{\mathscr{W}}, \tilde{A}_1) \ge p^{\text{elec}}(\mathbf{A}^{\mathscr{W}})$, the others $j \ne 1$ produce at least electricity for the allowances they have, $\varphi_j(\mathbf{A}_{-1}^{\mathscr{W}}, \tilde{A}_1) \ge \varphi_j(\mathbf{A}^{\mathscr{W}})$. Since the demand is decreasing we have $\varphi_1(\mathbf{A}_{-1}^{\mathscr{W}}, \tilde{A}_1) \le \varphi_1(\mathbf{A}^{\mathscr{W}})$.

Now, if $p^{\text{elec}}(\mathbf{A}_{-1}^{\mathscr{W}}, \tilde{A}_1) < p^{\text{elec}}(\mathbf{A}^{\mathscr{W}})$, the offer of Producer 1 based on his penalized marginal production cost is also greater than $p^{\text{elec}}(\mathbf{A}_{-1}^{\mathscr{W}}, \tilde{A}_1)$. Then $\varphi_1(\mathbf{A}_{-1}^{\mathscr{W}}, \tilde{A}_1) \le \frac{1}{e_1}\delta_1(\mathbf{A}_{-1}^{\mathscr{W}}, \tilde{A}_1) \le \varphi_1(\mathbf{A}^{\mathscr{W}})$. $\square$

Lemma 4. *Suppose* $\mathbf{A}$ *is such that* $p^{\text{CO}_2}(\mathbf{A}) < \tau^{\text{lower}}$. *Then* $\mathbf{A}$ *is not a Nash equilibrium.*

Proof. To prove this lemma we exhibit an unilateral favorable deviation of a producer.

(*a*) Assume first that at least one producer exists, say Producer 1, such that $\varphi_1(\mathbf{A}) < \kappa_1$ and there exists a tax value $\hat{\tau}_1$ such that $p^{\text{CO}_2}(\mathbf{A}) < \hat{\tau}_1 \le \tau^{\text{lower}}$ and, $\mathscr{W}_1(\tau) = e_1\kappa_1$ for any $\tau \in [p^{\text{CO}_2}(\mathbf{A}), \hat{\tau}_1]$.

This means that Producer 1 may sell κ_1, for any tax level τ in $[p^{\text{CO}_2}(\mathbf{A}), \hat{\tau}_1]$, and consequently we have $c_1 + \tau e_1 < p^{\text{elec}}(\tau)$ for τ in $[p^{\text{CO}_2}(\mathbf{A}), \hat{\tau}_1]$.

Consider a deviation $\tilde{A}_1$ of player 1, such that the resulting clearing price on CO_2 market, $p^{\text{CO}_2}(\mathbf{A}_{-1}, \tilde{A}_1) \in [p^{\text{CO}_2}(\mathbf{A}), \hat{\tau}_1]$. From Remark 3, we have

$$C_1 + \tau e_1 \le p^{\text{elec}}(p^{\text{CO}_2}(\mathbf{A}_{-1}, \tilde{A}_1)) \le p^{\text{elec}}(\mathbf{A}_{-1}, \tilde{A}_1).$$

This means that Producer 1 may sell its overall covered capacity: $\varphi_1(\mathbf{A}_{-1}, \tilde{A}_1) = \frac{1}{e_1}\delta_1(\mathbf{A}_{-1}, \tilde{A}_1)$.

Now we define $\tau \mapsto \tilde{A}_1(\tau)$ as follows, for $\varepsilon > 0$ arbitrarily small and $p^{\text{CO}_2}(\mathbf{A}) \le \tau$,

$$\tilde{A}_1(p^{CO_2}(\mathbf{A})) = e_1\kappa_1,$$

$$\tilde{A}_1(\tau) = \left(\Omega - \sum_{j>1} A_j(\tau) - \varepsilon\right) \mathbb{1}_{\{\sum_{\substack{j\neq 1}} A_j(\tau) + \delta_1(\mathbf{A}) \geq \Omega\}}$$

$$+ e_1\kappa_1 \mathbb{1}_{\{\sum_{\substack{j\neq 1}} A_j(\tau) + \delta_1(\mathbf{A}) < \Omega\}} \qquad \textbf{for } \tau \in (p^{CO_2}(\mathbf{A}), \hat{\tau}_1]$$

$$= A_1(\tau), \qquad\qquad\qquad \textbf{for } \tau > \hat{\tau}_1.$$

Note that $\tilde{A}_1(\tau) \geq A_1(\tau)$ for $p^{CO_2}(\mathbf{A}) \leq \tau \leq \hat{\tau}_1$, and consequently $p^{CO_2}(\mathbf{A}_{-1}, \tilde{A}_1) \geq p^{CO_2}(\mathbf{A})$.

If $p^{CO_2}(\mathbf{A}_{-1}, \tilde{A}_1) > p^{CO_2}(\mathbf{A})$, then $e_1\varphi_1(\mathbf{A}_{-1}, \tilde{A}_1)) = \delta(\mathbf{A}_{-1}, \tilde{A}_1) > \delta(\mathbf{A}) \geq e_1\varphi_1(\mathbf{A})$, and we get our favorable deviation.

If $p^{CO_2}(\mathbf{A}_{-1}, \tilde{A}_1) = p^{CO_2}(\mathbf{A})$, we observe that when $\Delta(A_1) \geq 0$, we also have $\Delta(\tilde{A}_1) = 0$. Then by the CO_2 market clearing mechanism, Producer 1 gets $e_1\kappa_1$ allowances instead of $\delta(\mathbf{A})$ and strictly improves its electricity market share. when $\Delta(A_1) < 0$, we have $\tilde{A}_1(p^{CO_2}(\mathbf{A})^+) > A_1(p^{CO_2}(\mathbf{A})^+)$, that also insures that Producer 1 increases $\delta(\mathbf{A}_{-1}, \tilde{A}_1) > \delta(\mathbf{A})$ (see (11)).

(b) Assume now that all producers are either such that $\varphi_j(\mathbf{A}) = \kappa_j$ or such that $\varphi_j(\mathbf{A}) < \kappa_j$ and $\mathcal{W}_j(p^{CO_2}(\mathbf{A})^+) < e_j\kappa_j$. Among the second category, there exists at least one producer (say Producer 1) such that $\varphi_1(\mathbf{A}) < \varphi_1(p^{CO_2}(\mathbf{A}))$ with $\varphi_1(p^{CO_2}(\mathbf{A})) > 0$ (unless to contradict $p^{CO_2}(\mathbf{A}) < \tau^{lower}$). Here we have used the notation (20) and (21). $\mathcal{W}_1(p^{CO_2}(\mathbf{A})^+) < e_1\kappa_1$ means that $c_1 + e_1 p^{CO_2}(\mathbf{A}) = p^{elec}(p^{CO_2}(\mathbf{A}))$ (as $p^{elec}(\cdot)$ is right-continuous).

A strictly favorable deviation $\tilde{A}_1$ of Producer 1, thus consists in increasing its ask at the price $p^{CO_2}(\mathbf{A})^+$, in order to increase its $\delta(\mathbf{A}_{-1}, \tilde{A}_1)$ (see (11)):

$$\tilde{A}_1(\tau) = \left(\Omega - \sum_{j>1} A_j(\tau) - \varepsilon\right) \mathbb{1}_{\{p^{CO_2}(\mathbf{A}) < \tau\}} + e_1\kappa_1 \mathbb{1}_{\{p^{CO_2}(\mathbf{A}) = \tau\}}.$$

Then $p^{CO_2}(\mathbf{A}_{-1}, \tilde{A}_1) = p^{CO_2}(\mathbf{A})$, $\tilde{A}_1(p^{CO_2}(\mathbf{A})) \geq A_1(p^{CO_2}(\mathbf{A}))$, but $\tilde{A}_1(p^{CO_2}(\mathbf{A})^+) > A_1(p^{CO_2}(\mathbf{A})^+)$, for ε sufficiently small. This last inequality guarantees that $\delta_1(\mathbf{A}_{-1}, \tilde{A}_1) > \delta_1(\mathbf{A})$ and finally $\varphi_1(p^{CO_2}(\mathbf{A})) \geq \varphi_1(\mathbf{A}_{-1}, \tilde{A}_1) > \varphi_1(\mathbf{A})$.   □

3.2.4 Higher Price Through Higher Price Strategy

Lemma 5. *Consider any strategy* $\mathbf{A}^{\overline{\mathcal{W}}} = (A_1^{\overline{\mathcal{W}}}, \cdots, A_J^{\overline{\mathcal{W}}})$ *such that*

$$A_j^{\overline{\mathcal{W}}}(\tau) = \begin{cases} \text{anything admissible,} & \text{for } \tau \leq \tau^{higher}, \\ \overline{\mathcal{W}}_j(\tau), & \text{for } \tau > \tau^{higher}. \end{cases} \qquad (27)$$

(i) $p^{CO_2}(\mathbf{A}^{\overline{\mathscr{W}}}) \leq \tau^{\text{higher}}$.

(ii) *There is no unilateral favorable deviation that clears the market at a CO_2 price higher than τ^{higher}.*

We call the *Higher price strategy* $(\overline{\mathscr{W}}_1, \ldots, \overline{\mathscr{W}}_J)$, as $p^{CO_2}((\overline{\mathscr{W}}_1, \ldots, \overline{\mathscr{W}}_J)) = \tau^{\text{higher}}$ by price definitions (10) and (25).

Proof. Point *(i)* follows directly from the definition of τ^{higher}.

To prove *(ii)*, suppose one producer, say Producer 1, chooses its strategy $\tilde{A}_1(\cdot)$ instead of $A_1^{\overline{\mathscr{W}}}(\cdot)$, and that the resulting CO_2 price is $\tilde{\tau} := p^{CO_2}(\mathbf{A}_{-1}^{\overline{\mathscr{W}}}, \tilde{A}_1) > \tau^{\text{higher}}$. Necessarily, due to the definition of $\mathbf{A}^{\overline{\mathscr{W}}}$, this means that $\overline{\mathscr{W}}_1(\tilde{\tau}) = 0$, which in turn means that $c_1 + \tilde{\tau}e_1 > p^{\text{elec}}(\tilde{\tau})$. To conclude, it is sufficient to notice that any Producer $j \neq 1$ obtains what he asks for, i.e. $\delta_j(\mathbf{A}_{-1}^{\overline{\mathscr{W}}}, \tilde{A}_1) = \overline{\mathscr{W}}_j(\tilde{\tau}^+)$, from which it follows that the *coupled* electricity price equals the *taxed* electricity price: $p^{\text{elec}}(\mathbf{A}_{-1}^{\overline{\mathscr{W}}}, \tilde{A}_1) = p^{\text{elec}}(\tilde{\tau})$, and then $\varphi_1(\mathbf{A}_{-1}^{\overline{\mathscr{W}}}, \tilde{A}_1) = \overline{\mathscr{W}}_i(\tilde{\tau}) = 0$ and the deviation of Producer 1 is not favorable.   □

A strategy $\mathbf{A}$ is said to be *effective* if all the producers that bought some allowances produce some electricity:

$$\forall j, \; \delta_j(\mathbf{A}) > 0 \Rightarrow \varphi_j(p^{CO_2}(\mathbf{A})) > 0.$$

Lemma 6.

(i) *Let $\mathbf{A}$ admissible such that $p^{CO_2}(\mathbf{A}) > \tau^{\text{higher}}$. Then $\mathbf{A}$ is not an effective strategy.*

(ii) *Let $\mathbf{A}$ admissible such that $p^{CO_2}(\mathbf{A}) > \tau^{\text{higher}}$. Then $\mathbf{A}$ is not a strong Nash equilibrium.*

As a consequence of this lemma, if a producer (or a set of producers) that does not produce electricity, tries to block the auction game of the carbon market by buying all the allowances he can, then there always exists a coalition with favorable deviation.

Proof.

(i). Effective means that for all producers such that $\delta_j(\mathbf{A}) > 0$, we have $\overline{\mathscr{W}}_j(p^{CO_2}(\mathbf{A})) = e_j\kappa_j$, which is clearly in contradiction with the definition of τ^{higher}.

(ii). Given $\mathbf{A}$, such that $p^{CO_2}(\mathbf{A}) > \tau^{\text{higher}}$, we consider the coalition of producers $\mathscr{K}$ such that for $j \in \mathscr{K}$, $\overline{\mathscr{W}}_j(p^{CO_2}(\mathbf{A})) = 0$. $\mathscr{K}$ is clearly non-empty by definition of τ^{higher}. Consider the following cooperating deviation of $\mathscr{K}$:

$$\tilde{A}_j(\cdot) = A_j^{\overline{\mathscr{W}}}(\cdot), \quad \text{for } j \in \mathscr{K}.$$

Then $p^{CO_2}(\mathbf{A}_{-\mathscr{K}}, \tilde{A}_{\mathscr{K}}) < p^{CO_2}(\mathbf{A})$, and at least for one member of the coalition $\mathscr{K}$, $\delta_j(\mathbf{A}_{-\mathscr{K}}, \tilde{A}_{\mathscr{K}}) > 0$ when $\overline{\mathscr{W}}_j(p^{CO_2}(\mathbf{A}_{-\mathscr{K}}, \tilde{A}_{\mathscr{K}})) > 0$. This means that $\varphi_j(\mathbf{A}_{-\mathscr{K}}, \tilde{A}_{\mathscr{K}}) > 0$ which is a strictly favorable deviation of j, whereas the situation is unchanged for the others in $\mathcal{K}$ that still produce nothing. Thus, we exhibit a coalition that allows a deviation from **A** that benefits to all of its members, and that benefits strictly to at least one. Then **A** is not a strong Nash equilibrium. □

3.2.5 Price Interval

From Lemmas 4 and 6, we have the following:

Corollary 1. *If* **A** *is a strong Nash equilibrium, or if it is an effective Nash equilibrium, then* $p^{CO_2}(\mathbf{A}) \in [\tau^{\text{lower}}, \tau^{\text{higher}}]$.

The interval in which the coupled carbon market Nash equilibria prices evolve is then $[\tau^{\text{lower}}, \tau^{\text{higher}}]$. This price range is generated by the existing gap between the functions $\mathscr{W}(\cdot)$ and $\overline{\mathscr{W}}(\cdot)$.

Thus a condition for a single unique carbon price is that this gap shrinks to zero: the equality between the two *willing-to-buy-allowances functionals* occurs i.e. for any value τ, and any producer i, the allowances needed to cover the global emissions generated by a player who has won electricity market shares and the allowances needed by a producer who has won electricity market shares and wants to cover its overall production capacity are the same. Clearly, this is very unlikely to happen.

It is worth of mentioning that the same lemmas apply when producers have an electricity production power plants portfolio, or when one modifies the maximal cap $\mathscr{E}_j$ of allowances that each producer j may buy while one redefines $\tau \mapsto \overline{\mathscr{W}}_j(\tau)$ by

$$\overline{\mathscr{W}}_j(\tau) = \mathscr{E}_j \mathbb{1}_{\{\varphi_j(\tau) > 0\}}.$$

Note that if one increases the maximal cap, τ^{higher} increases.

3.2.6 Intermediate Strategies

Consider any strategy profile $\mathbf{B} = (B_1, \cdots, B_J)$ satisfying the following:

$$B_j(\tau) = \begin{cases} \mathscr{W}_j(\tau^{\text{lower}}), & \text{for } \tau \leq \tau^{\text{lower}}, \\ \text{anything admissible}, & \text{for } \tau^{\text{lower}} < \tau \leq \tau^{\text{higher}}, \\ \overline{\mathscr{W}}_j(\tau), & \text{for } \tau > \tau^{\text{higher}}. \end{cases} \tag{28}$$

This is not in general an equilibrium, nevertheless we have the following properties:

Lemma 7.

(i) $p^{CO_2}(\mathbf{B}) \in [\tau^{\text{lower}}, \tau^{\text{higher}}]$.

(ii) *If there exists a favorable deviation from a producer, say Producer 1, that chooses $\tilde{B}_1$ instead of B_1, such that $p^{CO_2}(\mathbf{B}_{-1}, \tilde{B}_1) < \tau^{\text{lower}}$, then there exists* *another favorable deviation* $\hat{B}_1$ *defined by*

$$\hat{B}_1 = \begin{cases} \tilde{B}_1(\tau), & \text{for } \tau > \tau^{\text{lower}}, \\ \mathcal{W}_1(\tau^{\text{lower}}), & \text{for } \tau \leq \tau^{\text{lower}} \end{cases}$$

such that $p^{CO_2}(\mathbf{B}_{-1}; \hat{B}_1) = \tau^{\text{lower}}$, *and such that* $\varphi_1(\mathbf{B}_{-1}, \hat{B}_1) \geq \varphi_1(\mathbf{B}_{-1}, \tilde{B}_1)$.

Proof. Point *(i)* follows directly from Lemma 3–*(i)* and Lemma 5–*(i)*.

To prove *(ii)*, we first observe that, as producers $j \neq 1$ are served first on the carbon market,

$$\delta_1(\mathbf{B}_{-1}; \tilde{B}_1) = \Omega - \sum_{j \neq 1} \mathcal{W}_j(\tau^{\text{lower}}).$$

Moreover, we have $p^{CO_2}(\mathbf{B}_{-1}, \hat{B}_1) = \tau^{\text{lower}}$, and from the CO_2 market mechanism it follows that

$$\delta_1(\mathbf{B}_{-1}, \hat{B}_1) \geq \delta_1(\mathbf{B}_{-1}, \tilde{B}_1).$$

Since $\tilde{B}_j(p^{CO_2}(\mathbf{B}_{-1}, \tilde{B}_1)) = \tilde{B}_j(p^{CO_2}(\mathbf{B}_{-1}, \tilde{B}_1)^+) = \mathcal{W}_j(\tau^{\text{lower}})$ for any $j \neq 1$, it follows that $\delta_1(\mathbf{B}_{-1}, \tilde{B}_1) = \Omega - \sum_{j \neq 1} \mathcal{W}_j(\tau^{\text{lower}})$. Indeed, for strategy $(\mathbf{B}_{-1}, \hat{B}_1)$, the producers $j \neq 1$ such that $B_j(\tau^{\text{lower}+}) < \mathcal{W}_j(\tau^{\text{lower}})$ receive a quantity of quotas $\delta_j(\mathbf{B}_{-1}, \hat{B}_1) \leq \mathcal{W}_j(\tau^{\text{lower}})$, from which $\delta_1(\mathbf{B}_{-1}, \hat{B}_1) = \Omega - \sum_j \delta_j(\mathbf{B}_{-1}, \hat{B}_1) \geq \delta_1(\mathbf{B}_{-1}, \hat{B}_1)$. We also deduce that $\varphi_1(\mathbf{B}_{-1}, \hat{B}_1) = \frac{1}{e_1}\delta_1(\mathbf{B}_{-1}, \hat{B}_1)$. To conclude, it is sufficient to notice that $\varphi_1(\mathbf{B}_{-1}, \hat{B}_1) = \frac{1}{e_1}\delta_1(\mathbf{B}_{-1}, \hat{B}_1) \geq \frac{1}{e_1}\delta_1(\mathbf{B}_{-1}, \tilde{B}_1) \geq \varphi_1(\mathbf{B}_{-1}, \tilde{B}_1)$. $\qquad\square$

The following aims to characterize the form of effective Nash equilibria.

Corollary 2. *Let* $\mathbf{E}$ *be an effective Nash equilibrium (i.e* $p^{CO_2}(\mathbf{E}) \leq \tau^{\text{higher}}$*). Then the following* $\mathbf{E}'$ *is also an effective Nash equilibrium:*

$$E'_j(\tau) = \begin{cases} \mathcal{W}_j(\tau^{\text{lower}}), & \text{for } \tau \leq \tau^{\text{lower}}, \\ E_j(\tau), & \text{for } \tau^{\text{lower}} < \tau \leq \tau^{\text{higher}}, \\ \overline{\mathcal{W}}_j(\tau), & \text{for } \tau > \tau^{\text{higher}}. \end{cases} \tag{29}$$

Proof. From Lemmas 3 and 5, $p^{CO_2}(\mathbf{E}) \in [\tau^{\text{lower}}, \tau^{\text{higher}}]$. Consider a deviation that produces a bigger carbon price: Producer 1 deviates from E'_1 to $\tilde{E}'_1$ with $p^{CO_2}(\tilde{E}'_1, \mathbf{E}'_{-1}) > \tau^{\text{higher}}$. Then by definition of τ^{higher}, $\varphi_1(\tilde{E}'_1, \mathbf{E}'_{-1}) = 0$. Indeed, a deviation to this bigger price is possible only if $\overline{\mathcal{W}}_1(p^{CO_2}((\tilde{E}'_1, \mathbf{E}'_{-1})) = 0$.

Now if Producer 1 deviates from E'_1 to $\tilde{E}'_1$ with $p^{CO_2}((\tilde{E}'_1, \mathbf{E}'_{-1})) < \tau^{\text{lower}}$, and if we assume that this deviation is strictly favorable: $\varphi_1(\tilde{E}'_1, \mathbf{E}'_{-1}) > \varphi_1(\mathbf{E}')$. Then according to Lemma 7, we consider $\hat{E}'_1$ that gives $p^{CO_2}(\hat{E}_1, \mathbf{E}'_{-1}) = \tau^{\text{lower}}$. And we still have that $\varphi_1(\hat{E}'_1, \mathbf{E}'_{-1}) > \varphi_1(\mathbf{E}')$. But the deviation $(\hat{E}'_1, \mathbf{E}_{-1})$ from $\mathbf{E}$ produces the same price and shares than $(\hat{E}'_1, \mathbf{E}'_{-1})$). Since we also have $\varphi_1(\mathbf{E}') = \varphi_1(\mathbf{E})$, we get a strictly favorable deviation to $\mathbf{E}$ which gives the contradiction.

Same arguments apply when Producer 1 deviates from E'_1 to $\tilde{E}'_1$ with $p^{CO_2}(\tilde{E}'_1)$ in $[\tau^{lower}, \tau^{higher}]$. □

4 Conclusion

Once CO_2 is emitted into the atmosphere, it remains there for more than a century. Estimating its value is an essential indicator for efficiently defining policy. Carbon valuation is crucial for designing markets that foster emission reductions. In this paper, we established the links between an electricity market and a carbon auction market through an analysis of electricity producers' strategies. We proved that they lead to the interval where relevant Nash equilibria evolve, enabling the computation of equilibrium prices on both markets. For each producer, each equilibrium derives the level of electricity produced and the CO_2 emissions covered.

For a given design and set of players, the information provided by the interval may be interpreted as a diagnosis of market behavior in terms of prices and volume. Indeed, it enables the computation of the CO_2 emissions actually released, and opens the discussion of a relevant carbon market in terms of mitigation issues.

In addition to this analysis of the Nash equilibrium we plan to analyze the electricity production mix, with a particular focus on renewable shares that do not participate in emissions.

Acknowledgements This work was partly supported by Grant 0805C0098 from ADEME.

Appendix

Proof of Proposition 1

A. First We Prove the Dominance Property *(i)*

Suppose that one producer, let us say producer 1, deviates and chooses C_1 instead of s_1. We have to show that its market share cannot be reduced by this deviation. By definition of the admissibility (see (2)) we have

$$s_1(q) \geq C_1(q), \forall q \in [0, \kappa_1].$$

Hence the offer functions defined by (3) satisfy $\mathscr{O}(s_1; \cdot) \leq \mathscr{O}(C_1; \cdot)$. By adding the unchanged offers of the other producers

$$\mathcal{O}((\mathbf{s}_{-1}, s_1); \cdot) \le \mathcal{O}((\mathbf{s}_{-1}, C_1); \cdot), \tag{A.1}$$

where $(\mathbf{s}_{-1}, C_1)$ denotes the strategy profile that includes producer 1 deviation. The minimum market clearing price (5) for strategy profile $\mathbf{s}$ is

$$\underline{p}(\mathbf{s}) = \inf\{p, \ \mathcal{O}(\mathbf{s}; p) > D(p)\}.$$

The minimum market clearing price (5) for strategy profile $(\mathbf{s}_{-1}, C_1)$ is

$$\underline{p}(\mathbf{s}_{-1}, C_1) = \inf\{p, \ \mathcal{O}((\mathbf{s}_{-1}, C_1); p) > D(p)\}.$$

The inequality (A.1) together with the fact that the demand $D(\cdot)$ is a non-increasing function imply that $\underline{p}(\mathbf{s}_{-1}, C_1) \le \underline{p}(\mathbf{s})$, from which, with (6) we deduce that

$$p^{\text{elec}}(\mathbf{s}_{-1}, C_1) \le p^{\text{elec}}(\mathbf{s}).$$

Now let us show that Producer 1 does not reduce its market share by deviating from $s_1(\cdot)$ to $C_1(\cdot)$, that is $\varphi_1(\mathbf{s}_{-1}, C_1) \ge \varphi_1(\mathbf{s})$.

For the sake of clarity we adopt, in this paragraph, the following notation:

$$\begin{aligned} \underline{p}_{\mathbf{s}} &:= \underline{p}(\mathbf{s}) \\ p_{\mathbf{s}}^{\text{elec}} &:= p^{\text{elec}}(\mathbf{s}) \end{aligned} \quad \text{and} \quad \begin{aligned} \underline{p}_{\mathbf{s}C} &:= \underline{p}(\mathbf{s}_{-1}, C_1) \\ p_{\mathbf{s}C}^{\text{elec}} &:= p^{\text{elec}}((\mathbf{s}_{-1}, C_1)). \end{aligned}$$

We first consider the case where $\underline{p}_{\mathbf{s}C} < \underline{p}_{\mathbf{s}}$. By definition of the minimum clearing price $\underline{p}_{\mathbf{s}C}$, the fact that $D(\underline{p}_{\mathbf{s}}) \le D(\underline{p}_{\mathbf{s}C})$ and the fact that $\mathcal{O}((\mathbf{s}_{-1}, C_1); \cdot)$ is non-decreasing, we have

$$D(p_{\mathbf{s}}^{\text{elec}}) \le D(\underline{p}_{\mathbf{s}}) \le D(\underline{p}_{\mathbf{s}C}) \le \mathcal{O}((\mathbf{s}_{-1}, C_1); \underline{p}_{\mathbf{s}C}) \le \mathcal{O}((\mathbf{s}_{-1}, C_1); p_{\mathbf{s}C}^{\text{elec}}).$$

Hence,

$$\mathcal{O}((\mathbf{s}_{-1}, s_1), p_{\mathbf{s}}^{\text{elec}}) \wedge D(p_{\mathbf{s}}^{\text{elec}}) \le \mathcal{O}((\mathbf{s}_{-1}, C_1); p_{\mathbf{s}C}^{\text{elec}}) \wedge D(p_{\mathbf{s}C}^{\text{elec}}),$$

$$\mathcal{O}((\mathbf{s}_{-1}, s_1), \underline{p}_{\mathbf{s}}) \wedge D(\underline{p}_{\mathbf{s}}) \le \mathcal{O}((\mathbf{s}_{-1}, C_1); \underline{p}_{\mathbf{s}C}) \wedge D(\underline{p}_{\mathbf{s}C}).$$

From the market clearing (8) we get

$$\begin{aligned} &\varphi_1(\mathbf{s}_{-1}, s_1) - \varphi_1(\mathbf{s}_{-1}, C_1) \\ &= \mathcal{O}((\mathbf{s}_{-1}, s_1), p_{\mathbf{s}}^{\text{elec}}) \wedge D(p_{\mathbf{s}}^{\text{elec}}) - \mathcal{O}((\mathbf{s}_{-1}, C_1); p_{\mathbf{s}C}^{\text{elec}}) \wedge D(p_{\mathbf{s}C}^{\text{elec}}) \\ &\quad + \sum_{j>1} \left(\varphi_j(\mathbf{s}_{-1}, C_1) - \varphi_j(\mathbf{s}_{-1}, s_1) \right). \end{aligned}$$

According to Definition 2, let us denote

$$\mathcal{E}(\underline{p}_{\mathbf{s}}) = \left\{ j \in \{2, \ldots, J\} \text{ s.t. } \Delta^- \mathcal{O}(s_j; \underline{p}_{\mathbf{s}}) > 0 \right\}.$$

We have

$$\varphi_1(\mathbf{s}_{-1}, s_1) - \varphi_1(\mathbf{s}_{-1}, C_1)$$

$$= \mathcal{O}((\mathbf{s}_{-1}, s_1); p_\mathbf{s}^{\text{elec}}) \wedge D(p_\mathbf{s}^{\text{elec}}) - \mathcal{O}((\mathbf{s}_{-1}, C_1); p_{\mathbf{s}C}^{\text{elec}}) \wedge D(p_{\mathbf{s}C}^{\text{elec}})$$

$$+ \sum_{j>1, j \notin \mathscr{E}(\underline{p}_\mathbf{s})} \left(\varphi_j(\mathbf{s}_{-1}, C_1) - \mathcal{O}(s_j; p_\mathbf{s}^{\text{elec}}) \right)$$

$$+ \sum_{j>1, j \in \mathscr{E}(\underline{p}_\mathbf{s})} \left(\varphi_j(\mathbf{s}_{-1}, C_1) - \varphi_j(\mathbf{s}_{-1}, s_1) \right)$$

$$\leq \mathcal{O}((\mathbf{s}_{-1}, s_1); p_\mathbf{s}^{\text{elec}}) \wedge D(p_\mathbf{s}^{\text{elec}}) - \mathcal{O}((\mathbf{s}_{-1}, C_1); p_{\mathbf{s}C}^{\text{elec}}) \wedge D(p_{\mathbf{s}C}^{\text{elec}})$$

$$+ \sum_{j>1, j \notin \mathscr{E}(\underline{p}_\mathbf{s})} \left(\mathcal{O}(s_j; p_{\mathbf{s}C}^{\text{elec}}) - \mathcal{O}(s_j; p_\mathbf{s}^{\text{elec}}) \right)$$

$$+ \sum_{j>1, j \in \mathscr{E}(\underline{p}_\mathbf{s})} \left(\varphi_j(\mathbf{s}_{-1}, C_1) - \varphi_j(\mathbf{s}_{-1}, s_1) \right)$$

Since $p_{\mathbf{s}C}^{\text{elec}} \leq p_\mathbf{s}^{\text{elec}}$ we get

$$\varphi_1(\mathbf{s}_{-1}, s_1) - \varphi_1(\mathbf{s}_{-1}, C_1) \leq \sum_{j>1, j \in \mathscr{E}(\underline{p}_\mathbf{s})} \left(\varphi_j(\mathbf{s}_{-1}, C_1) - \varphi_j(\mathbf{s}_{-1}, s_1) \right).$$

But for any $j \in \mathscr{E}(\underline{p}_\mathbf{s})$, the quantity $\mathcal{O}(s_j; \underline{p}_\mathbf{s}^-) \leq \varphi_j(\mathbf{s}_{-1}, s_1)$. As $\mathcal{O}(\mathbf{s}_j; \cdot)$ is non-decreasing a since we have assumed $\underline{p}_{\mathbf{s}C} < \underline{p}_\mathbf{s}$, we get

$$\mathcal{O}(s_j; \underline{p}_{\mathbf{s}C}^-) \leq \mathcal{O}(s_j; \underline{p}_\mathbf{s}^-) \leq \varphi_j(\mathbf{s}_{-1}, s_1).$$

For such $j > 1$, we thus have

$$\varphi_j(\mathbf{s}_{-1}, C_1) - \varphi_j(\mathbf{s}_{-1}, s_1) \leq \varphi_j(\mathbf{s}_{-1}, C_1) - \mathcal{O}(s_j; p_\mathbf{s}^{\text{elec}^-})$$

$$\leq \varphi_j(\mathbf{s}_{-1}, C_1) - \mathcal{O}(s_j; p_{\mathbf{s}C}^{\text{elec}}) \leq 0,$$

from which it follows that $\varphi_1(\mathbf{s}_{-1}, s_1) - \varphi_1(\mathbf{s}_{-1}, C_1) \leq 0$.

Now consider the case where $\underline{p}_\mathbf{s} = \underline{p}_{\mathbf{s}C} := \underline{p}$. Due to the market rule (6), we necessarily have $p_\mathbf{s}^{\text{elec}} = p_{\mathbf{s}C}^{\text{elec}} := p^{\text{elec}}$.

- If $\mathcal{O}((\mathbf{s}_{-1}, s_1); p^{\text{elec}}) \leq \mathcal{O}((\mathbf{s}_{-1}, C_1); p^{\text{elec}}) \leq D(p^{\text{elec}})$, then by the market clearing

$$\varphi_1(\mathbf{s}_{-1}, s_1) = \mathcal{O}(s_1; p^{\text{elec}}) \leq \mathcal{O}(C_1; p^{\text{elec}}) = \varphi_1(\mathbf{s}_{-1}, C_1).$$

- If $\mathcal{O}((\mathbf{s}_{-1}, s_1); p^{\text{elec}}) \leq D(p^{\text{elec}}) \leq \mathcal{O}((\mathbf{s}_{-1}, C_1); p^{\text{elec}})$, then

$$\varphi_1(\mathbf{s}_{-1}, s_1) = \mathcal{O}(s_1; p^{\text{elec}}) \leq D(p^{\text{elec}}) - \sum_{j>1} \varphi_j(\mathbf{s}_{-1}, s_1) = D(p^{\text{elec}}) - \sum_{j>1} \mathcal{O}(s_j; p^{\text{elec}})$$

$$\leq D(p^{\text{elec}}) - \sum_{j>1} \varphi_j(\mathbf{s}_{-1}, C_1) = \varphi_1(\mathbf{s}_{-1}, C_1).$$

- If $D(p^{\text{elec}}) < \boldsymbol{\mathcal{O}}((\mathbf{s}_{-1}, s_1); p^{\text{elec}}) \leq \boldsymbol{\mathcal{O}}((\mathbf{s}_{-1}, C_1); p^{\text{elec}})$, by the market clearing we get

$$\varphi_1(\mathbf{s}_{-1}, s_1) - \varphi_1(\mathbf{s}_{-1}, C_1)$$
$$= \boldsymbol{\mathcal{O}}((\mathbf{s}_{-1}, s_1), p^{\text{elec}}) \wedge D(p^{\text{elec}}) - \boldsymbol{\mathcal{O}}((\mathbf{s}_{-1}, C_1); p^{\text{elec}}) \wedge D(p^{\text{elec}})$$
$$+ \sum_{j>1}(\varphi_j(\mathbf{s}_{-1}, C_1) - \varphi_j(\mathbf{s}_{-1}, s_1))$$
$$\leq \sum_{j>1}(\varphi_j(\mathbf{s}_{-1}, C_1) - \varphi_j(\mathbf{s}_{-1}, s_1))$$
$$\leq \sum_{j>1, j \in \mathcal{E}(\underline{p})}(\varphi_j(\mathbf{s}_{-1}, C_1) - \varphi_j(\mathbf{s}_{-1}, s_1)).$$

From (7), we have for $j \in \mathcal{E}(\underline{p})$

$$\varphi_j(\mathbf{s}_{-1}, s_1) = \mathcal{O}(s_j, p^-) + \Delta^- \mathcal{O}(s_j; \underline{p}) \frac{(D(\underline{p}) - \boldsymbol{\mathcal{O}}((\mathbf{s}_{-1}, s_1), p^-))}{\Delta^- \boldsymbol{\mathcal{O}}((\mathbf{s}_{-1}, s_1), \underline{p})}$$

and $\quad \varphi_j(\mathbf{s}_{-1}, C_1) = \mathcal{O}(s_j; \underline{p}^-) + \Delta^- \mathcal{O}(s_j; \underline{p}) \dfrac{(D(\underline{p}) - \boldsymbol{\mathcal{O}}((\mathbf{s}_{-1}, C_1); \underline{p}^-))}{\Delta^- \boldsymbol{\mathcal{O}}((\mathbf{s}_{-1}, C_1); \underline{p})}.$

Hence, if $\mathcal{E}(\underline{p})$ is non empty then at least one producer exists, $j \neq 1$ such that $\Delta^- \mathcal{O}(s_j; \underline{p}) > 0$. and from the desegregation of $\boldsymbol{\mathcal{O}}$ and definition of Δ^- it results that

$$\varphi_1(\mathbf{s}_{-1}, s_1) - \varphi_1(\mathbf{s}_{-1}, C_1)$$
$$= \sum_{j>1, j \in \mathcal{E}(\underline{p})} \Delta^- \mathcal{O}(s_j, \underline{p}) \left(\frac{(D(\underline{p}) - \boldsymbol{\mathcal{O}}(\mathbf{s}_{-1}, \underline{p}^-) - \mathcal{O}(C_1; \underline{p}^-))}{\boldsymbol{\mathcal{O}}((\mathbf{s}_{-1}, C_1); \underline{p}) - \boldsymbol{\mathcal{O}}(\mathbf{s}_{-1}, \underline{p}^-) - \mathcal{O}(C_1; \underline{p}^-)} \right.$$
$$\left. - \frac{(D(\underline{p}) - \boldsymbol{\mathcal{O}}(\mathbf{s}_{-1}, \underline{p}^-) - \mathcal{O}(s_1, \underline{p}^-))}{\boldsymbol{\mathcal{O}}((\mathbf{s}_{-1}, s_1), \underline{p}) - \boldsymbol{\mathcal{O}}(\mathbf{s}_{-1}, \underline{p}^-) - \mathcal{O}(s_1; \underline{p}^-)} \right).$$

We note that

$$0 < \boldsymbol{\mathcal{O}}((\mathbf{s}_{-1}, s_1); \underline{p}) - \boldsymbol{\mathcal{O}}(\mathbf{s}_{-1}, \underline{p}^-) - \mathcal{O}(C_1; \underline{p}^-)$$
$$< \boldsymbol{\mathcal{O}}((\mathbf{s}_{-1}, C_1); \underline{p}) - \boldsymbol{\mathcal{O}}(\mathbf{s}_{-1}, \underline{p}^-) - \mathcal{O}(C_1; \underline{p}^-),$$

and that $D(\underline{p}) - \boldsymbol{\mathcal{O}}((\mathbf{s}_{-1}, C_1); \underline{p}^-) > 0$ by definition of $\underline{p}$. Then

$$\varphi_1(\mathbf{s}_{-1}, s_1) - \varphi_1(\mathbf{s}_{-1}, C_1)$$

$$\leq \sum_{j>1, j \in \mathscr{E}(\underline{p})} \Delta^- \mathcal{O}(s_j; \underline{p}) \times \left(\frac{(D(\underline{p}) - \boldsymbol{\mathcal{O}}(\mathbf{s}_{-1}, \underline{p}_-) - \mathcal{O}(C_1; \underline{p}_-))}{\boldsymbol{\mathcal{O}}((\mathbf{s}_{-1}, s_1); \underline{p}) - \boldsymbol{\mathcal{O}}(\mathbf{s}_{-1}, \underline{p}^-) - \mathcal{O}(C_1; \underline{p}^-)} \right.$$

$$\left. - \frac{(D(\underline{p}) - \boldsymbol{\mathcal{O}}(\mathbf{s}_{-1}, \underline{p}_-) - \mathcal{O}(s_1; \underline{p}_-))}{\boldsymbol{\mathcal{O}}((\mathbf{s}_{-1}, s_1); \underline{p}) - \boldsymbol{\mathcal{O}}(\mathbf{s}_{-1}, \underline{p}^-) - \mathcal{O}(s_1; \underline{p}^-)} \right).$$

Since $D(\underline{p}) \leq \boldsymbol{\mathcal{O}}((\mathbf{s}_{-1}, s_1); \underline{p})$ and $\mathcal{O}(C_1; \underline{p}^-) \geq \mathcal{O}(s_1; \underline{p}^-)$, we can deduce that $\varphi_1(\mathbf{s}_{-1}, s_1) - \varphi_1(\mathbf{s}_{-1}, C_1) \leq 0$. This follows from the fact that when $A \leq B$, the map $x \mapsto \dfrac{A - x}{B - x}$ is decreasing on $[0, A)$.

A.0.0.7   B. We Prove the Uniqueness Property *(iii)*

All Nash equilibria induce the same electricity price and same quantities of electricity bought to each producer.

First, we state the following consequence of the dominance property *(i)*:

Lemma 8. *For any admissible strategy* $\mathbf{s} = (s_1, \ldots, s_J)$, *such that* $\underline{p}(\mathbf{s}) = \underline{p}(\mathbf{C})$, *if producer j is such that $s_j = C_j$, then*

$$\varphi_j(\mathbf{s}) \geq \varphi_j(\mathbf{C}).$$

Proof. As arguments are very similar to the proof of *(i)*, we just sketch them. Let $\mathbf{s}$ such that $\underline{p}(\mathbf{s}) = \underline{p}(\mathbf{C}) := \underline{p}$. Assume that Producer 1 is such that $s_1 = C_1$.

• If $\boldsymbol{\mathcal{O}}(\mathbf{s}; \underline{p}) \leq D(\underline{p})$, then by the market clearing

$$\varphi_1(\mathbf{s}) = \mathcal{O}(s_1; \underline{p}) = \mathcal{O}(C_1; \underline{p}) \geq \varphi_1(\mathbf{C}).$$

• If $D(\underline{p}) < \boldsymbol{\mathcal{O}}(\mathbf{s}; \underline{p}) \leq \boldsymbol{\mathcal{O}}(\mathbf{C}; \underline{p})$, by the market clearing we get

$$\varphi_1(\mathbf{s}_{-1}, C_1) = \mathcal{O}(C_1; \underline{p}^-) + \Delta^- \mathcal{O}(C_1; \underline{p}) \frac{(D(\underline{p}) - \boldsymbol{\mathcal{O}}((\mathbf{s}_{-1}; C_1); \underline{p}^-))}{\Delta^- \boldsymbol{\mathcal{O}}((\mathbf{s}_{-1}; C_1); \underline{p})}$$

and $$\varphi_1(\mathbf{C}_{-1}, C_1) = \mathcal{O}(C_1; \underline{p}^-) + \Delta^- \mathcal{O}(C_1; \underline{p}) \frac{(D(\underline{p}) - \boldsymbol{\mathcal{O}}((\mathbf{C}_{-1}; C_1), \underline{p}^-))}{\Delta^- \boldsymbol{\mathcal{O}}((\mathbf{C}_{-1}; C_1), \underline{p})}.$$

Thus,

$$\varphi_1(\mathbf{s}_{-1}, C_1) - \varphi_1(\mathbf{C}_{-1}, C_1)$$

$$= \Delta^{-}\mathscr{O}(C_1;\underline{p})\left(\frac{(D(\underline{p}) - \mathscr{O}((\mathbf{s}_{-1}, C_1);\underline{p}^{-}))}{\mathscr{O}((\mathbf{s}_{-1}, C_1);\underline{p}) - \mathscr{O}((\mathbf{s}_{-1}, C_1);\underline{p}^{-})}\right.$$

$$\left.- \frac{(D(\underline{p}) - \mathscr{O}((C_{-1}, C_1);\underline{p}^{-}))}{\mathscr{O}((C_{-1}, C_1);\underline{p}) - \mathscr{O}((C_{-1}, C_1);\underline{p}^{-})}\right).$$

Assuming that $\Delta^{-}\mathscr{O}(C_1;\underline{p}) > 0$, we note that

$$0 < \mathscr{O}((\mathbf{s}_{-1}, C_1);\underline{p}) - \mathscr{O}((\mathbf{s}_{-1}, C_1);\underline{p}^{-}) \le \mathscr{O}((C_{-1}, C_1);\underline{p}) - \mathscr{O}((\mathbf{s}_{-1}, C_1);\underline{p}^{-}).$$

Since $D(\underline{p}) - \mathscr{O}((C_{-1}, C_1);\underline{p}^{-}) > 0$ by definition of $\underline{p}$,

$$\varphi_1(\mathbf{s}_{-1}, C_1) - \varphi_1(C_{-1}, C_1)$$

$$\ge \Delta^{-}\mathscr{O}(C_1;\underline{p})\left(\frac{(D(\underline{p}) - \mathscr{O}((\mathbf{s}_{-1}, C_1);\underline{p}^{-}))}{\mathscr{O}((C_{-1}, C_1);\underline{p}) - \mathscr{O}((\mathbf{s}_{-1}, C_1);\underline{p}^{-})}\right.$$

$$\left.- \frac{(D(\underline{p}) - \mathscr{O}((C_{-1}, C_1);\underline{p}^{-}))}{\mathscr{O}((C_{-1}, C_1);\underline{p}) - \mathscr{O}((C_{-1}, C_1);\underline{p}^{-})}\right).$$

As $\mathscr{O}((\mathbf{s}_{-1}, C_1);\underline{p}^{-}) \le \mathscr{O}((C_{-1}, C_1);\underline{p}^{-})$, we get $\varphi_1(\mathbf{s}_{-1}, C_1) - \varphi_1(C_{-1}, C_1) \ge 0$.

$\square$

We prove that the quantities are the same for all Nash equilibria. Let $\mathbf{w}$ an other Nash equilibrium that differs from $\mathbf{C}$. On the global offers we always have $\mathscr{O}(\mathbf{w};\cdot) \le \mathscr{O}(\mathbf{C};\cdot)$ that implies $\underline{p}(\mathbf{w}) \ge \underline{p}(\mathbf{C})$. Note that when $\underline{p}(\mathbf{C}) = p_{\text{lolc}}$, all admissible strategies $\mathbf{s}$ are Nash as $\varphi_j(\mathbf{C}) = \varphi_j(\mathbf{s}) = \kappa_j$, for all j.

By the offers ordering, it is straightforward to show that

$$\sum_{j=1}^{J} \varphi_j(\mathbf{w}) \le \sum_{j=1}^{J} \varphi_j(\mathbf{C}).$$

Assume that the quantities are not the same, then there exists a producer, say Producer 1, such that $\varphi_1(\mathbf{w}) < \varphi_1(\mathbf{C})$. And we also have

$$\varphi_1(\mathbf{w}) < \varphi_1(\mathbf{C}) \le \mathscr{O}(C_1;p^{\text{elec}}(\mathbf{C})) \le \mathscr{O}(C_1;p^{\text{elec}}(\mathbf{w}_{-1}, C_1)).$$

If $\underline{p}(\mathbf{C}) = \underline{p}(\mathbf{w}_{-1}, C_1)$, then by Lemma 8, we have that $\varphi_1(\mathbf{w}_{-1}, C_1) \ge \varphi_1(\mathbf{C})$ and hence $\varphi_1(\mathbf{w}_{-1}, C_1) > \varphi_1(\mathbf{w})$. In other words, $\mathbf{w}$ has a strictly favorable deviation for Producer 1 that contradicts the assumption that $\mathbf{w}$ is a Nash equilibrium.

Now if $\underline{p}(\mathbf{C}) < \underline{p}(\mathbf{w}_{-1}, C_1)$, by (9),

$$\varphi_1(\mathbf{w}) < \varphi_1(\mathbf{C}) \le \mathscr{O}(C_1;\underline{p}(\mathbf{C})) \le \mathscr{O}(C_1;\underline{p}((\mathbf{w}_{-1}, C_1))^{-}) \le \varphi_1(\mathbf{w}_{-1}, C_1),$$

and the same conclusion follows.

We prove that the equilibrium best bid price is unique: $\bar{p}(\mathbf{w}) = \bar{p}(\mathbf{C})$, **for an other Nash equilibrium w.** Assume the contrary, $\bar{p}(\mathbf{w}) > \bar{p}(\mathbf{C})$. Then by the definition of $\bar{p}(\cdot)$, we have that $D(\underline{p}(\mathbf{w})) < D(\underline{p}(\mathbf{C}))$.

From (8) and (9),

$$\sum_{j=1}^{J} \varphi_j(\mathbf{w}) \le D(\underline{p}(\mathbf{w})) < D(\underline{p}(\mathbf{C})^+) \le D(\underline{p}(\mathbf{C})) \wedge \mathcal{O}(\mathbf{C}; \underline{p}(\mathbf{C})) = \sum_{j=1}^{J} \varphi_j(\mathbf{C})$$

that contradicts the fact that Nash equilibria have same clearing quantities.

Proofs of Lemmas 1 and 2

Proof of Lemma 1. Although the result of this lemma is intuitive, the proof is rather technical. This is due to our assumptions, in particular regarding demand, that allow the demand function to have discontinuity points and some non-elasticity areas (see Assumption 1).

More precisely, if we define the map $\tau \mapsto \mathcal{O}(\tau; p)$ by

$$\mathcal{O}(\tau; p) = \sum_{i=1}^{J} \mathcal{O}(C_j^\tau(\cdot); p) = \sum_{i=1}^{J} \kappa_i \mathbb{1}_{\{p \ge c_i + \tau e_i\}} = \sum_{i=1}^{J} \kappa_i \mathbb{1}_{\{\tau \le \frac{p - c_i}{e_i}\}},$$

then we can observe that, for any $p > 0$ far enough from the c_i, and any $\tau' \ge \tau$,

$$\mathcal{O}(\tau'; p) \le \mathcal{O}(\tau; p) \quad \text{and} \quad \lim_{\epsilon \to 0^+} \mathcal{O}(\tau + \epsilon; p) = \mathcal{O}(\tau; p).$$

We call $S_D = \{p_d; \lim_{\epsilon \to 0^+} D(p_d + \epsilon) < D(p_d)\}$, the set of discontinuity points of the Demand function.

We call $S_\kappa = \{p_c; D(p_c) = \sum \kappa_i\}$, the set of prices that make demand coincide with some accumulation of production capacities.

We observe that $p^{\text{elec}}(\tau) \in \{c_i + \tau e_i, i = 1, \ldots, j\} \cup S_D \cup S_\kappa$. In particular, from Definition 1, $\underline{p}(\tau) = \inf\{p > 0; \mathcal{O}(\tau; p) > D(p)\}$, and we obtain that $D(\underline{p}(\tau + \epsilon)) \le \mathcal{O}(\tau + \epsilon; \underline{p}(\tau + \epsilon)) \le \mathcal{O}(\tau; \underline{p}(\tau + \epsilon))$ from which we conclude that $\underline{p}(\tau + \epsilon) \ge \underline{p}(\tau)$.

Now we prove the right continuity of $\tau \mapsto \underline{p}(\tau)$. Let us fix a τ.

(i) **We first consider the case** $D(\underline{p}(\tau)) < \mathcal{O}(\tau; \underline{p}(\tau))$.

This means that $\underline{p}(\tau)$ is of the form $c_\ell + \tau \ell$, for a given ℓ. Then when $\epsilon > 0$ is small enough, we also have $\underline{p}(\tau + \epsilon) = c_\ell + (\tau + \epsilon)e_\ell$. Indeed, $D(c_\ell + (\tau + \epsilon)e_\ell) \le D(c_\ell + \tau e_\ell)$ and for a small enough ϵ,

$$\mathcal{O}(\tau; c_\ell + \tau e_\ell) = \kappa_\ell + \sum_{i \ne \ell} \kappa_i \mathbb{1}_{\{\tau \le \frac{c_\ell - c_i}{1 - e_i/e_\ell}\}} = \mathcal{O}(\tau + \epsilon; c_\ell + (\tau + \epsilon)e_\ell).$$

Thus, $D(c_\ell + (\tau + \epsilon)e_\ell) < \mathcal{O}(\tau + \epsilon; c_\ell + (\tau + \epsilon)e_\ell)$ which implies that $\underline{p}(\tau) + e_\ell \epsilon = c_\ell + (\tau + \epsilon)e_\ell \geq \underline{p}(\tau + \epsilon)$ and hence $e_\ell \epsilon \geq \underline{p}(\tau + \epsilon) - \underline{p}(\tau)$.

(ii) **We consider next the case** $D(\underline{p}(\tau)) > \mathcal{O}(\tau; \underline{p}(\tau))$.

This means that $\underline{p}(\tau) \in S_D$ is at a discontinuity point, say p_d of the demand, $\underline{p}(\tau) = p_d$. Then, for any $\delta > 0$,

$$D(\underline{p}(\tau) + \delta) < \mathcal{O}(\tau; \underline{p}(\tau) + \delta).$$

But,

$$\mathcal{O}(\tau; p_d + \delta) = \sum_{i=1}^{J} \kappa_i \mathbb{1}_{\{\tau \leq \frac{p_d + \delta - c_i}{e_i}\}},$$

and we can choose δ to be small enough so that $\tau \neq \frac{p_d + \delta - c_i}{e_i}$. Then, for a small enough ϵ,

$$D(\underline{p}(\tau) + \delta) < \mathcal{O}(\tau; \underline{p}(\tau) + \delta) = \mathcal{O}(\tau + \epsilon; \underline{p}(\tau) + \delta),$$

which implies that $\underline{p}(\tau) + \delta \geq \bar{p}(\tau + \epsilon)$, so we obtain $\delta \geq \underline{p}(\tau + \epsilon) - \underline{p}(\tau) \geq 0$.

(iii) **We consider now the case** $D(\underline{p}(\tau)) = \mathcal{O}(\tau; \underline{p}(\tau))$.

This means that $\underline{p}(\tau) \in S_\kappa$, say $\underline{p}(\tau) = p_c$ Then, for any $\delta > 0$,

$$D(\underline{p}(\tau) + \delta) < \mathcal{O}(\tau; \underline{p}(\tau) + \delta).$$

But,

$$\mathcal{O}(\tau; p_c + \delta) = \sum_{i=1}^{J} \kappa_i \mathbb{1}_{\{\tau \leq \frac{p_c + \delta - c_i}{e_i}\}}$$

and we can choose δ small enough such that $\tau \neq \frac{p_c + \delta - c_i}{e_i}$. Then, for ϵ small enough,

$$D(\underline{p}(\tau) + \delta) < \mathcal{O}(\tau; \underline{p}(\tau) + \delta) = \mathcal{O}(\tau + \epsilon; \underline{p}(\tau) + \delta)$$

which implies that $\underline{p}(\tau) + \delta \geq \underline{p}(\tau + \epsilon)$, so we get $\delta \geq \underline{p}(\tau + \epsilon) - \underline{p}(\tau) \geq 0$. The right-continuity of $\tau \mapsto \bar{p}(\tau)$ follows, by definition as $\bar{p}(\tau)$ is a continuous transformation of $\underline{p}(\tau)$. □

Proof of Lemma 2. The proof consists in a complete analysis of the entire combination of situations, but each situation is elementary.

Let us suppose the opposite, that is there exists $0 \leq t < t' \leq \mathfrak{p}$ such that the emission levels are $\mathcal{W}(t') > \mathcal{W}(t)$.

We define the function $\tau \mapsto I(\tau)$ valued in the subsets of $\{1, \ldots, J\}$ that lists the producers in the electricity market producing at tax level τ:

$$i \in I(\tau) \quad \text{if} \quad \varphi_i(\tau) > 0.$$

In particular we have for all $\tau \in [0, \mathfrak{p}]$,

$$\mathscr{W}(\tau) = \sum_{i \in I(\tau)} e_i \varphi_i(\tau).$$

(i) **We first examine the situation** $I(t') = I(t)$.

To shorten the expressions, we adopt the following notation

$$I(t) = I \quad \text{and} \quad I(t') = I'.$$

(i-a) If $\sum_{i \in I} \varphi_i(t) = D(t)$ then, from the demand constraint (DC) and the emission levels hypothesis (EH), we have

$$\sum_{i \in I} \varphi_i(t) = D(t) \geq D(t') \geq \sum_{i \in I'} \varphi_i(t') \tag{DC}$$

$$\sum_{i \in I} \varphi_i(t) e_i < \sum_{i \in I'} \varphi_i(t') e_i. \tag{EH}$$

We denote by $\hat{I}$ the subset of I of index such that $c_i + t e_i = \underline{p}(t)$. In particular, when $j \in I \setminus \hat{I}$, then $\varphi_j(t) = \kappa_j$.

Note that there exists at most one index (say ℓ) in the set $\hat{I} \cap \widehat{I'}$. If $j \in \hat{I} \setminus \widehat{I'}$ and $k \in \widehat{I'} \setminus \hat{I}$, then, by the definition of the sets

$$
\begin{aligned}
c_j + e_j t &= c_\ell + e_\ell t, & c_k + e_k t &< c_j + e_j t, \\
c_j + e_j t' &< c_\ell + e_\ell t', & c_k + e_k t' &= c_\ell + e_\ell t', \\
c_j + e_j t' &< c_k + e_k t, & c_k + e_k t &< c_\ell + e_\ell t,
\end{aligned}
$$

from which we easily deduce that

$$\max\{e_j, j \in \hat{I} \setminus \widehat{I'}\} < e_\ell < \min\{e_k, k \in \widehat{I'} \setminus \hat{I}\}. \tag{A.2}$$

Now we decompose the sets I and I' in the demand constraint (DC) and the emission levels hypothesis (EH) as follows:

$$\sum_{n \in I \setminus \hat{I} \cup \widehat{I'}} \kappa_n + \varphi_\ell(t) + \sum_{i \in \hat{I} \setminus \widehat{I'}} \varphi_i(t) + \sum_{k \in \widehat{I'} \setminus \hat{I}} \kappa_k$$

$$\geq \sum_{n \in I \setminus \hat{I} \cup \widehat{I'}} \kappa_n + \varphi_\ell(t') + \sum_{i \in \hat{I} \setminus \widehat{I'}} \kappa_i + \sum_{k \in \widehat{I'} \setminus \hat{I}} \varphi_k(t'), \tag{DC}$$

$$\sum_{n\in I\setminus\hat{I}\cup\widehat{I'}} e_n\kappa_n + e_\ell\varphi_\ell(t) + \sum_{i\in\hat{I}\setminus\widehat{I'}} e_i\varphi_i(t) + \sum_{k\in\widehat{I'}\setminus\hat{i}} e_k\kappa_k$$

$$< \sum_{n\in I\setminus\hat{I}\cup\widehat{I'}} e_n\kappa_n + e_\ell\varphi_\ell(t') + \sum_{i\in\hat{I}\setminus\widehat{I'}} e_i\kappa_i + \sum_{k\in\widehat{I'}\setminus\hat{i}} e_k\varphi_k(t'). \tag{EH}$$

After simplification, we obtain

$$\varphi_\ell(t) + \sum_{i\in\hat{I}\setminus\widehat{I'}} \varphi_i(t) + \sum_{k\in\widehat{I'}\setminus\hat{i}} \kappa_k \geq \varphi_\ell(t') + \sum_{i\in\hat{I}\setminus\widehat{I'}} \kappa_i + \sum_{k\in\widehat{I'}\setminus\hat{i}} \varphi_k(t'), \tag{DC}$$

$$e_\ell\varphi_\ell(t) + \sum_{i\in\hat{I}\setminus\widehat{I'}} e_i\varphi_i(t) + \sum_{k\in\widehat{I'}\setminus\hat{i}} e_k\kappa_k < e_\ell\varphi_\ell(t') + \sum_{i\in\hat{I}\setminus\widehat{I'}} e_i\kappa_i + \sum_{k\in\widehat{I'}\setminus\hat{i}} e_k\varphi_k(t').$$
$$\tag{EH}$$

Assume first that $\varphi_\ell(t) + \sum_{i\in\hat{I}\setminus\widehat{I'}} \varphi_i(t) \geq \varphi_\ell(t') + \sum_{i\in\hat{I}\setminus\widehat{I'}} \kappa_i$. Equivalently, we have

$$\varphi_\ell(t) - \varphi_\ell(t') \geq \sum_{i\in\hat{I}\setminus\widehat{I'}} (\kappa_i - \varphi_i(t))$$

and from (A.2),

$$e_\ell\left(\varphi_\ell(t) - \varphi_\ell(t')\right) \geq \sum_{i\in\hat{I}\setminus\widehat{I'}} e_i(\kappa_i - \varphi_i(t)).$$

By combining the above with the emission levels hypothesis (EH), we obtain the following contradiction: $\sum_{k\in\widehat{I'}\setminus\hat{i}} e_k\kappa_k < \sum_{k\in\widehat{I'}\setminus\hat{i}} e_k\varphi_k(t')$.

Assume now that $\varphi_\ell(t) + \sum_{i\in\hat{I}\setminus\widehat{I'}} \varphi_i(t) < \varphi_\ell(t') + \sum_{i\in\hat{I}\setminus\widehat{I'}} \kappa_i$. Multiplying the demand constraint (DC) by $\hat{e} := \min\{e_k, k \in \widehat{I'} \setminus \hat{I}\}$, we get

$$\sum_{k\in\widehat{I'}\setminus\hat{i}} e_k(\kappa_k - \varphi_k(t')) \geq \hat{e}\left(\varphi_\ell(t) - \varphi_\ell(t')\right) + \hat{e}\sum_{i\in\hat{I}\setminus\widehat{I'}} (\kappa_i - \varphi_i(t)).$$

But from (EH) and (A.2), we also have

$$\sum_{k\in\widehat{I'}\setminus\hat{i}} e_k(\kappa_k - \varphi_k(t')) < e_\ell\left(\varphi_\ell(t) - \varphi_\ell(t')\right) + e_\ell\sum_{i\in\hat{I}\setminus\widehat{I'}} (\kappa_i - \varphi_i(t)),$$

then

$$0 \geq (\hat{e} - e_\ell) \left(\varphi_\ell(t) - \varphi_\ell(t') \right) + (\hat{e} - e_\ell) \sum_{i \in \hat{I} \setminus \hat{I'}} (\kappa_i - \varphi_i(t)),$$

which contradicts our assumption.

(i-b) If $\sum_{i \in I} \varphi_i(t) < D(t)$ then, for all $i \in I$, $\varphi_i(t) = \kappa_i$ and (EH) is necessarily false.

(ii) **We examine the situation $I(t') \neq I(t)$.**

We add the following shortened notation: $I(t) \cap I'(t) = II'$.
We break down I and I' into the sets II', $I \setminus I'$ and $I' \setminus I$. We denote by $\hat{I}$ the set of index $i \in I$ such that $c_i + te_i = \underline{p}(t)$. In particular, when $j \in I \setminus \hat{I}$, then $\varphi_j(t) = \kappa_j$.

We first derive some generic relations between the emission rates for these.

Among the indexes in the set II', we observe that at most one index exists (say ℓ) in the set $\hat{I} \cap \hat{I'}$. If $j \in \hat{I} \setminus \hat{I'}$, if $k \in \hat{I'} \setminus \hat{I}$, then, by the definition of the sets

$$\begin{aligned} c_j + e_j t = c_\ell + e_\ell t, \qquad & c_k + e_k t < c_j + e_j t, \\ c_j + e_j t' < c_\ell + e_\ell t', \qquad & c_k + e_k t' = c_\ell + e_\ell t', \\ c_j + e_j t' < c_k + e_k t, \qquad & c_k + e_k t < c_\ell + e_\ell t, \end{aligned}$$

from which, we easily deduce that

$$\hat{e} := \max \left\{ e_j, j \in II' \cap \left(\hat{I} \setminus \hat{I'} \right) \right\} < e_\ell < \min \left\{ e_k, k \in II' \cap \left(\hat{I'} \setminus \hat{I} \right) \right\} := \hat{e'}. \quad \text{(A.3)}$$

For $j \in I \setminus I'$ and $k \in I' \setminus I$, we have

$$c_j + e_j t < c_k + e_k t \quad \text{and} \quad c_j + e_j t' > c_k + e_k t'$$

from which, we also easily deduce that

$$\max\{e_k, k \in I' \setminus I\} < \min\{e_j, j \in I \setminus I'\}. \quad \text{(A.4)}$$

For the same j and k, for $(\hat{c}, \hat{e})$ representative of index in $II' \cap \hat{I} \setminus \hat{I'}$, and $(\hat{c'}, \hat{e'})$ representative of index in $II' \cap \hat{I'} \setminus \hat{I}$, we also have

$$\begin{aligned} c_j + e_j t \leq \hat{c} + \hat{e} t \\ c_j + e_j t' > \hat{c} + \hat{e} t' \end{aligned} \quad \text{and} \quad \begin{aligned} c_k + e_k t > \hat{c'} + \hat{e'} t \\ c_k + e_k t' \leq \hat{c'} + \hat{e'} t' \end{aligned}$$

from which, we deduce that

$$\begin{aligned} \min\{e_j, j \in I \setminus I'\} > (e_\ell, \hat{e}) \vee \max\{e_k, k \in I' \setminus I\} \\ \max\{e_k, k \in I' \setminus I\} < (e_\ell, \hat{e'}) \wedge \min\{e_j, j \in I \setminus I'\}. \end{aligned} \quad \text{(A.5)}$$

We divide the analysis in cases. In the first one the demand is fully satisfied for the price $p^{\text{elec}}(t)$.

(ii-a) If $\sum_{i \in I} \varphi_I(t) = D(p^{\text{elec}}(t))$,

$$\sum_{i \in I \setminus I'} \varphi_I + \sum_{i \in II'} \varphi_i(t) = D(p^{\text{elec}}(t)) \geq D(p^{\text{elec}}(t')) \geq \sum_{i \in II'} \varphi_i(t') + \sum_{i \in I' \setminus I} \varphi_i(t'),$$

(DC)

$$\sum_{i \in I \setminus I'} \varphi_i(t)e_i + \sum_{i \in II'} \varphi_i(t)e_i < \sum_{i \in II'} \varphi_i(t')e_i + \sum_{i \in I' \setminus I} \varphi_i(t)e_i.$$

(EH)

We must then examine the following two subcases, relative to the situations where the demand is satisfied or not at the price $p^{\text{elec}}(t')$.

(ii-a-1) If $\sum_{i \in I'} \varphi_i(t') < D(p^{\text{elec}}(t'))$, then $\varphi_i(t') = \kappa_i$ for all $i \in I'$ and

$$\sum_{j \in I \setminus I'} \varphi_j(t) + \sum_{i \in II'} \varphi_i(t) > \sum_{i \in II'} \kappa_i + \sum_{k \in I' \setminus I} \kappa_k,$$

(DC)

$$\sum_{j \in I \setminus I'} \varphi_j(t)e_j + \sum_{i \in II'} \varphi_i(t)e_i < \sum_{i \in II'} \kappa_i e_i + \sum_{k \in I' \setminus I} \kappa_k e_k.$$

(EH)

As $\varphi_i(t) = \kappa_i$ when $i \in (I \setminus \hat{I}) \cap II'$, we can simplify the two sides of (DC) and (EH) by the sum over $(I \setminus \hat{I}) \cap II'$. The remaining part of II' is $\{\ell\} \cup \left(\hat{I} \setminus \widehat{I'} \cap II' \right)$:

$$\sum_{j \in I \setminus I'} \varphi_j(t) + \varphi_\ell + \sum_{i \in \hat{I} \setminus \widehat{I'} \cap II'} \varphi_i(t) > \kappa_\ell + \sum_{i \in \hat{I} \setminus \widehat{I'} \cap II'} \kappa_i + \sum_{k \in I' \setminus I} \kappa_k,$$

(DC)

$$\sum_{j \in I \setminus I'} e_j \varphi_j(t) + e_\ell \varphi_\ell + \sum_{i \in \hat{I} \setminus \widehat{I'} \cap II'} e_i \varphi_i(t) < e_\ell \kappa_\ell + \sum_{i \in \hat{I} \setminus \widehat{I'} \cap II'} e_i \kappa_i + \sum_{k \in I' \setminus I} e_k \kappa_k.$$

(EH)

Then we multiply (DC) by $\bar{e} := (e_\ell, \hat{e}) \vee \max\{e_k, k \in I' \setminus I\}$, and we obtain by (A.5)

$$\sum_{j \in I \setminus I'} e_j \varphi_j(t) + \bar{e} \varphi_\ell + \bar{e} \sum_{i \in \hat{I} \setminus \widehat{I'} \cap II'} \varphi_i(t) > \bar{e} \kappa_\ell + \bar{e} \sum_{i \in \hat{I} \setminus \widehat{I'} \cap II'} \kappa_i + \sum_{k \in I' \setminus I} e_k \kappa_k.$$

We subtract with (EH) :

$$(\bar{e} - e_\ell)\varphi_\ell + \sum_{i \in \hat{I} \setminus \widehat{I'} \cap II'} (\bar{e} - e_i)\varphi_i(t) > (\bar{e} - e_\ell)\kappa_\ell + \sum_{i \in \hat{I} \setminus \widehat{I'} \cap II'} (\bar{e} - e_i)\kappa_i.$$

But $\bar{e} \geq e_\ell$ when ℓ exists, and $\bar{e} \geq \hat{e} \geq e_i$ for $i \in \hat{I} \backslash \widehat{I'} \cap II'$. So we obtain our contradiction.

(ii-a-2) If $\sum_{i \in I'} \varphi_i(t') = D(p^{\text{elec}}(t'))$, then

$$\sum_{j \in I \backslash I'} \varphi_j(t) + \sum_{i \in II'} \varphi_i(t) > \sum_{i \in II'} \varphi_i(t') + \sum_{k \in I' \backslash I} \varphi_k(t'), \tag{DC}$$

$$\sum_{j \in I \backslash I'} \varphi_j(t) e_j + \sum_{i \in II'} \varphi_i(t) e_i < \sum_{i \in II'} \varphi_i(t') e_i + \sum_{k \in I' \backslash I} \varphi_k(t') e_k. \tag{EH}$$

We decompose $I \backslash I' = \left(I \backslash (I' \cup \hat{I}) \right) \cup \hat{I} \backslash I'$ and $I' \backslash I = \left(I' \backslash (I \cup \widehat{I'}) \right) \cup \widehat{I'} \backslash I$:

$$\sum_{j \in I \backslash (I' \cup \hat{I})} \kappa_j + \sum_{j \in \hat{I} \backslash I'} \varphi_j(t) + \sum_{i \in II'} \varphi_i(t) > \sum_{i \in II'} \varphi_i(t') + \sum_{k \in \widehat{I'} \backslash I} \varphi_k(t') + \sum_{k \in I' \backslash (I \cup \widehat{I'})} \kappa_k, \tag{DC}$$

$$\sum_{j \in I \backslash (I' \cup \hat{I})} e_j \kappa_j + \sum_{j \in \hat{I} \backslash I'} e_j \varphi_j(t) + \sum_{i \in II'} e_i \varphi_i(t)$$
$$< \sum_{i \in II'} e_i \varphi_i(t') + \sum_{k \in \widehat{I'} \backslash I} e_k \varphi_k(t') + \sum_{k \in I' \backslash (I \cup \widehat{I'})} e_k \kappa_k. \tag{EH}$$

We also break down the set $II' = (I \cap I')$:

$$II' = \left(II' \cap \{\ell\} \right) \cup \left(II' \cap \hat{I} \widehat{I'} \right) \cup \left(II' \cap \widehat{I'} \backslash \hat{I} \right) \cup \left(I \backslash \hat{I} \cap I' \backslash \widehat{I'} \right).$$

$$\sum_{j \in I \backslash (I' \cup \hat{I}))} \kappa_j + \sum_{j \in \hat{I} \backslash I'} \varphi_j(t) + \varphi_\ell(t) + \sum_{i \in \hat{I} \backslash \widehat{I'} \cap II'} \varphi_i(t) + \sum_{i \in \widehat{I'} \backslash \hat{I} \cap II'} \varphi_i(t)$$
$$> \varphi_\ell(t') + \sum_{i \in \hat{I} \backslash \widehat{I'} \cap II'} \varphi_i(t') + \sum_{i \in \widehat{I'} \backslash \hat{I} \cap II'} \varphi_i(t') + \sum_{k \in \widehat{I'} \backslash I} \varphi_k(t') + \sum_{k \in I' \backslash (I \cup \widehat{I'})} \kappa_k, \tag{DC}$$

$$\sum_{j \in I \backslash (I' \cup \hat{I})} e_j \kappa_j + \sum_{j \in \hat{I} \backslash I'} e_j \varphi_j(t) + e_\ell \varphi_\ell(t) + \sum_{i \in \hat{I} \backslash \widehat{I'} \cap II'} e_i \varphi_i(t) + \sum_{i \in \widehat{I'} \backslash \hat{I} \cap II'} e_i \varphi_i(t)$$
$$< e_\ell \varphi_\ell(t') + \sum_{i \in \hat{I} \backslash \widehat{I'} \cap II'} e_i \varphi_i(t') + \sum_{i \in \widehat{I'} \backslash \hat{I} \cap II'} e_i \varphi_i(t') + \sum_{k \in \widehat{I'} \backslash I} e_k \varphi_k(t') + \sum_{k \in I' \backslash (I \cup \widehat{I'})} e_k \kappa_k. \tag{EH}$$

For index i in the last subset $(I\backslash\hat{I} \cap I'\backslash\widehat{I'})$, we have $\varphi_i(t) = \kappa_i$ and $\varphi_i(t') = \kappa_i$, so we simplify (DC) and (EH) from this last subset. Thus,

$$\sum_{j\in I\backslash(I'\cup\hat{I})} \kappa_j + \sum_{j\in\hat{I}\backslash I'} \varphi_j(t) + \varphi_\ell(t) + \sum_{i\in\hat{I}\backslash\widehat{I'}\cap II'} \varphi_i(t) + \sum_{i\in\widehat{I'}\backslash\hat{I}\cap II'} \kappa_i$$

$$> \varphi_\ell(t') + \sum_{i\in\hat{I}\backslash\widehat{I'}\cap II'} \kappa_i + \sum_{i\in\widehat{I'}\backslash\hat{I}\cap II'} \varphi_i(t') + \sum_{k\in\widehat{I'}\backslash I} \varphi_k(t') + \sum_{k\in I'\backslash(I\cup\widehat{I'})} \kappa_k,$$

$$\text{(DC)}$$

$$\sum_{j\in I\backslash(I'\cup\hat{I})} e_j\kappa_j + \sum_{j\in\hat{I}\backslash I'} e_j\varphi_j(t) + e_\ell\varphi_\ell(t) + \sum_{i\in\hat{I}\backslash\widehat{I'}\cap II'} e_i\varphi_i(t) + \sum_{i\in\widehat{I'}\backslash\hat{I}\cap II'} e_i\kappa_i$$

$$< e_\ell\varphi_\ell(t') + \sum_{i\in\hat{I}\backslash\widehat{I'}\cap II'} e_i\kappa_i + \sum_{i\in\widehat{I'}\backslash\hat{I}\cap II'} e_i\varphi_i(t') + \sum_{k\in\widehat{I'}\backslash I} e_k\varphi_k(t') + \sum_{k\in I'\backslash(I\cup\widehat{I'})} e_k\kappa_k.$$

$$\text{(EH)}$$

We multiply (DC) by $\bar{e} := (e_\ell, \hat{e}) \vee \max\{e_k, k \in I'\backslash I\}$, we get by (A.5)

$$\sum_{j\in I\backslash(I'\cup\hat{I})} e_j\kappa_j + \sum_{j\in\hat{I}\backslash I'} e_j\varphi_j(t) + \bar{e}\varphi_\ell(t) + \bar{e}\sum_{i\in\hat{I}\backslash\widehat{I'}\cap II'} \varphi_i(t) + \bar{e}\sum_{i\in\widehat{I'}\backslash\hat{I}\cap II'} \kappa_i$$

$$> \bar{e}\varphi_\ell(t') + \bar{e}\sum_{i\in\hat{I}\backslash\widehat{I'}\cap II'} \kappa_i + \bar{e}\sum_{i\in\widehat{I'}\backslash\hat{I}\cap II'} \varphi_i(t') + \sum_{k\in\widehat{I'}\backslash I} e_k\varphi_k(t') + \sum_{k\in I'\backslash(I\cup\widehat{I'})} e_k\kappa_k.$$

We subtract (EH)

$$(\bar{e} - e_\ell)\varphi_\ell(t) + \sum_{i\in\hat{I}\backslash\widehat{I'}\cap II'} (\bar{e} - e_i)\varphi_i(t) + \sum_{i\in\widehat{I'}\backslash\hat{I}\cap II'} (\bar{e} - e_i)\kappa_i$$

$$> (\bar{e} - e_\ell)\varphi_\ell(t') + \sum_{i\in\hat{I}\backslash\widehat{I'}\cap II'} (\bar{e} - e_i)\kappa_i + \sum_{i\in\widehat{I'}\backslash\hat{I}\cap II'} (\bar{e} - e_i)\varphi_i(t').$$

We arrange the terms

$$(\bar{e} - e_\ell)\varphi_\ell(t) + \sum_{i\in\hat{I}\backslash\widehat{I'}\cap II'} (\bar{e} - e_i)\varphi_i(t) + \sum_{i\in\widehat{I'}\backslash\hat{I}\cap II'} (\bar{e} - e_i)\kappa_i$$

$$> (\bar{e} - e_\ell)\varphi_\ell(t') + \sum_{i\in\hat{I}\backslash\widehat{I'}\cap II'} (\bar{e} - e_i)\kappa_i + \sum_{i\in\widehat{I'}\backslash\hat{I}\cap II'} (\bar{e} - e_i)\varphi_i(t').$$

If ℓ exists, then $\bar{e} = e_\ell$ and

$$
\sum_{i \in \widehat{I'} \setminus \hat{I} \cap II'} (e_\ell - e_i) \left(\kappa_i - \varphi_i(t') \right) > \sum_{i \in \hat{I} \setminus \widehat{I'} \cap II'} (e_\ell - e_i) \left(\kappa_i - \varphi_i(t) \right),
$$

$$
\sum_{i \in \widehat{I'} \setminus \hat{I} \cap II'} (e_\ell - \hat{e}') \left(\kappa_i - \varphi_i(t') \right) > \sum_{i \in \hat{I} \setminus \widehat{I'} \cap II'} (e_\ell - \hat{e}) \left(\kappa_i - \varphi_i(t) \right).
$$

(A.6)

But $\hat{e} < e_\ell < \hat{e}'$, and the contradiction follows.
If ℓ does not exist, then $\bar{e} = \hat{e} \vee \max\{e_k, k \in I' \setminus I\}$

$$
\sum_{i \in \widehat{I'} \setminus \hat{I} \cap II'} (\bar{e} - e_i) \left(\kappa_i - \varphi_i(t') \right) > \sum_{i \in \hat{I} \setminus \widehat{I'} \cap II'} (\bar{e} - e_i) \left(\kappa_i - \varphi_i(t) \right),
$$

$$
\sum_{i \in \widehat{I'} \setminus \hat{I} \cap II'} (\bar{e} - \hat{e}') \left(\kappa_i - \varphi_i(t') \right) > \sum_{i \in \hat{I} \setminus \widehat{I'} \cap II'} (\bar{e} - \hat{e}) \left(\kappa_i - \varphi_i(t) \right).
$$

(A.7)

But $\max\{e_k, k \in I' \setminus I\} < \hat{e}'$, and the contradiction follows.

(ii-b) If $\sum_{i \in I} \varphi_i(t) < D(p^{\text{elec}}(t))$ then for all $i \in I$, $\varphi_i(t) = \kappa_i$.

(ii-b-1) If $\sum_{i \in I'} \varphi_i(t') < D(p^{\text{elec}}(t'))$, then $\varphi_i(t') = \kappa_i$ for all $i \in I'$. Moreover, we have that $\mathcal{O}(t, \underline{p}(t)) \geq D(\underline{p}(t)) + \varepsilon) \geq D(\underline{p}(t')) > \mathcal{O}(t', \underline{p}(t'))$ and (DC)–(EH) becomes

$$
\sum_{j \in I \setminus I'} \kappa_j > \sum_{k \in I' \setminus I} \kappa_k,
$$

(DC)

$$
\sum_{j \in I \setminus I'} e_j \kappa_j < \sum_{k \in I' \setminus I} e_k \kappa_k.
$$

(EH)

Then, we multiply (DC) by $\min\{e_j; j \in I \setminus I'\} \geq \max\{e_k; k \in I' \setminus I\}$, and we obtain a contradiction with (EH).

(ii-b-2) If $\sum_{i \in I'} \varphi_i(t') = D(p^{\text{elec}}(t'))$, we go back to the analysis of the case **(ii-a-2)**, with the main difference that all quantities $\varphi_i(t)$ are now equal to κ_i. We go to inequalities (A.6) and (A.7) which are simplified as the right-had sides are now zero. The contradiction follows with the same arguments $\qquad\square$

References

1. Başar, T., Olsder, G.J.: Dynamic Noncooperative Game Theory. Classics in Applied Mathematics, vol. 23. Society for Industrial and Applied Mathematics (SIAM), Philadelphia, PA (1999). Reprint of the second edn (1995)
2. Bossy, M., Maïzi, N., Olsder, G.J., Pourtallier, O., Tanré, E.: Electricity prices in a game theory context. In: Dynamic Games: Theory and Applications. GERAD 25th Anniversary Series, pp. 135–159, vol. 10. Springer, New York (2005)
3. Carmona, R., Coulon, M., Schwarz, D.: The valuation of clean spread options: linking electricity, emissions and fuels. Quant. Finan. 12(12), 1951–1965 (2012)
4. Carmona, R., Delarue, F., Espinosa, G.E., Touzi, N.: Singular forward-backward stochastic differential equations and emissions derivatives. Ann. Appl. Probab. 23(3), 1086–1128, 06 (2013)
5. Chiesa, G., Denicolò, V.: Trading with a common agent under complete information: a characterization of nash equilibria. J. Econ. Theory 144, 296–311 (2009)
6. Hortaçsu, A.: Recent progress in the empirical analysis of multi-unit auctions. J. Ind. Organ. 29, 345–349 (2011)
7. Hortaçsu, A., Puller, S.L.: Understanding strategic bidding in multi-unit auctions: a case study of the texas electricity spot market. RAND J. Econ. 39(1), 86–114 (2008)

Dynamic Cournot Models for Production of Exhaustible Commodities Under Stochastic Demand

Michael Ludkovski and Xuwei Yang

Abstract We extend the dynamic Cournot model of Ludkovski and Sircar (2011) by considering stochastic demand. We analyze a duopoly between an exhaustible producer and a "green" competitor. Both producers dynamically make decisions regarding their production rates; in addition the exhaustible producer optimizes search for new reserves. The aggregate price earned by the producers switches between high and low demand regimes with exogenously given holding rates. We study how the regime changes and the relative cost of production, which is a proxy for market competitiveness, affect game equilibria, and compare with the case of deterministic demand. A novel feature driven by stochasticity of demand is that production may shut down during low demand to conserve reserves.

1 Introduction

Dynamic models of competition have been a major application of game theory in finance and industrial organization [16]. Beyond the traditional optimization settings that feature a single economic agent maximizing her economic value (in terms of terminal wealth, consumption, investment, etc.), game theoretic frameworks consider interaction of multiple players, replacing the notion of optimality by equilibrium. For commodities markets, exemplified by natural resources such as energy, agricultural, and metals commodities, two basic notions of a competitive equilibrium are given by the Cournot and Bertrand models. Both models were first proposed back in the nineteenth century and focus on producers competing against each other via a non-cooperative game. In a Cournot game, producers choose their supply levels (i.e. quantities of goods produced) and the common clearing price is determined by an exogenous inverse demand curve. In a Bertrand game producers set prices and quantities are determined from the resulting aggregate demand. In both models, the main mechanism through which players interact is the aggregate

M. Ludkovski (✉) • X. Yang
Department of Statistics and Applied Probability, University of California Santa Barbara, South Hall, Santa Barbara, CA 93106-3110, USA
e-mail: ludkovski@pstat.ucsb.edu; yangx@pstat.ucsb.edu

© Springer Science+Business Media New York 2015

R. Aïd et al. (eds.), *Commodities, Energy and Environmental Finance*, Fields Institute Communications 74, DOI 10.1007/978-1-4939-2733-3_14

demand curve which translates supply into price (and vice-versa). Generally, this curve is assumed to be given; the agents can only strategically shift the supply.

Classical game theory deals with deterministic setups where game payoffs are fully prescribed by the agents' joint strategy. To address the observed "noise" in empirical settings (arising from fluctuations in any number of game parameters), stochastic dynamic models have become increasingly popular. Such models typically introduce a stochastic state variable that drives behavior of the agents and is modeled as a (controlled) stochastic process. The state variable could be a macroeconomic indicator, price level of a reference asset, size/capacity of the players, etc. In the context of commodities and environmental finance a common choice is to model the reserves. If the commodity is in finite supply (exhaustible) then its total level is important since there is the exhaustibility constraint on production. Reserves introduce a long-term memory effect into the competition and force agents to pay attention not just to their nominal profits but also to the marginal shadow cost of having fewer reserves. The presence of reserves necessarily makes players non-symmetric since each agent has their own (private) shadow cost. Study of non-symmetric stochastic dynamic games remains rudimentary. The resulting explosion in dimensionality of the problem (officially requiring at least as many dimensions as the number of players) places major constraints on what is tractable. For these reasons, while realism demands analysis of several (typically at least a half dozen) agents, we will restrict our attention to a duopoly.

There is a long literature on optimal economic behavior of a natural resource monopolist extracting non-renewable resources dating back to [10]. We refer to [4, 7] for modern textbook treatment of this field. Nevertheless, to our knowledge, the first paper that rigorously treated a dynamic noncooperative model for exhaustible resource extraction was published less than 5 years ago by Harris et al. [9]. That work studied a deterministic Cournot market focusing on the impact of exhaustibility and subsequent entry of "green" producers that always have infinite reserves. Subsequently, [13] studied a related model that allowed for stochastic evolution of reserves by considering *exploration* that can lead to discovery of new reserves. This analysis was motivated by the oil market where E&P (exploration and production) efforts total many billions of dollars a year. In that sense, while oil is exhaustible, it is also replenishable since there is a difference between total abstract reserves on Earth, and what is actually commercially "proven" and drives production decisions. With exploration, players have two complementary choices regarding running down existing reserves and expending effort in the hopes of finding new reserves. In particular, players may never fully "leave" the game since they can periodically resurrect themselves by ongoing discoveries.

In the present paper we further extend [13] to consider stochastic demand. Because demand is a fundamental ingredient of both Cournot and Bertrand markets, it ought to be modeled for any degree of realism. Empirically, commodity producers have a major exposure to the business cycle both in terms of profit and E&P activities. For instance, during the worldwide financial crisis of 2008–09, commodity demand (and subsequently prices) had a major downward leap, triggering significant shrinkage of E&P expenditures by oil producers. This heightened link between demand and exploration motivates us to incorporate demand uncertainty into the game model. We maintain the basic setting of [9] and [13] which considers a heterogenous duopoly between an exhaustible old-school producer, and an inexhaustible "green" producer. This model maintains some analytic tractability and allows to explore the switch between energy production mixes within a game-theoretic framework. A related concept would be to consider stochastic production *costs*, or more precisely the relative costs of the two agents that drive the original asymmetry in their profits. Indeed, commodity production frequently requires massive other investments (e.g. labor, energy use, favorable government regulation) whose cost may rise or fall over time. Therefore, stochastic costs could be a proxy for shifting differences among the players. We also mention [12] for analysis of cost asymmetries in Bertrand models and [14] for a survey of other stochastic models of Cournot markets.

The rest of this paper is organized as follows. After a brief overview of the model considered, Sect. 2 sets up the full mathematical model and derives the HJB-I equations and their basic properties. Section 3 focuses on the impact of stochastic demand compared to the earlier analysis in [13]. Section 4 mentions related extensions or modifications of the basic dynamic Cournot duopoly. Finally, Sect. 5 presents conclusions of our investigations.

1.1 Model Overview

Our model features two players: producer 1 that extracts a non-renewable resource (oil) and has to worry about reserves; and producer 2 who extracts a renewable resource (green energy) and therefore has infinite reserves. We also consider two stochastic state variables: demand D_t and current reserves X_t of the exhaustible player 1. Moreover, agents have a total of three controls, namely production rates for producers 1 and 2, as well as exploration effort for producer 1.

The two producers compete through a Cournot framework, in which producers choose quantities of energy to produce and receive profit based on a single market price determined through aggregate supply. Costs of the exhaustible player are driven solely by the costly (convex) exploration efforts; her production costs are taken to be zero. On the contrary, the green producer has a positive marginal cost of production but inexhaustible resources so his additional shadow marginal costs are zero. These production costs of player 2 are a proxy for the amount of competition. They also reflect the present reality of non-renewable energy production as being the cheaper incumbent against the new renewable entrants. The Cournot framework is used because at the macro level, energy is perfectly substitutable and so the two producers' products are in direct competition.

Our aim is to study this duopoly of the exhaustible resources producer with a green producer in terms of dynamic Nash equilibria. The model is cast in continuous-time so as to allow use of the well-understood Hamilton-Jacobi-Bellman-Isaacs (HJB-I) methodology that reduces computational analysis to study of coupled systems of differential equations. In order to simplify the mathematics as much as possible while maintaining dynamic effects, we keep the dynamics of (D_t) and (X_t) stylized. To this end, reserves (X_t) follow piecewise deterministic trajectories, smoothly decreasing due to production and experiencing constant-size jumps upon new reserves discovery. These upward jumps of fixed size δ mimic discrete discoveries of new oil fields or new oil recovery technologies that take place abruptly. Such "Poissonian" dynamics date back to work of Arrow and Chang [1] and are arguably more fidel to realistic reserves evolution. The demand level (D_t) is modeled as $D_t = D(M_t)$ where M_t is a finite-state Markov chain; this is meant to evoke the popular regime-switching models that are frequently used in financial mathematics to model the business cycle fluctuations. In the context of single-agent optimization, a related model of resource extraction (with an exogenous discovery process) within a random environment was studied in [6].

The main setting just takes D_t to have two possible levels L, H. The demand level modulates the common price obtained by the producers for a fixed supply level. In a toy setting this occurs linearly. Both state variables are modeled by stationary processes, leading to an infinite-horizon discounted game. With the Markovian dynamics, this allows to reduce equilibrium behavior to Markov feedback (closed-loop) strategies. A drawback is that agents are infinitely long-lived (i.e. never leave or enter the game) and no off-equilibrium behavior is modeled.

The combination of the above choices, in particular lack of any diffusive stochastic factors, keeps the overall state-space as simple as possible and makes the HJB-I equations first-order only, removing many of the analytical difficulties arising in second-order equations (for example [9] found that these equations can sometimes be hyperbolic rather than parabolic, causing unstable analytic and numeric behavior). Moreover, since only a single agent has reserves, there is just one continuous state-variable, effectively allowing us to deal only with ordinary differential equations, rather than pde's. We however stress that exploration necessarily introduces additional subtleties; in our model it brings in a *non-local* term that requires careful treatment even at the implementation level.

2   Model

Two players (named 1 and 2) produce perfectly substitutable goods at rates q^1, q^2. They sell into the same market which features the inverse demand curve $p \mapsto D(p)$. The clearing market price p is determined from the global supply-demand equilibrium,

$$p = D^{-1} \left(\sum_\ell q^\ell \right).$$

In the simplest case of a linear demand curve, we have that $p = \bar{D} - q^1 - q^2$ where $\bar{D}$ is the maximum (finite) *choke* price under zero supply. Players interact with each other through this price mechanism that is driven by aggregate supply (and affects players' profits) leading to a noncooperative game.

In the present paper we consider the situation of *stochastic demand*, whereby market demand exhibits exogenous fluctuations over time. All variables are thereafter continuously indexed by $t \in \mathbb{R}_+$. We model stochastic demand by making $\bar{D}$ non-constant, modulated by an exogenous factor (M_t), namely

$$p_t \equiv p(q^1, q^2, M_t) = M_t - q^1 - q^2. \tag{1}$$

We assume that (M_t) is a finite-state stationary Markov chain with state space E and generator $\Lambda \equiv (\lambda_{ij})$. Thus, a larger value of M_t means stronger demand and therefore a higher price p_t for the same level of supply. For illustrative purposes we shall focus on the case where (M_t) is a two-state Markov chain with state space $\{D_0, D_1\} \equiv \{L, H\}$, where $0 < L \leq H$. In that case we label the time-homogeneous switching rates between the two regimes of (M_t) as λ_{01} and λ_{10} respectively.

Player 1 extracts a non-renewable resource that may become exhausted. His reserves at time t are denoted by $X_t \geq 0$. Reserves decrease through production but can be replenished via exploration. Without any reserves, the player may not produce but may continue to search for replenishments. Denote by $a_t \geq 0$ the exploration effort at time t, and let (N_t) be a point process for counting discoveries of new resources. Then (N_t) has controlled intensity λa_t, i.e.

$$\mathbb{P}(\tau_{n+1} > \tau_n + t) = \exp\left(-\int_0^t \lambda a_{s+\tau_n} ds\right),$$

where τ_n's are the arrival times of (N_t). The unit amount of new discovery is a fixed $\delta > 0$ (see Remark 3 for discussion of making discovery amounts random). Overall, the reserve process (X_t) of producer 1 follows

$$dX_t = -q_t^1 \mathbf{1}_{\{X_t > 0\}} dt + \delta dN_t,$$

where q_t^1 is the production rate. Exploration is costly and generates costs at rate $C(a_t)$ per unit time. The cost function C is convex with $C(0) = 0$, a typical example (see [13]) is

$$C(a) = \frac{1}{\beta} a^\beta + \kappa a, \qquad \beta > 1, \kappa \geq 0. \tag{2}$$

Producer 2 always has infinite resources, but faces positive fixed production costs $c \geq 0$. The parameter c is a proxy for the amount of competition between the producers; large c makes Producer 2 uncompetitive. It is possible for the controls to be zero in which case there is no production (reserves remain constant) or no exploration (i.e. discovery rate is zero).

Players aim to maximize their total discounted profit, which is equal to the instantaneous revenue $p_t \cdot q_t$, minus the production and exploration costs, integrated and discounted (using continuous discount rate $r > 0$) on the infinite time horizon.

To analyze the game equilibria we use the notion of Markov Nash equilibria. Thus, player strategies are assumed to be in closed-loop feedback form, $q_t^\ell = q^\ell(X_t, M_t)$, $\ell = 1, 2$ and $a_t = a(X_t, M_t)$. Given an equilibrium $(q^{\ell,*}(X_t, M_t), a^*(X_t, M_t))$ we denote the corresponding game functions of producer 1 by $v_L(x)$ and $v_H(x)$; and the game functions of producer 2 by $g_L(x)$ and $g_H(x)$. Here the subscript indicates the initial value $M_0 \in \{L, H\}$ of the Markov chain. These game values are the discounted cumulative expected profits starting with $X_0 = x, M_0 = i$,

$$v_i(x) = \mathbb{E}\left[\int_0^\infty e^{-rt} \left(q_t^{1,*} p(q_t^{1,*}, q_t^{2,*}, M_t) - C(a_t^*) \right) dt \middle| X_0 = x, M_0 = i \right];$$

$$g_i(x) = \mathbb{E}\left[\int_0^\infty e^{-rt} q_t^{2,*} \left(p(q_t^{1,*}, q_t^{2,*}, M_t) - c \right) dt \middle| X_0 = x, M_0 = i \right],$$

and must satisfy the Nash optimality conditions

$$v_i(x) = \sup_{q^1, a} \mathbb{E}\left[\int_0^\infty e^{-rt} \left(q_t^1 p(q_t^1, q_t^{2,*}, M_t) - C(a_t) \right) \mathbf{1}_{\{X_t > 0\}} \, dt \middle| X_0 = x, M_0 = i \right]$$

$$\tag{3}$$

$$g_i(x) = \sup_{q^2} \mathbb{E}\left[\int_0^\infty e^{-rt} q_t^2 \left(p(q_t^{1,*}, q_t^2, M_t) - c \right) dt \middle| X_0 = x, M_0 = i \right], \quad i = L, H.$$

Thus, given the other player's equilibrium strategy, each player chooses optimal strategies for her own production (and exploration). To analyze (3), we employ the Hamilton-Jacobi-Bellman-Isaacs framework that aims to express game values through a system of coupled differential equations. Define $\Delta f(x) := f(x + \delta) - f(x)$. We also index a generic regime by $i = 1, 2$ and the *other* regime by j. Assuming the functional forms of demand and exploration costs in (1)–(2), the HJB-I ODEs of v_L, v_H and g_L, g_H are

$$\sup_{q_i^1} \left[q_i^1(x) \left(D_i - q_i^1(x) - q_i^{2,*}(x) \right) - v_i'(x) q_i^1(x) \right] + \sup_{a_i} \left[a_i \lambda \Delta v_i(x) - C(a_i) \right]$$

$$+ \lambda_{ij}(v_j(x) - v_i(x)) - rv_i(x) = 0,$$

$$\tag{4}$$

$$\sup_{q_i^2} \left[q_i^2(x) \left(D_i - q_i^{1,*}(x) - q_i^2(x) - c \right) \right] - g_i'(x) q_i^{1,*}(x) + a_i^*(x) \lambda \Delta g_i(x)$$

$$+ \lambda_{ij}(g_j(x) - g_i(x)) - rg_i(x) = 0.$$

$$\tag{5}$$

Upon exhaustion of reserves $X_t = 0$, player 1 can no longer produce, yielding a temporary monopoly for player 2. However, player 1 remains in the game and may continue to explore for reserves (financing exploration by borrowing against future earnings). Fix $X_0 = 0$ and denote by $\tau \equiv \tau_1$ the time of the first discovery of new reserves (so that $X_t = 0$ on $[0, \tau)$ and $X_\tau = \delta$) and by σ the first transition time of the Markov chain (M_t). Then by conditioning on τ and σ we have

$$
v_i(0) = \sup_{a_i \geq 0} \mathbb{E} \left\{ \mathbf{1}_{\{\sigma < \tau\}} \left[e^{-r\sigma} v_j(0) - \int_0^\sigma e^{-rt} C(a_i)\, dt \right] \right.
$$

$$
\left. + \mathbf{1}_{\{\tau \leq \sigma\}} \left[e^{-r\tau} v_i(\delta) - \int_0^\tau e^{-rt} C(a_i)\, dt \right] \,\Big|\, X_0 = x, M_0 = i \right\} \vee 0. \qquad (6)
$$

By stationarity of (M_t) it follows that the optimal exploration rate a_i is constant until $\tau \wedge \sigma$ and hence $\tau \wedge \sigma \sim Exp(\lambda a + \lambda_{ij})$ has an exponential distribution. Using the fact that $\mathbb{P}(\tau < \sigma) = \frac{\lambda a_i}{\lambda a_i + \lambda_{ii}}$ then leads to

$$
v_i(0) = \sup_{a_i \geq 0} \frac{v_j(0)\lambda_{ij} + v_i(\delta)\lambda a_i - C(a_i)}{r + \lambda_{ij} + \lambda a_i} \vee 0, \qquad (7)
$$

yielding an implicit condition linking $v_i(0)$, $v_i(\delta)$ and $v_j(0)$.

Optimizing for the production rates q^ℓ which must be non-negative in (4)–(5) yields that the candidate equilibrium strategies are given by

$$
\begin{cases}
q_i^{1,*}(x) = \dfrac{1}{2} \max \left(D_i - q_i^{2,*}(x) - v_i'(x), 0 \right), \\[2mm]
q_i^{2,*}(x) = \dfrac{1}{2} \max \left(D_i - q_i^{1,*}(x) - c, 0 \right).
\end{cases} \qquad (8)
$$

For simpler notation, we write $z^+ \equiv \max(z, 0)$. Figure 1 illustrates how (8) is used to determine the equilibrium given a fixed value of say $v_H'(x)$. Assuming (2), the candidate optimal exploration rate is similarly

$$
a_i^*(x) = [(\lambda \Delta v_i(x) - \kappa)^+]^{\gamma-1}, \qquad \gamma = \frac{\beta}{\beta - 1}. \qquad (9)
$$

Equation (9) holds also for $x = 0$ since the exhaustibility constraint does not apply to exploration.

We observe that (4) yields two coupled equations for $v_L(x)$ and $v_H(x)$ which are however autonomous from $g_i(x)$. This is due to the state variable (X_t) being completely controlled by player 1. The system (4) features only a first-order differential of $v_i(x)$ due to the continuous decrease in (X_t); it also has two non-local effects, the term $\Delta v_i(x)$ arising from jumps induced by exploration successes, and the term $v_j(x) - v_i(x)$ due to the regime-shifts in (M_t). Therefore, overall (4) is

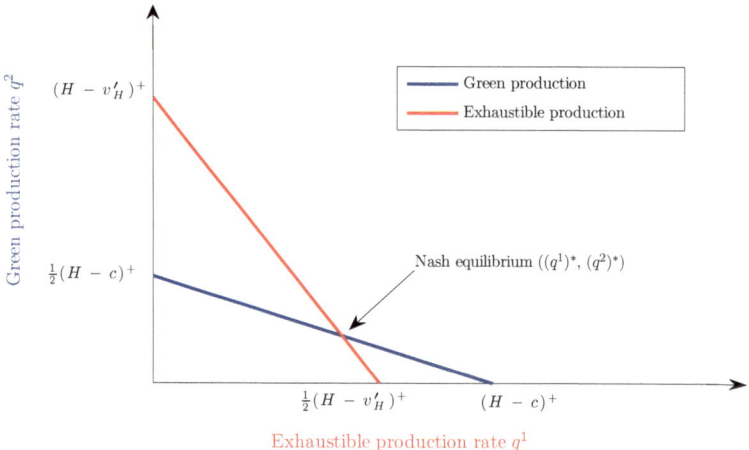

Fig. 1 Nash equilibrium of the Cournot duopoly in high-demand regime. The two piecewise linear curves show optimal production rates of player 1 and player 2 given $v'_H(x)$ and the production rate of the other player (e.g. $q^{1,*}(x; q^2(x), v'_H(x))$) as defined in (8). Equilibrium is achieved when the curves cross

a system of first-order nonlinear (forward)-delay ODEs in x. Respectively, (5) leads to a first order linear delay-ODE for $g_L(x)$ and $g_H(x)$ in terms of $v_L(x)$ and $v_H(x)$. However, as already written in (8), all the equilibrium production and exploration strategies depend only on $v_i(x)$, and so we will not deal much with the g_i equations, focusing mostly on (4).

2.1 Game Stages

The maximizers in (4)–(5) intuitively determine the equilibrium strategies of the players. Several different equilibrium types are possible due to the constraints $q^1 \geq 0$, $q^2 \geq 0$ and $a \geq 0$ that can be binding. These situations can be seen through the piecewise nature of (4)–(5) that arises from the $\max(\cdot, 0)$ terms. They can also be imagined through Fig. 1: if the two piecewise linear curves for $q_i^{1,*}, q_i^{2,*}$ do not cross in the interior then we have a boundary solution on one of the axes.

For each L, H, and c fixed, the strategies of the two players depend on the shadow reserves cost $v'_i(x), i = L, H$, which determines the type of equilibrium at $X_t = x$. Depending on the shadow costs, the alternative game types are:

Type I: *Interior* solution where both players are active: $q^1 > 0, q^2 > 0$. This case arises when $v'_i(x)$ satisfies $2c - D_i < v'_i(x) < \frac{c+D_i}{2}$.

Type M1:   The exhaustible player 1 has a monopoly because $q^2 = 0$. This occurs when $v'_i(x) \leq 2c - D_i$.

Type M2:   The green player 2 has a monopoly while $q^1 = 0$. This occurs when $v_i'(x) \geq \frac{c+D_i}{2}$.

In each case we further can have either $a_i(x) > 0$, or $a_i(x) = 0$ (no exploration, i.e. "saturation" of reserves) depending on the positivity of the term $\lambda \Delta v_i(x) - \kappa$ in (9).

In the interior game Type I, we have

$$q_i^1 = \frac{1}{3}(D_i + c - 2v_i'(x)) \qquad q_i^2 = \frac{1}{3}(D_i + v_i'(x) - 2c), \quad x > 0, \tag{10}$$

and the ODE of v_i reduces to

$$v_i'(x) = \frac{D_i + c}{2} - \frac{3}{2}\left[(\lambda_{10} + r)v_i(x) - \lambda_{ij}v_j(x) - \frac{1}{\gamma}((\lambda \Delta v_i(x) - \kappa)^+)^\gamma\right]^{\frac{1}{2}}. \tag{11}$$

In Type M1 equilibrium, exhaustible producer 1 monopolizes the market,

$$q_i^1 = \frac{1}{2}\left(D_i - v_i'(x)\right), \quad q_i^2 = 0, \tag{12}$$

and the ODE of v_i is given by the monopoly equation

$$\frac{1}{2}(D_i - v_i'(x))^2 + \frac{1}{\gamma}((\lambda \Delta v_i(x) - \kappa)^+)^\gamma + \lambda_{ij}(v_j(x) - v_i(x)) - rv_i(x) = 0. \tag{13}$$

In Type M2 equilibrium, the green producer 2 monopolizes the market, $q_i^1 = 0$, giving $q_i^2 = \frac{1}{2}(D_i - c)$, and $v_i(x)$ is determined through the nonlinear equation

$$\frac{1}{\gamma}((\lambda \Delta v_i(x) - \kappa)^+)^\gamma + \lambda_{ij}v_j(x) - (r + \lambda_{ij})v_i(x) = 0. \tag{14}$$

Intuitively, $v_i(x)$ is concave, so that $x \mapsto v_i'(x)$ is decreasing. Thus, Type M2 equilibrium arises for small x (when the shadow cost of exhaustibility is very large, driving player 1 to sit out); Type I equilibrium arises for moderate x and Type M1 equilibrium arises for large x where the shadow cost is negligible. From the conditions for each equilibrium, it is clear that larger Player 2 costs c make Type M1 game more likely, i.e. increase the region where the exhaustible player is a monopolist. We also note that in high demand regime, exhaustible resources production is always positive for any $x > 0$ since player 1 cannot expect higher profit by holding on to reserves. Therefore Type M2 equilibrium may only arise for small x and $M_t = L$.

Combining (11)–(14), the HJB ODEs of $v_i(x)$ can be written as the single piecewise-defined equation

$$\left[\frac{2}{3}\left(\frac{D_i + c}{2} - v_i'(x)\right)^+ - \frac{1}{6}\left(2c - D_i - v_i'(x)\right)^+\right]^2 + \frac{1}{\gamma}\left[(\lambda \Delta v_i(x) - \kappa)^+\right]^\gamma$$

$$+ \lambda_{ij}v_j(x) - (\lambda_{ij} + r)v_i(x) = 0, \quad i,j = L, H. \tag{15}$$

For the remainder of the paper we assume existence-uniqueness of a solution to (15). As explained below, numerically we in fact solve an approximation to (15) that is guaranteed to be well-posed, so the above issue is of theoretic concern only.

Remark 1. Within a single-agent optimization setting, [6] showed the well-posedness of the system (11) and associated implicit boundary condition (7). In particular, using the functional equations methods of [15], Deshmukh and Pliska [6] prove that (11) has a unique bounded differentiable solution and the optimal controls are indeed (10). They moreover show that v_i is strictly concave, i.e. marginal cost of reserves is decreasing in x. However, these analytic tools become unavailable in the presence of free boundaries that arise due to game effects such as blockading or production shut-down (i.e. other equilibria types beyond Type I). In particular, the interaction of the non-local term $\Delta v_i(x)$ with the piecewise-defined functional form of $v_i'(x)$ in (15) (and the additional coupling among $v_i(x)$ and $v_j(x)$) poses a major challenge. Additionally, the possibility that both terms involving $v_i'(x)$ are zeroed-out, which happens under Type M2 game, means that in the latter region $v_L(x)$ satisfies an algebraic relation with $v_H(x)$ (rather than an ODE), i.e. there is possibly an algebraic free-boundary embedded in (15). Due to these difficulties, we are not able to analytically establish smoothness of (15) in the duopoly model.

2.2 Numerical Scheme

Due to the challenge of implicit boundary condition and the presence of a "forward" delay term on the semi-infinite domain $\mathbb{R}_+$, numerically solving the system (4) (or (15)) is nontrivial. We note that the equations have a non-local term and several free boundaries indicated by the critical values of x that trigger the $(\cdot)^+$ terms to be zero. In particular, there is x_i^{start} which separates Type M2 and Type I equilibria; x_i^{block} that separates Type I and Type M1 equilibria; and x_i^{sat} that separates regions where $a_i(x) > 0$ and $a_i(x) = 0$. The meaning of these quantities is the production start level (below x_i^{start} production is shut-down); the blockading level (above x_i^{block} player 2 is blockaded) and the reserves saturation level (above x_i^{sat} no exploration takes place).

To solve the HJB-I ODE system we use the iterative scheme in [13] that originates in the ideas of Davis [5]. Let $v_L^{(0)}(x)$ and $v_H^{(0)}(x)$ be the game values corresponding to the case where resources are completely non-replenishable, so that no new resource discoveries are possible. Similarly to (15) (and removing the exploration-related term), we have that

$$
\left[\frac{2}{3}\left(\frac{D_i+c}{2}-(v_i^{(0)})'(x)\right)^+ - \frac{1}{6}\left(2c-D_i-(v_i^{(0)})'(x)\right)^+\right]^2
$$

$$
+ \lambda_{ij}v_j^{(0)}(x) - (\lambda_{ij}+r)v_i^{(0)}(x) = 0, \quad i,j = L,H, \tag{16}
$$

with boundary conditions $v_L^{(0)}(0) = v_H^{(0)}(0) = 0$, since starting with no reserves $(x = 0)$ and no possibility of discoveries, player 1 will never have any revenue. We then define inductively for $n \geq 1$ the functions $v_i^{(n)}$ via

$$
\left[\frac{2}{3}\left(\frac{D_i+c}{2}-(v_i^{(n)})'(x)\right)^+ - \frac{1}{6}\left(2c-D_i-(v_i^{(n)})'(x)\right)^+\right]^2
$$

$$
+ \frac{1}{\gamma}\left[\left(\lambda(v_i^{(n-1)}(x+\delta)-v_i^{(n)}(x))-\kappa\right)^+\right]^\gamma + \lambda_{ij}v_j^{(n)}(x) - (\lambda_{ij}+r)v_i^{(n)}(x) = 0,
$$

$$
\tag{17}
$$

with boundary conditions

$$
v_i^{(n)}(0) = \sup_{a_i \geq 0} \frac{v_j^{(n)}(0)\lambda_{ij} + v_j^{(n-1)}(\delta)\lambda a_i - C(a_i)}{r + \lambda_{ij} + \lambda a_i}, \quad i,j = L,H. \tag{18}
$$

Note that (17)–(18) partially uncouple the original equations by making the non-local term a source term instead. Therefore, (17)–(18) are now a standard system of nonlinear first order ODE with an implicit boundary condition at $x = 0$ and hence well-posed. We start the computation with no discovery case $n = 0$ by solving the system (16). For $n \geq 1$, we solve the system (17) by using the data from the $(n-1)$-st iteration for the forward delay term in the n-th iteration. In numerical computation, at each $x > 0$ we start with assuming $v_i'(x)$ satisfy any one of the three equilibrium types and then solve for $v_i(x)$ and $v_i'(x)$. If the computed $v_i'(x)$ is consistent with the assumed range of values, the assumption and the results are correct, otherwise we switch to another assumption to compute $v_i(x)$ and $v_i'(x)$. To handle Type M2 equilibrium whereby the terms in (17) involving $(v_i^{(n)}(x))'$ are zero, we formally differentiate (14) to obtain an expression for $(v_L^{(n)})'(x)$ in terms of $(v_H^{(n)})'(x)$ and $(v_L^{(n-1)})'(x+\delta)$.

Remark 2. Equation (17) is an implicit quadratic in v'. To use a time-stepping ODE solver then requires inverting this to directly express $v'(x)$ in terms of $(x, v(x))$. Due to the special form above, one obtains $v'(x) = \sqrt{F(x, v(x))}$ where only the positive square root is relevant since $v(x)$ must be increasing. Standard tools such as Runge-Kutta methods can be applied to solve the ODEs. We used the Matlab fourth-order ODE solver `ode45` in our numerical experiments.

Informally, $v^{(n)}$ represents the game value assuming a horizon of τ_n, where τ_n is the time of n-th resource discovery. Thus, $v^{(n)}$ is an upper bound for equilibrium profit from time period $[0, \tau_n]$. Because total discounted profits are uniformly bounded by $\max_q r^{-1} p \cdot q$, it follows that the absolute error $|v_i^{(n)}(x) - v_i(x)|$ is bounded by $C\mathbb{E}[e^{-r\tau_n}]$. In turn, using the fact that exploration rates are uniformly bounded (in particular $a_i^*(x) \le a_H^*(0) \forall x$) we have by a simple coupling argument that $\tau_n > e_n$ (in the sense of first-order stochastic dominance) where $e_n \sim Gamma(n, \lambda a_H^*(0))$, so that $\mathbb{E}[e^{-r\tau_n}] = \mathcal{O}(C^n)$ for some constant $C < 1$. It follows that $v^n \to v$ exponentially fast. Practically, we have observed convergence after 10–20 iterations in all examples presented.

Remark 3. The size δ of each new discovery can be random in general. Namely, we may model discovery amounts via a stochastic sequence δ_n, $n = 1, 2, \ldots$, where each δ_n is identically distributed with some distribution $F_\delta(\cdot)$ and independent of everything else in the model (in particular of other δ_n's and of past controls). Introducing F_δ entails replacing $v_i(x + \delta)$ in the HJB-I equations (4)–(5) and boundary condition (7) with $I_\delta[v_i](x) := \int_0^\infty v_i(x + u) F_\delta(du)$. Similarly, one would apply this integral operator I_δ to $v^{(n-1)}$ in (17)–(18), which is straightforward to handle numerically.

2.3 *Illustration*

Figure 2 illustrates the obtained solution of (4)–(5). We show the game values, production rates and exploration rates for two different values of player 2 costs c so as to illustrate the basic impact of competition. Smaller c makes player 2 more competitive, while larger c gives extra preference to player 1. As expected, $v_i(x)$ are concave increasing; $q_i^1(x)$ are also concave increasing; $q_i^2(x)$ are convex decreasing, and $a_i(x)$ is convex decreasing (note that in general the control mappings will not be differentiable in x across the free boundaries so the above characterization is heuristic). Note that the impact of c is ambiguous. While lower c raises competition, it may also spur higher exploration efforts since the marginal cost of reserves could rise. Hence $c \mapsto a_i(x; c)$ may be non-monotone. In contrast, competition unambiguously lowers production rates $q_i^1(x)$ of the exhaustible player and her game value $v_i(x)$.

In the far-field limit $x \to \infty$, dependence on reserves vanishes and all quantities have a limit that can be directly computed via a one-stage static game. In particular, we have

$$\lim_{x \to \infty} v_L(x) = \frac{1}{r} \frac{(\lambda_{10} + r)\left(\frac{L+c}{2} - \frac{1}{6}(2c - L)^+\right)^2 + \lambda_{01}\left(\frac{H+c}{2} - \frac{1}{6}(2c - H)^+\right)^2}{r + \lambda_{01} + \lambda_{10}};$$

$$\lim_{x \to \infty} v_H(x) = \frac{1}{r} \frac{\lambda_{10}\left(\frac{L+c}{2} - \frac{1}{6}(2c - L)^+\right)^2 + (\lambda_{01} + r)\left(\frac{H+c}{2} - \frac{1}{6}(2c - H)^+\right)^2}{r + \lambda_{01} + \lambda_{10}}.$$

Similarly, the limiting values of the controls admit explicit solutions, including $\lim_{x \to \infty} a_i(x) = 0$.

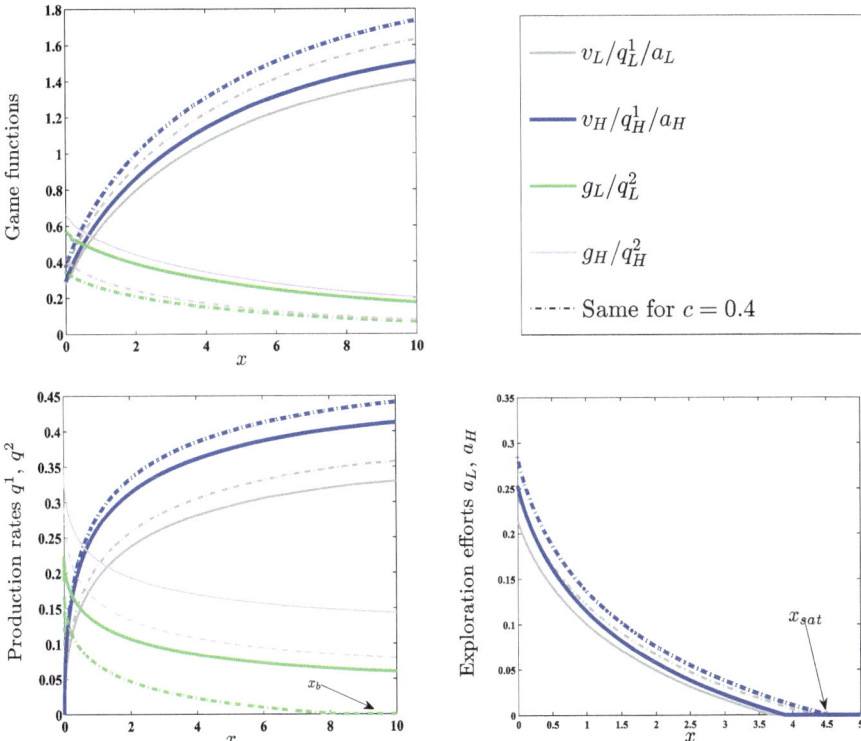

Fig. 2 Duopoly axis game solutions for two different levels of production cost c: *solid curves* are for $c = 0.3$; *dashed* for $c = 0.4$. *Top left panel*: Game functions $v_i(x), g_i(x)$. *Bottom left panel*: Production rates $q_i^\ell(x)$, $\ell = 1, 2$. *Bottom right panel*: Exploration efforts $a_i(x)$ of exhaustible producer 1. We take linear inverse demand with $L = 0.75, H = 1$, and switching rates $\lambda_{01} = \frac{1}{3}, \lambda_{10} = \frac{1}{5}$. Exploration costs are $C(a) = 0.1a + a^2/2$

3 Effects of Stochastic Demand on Production and Exploration

Stochastic demand is one of the key features of our model which is determined by two characteristics: the demand levels D_i and stationary distribution (π_L, π_H) of the demand regimes that is driven by the switching rates λ_{ij}. We analyze the effects of changes in these parameters on the equilibrium strategies of production and exploration.

The regime-switching stochastic demand is meant to mimic the macroeconomic business cycle. When the macroeconomy is running low, the demand for energy is low; when the macroeconomy is running high, the demand for energy is high and therefore the price function moves up to the high regime. In general, higher demand (or better opportunities for profit), lead to higher value for the producer and induce both higher production and higher exploration. This is because the marginal value of reserves rises, stimulating both short-term production and longer-term extraction. This basic feature is present in all single-parameter comparative statics we tried, such as unilaterally changing the rates $\lambda_{01}, \lambda_{10}$ or increasing/decreasing the levels of market demand L, H. Similarly, due to the intuitive ordering of the two regimes, we expect that changing the regime to high while keeping all other parameters constant should also increase expected profit, increase production and increase exploration. Indeed, by a coupling argument it is easy to show that $v_H(x) > v_L(x)$. The argument regarding $q_H^1(x) > q_L^1(x)$ is more intricate. Indeed, while production rates are tied to shadow costs of reserves and $v_H'(x) > v_L'(x)$, one must also establish that $v_H'(x) - v_L'(x) < \frac{H-L}{2}$ under Type I equilibrium (cf. (10)) and $v_H'(x) - v_L'(x) < H - L$ under Type M1 equilibrium. Numerical experiments suggest that $x \mapsto v_H'(x) - v_L'(x)$ is decreasing and clearly $\lim_{x \to \infty} v_H'(x) = \lim_{x \to \infty} v_L'(x) = 0$. Thus, the upper bounds above are only relevant for small x. In the case of full monopoly (i.e. $c > H$), whence only Type M1 occurs, one can indeed show that $v_H'(0) - v_L'(0) \leq H - L$, but the situation when all three equilibria are present for different x remains open. Nevertheless, based on numerical evidence we make the following conjecture.

Conjecture 1. Suppose that the state space of (M_t) is $E = \{L, H\}$ and $c < H$. Then for all x,

$$0 \leq v_H'(x) - v_L'(x) \leq \frac{H-L}{2}. \tag{19}$$

Corollary 1. *Suppose Conjecture 1 holds. Then for all $x \geq 0$, $0 \leq q_H^1(x) - q_L^1(x) \leq \frac{H-L}{2}$ and $a_H(x) > a_L(x)$.*

Proof. Substituting the conjectured relationship of v_L' and v_H' into the equations for $q_i^1(x)$ in (10)–(12) gives the stated inequalities. Similarly, we have

$$\Delta v_L(x) = v_L(x + \delta) - v_L(x) = \int_0^\delta v_L'(x + u)du$$

$$\leq \int_0^\delta v_H'(x + u)du = v_H(x + \delta) - v_H(x) = \Delta v_H(x),$$

and therefore $a_L^*(x) = [(\lambda \Delta v_L(x) - \kappa)^+]^{\gamma-1} \leq [(\lambda \Delta v_L(x) - \kappa)^+]^{\gamma-1} = a_H^*(x)$.

The Corollary can be observed in the panels of Fig. 2 showing impact of demand regimes on equilibrium $q^1(x)$ and $a(x)$. We have observed Conjecture 1 holding in all parameter settings we tried; but in general it is known that comparison of stochastic regimes is extremely difficult (see an extended discussion on this issue in [6]). Finally, we note that Conjecture 1 only covers two regimes and the situation where (M_t) modulates $p(\mathbf{q}_t)$ only. See Sect. 4.1 for counterexamples in more general situations.

3.1 Limiting Cases

The stationary distribution of the Markov chain (M_t) driving demand regimes is

$$\pi_L = \frac{\lambda_{10}}{\lambda_{01} + \lambda_{10}}, \quad \pi_H = \frac{\lambda_{01}}{\lambda_{01} + \lambda_{10}}. \tag{20}$$

The game value functions v_L, v_H of producer 1 under stochastic demand are bounded above by the value function under the high regime demand $D_t \equiv H$, which we denote by $\tilde{u}_H$. They are also bounded below by the value function under the deterministic low-regime demand $D_t \equiv L$, which we denote by $\tilde{u}_L$. As a result, v_L, v_H are "averages" of $\tilde{u}_L$ and $\tilde{u}_H$. The following two Lemmas clarify this interpretation by considering the extreme cases.

Lemma 1. *We have* $\lim_{\lambda_{01} \to +\infty} v_L(x) = \lim_{\lambda_{01} \to +\infty} v_H(x) = \tilde{u}_H$, *the game value of the model with fixed demand* $D_t \equiv H$.

Proof. See the Appendix.

Lemma 2. *Let* $\lambda_{01} = \frac{c_{01}}{\epsilon}, \lambda_{10} = \frac{c_{10}}{\epsilon}$. *As* $\epsilon \to 0$, *the game value functions* $v_L(x)$ *and* $v_H(x)$ *both converge to* $\bar{v}(x)$, *which is determined by the following delay ODE*

$$\pi_L \left[\frac{2}{3} \left(\frac{L+c}{2} - \bar{v}'(x) \right)^+ - \frac{1}{6} \left(2c - L - \bar{v}'(x) \right)^+ \right]^2$$

$$+ \pi_H \left[\frac{2}{3} \left(\frac{H+c}{2} - \bar{v}'(x) \right)^+ - \frac{1}{6} \left(2c - H - \bar{v}'(x) \right)^+ \right]^2$$

$$+ \frac{1}{\gamma} [(\lambda \Delta \bar{v}(x) - \kappa)^+]^\gamma - r \bar{v}(x) = 0, \quad x > 0 \tag{21}$$

with boundary condition

$$\bar{v}(0) = \frac{\lambda \bar{a}(0) \bar{v}(\delta) - C(\bar{a}(0))}{r + \lambda \bar{a}(0)}, \quad \bar{a}(0) = [(\lambda \Delta \bar{v}(0) - \kappa)^+]^{\gamma-1}, \tag{22}$$

where π_i *are given in* (20).

Proof. See the Appendix.

Equation (21) can have as many as 5 free boundaries due to the multiple piecewise defined terms $(\cdot)^+$. While we have the basic ordering $(L + c)/2 < (H + c)/2$ and $2c - H < 2c - L$, the relationship between say $(H + c)/2$ and $2c - H$ depends on c. Therefore, the order of all the potential solution pieces in terms of x is parameter dependent. Moreover, while $\bar{v}(x)$ is expected to be concave the nonlocal term $\Delta \bar{v}(x)$ does not allow to guarantee this and therefore the a priori ordering of the pieces cannot be fully determined. Compared to the basic (15), (21) is of the form $(A_1 - v')^2 + (A_2 - v')^2 = rv$ so solving the quadratic for v' it is possible to obtain multiple positive roots. Determining the correct root is then done by enforcing the C^1 continuity of v', i.e. making sure that $x \mapsto v'(x)$ is continuous (and decreasing).

Lemma 2 illustrates what happens when the macroeconomic environment becomes more volatile. The parameter ϵ can be thought of as a proxy for volatility. The lemma shows the homogenization arising as $\epsilon \to 0$. Indeed, increasing ϵ can be viewed as increase in market volatility. Figure 3 illustrates the behavior of the value functions and controls in terms of ϵ. The value and marginal value of reserves decreases in high regime and increases in low regime. Therefore the production rate increases in high regime and decreases in low regime, since holding reserves becomes more valuable in high regime and less valuable in low regime. Similarly, as $\epsilon \searrow 0$, the exploration effort decreases in high regime due to decreased marginal value of a new discovery and increases in low regime.

3.2 Production Shut-Down in Low-Demand Regime

Due to fluctuating profit levels across the macroeconomic regimes, there may arise situations in which player 1 voluntarily shuts down production in the low demand regime when reserves level x is small. We define $x_{start} := \inf\{x > 0 : q_L(x) > 0\}$, the critical reserves level below which production stops. Heuristically, below x_{start} marginal value of reserves is so high relative to the low price offered that $v'_L(x) > L$ leading to $q^1_L(x) = 0$ in (8).

As mentioned before, such a shutdown (M2-equilibrium) can only happen in the low regime and is driven by the expectation of collecting higher revenue

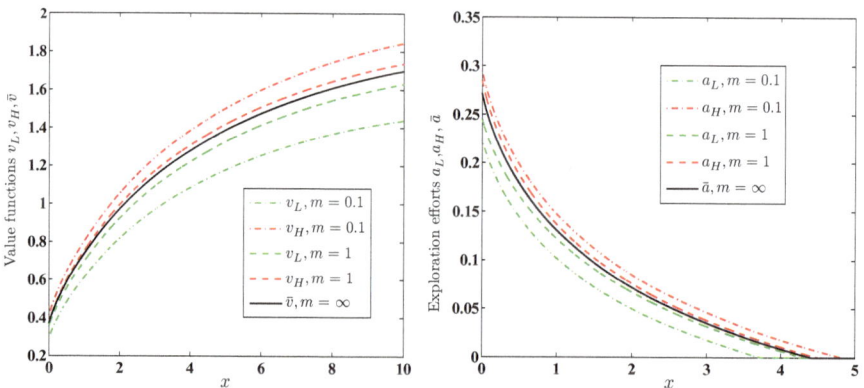

Fig. 3 *Left panel*: convergence of the game functions $v_i(x) \to \bar{v}(x)$ as $\lambda_{01}, \lambda_{10} \to \infty$ together. *Right panel*: convergence of the exploration effort $a_i(x) \to \bar{a}(x)$. We take $\lambda_{01} = m/3, \lambda_{10} = m/5$, with $m = 0.1, 1$ as well as the limiting solution defined in Lemma 2

once demand reverts to the high regime. Typically, x_{start} is quite low, so M2-equilibrium arises just before total exhaustion of reserves. There are two motivations for production shut-down. The *quantity effect* reflects the extra profits available under high demand. Thus, a sufficiently large difference in demand levels across regimes is required for an M2 equilibrium to appear. The *time effect* reflects the anticipated waiting time until high demand which is important due to the present value discounting involved. Thus, for M2 equilibrium it is necessary that the holding time in the low regime is sufficiently short relative to the discount rate r.

Figure 4 illustrates these phenomena. The left panel of Fig. 4 shows that x_{start} is increasing in λ_{01}. This is the time effect: as λ_{01} increases, the exhaustible producer is anticipating imminent higher profits and is more willing to shut down production to save reserves for that purpose. Asymptotically, as $\lambda_{01} \rightarrow \infty$, x_{start} converges to $\tilde{x}_{start} := \inf\{x > 0 : L - \tilde{v}_H(x) > 0\}$, where $\tilde{v}_H(x)$ is the game function corresponding to the constant-high-demand case, see Lemma 1. The middle panel of Fig. 4 shows that when H is close to L there is no production shutdown, whereas as H increases, x_{start} increases in H unboundedly. This is the quantity effect: larger H increases the marginal value of reserves which makes the situation $v'_L(x) > L$ more likely.

Finally, the right panel of Fig. 4 shows x_{start} as a function of green production cost c. We observe an ambiguous effect of competition on voluntary shutdown. Because shutdown only happens with very low reserves, it takes place when the green producer 2 is generally the "leader" of the market and hence the equilibrium rates are very sensitive to the green leader's costs. When c is very small, competition lowers the value of reserves and makes the marginal value large. Therefore, x_{start} is large when c is small. At moderate c, the competition is alleviated and the game becomes in favor of the exhaustible producer 1, thus the exhaustible production is expanded due to decreased marginal value of reserves. When c is close to L, green production is very low or even blockaded in low regime, causing producer 1 to *raise* production under low demand, and eschew shutdown $x_{start} = 0$. As c increases beyond $c > L$, the exhaustible producer begins to lead the market under

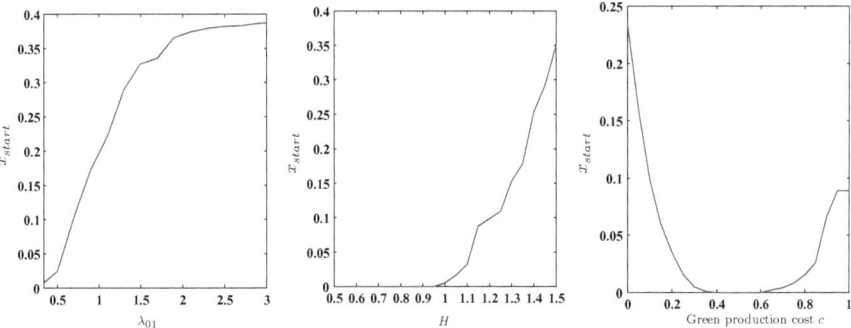

Fig. 4 *Left panel*: x_{start} as a function of $\lambda_{01} \in [1/3, 3]$. *Middle panel*: x_{start} as a function of $H \in [L, 1.5]$. *Right panel*: x_{start} as a function of green production cost $c \in [0, H]$. The default parameters are $L = 0.5, H = 1, c = 0.3, \lambda_{01} = \frac{1}{3}, \lambda_{10} = \frac{1}{5}$

both regimes, and is driven by the quantity/time effects to shutdown production. Thus, x_{start} increases in c when $L < c < H$. When $c = H$, the green producer 2 is completely blockaded in both the low and high regimes, thus the model reduces to exhaustible production monopoly. It also deserves our attention that in the setting $c \approx 0$ that is most favorable to the green producer, x_{start} is significantly larger than in a player 1 monopoly market ($c = H$). This is because competition encourages producer 1 to conserve reserves in low regime, much more so than in the monopoly setting.

4 Extensions

4.1 Multiple Regimes

We could similarly analyze the situation where the market demand switches among $n > 2$ macroeconomic regimes such that the demand function is given by

$$p_t = M_t - q^1 - q^2, \quad M_t \in \{D_1, D_2, \ldots, D_n\}, \quad 0 < D_1 < D_2 < \ldots < D_n.$$

Denote the generator of the Markov chain (M_t) as

$$\Lambda = \begin{array}{c} \\ D_1 \\ D_2 \\ \vdots \\ D_n \end{array} \begin{pmatrix} -\sum_{j\neq 1} \lambda_{1j} & \lambda_{12} & \cdots & \lambda_{1n} \\ \lambda_{21} & -\sum_{j\neq 2} \lambda_{2j} & \cdots & \lambda_{2n} \\ \vdots & \vdots & \ddots & \vdots \\ \lambda_{n1} & \lambda_{n2} & \cdots & -\sum_{j\neq n} \lambda_{nj} \end{pmatrix}.$$

$$\begin{array}{cccc} D_1 & D_2 & \cdots & D_n \end{array}$$

Then (4) is generalized to

$$\sup_{q_i^1} \left[q_i^1(x) \left(D_i - q_i^1(x) - (q_i^2)^*(x) \right) - v_i'(x) q_i^1(x) \right] + \sup_{a_i} \left[a_i \lambda \Delta v_i(x) - C(a_i) \right]$$

$$+ \sum_{j\neq i} \lambda_{ij} \left(v_j(x) - v_i(x) \right) - r v_i(x) = 0, \quad (23)$$

and similarly for $g_i(x)$, and the boundary conditions are

$$v_i(0) = \sup_{a_i \geq 0} \frac{\sum_{j\neq i} \lambda_{ij} v_j(0) + v_i(\delta) \lambda a_i - C(a_i)}{r + \lambda a_i + \sum_{j\neq i} \lambda_{ij}}, \quad i = 1, 2, \ldots, n.$$

The candidate optimizers $a_i^*(x), q_i^{\ell,*}(x), i = L, H, \ell = 1, 2$, remain as in (8)–(9).

With multiple regimes, some of the intuitive comparative statics become unavailable. For example, Conjecture 1 states that with only two regimes the production

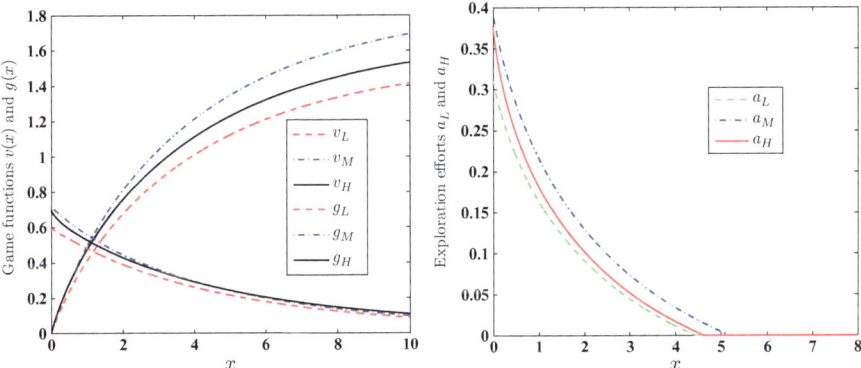

Fig. 5 Duopoly with three demand regimes, $D_1 = L, D_2 = M, D_3 = H$. *Left panel*: Game functions $v_i(x)$ and $g_i(x)$, $i = L, M, H$ of the two producers. *Right panel*: exploration efforts $a_i(x)$, $i = L, M, H$. We take $L = 0.5, M = 1, H = 1.5$ and $\lambda_{12} = \frac{1}{2}, \lambda_{23} = \frac{1}{4}, \lambda_{31} = 1$ in (24)

rates (and hence the shadow marginal costs of reserves) appear to be always ordered. Namely, production and exploration efforts are larger under higher demand. For a generic chain (M_t), such monotonicity no longer necessarily holds.

As explained, the game values can be thought of as "averages" of the corresponding rewards under fixed demand regimes. The averaging is done in terms of the expected discounted time spent in each regime given M_0. With more than two regimes, this averaging is non-trivial: even if M_0 is high today, the future prospects could be worse compared to a lower starting point. For instance, consider the case where the market demand switches cyclically among three levels $D_1 < D_2 < D_3$, with a generator of the form

$$\Lambda = \begin{pmatrix} -\lambda_{12} & \lambda_{12} & 0 \\ 0 & -\lambda_{23} & \lambda_{23} \\ \lambda_{31} & 0 & -\lambda_{31} \end{pmatrix}. \tag{24}$$

Thus, (M_t) moves cyclically along $D_1 \rightarrow D_2 \rightarrow D_3 \rightarrow D_1$. If the proportion of time spent in regime 2 is significantly longer than in regime 3 (namely $\lambda_{12} \ll \lambda_{23} \ll \lambda_{31}$), then regime 3 could be a worse starting point than regime 2. This situation is illustrated in Fig. 5 that shows that $v_i(x)$ are no longer monotone (and neither are $q_i^1(x)$ or $a_i(x)$) in i. Therefore, the original ordering of D_i gets shuffled due to the influence of the transition rates λ_{ij}.

A further generalization would be to consider continuous fluctuations of market demand, for instance taking (M_t) as an Itô diffusion modulating the price function. Some of the standard choices could include a (Geometric) Brownian motion factor to model for example the evolution of total economy GDP, or a stationary Ornstein-Uhlenbeck factor to model the macroeconomic business cycle relative to the long-run baseline. Such a model would require replacing the current HJB-I equations (4)–(5) by a partial differential equation since the new game values would be a function $v(x, m)$ of both the reserves level x and the demand level $M_0 = m$. If M_t is a diffusion, this will lead to a parabolic HJB-I system. The present theory of such systems, including smoothness of the game values, existence of equilibrium, etc. is not well-developed. In our context some of the additional difficulties include (i) non-standard implicit boundary conditions at $x = 0$; (ii) non-local terms arising from discrete exploration discoveries; (iii) degeneracy in the x variable that has only first-order dynamics (no volatility) and is fully endogenized by the player strategies. Overcoming all these challenges (that extend beyond the analytic properties to numerics and economic interpretation) is well beyond the scope of this paper. Indeed, we believe that the extra realism gained from incorporating (M_t) as a continuous factor is not worth the additional model complexity. Our take is that all models are stylized and are aiming at basic insights rather than complete practicality.

4.2 Stochastic Production Costs

One may use the stochastic factor to modulate other game parameters. For example, the production costs of player 2 (the green producer) might be changing over time. Such fluctuations could be due to varying technology costs; changing government policies such as renewable energy subsidies; or non-constant financing costs. To capture this setup we could then assume that $c = c(M_t)$ is modulated by the chain (M_t) whereas the demand is for simplicity now fixed at some $\bar{D}$.

The resulting game values would solve HJB-I equations essentially matching (4)–(5), except that player 2 production costs $c_i = c(D_i)$ now differ across regimes. Figure 6 illustrates the solution assuming a two-state (M_t). The parameters are similar to those in Fig. 2 allowing a degree of comparison. As before, green production rises when costs are low (regime 1) and falls when costs become larger. In particular, the game can switch between Type I and type M1 equilibria due to regime change. Moreover, because exploration efforts of player 1 are not monotone in c in the setup of Sect. 2, they are also non-monotone here, see right panel of Fig. 6. Thus, for some reserve levels x, a drop in competitor's production costs may induce increased motivation to explore.

Remark 4. One could also imagine fluctuating production costs of player 1. In fact, with linear inverse demand, one may interpret the choke level D in $p = D - q^1 - q^2$ as the *net* difference between demand level and production costs for the exhaustible producer (recall that in Sect. 2 we took these fixed production costs to be zero for convenience). Hence, the original model that takes $D = D(M_t)$ is equivalent to assuming that $c^1 = c^1(M_t)$ with a baseline case $c^1(D_1) = 0$.

One could also modulate other parts of the model, such as the exploration costs $C(a; M_t)$ (indeed, there was plenty of evidence that E&P costs in the oil industry rose sharply during the bull oil market of 2006–08 as increased demand spurred all companies to replenish reserves). Another idea that was suggested in [6] is

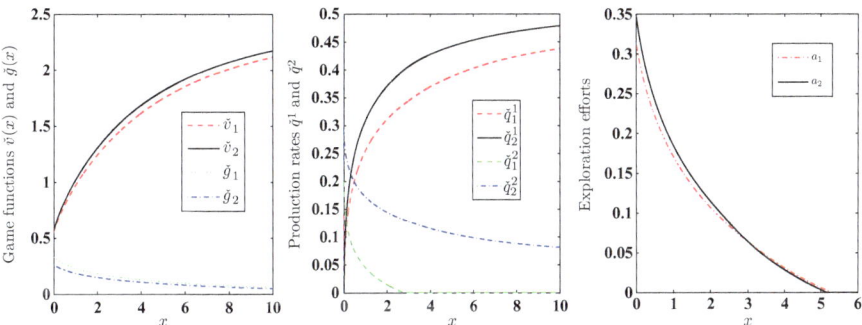

Fig. 6 Duopoly with regime-switching green production costs. *Left panel*: Game functions $\breve{v}_i(x)$ and $\breve{g}_i(x)$, $i = 1, 2$ of the two producers. *Middle panel*: Production rates $q_i^\ell(x)$ of the two producers $\ell = 1, 2$. *Right panel*: Exploration efforts $a_i(x)$ of the exhaustible producer. We take $\bar{D} = 1, c_1 = 0.4, c_2 = 0.6$, and switching rates $\lambda_{01} = \frac{1}{3}, \lambda_{10} = \frac{1}{5}$. Exploration costs are $C(a) = 0.1a + a^2/2$

to modulate the discovery rate λ (so as to capture for instance approach of the global exhaustibility of the resource). In general, one could mix and match above features, taking both $c(M_t)$ and $D(M_t)$ to be dependent on the macroeconomy. In that situation, there is no longer a clear ordering of the regimes. For example, it could be the case that in regime 1, demand is high but so is the competition with green production; in regime 2 demand is lower but green producer is blockaded. Our above investigations can serve as a guide to disentangle the opposite effects that would then be induced on $q^\ell(x)$ and $a(x)$.

4.3 More Players

Our main model featured two players. These are meant to be representative of the exhaustible and renewable producers, for example oil and "green" energy industries competing on the electricity market. By having only a single player with reserves, our continuous state variable X_t is one-dimensional which greatly facilitates the analysis. It would of course be more realistic to include multiple players with reserves (e.g. different type of conventional fossil fuels). An immediate extension would be to analyze the duopoly among two exhaustible producers. The difference in their reserves X_t^1, X_t^2 would determine the asymmetry in the game and decide who is the leader based on the respective shadow reserve costs. A version of such a model without exploration or stochastic demand was treated in [9]. Given the major challenges encountered there (in particular regularity issues for the game functions), this is another extension that we are not able to fully address here.

Beyond two players, the theory of Cournot oligopolies is rudimentary. See [11] for analysis of Bertrand oligopolies in deterministic markets (with no exploration nor stochastic factor). Consideration of more than two non-symmetric players will necessarily be challenging due to the exploding dimensionality. Moreover, it raises thorny questions regarding private and public information, namely whether all players can be fully informed about all other players given the complexity of the game. Since real life markets actually feature hundreds of agents, another useful approximation would be to study an infinity of players using the framework of mean field games [2]. Namely, one might consider the strategic interaction among a continuum of exhaustible producers with exploration. Related models without exploration were analyzed in the deterministic case by [8] and in the stochastic case (where reserves X_t receive Brownian shocks) by [3].

5   Conclusion

We studied the effect of exploration and stochastic demand in dynamic Cournot games. In the model of [13], players competed in a dynamic noncooperative game as their reserves of an exhaustible resource depleted, simultaneously exploring new reserves. The only stochastic aspect was (Poissonian) randomness in reserve discoveries, making the overall game to be piecewise deterministic. The stochastic demand in our research adds a further feature of fluctuating market prices, introducing further dynamic aspects into the duopoly. We modelled this feature through a regime-switching price (inverse demand) function, which represented a random environment under which the producers make strategies of production and exploration.

Stochastic demand creates the possibility of a new (dubbed Type M2) equilibrium whereby the exhaustible producer may opportunistically shutdown production in hopes of higher profits in the future. This happens when reserves are low and their shadow marginal cost is high enough. Additionally, the non-monotonic impact of competition on exploration efforts already observed in [13] continues to occur in our model and leads to interesting phenomena herein.

Acknowledgements We thank an anonymous referee for helpful questions and comments that have improved our final version.

Appendix

Proof of Lemma 1

Proof. To derive the asymptotic game functions as $\lambda_{01} \rightarrow +\infty$, we set $\lambda_{01} = \frac{1}{\epsilon}$. Without loss of generality, we assume that the asymptotic expansion of $v_M(x)$ with respect to ϵ is

$$v_i^\epsilon = v_i^0 + g(\epsilon)v_i^1 + o(g(\epsilon)), \qquad i = L, H \tag{25}$$

where $g(\epsilon) \rightarrow 0$, as $\epsilon \rightarrow 0$.

We substitute the v_L and v_H in the HJB ODEs with their asymptotic expansion to obtain

$$\left[\frac{2}{3}\left(\frac{L+c}{2} - (v_L^\epsilon)'(x)\right)^+ - \frac{1}{6}\left(2c - L - (v_L^\epsilon)'(x)\right)^+\right]^2 + \frac{1}{\gamma}[(\lambda\Delta v_L^\epsilon(x) - \kappa)^+]^\gamma$$

$$+ \frac{1}{\epsilon}\left(v_H^0(x) - v_L^0(x)\right) + \frac{1}{\epsilon}\left(g(\epsilon)[v_H^1(x) - v_L^1(x)] + o(g(\epsilon))\right) - rv_L^\epsilon(x) = 0,$$

$$(26)$$

$$\left[\frac{2}{3}\left(\frac{H+c}{2} - (v_H^\epsilon)'(x)\right)^+ - \frac{1}{6}\left(2c - H - (v_H^\epsilon)'(x)\right)^+\right]^2 + \frac{1}{\gamma}[(\lambda\Delta v_H^\epsilon(x) - \kappa)^+]^\gamma$$

$$+ \lambda_{10}\left(v_L^0(x) - v_H^0(x)\right) + g(\epsilon)\lambda_{10}(v_L^1(x) - v_H^1(x)) - rv_H^\epsilon(x) + o(g(\epsilon)) = 0.$$

$$(27)$$

We must have that $\lim_{\epsilon\to 0} v_L^\epsilon = \lim_{\epsilon\to 0} v_L^\epsilon$, i.e. $v_L^0 = v_H^0 =: \tilde{v}$, otherwise the term $\epsilon^{-1}(v_H^0 - v_L^0)$ will explode as $\epsilon \to 0$. Making that simplification, multiplying (26) by $\epsilon\lambda_{10}$ and adding (27) we obtain

$$0 = \epsilon\lambda_{10}\left[\frac{2}{3}\left(\frac{L+c}{2} - (v_L^\epsilon)'(x)\right)^+ - \frac{1}{6}\left(2c - L - (v_L^\epsilon)'(x)\right)^+\right]^2$$

$$+ \left[\frac{2}{3}\left(\frac{H+c}{2} - (v_H^\epsilon)'(x)\right)^+ - \frac{1}{6}\left(2c - H - (v_H^\epsilon)'(x)\right)^+\right]^2$$

$$+ \frac{\epsilon\lambda_{10}}{\gamma}[(\lambda\Delta v_L^\epsilon(x) - \kappa)^+]^\gamma + \frac{1}{\gamma}[(\lambda\Delta v_H^\epsilon(x) - \kappa)^+]^\gamma - r\left(\lambda_{10}\epsilon v_L^\epsilon(x) + v_H^\epsilon(x)\right).$$

$$(28)$$

One can now take $\epsilon \to 0$ which reduces to a regular perturbation of the following ODE for $\tilde{v}(x)$ (note that the first term involving L vanishes):

$$\left[\frac{2}{3}\left(\frac{H+c}{2} - (\tilde{v}_H)'(x)\right)^+ - \frac{1}{6}\left(2c - H - (\tilde{v}_H)'(x)\right)^+\right]^2$$

$$+ \frac{1}{\gamma}[(\lambda\Delta\tilde{v}_H(x) - \kappa)^+]^\gamma - r\tilde{v}_H(x) = 0,$$

which matches the solution of an exploration duopoly game studied in [13] with linear inverse demand $p_t = H - q_t^1 - q_t^2$.

For the boundary conditions, we re-write (6) as

$$\left(v_H^\epsilon(0) - v_L^\epsilon(0)\right)\left(\frac{1}{\epsilon}\right) + v_L^\epsilon(\delta)\lambda a_L^\epsilon(0) - C(a_L^\epsilon(0)) - \left(r + \lambda a_L^\epsilon(0)\right)v_L^\epsilon(0) = 0,$$
(29)

$$\left(v_L^\epsilon(0) - v_H^\epsilon(0)\right)\lambda_{10} + v_H^\epsilon(\delta)\lambda a_H^\epsilon(0) - C(a_H^\epsilon(0)) - \left(r + \lambda a_H^\epsilon(0)\right)v_H^\epsilon(0) = 0.$$
(30)

We multiply (29) by $\epsilon\lambda_{10}$ and add to (30) to obtain

$$\epsilon\lambda_{10}\left[v_L^\epsilon(\delta)\lambda a_L^\epsilon(0) - C(a_L^\epsilon(0)) - \left(r + \lambda a_L^\epsilon(0)\right)v_L^\epsilon(0)\right]$$
$$+ \left[v_H^\epsilon(\delta)\lambda a_H^\epsilon(0) - C(a_H^\epsilon(0)) - \left(r + \lambda a_H^\epsilon(0)\right)v_H^\epsilon(0)\right] = 0.$$
(31)

Letting $\epsilon \to 0$ removes the first terms and we are left with

$$\tilde{v}(\delta)\lambda\tilde{a}(0) - C(\tilde{a}(0)) - (r + \lambda\tilde{a}(0))\,\tilde{v}(0) = 0,$$

which is equivalent to

$$\tilde{v}(0) = \frac{\tilde{v}(\delta)\lambda\tilde{a}(0) - C(\tilde{a}(0))}{(r + \lambda\tilde{a}(0))} = \sup_a \frac{\tilde{v}(\delta)\lambda a - C(a)}{(r + \lambda a)},$$
(32)

again matching the corresponding boundary condition in the deterministic demand setting.

Proof of Lemma 2

Proof. We set $\lambda_{01} = \frac{b_L}{\epsilon}, \lambda_{10} = \frac{b_H}{\epsilon}$, where b_L, b_H are some constants, and $\epsilon > 0$ can be arbitrarily small. The stationary distribution $\boldsymbol{\pi}$ given by $\pi_L = \frac{\lambda_{10}}{\lambda_{01}+\lambda_{10}} = \frac{b_H}{b_L+b_H}$, $\pi_H = \frac{b_L}{b_L+b_H}$ is unchanged as $\epsilon \to 0$.

We consider the asymptotic expansions of v_M in terms of ϵ:

$$v_i^\epsilon = v_i^0 + f(\epsilon)v_i^1 + o(f(\epsilon)), \qquad i = L, H,$$
(33)

where $f(\epsilon) \to 0$, as $\epsilon \to 0$. Substituting (33) into (15) yields

$$\left[\frac{2}{3}\left(\frac{L+c}{2} - (v_L^\epsilon)'(x)\right)^+ - \frac{1}{6}\left(2c - L - (v_L^\epsilon)'(x)\right)^+\right]^2 + \frac{1}{\gamma}[(\lambda\Delta v_L^\epsilon(x) - \kappa)^+]^\gamma$$

$$+ \frac{b_L}{\epsilon}\left(v_H^0(x) + f(\epsilon)v_H^1(x) - v_L^0(x) - f(\epsilon)v_L^1(x)\right) - rv_L^\epsilon(x) + o(f(\epsilon)) = 0;$$
(34)

$$
\left[\frac{2}{3} \left(\frac{H+c}{2} - (v_H^\epsilon)'(x) \right)^+ - \frac{1}{6} \left(2c - H - (v_H^\epsilon)'(x) \right)^+ \right]^2 + \frac{1}{\gamma} [(\lambda \Delta v_H^\epsilon(x) - \kappa)^+]^\gamma
$$

$$
+ \frac{b_H}{\epsilon} \left(v_L^0(x) + f(\epsilon) v_L^1(x) - v_H^0(x) - f(\epsilon) v_H^1(x) \right) - r v_H^\epsilon(x) + o(f(\epsilon)) = 0.
$$

$$(35)$$

We must have that $\lim_{\epsilon \to 0} v_L^\epsilon = \lim_{\epsilon \to 0} v_H^\epsilon$, i.e. $v_L^0 = v_H^0 = \bar{v}$, otherwise the terms $\frac{b_L}{\epsilon} \left(v_H^\epsilon - v_L^\epsilon \right)$ and $\frac{b_H}{\epsilon} \left(v_L^\epsilon - v_H^\epsilon \right)$ above would explode as $\epsilon \to 0$. Indeed, it is clear that $|v_L(x) - v_H(x)| \to 0$ as $\epsilon \to 0$ due to the fast switching of the regimes, making the initial macroeconomic conditions irrelevant.

Canceling the terms $v_L^0 - v_H^0 \equiv 0$ in (34)–(35), multiplying (34) by $b_H/(b_L + b_H)$, (35) by $b_L/(b_L + b_H)$, and adding them up we obtain

$$
0 = \pi_L \left[\frac{2}{3} \left(\frac{L+c}{2} - (v_L^\epsilon)'(x) \right)^+ - \frac{1}{6} \left(2c - L - (v_L^\epsilon)'(x) \right)^+ \right]^2
$$

$$
+ \pi_H \left[\frac{2}{3} \left(\frac{H+c}{2} - (v_H^\epsilon)'(x) \right)^+ - \frac{1}{6} \left(2c - H - (v_H^\epsilon)'(x) \right)^+ \right]^2
$$

$$
+ \frac{\pi_L}{\gamma} [(\lambda \Delta v_L^\epsilon(x) - \kappa)^+]^\gamma + \frac{\pi_H}{\gamma} [(\lambda \Delta v_H^\epsilon(x) - \kappa)^+]^\gamma
$$

$$
- r \left(\pi_L v_L^\epsilon(x) + \pi_H v_H^\epsilon(x) \right) + o(f(\epsilon)).
$$

Note that all the terms involving ϵ^{-1} have cancelled out. Once again plugging in (33) we can now take $\epsilon \to 0$ since this just amounts to a regular perturbation; the result is precisely (21).

For the boundary conditions, we re-write the original

$$
v_i^\epsilon(0) = \frac{v_j^\epsilon(0)(\frac{b_i}{\epsilon}) + v_i^\epsilon(\delta) \lambda a_i^\epsilon(0) - C(a_i^\epsilon(0))}{r + \frac{b_i}{\epsilon} + \lambda a_i^\epsilon(0)}, \quad i, j = L, H
$$

as

$$
\left(v_H^\epsilon(0) - v_L^\epsilon(0) \right) \left(\frac{b_L}{\epsilon} \right) + v_L^\epsilon(\delta) \lambda a_L^\epsilon(0) - C(a_L^\epsilon(0)) - \left(r + \lambda a_L^\epsilon(0) \right) v_L^\epsilon(0) = 0,
$$

$$(36)$$

$$
\left(v_L^\epsilon(0) - v_H^\epsilon(0) \right) \left(\frac{b_H}{\epsilon} \right) + v_H^\epsilon(\delta) \lambda a_H^\epsilon(0) - C(a_H^\epsilon(0)) - \left(r + \lambda a_H^\epsilon(0) \right) v_H^\epsilon(0) = 0.
$$

$$(37)$$

Once again multiplying (36) by $b_H/(b_L + b_H)$ and (37) by $b_L/(b_L + b_H)$, and summing produces

$$\pi_L v_L^\epsilon(\delta)\lambda a_L^\epsilon(0) + \pi_H v_H^\epsilon(\delta)\lambda a_H^\epsilon(0) - \pi_L C(a_L^\epsilon(0)) - \pi_H C(a_H^\epsilon(0))$$

$$- \pi_L\left(r + \lambda a_L^\epsilon(0)\right)v_L^\epsilon(0) - \pi_H\left(r + \lambda a_H^\epsilon(0)\right)v_H^\epsilon(0) = 0. \tag{38}$$

As $\epsilon \to 0$, $a_M^\epsilon(0) = [(\lambda \Delta v_M^\epsilon(0) - \kappa)^+]^{\gamma-1} \to [(\lambda \Delta \bar{v}(0) - \kappa)^+]^{\gamma-1} = \bar{a}(0)$, and we find $\bar{v}(\delta)\lambda\bar{a}(0) - C(\bar{a}(0)) - (r + \lambda\bar{a}(0))\bar{v}(0) = 0$, which is equivalent to (22).

References

1. Arrow, K.J., Chang, S.: Optimal pricing, use, and exploration of uncertain resource stocks. J. Environ. Econ. Manag. **9** (1), 1–10 (1982)
2. Bensoussan, A., Frehse, J., Yam, P.: Mean Field Games and Mean Field Type Control Theory. Springer, New York (2013)
3. Chan, P., Sircar, R.: Bertrand & Cournot mean field games. Technical Report SSRN (2014). http://papers.ssrn.com/sol3/papers.cfm?abstract_id=2387226
4. Conrad, J.: Resource Economics. Cambridge University Press, Cambridge (1999)
5. Davis, M.H.A.: Markov Models and Optimization. Chapman & Hall, London (1993)
6. Deshmukh, S.D., Pliska, S.R.: Optimal consumption of a nonrenewable resource with stochastic discoveries and a random environment. Rev. Econ. Stud. **50**(3), 543–554 (1983)
7. Dockner, E.J., Jørgensen, S., Long, N.V., Sorger, G.: Differential Games in Economics and Management Science. Cambridge University Press, Cambridge (2000)
8. Guéant, O., Lasry, J.M., Lions, P.L.: Mean field games and applications. In: Paris-Princeton Lectures on Mathematical Finance 2010, pp. 205–266. Springer, Berlin (2011)
9. Harris, C., Howison, S., Sircar, R.: Games with exhaustible resources. SIAM J. Appl. Math. **70**, 2556–2581 (2010)
10. Hotelling, H.: The economics of exhaustible resources. J. Polit. Econ. **39** (2), 137–175 (1931)
11. Ledvina, A., Sircar, R.: Dynamic Bertrand oligopoly. Appl. Math. Optim. **63**(1), 11–44 (2011)
12. Ledvina, A., Sircar, R.: Oligopoly games under asymmetric costs and an application to energy production. Math. Finan. Econ. **6**, 1–33 (2012)
13. Ludkovski, M., Sircar, R.: Exploration and exhaustibility in dynamic Cournot games. Eur. J. Appl. Math. **23**(3), 343–372 (2011)
14. Ludkovski, M., Sircar, R.: Game models for exhaustible resources. In: Aid, R., Ludkovski, M., Sircar, R. (eds.): Energy, Commodities and Environmental Finance, Fields Insititute Communications. Springer, Heidelberg (2014)
15. Pliska, S.R.: On a functional differential equation that arises in a Markov control problem. J. Differ. Equ. **28**(3), 390–405 (1978)
16. Vives, X.: Oligopoly Pricing: Old Ideas and New Tools. MIT press, Cambridge (2001)

Variable Costs in Dynamic Cournot Energy Markets

Anirudh Dasarathy and Ronnie Sircar

Abstract We study a game-theoretic model for energy markets. Our framework is an N-player stochastic dynamic Cournot game where one producer has a reserve (or stock) that depletes over time, while the others can produce indefinitely with no such quantity restriction. We think of the first player as producing energy from a fossil fuel such as oil, which is an exhaustible resource, while the others are producing from renewables. All players have costs of production that evolve over time, and the exhaustible player can choose to invest in R&D (research and development, including exploration) which may yield increases in stock probabilistically over time. The assumption that the players have heterogeneous and time-varying costs requires a reexamination and extension of previous literature which has typically considered homogeneous costs. We also study how this model may be applied to energy policy, comparing when it is optimal to consider taxing oil producers, opposed to subsidizing green energy, as a matter of public policy.

1 Introduction

The motivation behind the work presented in this paper is to present a model for energy markets which may be considered as oligopolies, where a small number of different producers compete against each other to maximize profits. The initial work in the economics literature on oligopolistic competition was by Cournot [2] in 1838, who introduced the idea of competition through production output. This work was re-envisioned by Bertrand [1] in 1883, who framed competitions in terms of prices. More recently, however, energy markets have been modeled through dynamic, as opposed to the static games that Cournot and Bertrand considered. For more modern interpretations of oligopolistic competition, we recommend Friedman [4], Vives [10], or for dynamic models, Dockner *et al.* [3].

A. Dasarathy • R. Sircar (✉)

ORFE Department, Princeton University, Sherrerd Hall, Princeton, NJ 08544, USA

e-mail: anirudhd@princeton.edu; sircar@princeton.edu

© Springer Science+Business Media New York 2015

R. Aïd et al. (eds.), *Commodities, Energy and Environmental Finance*,
Fields Institute Communications 74, DOI 10.1007/978-1-4939-2733-3_15

Additionally, energy markets often have two distinct types of players: some, like oil, depend on a fixed reserve, while others, like renewable players such as wind and solar, have effectively infinite or inexhaustible resources. Study of the impact of exhaustibility of resources was initiated by Hotelling [6], within a monopoly. In [5], energy production is modeled as a dynamic Cournot game, where certain players depend on stock remaining to produce, while others have a higher extraction cost but can produce indefinitely.

Previous work to model the behavior we consider in energy market production has made certain assumptions that we relax here. For example, [8] assumes that there is no R&D or exploration, while [9] assumes that the only cost oil producers incur are their research costs. We expand such work in two ways. First, we relax the assumption that the oil producer has zero cost of extraction. Then we allow for evolving costs of energy production: costs over time are realistically not constant. In particular, as stock begins to run out, costs for oil producers often increase (deeper drilling, more expensive extraction technology required), whereas costs for green energy often decrease due to external investment and financing that leads to more efficient technology. For instance "the price of solar panels has fallen more than 75 percent just since 2008" [7].

The outline of this paper is as follows. In Sect. 2, we analytically solve a special case of constant costs, which requires no numerical calculations. We define the notion of "blockading" and find the Markov perfect equilibrium for the case where all costs are positive, yet constant. In Sect. 3, we provide a partially analytical solution for the case without exploration. Then, in Sect. 4, we solve the full model which incorporates exploration. This extends the stochastic model in [9] to allow for varying costs. Finally, in Sect. 5, we demonstrate how our model may be applied to the setting of energy policy. We provide two examples, the latter of which compares taxing finite resource producers and subsidizing green energy. We then conclude and suggest methods by which this work may be extended.

2 Dynamic Game Model with Constant Costs

We begin with the case of constant production costs and no exploration where we can establish analytical results.

2.1 Preliminary Notation

We consider an N-player oligopoly game that models energy markets. The first $k < N$ of the energy players have exhaustible stocks (or reserves) $\{x_1(t), \ldots, x_k(t)\}$. They are active whenever their stock $x_i(t) > 0$ and are "eliminated" or exit production when their reserve $x_i(t)$ hits 0. The remaining players can be considered "renewable," and have infinite (or inexhaustible) reserves. Each of the N players has a marginal cost function s_i that is associated with the ith player (so it costs the ith player s_i units of money to produce one unit of output), and they compete through Cournot (quantity-setting) competition. We will see that all players will participate provided that their cost is low enough compared to the other active players.

The case where $k = N \geq 2$ (that is, there are only exhaustible players who compete against each other), both with zero extraction costs for all time, was examined in [5], and involved the analysis of a coupled system of nonlinear partial differential equations (PDEs) which are typically difficult to solve even numerically. We will thus consider the case where $k = 1$, so there is only one exhaustible player. Since the motivation behind this paper is to provide a framework within which energy policy comparisons can be made, we will constrain ourselves to this case.

Notationally, when we consider an N-player game, we will let player 0 be the exhaustible player. Given the evident link to energy markets, we will interchangeably refer to player 0 as the "exhaustible" player, the "stock" player, and the "oil" player. Then, players $1, \ldots, N - 1$ shall be our renewable players. Since there is only one exhaustible player in our consideration, we denote by $x(t) = x_0(t)$, the remaining stock of player 0. We let the costs also vary in the amount $x(t)$ remaining: $s_i(x)$. In general, costs evolve as stock begins to run out and this generalization allows us to encapsulate the meaning of such changes. Further, to have a meaningful model that can be applied to policy, we must be able to capture evolving costs, as most elements of energy policy affect the perceived price of various players in the market. The case where $s_0(x) = 0$ and $s_i(x) = s_i$, that is, the oil player has zero extraction costs and the renewable players have constant cost has been evaluated analytically in [8].

As assumed there, we will have a representative market for the energy that is produced by these players. In particular, we assume that the utility function represented by market demand makes no differentiation between the energy produced by all N players; that is, the individual consumer cares only about prices and quantities produced, but not the actual source fuel (or technology) itself. We can, in general, assume that the market demand function follows a constant prudence price curve; that is, if we let Q denote the total market output, then the inverse demand curve is

$$P(Q) = \begin{cases} \dfrac{1 - Q^{1-\rho}}{1 - \rho} & \text{if } \rho \neq 1 \\ -\log Q & \text{if } \rho = 1, \end{cases} \tag{1}$$

where we restrict $0 \leq Q \leq 1$. Here, we will consider the case where $\rho = 0$, so the inverse market demand curve reduces to $P(Q) = 1 - Q$.

We first review the static (one-period) Cournot game with constant costs.

2.2 Static Game

In the static (or one-period or stage) Cournot game with N players who have constant costs $s_0, \ldots, s_{N-1}$, each player i chooses a production quantity $q_i \geq 0$ to maximize revenue, that is price minus cost multiplied by quantity produced:

$$\pi(q_i, q_{-i}) = q_i \left(1 - q_i - \sum_{j=0, j \neq i}^{N-1} q_j - s_i \right),$$

where $q_{-i} = (q_0, \cdots, q_{i-1}, q_{i+1}, \cdots, q_{N-1})$. Here there is no distinction between player 0 and the others as there is no dynamic component whereby things change over time.

We say that $(q_0^*, \cdots, q_{N-1}^*)$ is a Nash equilibrium if $\pi(q_i^*, q_{-i}^*) \geq \pi(q_i, q_{-i}^*)$ for any $q_i \geq 0$ and all $i = 0, 1, \cdots, N-1$. That is q_i^* maximizes revenue for player i when all the other players play their Nash equilibrium strategies. For this problem, the Nash equilibrium is as follows.

Proposition 1. *In an N-player static game with constant costs $0 \leq s_0 < s_1 < \cdots < s_{N-1} < 1$, we let*

$$P_i = \frac{1}{i+1} \left(1 + \sum_{j=0}^{i-1} s_j \right), \qquad i = 1, \ldots, N,$$

and $\bar{P} = \min\{P_i \mid i = 1, \ldots, N\}$. The number of active players is given by

$$n = \min\{i \mid P_i = \bar{P}, i = 1, \ldots, N\}.$$

Then players $\{0, \ldots, n-1\}$ are active, and players $\{n, \ldots, N-1\}$ do not produce. The unique Nash equilibrium is given by

$$q_i^* = \frac{1}{n+1} \left(1 - n s_i + \sum_{j=0, j \neq i}^{n-1} s_j \right), \qquad i \in \{0, \ldots, n-1\},$$

and $q_i^* = 0$ for $i \in \{n, \ldots, N-1\}$.

Proof. See [5, Proposition 2.9]. $\quad\blacksquare$

Remark. We chose to have the costs to be strictly increasing; if costs of players are allowed to be the same, then the blockading points for the dynamic game (defined in Sect. 2.3) might coincide. For purposes of consistency, we use the same assumption here. The result in Proposition 1 would hold even if the sequence of costs is only weakly increasing and the calculations presented in Sect. 2.3 and beyond can be reproduced for weakly increasing costs. For simplicity and to highlight the focus of our results, we use a strictly increasing costs assumption.

When the oil producer is inactive (as will occur in the dynamic game when his reserves are exhausted) and only players $1 \le i \le N - 1$ are in the market, the static game Nash equilibrium is

$$
q_i^* = \begin{cases} \frac{1}{n+1} \left(1 - ns_i + \sum_{j=1; j \neq i}^n s_j \right), & i \in \{1, \ldots, n\} \\ 0 & i \in \{n+1, \ldots, N-1\}, \end{cases} \tag{2}
$$

where $n = \min\{i \mid P_i = \bar{P}, i = 1, \ldots, N-1\}$ and

$$
P_i = \frac{1}{i+1} \left(1 + \sum_{j=1}^i s_j \right), \quad i = 1, \ldots, N-1, \qquad \bar{P} = \min\{P_i \mid i = 1, \ldots, N-1\}. \tag{3}
$$

We denote by $S^{(k)} = \sum_{i=1}^k s_i$, the cumulative cost of the first k renewable players, and we also define

$$
\rho_i = \frac{1 + S^{(i-1)}}{i}, \qquad i \in 2, \ldots, N, \tag{4}
$$

with $\rho_1 = 1$. Further, we make the following assumption.

Assumption 1. *We assume that $s_{N-1} < \rho_{N-1}$.*

Remark. This assumption is necessary to guarantee that when the oil producer is not present, the costs of the renewable producers are low enough that all participate in equilibrium. This follows from the fact that $P_{i-1} = \rho_i$, where P_i is given in (3). Then a calculation shows that

$$
\rho_N < \rho_{N-1} \quad \Longleftrightarrow \quad s_{N-1} < \rho_{N-1},
$$

and so under Assumption 1, $P_{N-1} < P_{N-2}$. Therefore $\bar{P} = P_{N-1}$ and $n = N - 1$.

We also have:

Lemma 1. *Under Assumption 1, we have that $s_j < \rho_j$ for all $j \in 1, \ldots, N-1$.*

Proof. For player $N - 2$:

$$
s_{N-2} - \rho_{N-2} = \frac{(N-1)s_{N-2} - (1 + S^{(N-2)})}{N-2} = \frac{N-1}{N-2}(s_{N-2} - \rho_{N-1})
$$

$$
\le \frac{N-1}{N-2}(s_{N-1} - \rho_{N-1}) < 0,
$$

where in the second to last step, we used $s_j < s_{j+1}$ for all $j \in \{1, \ldots, N-2\}$. The implication $s_j < \rho_j$ can be shown inductively from here.

Lemma 2. *When s_i is increasing in i and Assumption 1 holds, we have that ρ_i is decreasing in i.*

Proof. This follows from

$$\rho_i - \rho_{i-1} = \frac{1 + S^{(i-1)}}{i} - \frac{1 + S^{(i-2)}}{i-1} = \frac{is_{i-1} - S^{(i-1)} - 1}{i(i-1)} < 0,$$

since

$$s_{i-1} < \rho_{i-1} \implies s_{i-1} < \frac{1 + S^{(i-1)}}{i} \implies is_{i-1} < 1 + S^{(i-1)}.$$

Next, we introduce the dynamic N player Cournot game, where player 0 is our "oil" producer with exhaustible resources, and the remaining players $1, \ldots, N-1$ are renewable energy producers.

2.3 Dynamic Game

Energy is produced from different sources by players $0, \ldots, N-1$, who have constant costs of production $(s_0, \ldots, s_{N-1})$. The case where $s_0 = 0$ has been solved analytically in [8]. The general case where s_0 is allowed to be a constant that is in $[0, 1)$ can also be solved completely analytically and will be presented below. We will study the game when costs vary as time goes on and oil runs out in Sect. 3.

Player 0 is our oil producer who plays when his stock $x(t) > 0$ and has an extraction cost $s_0 > 0$. The stock $x(t)$ evolves according to the flow equation

$$\frac{dx(t)}{dt} = -q_0(x(t))\mathbb{1}_{\{x(t)>0\}},$$

where q_0 is the extraction strategy of player 0, and his initial reserve is $x(0)$. Players $i = 1, \ldots, N-1$ are renewable producers and have a fixed marginal cost of production $s_i > 0$. We order the players such that $s_1 \leq s_2 \leq \cdots \leq s_{N-1}$ and further require that Assumption 1 holds. They produce energy at the rates q_i.

Each player has an infinite time horizon objective value function that is determined by future profits discounted at rate $r > 0$. In particular, the Nash equilibrium $(q_0^*(\cdot), q_1^*(\cdot), \ldots, q_{N-1}^*(\cdot))$ are given by the arguments of the following suprema:

$$v(x) = \sup_{q_0} \int_0^\tau e^{-rt} q_0(x(t)) \left(1 - q_0(x(t)) - \sum_{j=1}^{N-1} q_j^*(x(t)) - s_0 \right) dt \qquad (5)$$

$$w_i(x) = \sup_{q_i} \int_0^\tau e^{-rt} q_i(x(t)) \left(1 - q_0^*(x(t)) - \sum_{j=1, j \neq i}^{N-1} q_j^*(x(t)) - q_i(x(t)) - s_i\right) dt$$

$$+ \frac{1}{r} e^{-r\tau} G_i, \tag{6}$$

where $i = 1, \cdots, N - 1$. Here τ is the exhaustion time $\tau = \inf\{t \mid x(t) = 0\}$, and G_i is the equilibrium profit of player i in the static game with only players $1, \ldots, N-1$ (who all participate under Assumption 1):

$$G_i = q_i^* \left(1 - \sum_{j=1}^{N-1} q_j^* - s_i\right) = \left(\frac{1}{N}(1 - Ns_i + S^{(N-1)})\right)^2,$$

where we have used (2) for q_i^* with $n = N - 1$. The admissible Markov strategies $q_i(x)$ are such that $q_i \geq 0$ and the $q_i(x)$ are Lipschitz continuous.

2.4 Blockading of Renewable Producers

Under some conditions, some subset of the players are blockaded from production because their costs are too high to generate a profit given the competition from players with lower costs. In the context of the renewable players $i \in \{1, \ldots, N-1\}$, this will be denoted by a point x_b^i such that

$$x_b^i = \inf\{x > 0 : q_i^*(x) = 0\}.$$

That is, for all points $x < x_b^i$, player i produces and participates in the game, but for $x \geq x_b^i$, the supply of cheap oil makes the market energy price too low for him to participate. We define this to be the blockading point for player i. In the case where $q_i^*(x) > 0$ for all x, we set $x_b^i = \infty$. In such an instance, we say that player i is never blockaded. We also identify $x_b^N = 0$ and $x_b^0 = \infty$.

Additionally, unlike the $s_0 = 0$ case, since s_0 is now positive, there is also the chance that the oil player may be blocked from playing if his extraction cost s_0 is too high compared to the renewable players. In other words, it may be possible that the oil player is inactive even if $x(t) > 0$. We will defer this case to Sect. 2.5. The intuition behind the case where s_0 is sufficiently low so that the oil player plays whenever he has remaining stock is presented in [8] and reproduced in Fig. 1, where the blocking times t_b^i are defined by $x(t_b^i) = x_b^i$.

In the region $x \in [x_b^n, x_b^{n-1})$, there are $n \leq N$ active players *including the oil player*. Shifting the variable x, we write $v(x_b^n + x) = v^{(n)}(x)$ for $x \in (0, x_b^{n-1} - x_b^n)$. A straightforward extension of [8, Proposition 5.3] to incorporate the cost s_0 shows that $v^{(n)}$ solves the Hamilton-Jacobi-Bellman (HJB) equation

$$rv^{(n)} = \frac{1}{(n+1)^2}\left(1 - n(v^{(n)'} + s_0) + \sum_{j=1}^{n-1} s_j\right)^2, \qquad n = 1,\ldots,N. \qquad (7)$$

These are solved with the boundary conditions $v^{(N)} = 0$ and $v^{(n)}(0) = v^{(n+1)}(x_b^n - x_b^{n+1})$ for continuity of the value function $v(x)$. The equilibrium strategies in the region $x \in [x_b^n, x_b^{n-1})$ are given by the formula in Proposition 1 with the replacement $s_0 \mapsto s_0 + v^{(n)'}(x - x_b^n)$:

$$q_0^*(x) = \frac{1}{n+1}\left(1 - n(s_0 + v^{(n)'}(x - x_b^n)) + \sum_{j=1}^{n-1} s_j,\right) \qquad (8)$$

$$q_i^*(x) = \frac{1}{n+1}\left(1 - ns_i + (s_0 + v^{(n)'}(x - x_b^n)) + \sum_{j=1;j\neq i}^{n-1} s_j\right), \qquad (9)$$

for $i = 1,\ldots,n-1$.

Further, since at any given point x, $q_j^*(x) > q_{j+1}^*(x)$ (that is, higher cost renewable players produce less), and since costs are constant in x, we have that $x_b^j > x_b^{j+1}$; so, it takes more oil to run out before player j enters compared to player $j+1$, as indicated in Fig. 1. To solve (7) analytically, the following Lemma shall be useful:

Lemma 3. *The solution to the ODE*

$$(\alpha - v')^2 = \kappa v, \qquad (10)$$

where $v_0 = v(0) \geq 0$ and $\alpha, \kappa > 0$ is

$$v(x) = \frac{\alpha^2}{\kappa}\left(1 + \mathbf{W}\left(\theta(x)\right)\right)^2,$$

where $\mathbf{W}(\cdot)$ is the Lambert-W function, satisfying $Z = \mathbf{W}(Z) e^{\mathbf{W}(Z)}$ restricted to $Z \geq -e^{-1}$. Further, $\theta(x) = \beta e^\beta e^{-\kappa x/(2\alpha)}$ and $\beta = -1 + \dfrac{\kappa v_0}{\alpha}$.

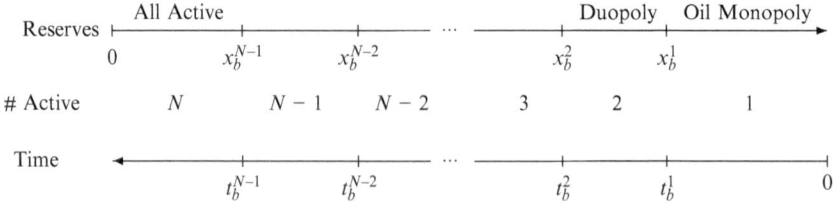

Fig. 1 Blockading intuition

Proof. The result follows by direct evaluation, which is explained in detail in [8].

Writing (7) in the form (10), we see that $\alpha = \dfrac{1 - ns_0 + S^{(n-1)}}{n+1} > 0$ by Assumption 1. Consequently, the closed form solution to (7) is

$$v^{(n)}(x) = \frac{1}{r}\left(\frac{1 - ns_0 + S^{(n-1)}}{n+1}\right)^2 (1 + \mathbf{W}(\theta(x)))^2, \qquad (11)$$

where $\theta(x) = \beta e^\beta e^{-\kappa x/(2\alpha)}$ with $\beta = -1 + \dfrac{\kappa v^{(n)}(0)}{\alpha}$ and $\kappa = r$. Taking the derivative, we have

$$v^{(n)'}(x) = -\frac{1}{n}\mathbf{W}(\theta(x))(1 - ns_0 + S^{(n-1)}) = -\mathbf{W}(\theta(x))(\rho_n - s_0), \qquad (12)$$

where ρ_n was defined in (4). Assumption 1 and Lemma 1 guarantee that $v^{(n)'}(x) > 0$. We define:

$$\hat{\rho}_n = \frac{1 + S^{(n-1)} - ns_0}{n} = \rho_n - s_0 \qquad (13)$$

for $n = 2, \ldots, N$. When $\hat{\rho}_n < 0$ for all n, the oil producer does not play in our game, and we have a perpetually repeated static game with $N - 1$ players, meaning v and w_i ($i = 1, \ldots, N - 1$) are given by (5)–(6) with $\tau = 0$.

It is straightforward to show from the concavity of $v(x)$ that whenever the stock for the oil producer is greater than the blockading point x_b^n, player n produces nothing. The blockading point is thus a threshold, below which the player enters the market, and above which, the player does not enter the market.

2.5   Blockading of the Oil Producer

When we allow s_0 to be greater than zero, we cannot assume as in [8] that the oil producer will always participate should $x(t) > 0$. Indeed, should s_0 be sufficiently high, it is possible that the oil producer does not play because his extraction cost is too high.

Heuristically, we can view the "cost" for the oil producer as $s_0 + v'(x)$ at any given stock $x(t)$ (the sum of extraction and shadow costs). This is bounded below by s_0, so if s_0 is sufficiently high, it may be too expensive for him to produce and he may be forced to exit the market. This shall be referred to as blockading of the oil player.

We define a blockading point for our oil producer as the point

$$x_b^* = \sup\{x > 0 : q_0^*(x) = 0\}.$$

For all $x > x_b^*$, the oil producer will produce, whereas for all $x < x_b^*$, the oil producer does not participate. If there is no such $x > 0$ such that $q_0^*(x) = 0$, then we say the oil producer is never blockaded and let $x_b^* = 0$.

We will first demonstrate that provided that the oil producer always plays when $x(t) > 0$ provided s_0 is small enough.

Proposition 2. *If $s_0 < \rho_n$ for all $n \in \{2, \ldots, N\}$, then the oil producer is never blockaded.*

By Lemma 2, this condition is equivalent to $s_0 < \rho_N$, as ρ_n is decreasing in n.

Proof. In the interval $[0, x_b^{N-1})$, the candidate value function is $v(x) = v^{(N)}(x)$ which solves the ODE (7) with boundary condition $v^{(N)}(0) = 0$, and the corresponding equilibrium oil production is

$$ q_0^*(x) = \frac{N}{N+1} \left(\rho_N - (s_0 + v^{(N)'}(x)) \right), $$

following from formula (8). Then $q_0^*(x) \geq 0$ as long as $v^{(N)'}(x) \leq \rho_N - s_0$, where the bound is positive by hypothesis. It follows from the ODE (7) that $v^{(N)'}(0) = \rho_N - s_0$, and it is easily verified from the formula (11) for $v^{(N)}$ that $v^{(N)}(x)$ is strictly concave and so $v^{(N)'}(x) \leq \rho_N - s_0$ for all $x \in [0, x_b^{N-1})$. Therefore the candidate solution in which player 0 is not blockaded hold in the first interval as the unique Markov perfect equilibrium. A similar argument hold in the other intervals (in which the shadow cost v' of the oil producer is even lower). ∎

This is consistent for the $s_0 = 0$ limiting case, since then it is always true that $s_0 \leq \rho_N$ and thus there are no blockading points for player 0. In this case, we can evaluate the blockading points explicitly. We first define

$$ \delta_n = (n+1)s_n - (1 + s_0 + S^{(n-1)}). $$

Then, the following proposition holds:

Proposition 3. *Provided that $s_0 < \rho_N$, the blockading point for the kth player is finite if $\delta_k > 0$. Let $i = \min\{k : \delta_k > 0\}$, or the lowest cost player who is blockaded at some point. Players $\{i, \ldots, N-1\}$ are blockaded to the right of their blockading points which are determined recursively by the equations:*

$$ x_b^{N-1} = \frac{1}{\mu_N} \left[-1 + \frac{\delta_{N-1}}{\hat{\rho}_N} - \log\left(\frac{\delta_{N-1}}{\hat{\rho}_N} \right) \right] \tag{14} $$

$$ x_b^{n-1} = x_b^n + \frac{1}{\mu_n} \left[\log\left(\frac{\delta_n}{\delta_{n-1}} \right) - \frac{(n+1)(s_n - s_{n-1})}{\hat{\rho}_n} \right], \tag{15} $$

where, for $n \in \{i, \ldots, N-2\}$,

$$\mu_n = \frac{2r}{\hat{\rho}_n} \left(\frac{1}{2} + \frac{1}{2n} \right)^2,$$

and $\hat{\rho}_n$ was defined in (13).

Proof. To find x_b^{N-1}, we look for a solution of $q_{N-1}^*(x_b^{N-1}) = 0$, and thus, from (9), we require that

$$v^{(N)'}(x_b^{N-1}) = N s_{N-1} - (1 + s_0 + S^{(N-2)}) = \delta_{N-1}.$$

Since, from (11), $v^{(N)'} = -\hat{\rho}_N \mathbf{W}(\theta(x))$, we have that

$$\mathbf{W}\left(\theta(x_b^{N-1})\right) = -\frac{\delta_{N-1}}{\hat{\rho}_N}.$$

This equation only has a positive solution for x_b^{N-1} for $\delta_{N-1} > 0$, which is given by (14). The recursion formula (15) follows from shifting the axes left by x_b^{N-1} and proceeding with similar analysis to that for the $s_0 = 0$ case in the proofs of Propositions 5.2 and 5.3 in [8]. $\qquad \blacksquare$

The following proposition solves the case where $s_0 \geq \rho_N$:

Proposition 4. *If $s_0 \geq \rho_N$, then the Markov perfect equilibrium is that the oil producer does not play and the renewables play an $N - 1$ player Cournot game as given by (2) with $n = N - 1$:*

$$q_i^*(x(t)) = \frac{1}{N}\left(1 - N s_i + S^{(N-1)}\right), \qquad i = 1, \ldots, N-1. \qquad (16)$$

Proof. In the candidate $N - 1$ player equilibrium (16), the total output is

$$Q = \sum_{i=1}^{N-1} q_i^* = 1 - \rho_N,$$

and so the market price is $P = 1 - Q = \rho_N$. Since $s_0 \geq \rho_N$, player 0 will not enter the market and his best response is $q_0^* = 0$. Therefore (16) gives the Nash equilibrium in this case. $\qquad \blacksquare$

Finally, in Fig. 2, we plot $v(x)$ for two constant costs: $s_0 = 0.2$ and $s_0 = 0.4$. For the lower cost, note that the value function is strictly higher, as expected, because profits are greater. On the right, however, we note that the oil producer lasts longer in time t with a higher cost than with a lower cost. The additional cost produces an amplified incentive to save until tomorrow and hence the oil producer lasts longer. This is reflected in the market price as well, as there is a greater market price with a higher cost, but the jump in market price is not as significant when oil runs out. In other words, the higher fixed cost of extraction results in price stability over time at the cost of higher market prices even when the oil player produces.

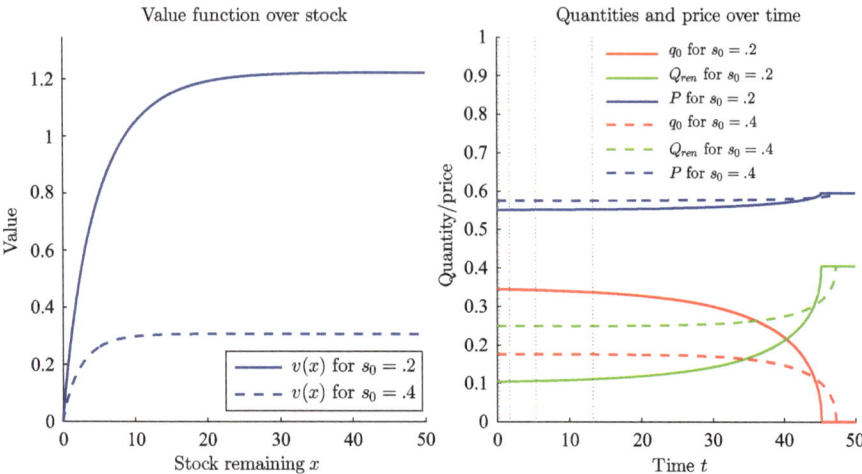

Fig. 2 Constant cost dynamic game based on analytical solution for $s_0 = 0.2$ and $s_0 = 0.4$ for $s_1 = 0.51, s_2 = 0.52, \ldots, s_9 = 0.59$. Notationally, $Q_{ren} = \sum_{i=1}^{9} q_i$ (total renewable production). In this case $\rho_N = 0.595$, and so, as $s_0 < 0.595$, the oil producer always produces

3 Varying Costs

Now that we have considered the one subcase in which we can derive a fully analytical Markov strategy, we consider the general case, where costs vary for each of the players. We associate to each player i in our N player game a cost function $s_i(x)$, where x is the remaining stock of player 0, the only exhaustible player. To understand how this generalizes the earlier result, we first consider a base case, where $s_1, \ldots, s_{N-1}$ are constant.

In this section, we do not yet allow discovery of new reserves. Since disabling discovery of new reserves enables us to have partially analytical solutions, we present these calculations here and resort to a numerical approach to the full discovery problem in Sect. 4.

3.1 Holding Renewable Costs Constant

The array of costs for this base case is such that $s_i(x) = s_i \in [0, 1]$ for each $i \in \{1, \ldots, N-1\}$, while $s_0(x)$ is a decreasing function. The following assumptions are made:

Assumption 2. *We retain the assumption that $s_i < \rho_i$ for the renewable players $i = 1, \ldots, N-1$. This implies that should the oil producer exit, the appropriate Cournot solution among the renewable players, given in formula (2), entails that all players are active.*

Assumption 3. *We assume that the cost function of the oil producer $s_0(x) \in \mathcal{C}^1$ and $s_0'(x) < 0$, so cost increases as remaining reserves are depleted. This assumption can be justified, as oil producers typically delay extraction from their most expensive ores, so as reserves begin to run out, costs are increased.*

When there are $n-1$ active renewable players, and hence n total players, should the oil producer play, the Hamilton-Jacobi equation for his value function $v(x)$ is

$$rv = \frac{1}{(n+1)^2}\left(1 - n(v'(x) + s_0(x)) + S^{(n-1)}\right)^2, \qquad x \in [x_b^n, x_b^{n-1}), \qquad (17)$$

analogous to (7) with the constant s_0 replaced by $s_0(x)$. The oil producer's strategy is

$$q_0^*(x) = \frac{1}{(n+1)}\left(1 - n(v'(x) + s_0(x)) + S^{(n-1)}\right) = \frac{n}{(n+1)}\left(\rho_n - (v'(x) + s_0(x))\right), \tag{18}$$

where ρ_n was defined in (4). We will also see that, while the effective cost for the oil player is always decreasing in x, even when Assumption 3 does not hold, the shadow cost $v'(x)$ may be increasing or decreasing in x, depending on the properties of $s_0(x)$.

Proposition 5. *The sum $v'(x) + s_0(x)$ is strictly decreasing, but $v'(x)$ is not necessarily decreasing. In particular, if we define*

$$T(x, n) = -\frac{(n+1)^2 rv'(x)}{2nq_0^*(x)},$$

then if $|s_0'(x)| \geq |T(x,n)|$, when there are n players total including the oil producer, then $v(x)$ is convex ($v'(x)$ increasing), while $|s_0'(x)| \leq |T(x,n)|$ implies that $v(x)$ is concave, with $v'(x)$ decreasing.

Proof. The result follows immediately from taking the derivative of Eq. (17) with respect to x:

$$v''(x) + s_0'(x) = -T(x, n),$$

so if $|s_0'(x)| \geq |T(x,n)|$, then $v''(x) > 0$ and hence $v'(x)$ is increasing, and similarly for the other case.

Note that $q_0^*(x)$ is increasing in x since the effective cost $v'(x) + s_0(x)$ is decreasing; further, as $x \to 0$, $q_0^*(x) \to 0$, but v' is bounded above ($v'(0) + s_0(0) = \rho_N$). Assuming that $s_0(x)$ is bounded above as well, then in a neighborhood around 0, $v(x)$ is concave. If $s_0(x)$ decreases sufficiently quickly in some neighborhood of a point $x > 0$, then $v(x)$ becomes convex. Figure 3 illustrates with two different cost functions $s_0(x)$.

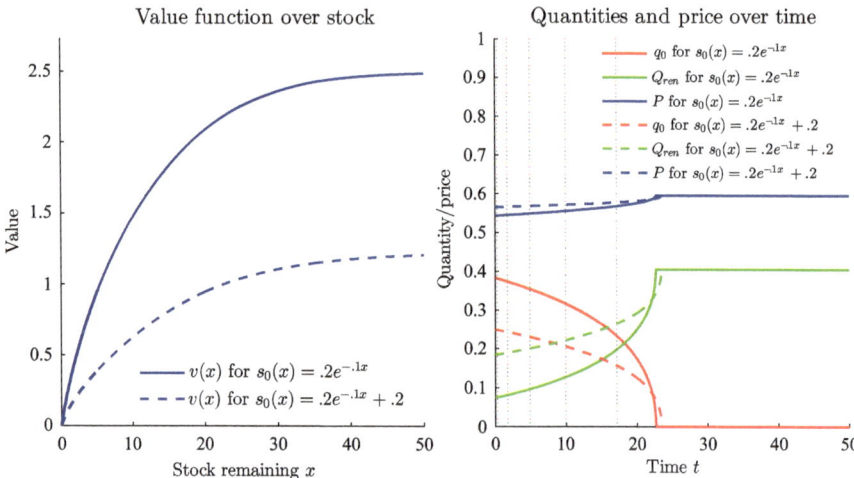

Fig. 3 Compares value functions as well as quantity and price evolution over time for 9 renewable players with costs $0.51, \ldots, 0.59$, while we adjust the cost of extraction for the oil producer from $s_0(x) = .2e^{-.1x}$ (*solid*) to $s_0(x) = .2e^{-.1x} + .2$ (*dashed*)

We now let $x_b^* = \sup\{x > 0 : q_0^*(x) = 0\}$. Then, it is evident that since $v'(x) + s_0(x)$ is decreasing that for all $x > x_b^*$, the oil producer participates and for all $x < x_b^*$, the oil producer does not participate. In the region the oil producer does not participate, the assumption $s_n < \rho_n$ for renewable players implies the quantities and market price is determined by an $N - 1$ player Cournot static game among the renewable players. The following establishes that the condition upon which the oil producer plays is solely dependent on extracting costs and not remaining stock:

Proposition 6. *The blockading point x_b^* for player 0 is given by*

$$x_b^* = \sup\{x > 0 : q_0^*(x) = 0\} = \inf\{x > 0 : s_0(x) \leq \rho_N\}.$$

Proof. Let $I = \inf\{x > 0 : s_0(x) \leq \rho_N\}$. It suffices to show that when $x < I$, the oil producer does not play and when $x > I$, the oil producer does. When $x < I$, we have that $s_0(x) > \rho_N$; assume the oil producer plays. Prior to this assumption, the equilibrium in this region would have been an $N - 1$ player Markov game, so for the assumption to hold, the quantity produced in an N player game once the oil producer enters should be positive. That is, from (18) with $n = N$, this requires $v'(x) + s_0(x) < \rho_N$, but this forces $v' < 0$, a contradiction. So, the oil producer does not play on $x < I$.

Similarly, the oil producer must play when $x > I$. Assume he doesn't; then it must not be profitable for him to play. Hence, if $q_0 > 0$ when $x > I$, the oil player plays. This is evident, since if $x > I$, $s_0(x) \leq \rho_N$, so $v'(I) \leq 0$ and the Hamilton-Jacobi equation is well-defined and evolves to give positive q_0 and hence positive profit. Since $s(x)$ is decreasing and continuous, the result follows. $\qquad\square$

Remark. The first point coming from the left at which $s(x) = \rho_N$ (and also the only point since $s(x)$ is decreasing) is the threshold point at which the oil producer enters. The mechanism here is not that the oil producer's shadow cost is too high. Rather, the oil producer's cost of extraction becomes too prohibitive in comparison to the other alternatives. To the left of this point, since the assumption $s_n < \rho_n$ for the renewable players binds, only the renewable players remain and they play indefinitely in a static game. To the right of this point, blockading for the renewable players once again can occur; that is, there may be associated threshold values for each of the renewable players to the left of which they do play and to the right of which they do not. In particular, the following proposition highlights when the renewable players are blockaded.

3.2   Numerical Examples with Renewable Costs Held Constant

Figures 3 and 4 depict numerically evaluated solutions for the case of decreasing $s_0(x)$. In particular, in Fig. 3, we note that if $s_0'(x) < s_0(x)$ for all x, then the $v(x)$ that corresponds to $s_0'(x)$ is less than or equal to than the $v(x)$ for $s_0(x)$. In particular, we note that similar to earlier, increasing $s_0(x)$ by a fixed cost leads to the oil player staying in for a longer period of time. In Fig. 3, the vertical dotted lines correspond to blockading points x_b^i for the oil producers for $s_0(x) = .2e^{-.1x}$. That is, to the left of these lines, additional players enter since the reserves have depleted

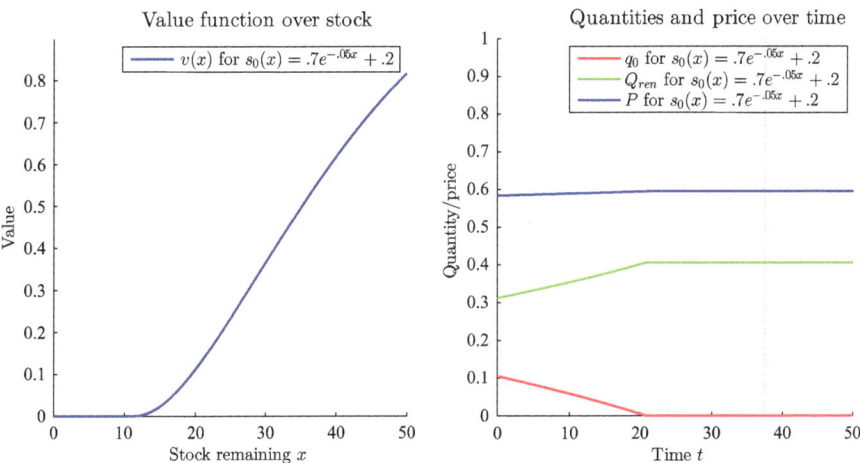

Fig. 4 Demonstrates evolution of game when $s_0(x)$ is high enough for small enough x so that the oil player is effectively blockaded for low x. Notationally, $Q_{ren} = \sum_{i=1}^{9} q_i$. Also demonstrates case where Lipschitz constant of $s_0(x)$ is so large to cause concavity of $v(x)$ on certain regions of x. The dotted vertical line reflects blockading of the 9th renewable player

sufficiently. When we increase $s_0(x)$, the blockading points are shifted to the left, since the relative cost of oil is higher, so the threshold for the renewables entering is lower. Hence, when simulating such a game over time, the renewable players come in earlier.

Figure 4 depicts the case where for some $x < x^*$, we have that $s_0(x) > \rho_N$. In particular, note that until $x = 11.6$, the oil producer does not produce. This is reflected in the graph on the left. At the time when the oil producer leaves, there are still reserves left that have yet to be drilled (specifically, 11.6 units). Further, Fig. 4 depicts a situation that is consistent with that explained in Proposition 8. That is, $s_0'(x)$ is greater than the threshold value $T(x, n)$ for a subset of $x \in [0, 50]$, resulting in a shift from concavity to convexity. This is due to the high Lipschitz constant that bounds the value of $s_0'(x)$ from the above and continuity, which ensures concavity in a neighborhood where the derivative is maximized.

3.3 Varying Renewable Costs: Analytic Results

The above problem can be solved partially analytically for N players, all of whom have varying costs that obey the following assumption:

Assumption 4. *In an N player game, let player 0 be the exhaustible player and players $1, \ldots, N - 1$ be renewable players. Let $s_i(x)$ be the cost of player i. We assume that $s_0'(x) < 0$ and $s_i'(x) > 0$ for all $i \in \{1, \ldots, N - 1\}$.*

This assumption is justified because, as oil tends to deplete, governments tend to subsidize green energy, thus reducing costs for the renewable players as x approaches 0.

The following assumption is made so that players are not forced out of the game due solely to their absolute extraction cost, but rather only choose not to play because of relative extraction cost:

Assumption 5. *For all $i \in \{0, 1, \ldots, N - 1\}$, we require $s_i(x) \in [0, 1]$ for all $x \geq 0$.*

We now let $N = 2$ for a couple of reasons: first, the structure of the game is not substantively different with $N > 2$, and second, $N = 2$ simplifies numerical computation significantly.

Proposition 7. *We define*

$$\rho_1(x) = \frac{1 + s_1(x)}{2}.$$

The dominant strategy for the oil producer is not to produce when $s_0(x) > \rho_1(x)$.

Proof. Assume that the oil producer will want to produce; then there must be some neighborhood around the point x by continuity of $s_0(x)$ and $s_1(x)$ such that both players will participate with positive quantities. However, in such a case,

$$q_0^*(x) = \frac{1}{3}\left(1 - 2v' - 2s_0(x) + s_1(x)\right) > 0,$$

implying that $v' < \rho_1(x) - s_1(x) < 0$, a contradiction to the obvious fact that giving a player extra resources does not make him worse off.

Remark. When $s_1(x)$ is held constant, then $\rho_1(x)$ is also constant, so this result is identical with that of the fixed cost case discussed previously.

In particular, since $s_1(x)$ is decreasing, so must $\rho_1(x)$. However, at the same time, $s_0(x)$ is decreasing, leading to the following corollary:

Corollary 1. *It cannot be the case that at some point x_1, the oil producer produces, but for some $x_2 > x_1$ the oil producer does not produce.*

Thus, we define the point

$$x_b^* = \inf\{x > 0 : q_0^*(x) \geq 0\}.$$

For all points $x < x_b^*$, the oil producer will not produce. In particular, it is easy to see that

$$\sup\{x > 0 : q_0^*(x) = 0\} = \inf\{x > 0 : q_0^*(x) \geq 0\}.$$

Proposition 8. *If there exists a point x such that $s_0(x) < \rho_1(x)$, then the oil producer will participate.*

Proof. Assume for sake of contradiction that the oil player does not play. Then, should he play, the cost must be too high. In other words, $v'(x) + s_0(x) > \rho_1(x)$ must hold, requiring $v'(x) > \rho_1(x) - s_0(x) > 0$, so the value function is increasing at the point x. However, then, there must be a neighborhood to the left of x on which the value function is also rising; since the value function measures objective utility, the player must produce nonzero quantity. If the player were to produce zero quantity, then his utility could not increase. Thus, there must be a neighborhood to the left of x on which the oil player produces. However, Corollary 1 then implies that at x, the oil player must participate, contradicting the assumption.

It thus follows immediately that there are three cases for the form of $s_0(x)$

CI: It could be that $s_0(0) < \rho_1(0)$, which implies that $s_0(x) < \rho_1(x)$ for all x, in which case the oil producer always plays.

CII: It could be that $s_0(x) \geq \rho_1(x)$ for all x, in which case the oil producer never plays.

CIII: It could be that $s_0(0) \geq \rho_1(0)$, but there is some later $x' > 0$ such that $s_0(x') < \rho_1(x')$. By continuity, it is evident that the point x_b^* can be expressed as

$$x_b^* = \inf\{s_0(x) < \rho_1(x)\}.$$

In the third case we were examining, continuity and closure of $[0, x']$ requires that $x_b^* \in [0, x']$. In particular, continuity of $s_0(x)$ and $s_0(x)$ decreasing implies that in this case, x_b^* is the unique point that satisfies $s_0(x_b^*) = \rho_1(x_b^*)$.

The following Proposition then establishes the Markov perfect equilibrium:

Proposition 9. *The Markov perfect equilibrium for the game is that for all points $x < x_b^*$, the renewable player is the only player and the strategies are*

$$(q_0^*, q_1^*) = \left(0, \frac{1 - s_0(x)}{2}\right).$$

For all $x > x_b^*$,

$$(q_0^*, q_1^*) = \left(1 - 2v'(x) - 2s_0(x) + s_1(x), 1 - 2s_1(x) + v'(x) + s_0(x)\right).$$

Proof. Should we show that this holds for Case III, we can merely take the limits $x_b^* \to 0$ for Case I and $x_b^* \to \infty$ for Case II, so proving that this is the relevant equilibrium for Case III suffices. Assume for contradiction that this is not a Markov equilibrium. Then, either there is some strategy q_0 such that profit is larger for the oil player holding q_1^* constant (Case A), or there is some strategy q_1 such that profit is larger for the renewable player holding q_0^* constant (Case B). Assume Case A holds; then, Proposition 8 implies optimality of q_0^*. Assume Case B holds; then, when $x < x_b^*$, the renewable player has a monopoly on the market, which implies q_1^* is optimal. For $x > x_b^*$, the optimal Cournot equilibrium is described by q_1^* and hence is already optimal. Thus, the assumption fails to hold and we have a contradiction, implying the result.

3.4 Varying Costs: Numerical Calculations

We now investigate the problem numerically. We first consider a simple case, shown in Fig. 5; the oil producer shall have an exponentially decreasing cost in stock, that is $s_0(x) = .6e^{-x}$ and player 1, our renewable player, shall have an exponentially increasing cost in stock, that is, $s_1(x) = .6(1 - e^{-x})$. The left plot suggests a result that follows directly from the analytical propositions above. That is, for sufficiently low stock, that is for all stock $x < x^*$, where x^* satisfies $s_0(x^*) = \rho_1(x^*)$, the oil producer has a cost too high for him to ever produce. For all $x > x^*$, his cost is lower than the necessary threshold. This is reflected in the value functions $v(x)$ which satisfy $v(x) = 0$ for all $x \in [0, x^*]$.

Although not explicitly depicted, for large reserves of oil, the renewable players' cost increases to the point where he must drop out, leading to an oil monopoly for sufficiently large x. In particular, we note that the total quantity produced is not smooth at these two points. This is as expected; if we let x' be the point at which the oil player has a monopoly for all $x > x'$, the intervals $[0, x^*]$, (x^*, x'), and (x', ∞) correspond to fundamentally different games.

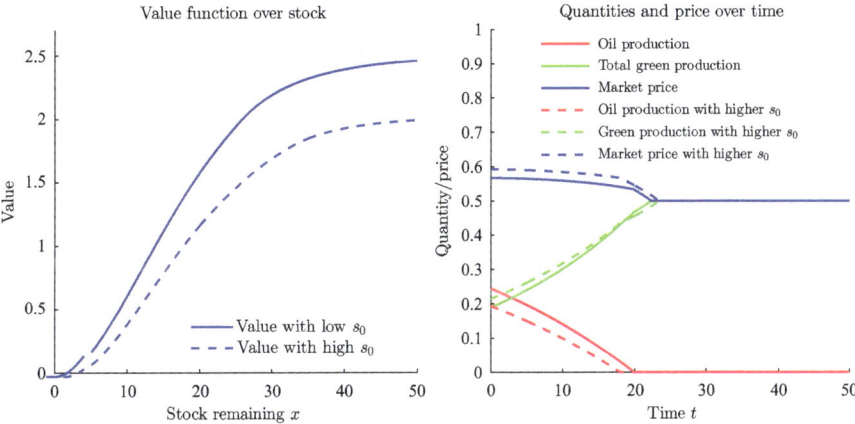

Fig. 5 The "low" cost corresponds to $s_0(x) = .6e^{-.1x}$ and $s_1(x) = .6(1 - e^{-.1x})$, while the "high" cost case corresponds to $s_0(x) = .6e^{-.1x} + .1$ and $s_1(x) = .6(1 - e^{-.1x})$

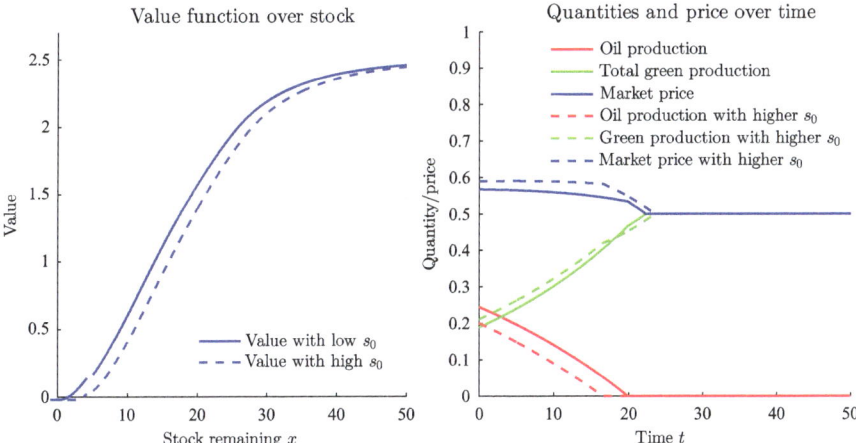

Fig. 6 The "low" cost corresponds to $s_0(x) = .6e^{-.1x}$ and $s_1(x) = .6(1 - e^{-.1x})$, while the "high" cost case corresponds to $s_0(x) = .8e^{-.1x}$ and $s_1(x) = .6(1 - e^{-.1x})$

Finally, Fig. 6 plots two value functions; this time, we once again fix $s_1(x) = .6(1 - e^{-.1x})$. However, the low $s_0(x)$ case corresponds to $s_0(x) = .6e^{-.1x}$, while the high $s_0(x)$ case corresponds to $s_0(x) = .8e^{.1x}$. While the right plot exhibits similar features to those discussed earlier, the left plot is of particular interest. If we denote $v_L(x)$ to correspond to the low cost case and $v_H(x)$ to correspond to the high cost, we note that as $x \to \infty$, $\|v_H(x) - v_L(x)\| \to 0$. This is a consequence of the lower and higher costs both converging to 0, which reduces the difference in oil value over time. This suggests that as stock is increased to arbitrarily large amounts, the differences in discounted profits for the oil producer becomes negligible as long as the costs over time converge to the same value.

4 Resource Discovery

We now allow that the oil producer may invest in research and development, choosing to invest in exploration that yields a possibility of discovering new reserves. In particular, the oil producer may, at any given point, explore with intensity a at a cost $\mathcal{C}(a)$. We choose $\mathcal{C}(a)$ to be increasing in a. As intensity increases, the probability of finding additional reserves also increases. In particular, we let reserves at any time t evolve according to

$$dX_t = -q_0(X_t)\mathbb{1}_{\{X_t>0\}}\,dt + \delta\,dN_t.$$

Here, N_t is a point process with intensity λa_t, where λ is determined exogenously. The probability of increasing one's reserves by a fixed quantity δ at a given time t is thus $\lambda a_t dt$. This is the situation considered in [9] for constant costs.

4.1 Two Players

We first consider a scenario similar to above where there are two players, an oil player (denoted player 0) and a renewable player (denoted player 1). The oil producer's strategy will depend on his stock at any given time and produces at $q_0(X_t)$, where q_0 is part of a Markov strategy with player 1.

We let both players have a cost of extraction $s_0(x)$ and $s_1(x)$, where $s_0'(x) < 0$. This reflects that as oil runs out, the oil producer's cost of extraction also goes up. Meanwhile, we let $s_1'(x) > 0$, since as x decreases, investment and the like in renewable resources increasing.

We then can write the value functions for both players, where (q_0^*, q_1^*) is a Markov equilibrium:

$$v(x) = \sup_{q_0,a} \mathbb{E}\left[\int_0^\infty e^{-rt}\left(q_0(X_t)(1 - q_0(X_t) - q_1^*(X_t) - s_0(X_t)) - \mathcal{C}(a_t)\right)dt\Big|X_0 = x\right]$$

$$w(x) = \sup_{q_1} \mathbb{E}\left[\int_0^\infty e^{-rt}q_1(X_t)(1 - q_0^*(X_t) - q_1(X_t) - s_1(X_t))\mathbb{1}_{\{X_t>0\}}dt\right.$$

$$\left. + \int_0^\infty e^{-rt}\frac{1}{4}(1 - s_1(X_t))^2\mathbb{1}_{\{X_t=0\}}dt\Big|X_0 = x\right].$$

As earlier, we will note that $v_0(x)$ will be the only value function necessary for a description of the strategies. From just the value functions, we may note that the oil player has to optimize over two variables, q_0 and a. In other words, in certain intervals it may be optimal for player 1 to defer production by investing in exploration (R&D) while at other times, he may choose to not invest in R&D at all.

The corresponding Hamilton-Jacobi equation for the oil producer is

$$rv(x) = \sup_{q_0,a} \left[(1 - q_0 - q_1^*)q_0 - q_0(v'(x) + s_0(x)) - \mathcal{C}(a) + a\lambda \Delta v(x) \right].$$

Here, we have that $\mathcal{C}(a)$ is the cost of exploring with intensity a and $\Delta v(x)$ is a jump term defined by

$$\Delta v(x) = v(x + \delta) - v(x).$$

Since q_1 and a are additively separable, we can simplify the above Hamilton-Jacobi equation to

$$rv(x) = \sup_{q_0} \left\{ (1 - q_0 - q_1^*)q_0 - q_0(v'(x) + s_0(x)) \right\} + \sup_a \left\{ -\mathcal{C}(a) + a\lambda \Delta v(x) \right\}.$$

In particular, the optimum exploration intensity at any given x is given by

$$a^* = \operatorname*{argsup}_{a \geq 0} -\mathcal{C}(a) + a\lambda \Delta v(x),$$

the Legendre transform of the exploration cost function evaluated at $\lambda \Delta v(x)$.

As is done in [9], we will take the cost of exploration to be

$$\mathcal{C}(a) = \frac{1}{\beta} a^\beta + \kappa a, \tag{19}$$

with $\beta > 1$ and $\kappa \geq 0$. We define a saturation point x_{sat} to be the point where the oil producer stops exploring. In particular,

$$x_{sat} = \inf\{a^*(x) = 0 : x > 0\}.$$

Since we require $\mathcal{C}(a)$ to be increasing in a, it is evident from $\lambda > 0$ and $\Delta v(x) > 0$ ($v(x)$ non-decreasing is immediate as additional reserves cannot make one worse off), we have that for all $x > x_{sat}$, $a^*(x) = 0$ and for all $x < x_{sat}$, $a^*(x) > 0$. It is immediately seen that

$$a^*(x) = [\max(\lambda \Delta v(x) - \kappa, 0)]^{\gamma - 1}, \qquad \text{where} \qquad \gamma = \frac{\beta}{\beta - 1}. \tag{20}$$

4.1.1 Structure of Solution

In particular, there are three possibilities. In regions where $s_0(x) > \dfrac{1 + s_1(x)}{2}$, the oil producer is blockaded and does not produce. On this interval, the above Hamilton-Jacobi equation reduces to

$$v(x) = -\mathcal{C}(a^*) + a^*\lambda \Delta v(x).$$

When both the oil and renewable players produce, then the Hamilton-Jacobi equation reduces to

$$rv = \frac{1}{9}\left(1 - 2s_0(x) - 2v'(x) + s_1(x)\right)^2 + \frac{1}{\gamma}[\max(\lambda\,\Delta v(x) - \kappa, 0)]^{\gamma}.$$

If the renewable player is blockaded, in the sense that there exists a finite x_b^1 such that

$$x_b^1 = \inf\{q_1^*(x) = 0 : x > 0\},$$

then for all $x \geq x_b^1$, the Hamilton-Jacobi equation is

$$rv = \frac{1}{4}(1 - s_0(x) - v'(x))^2 + \frac{1}{\gamma}[\max(\lambda\,\Delta v(x) - \kappa, 0)]^{\gamma}.$$

The case where $s_0(x) = 0$ has been asymptotically evaluated when $\lambda < \epsilon$ for small ϵ in [9].

Finally, to compute the boundary condition, we note that at $x = 0$, the oil producer cannot produce; that is, we have $q_0(0) = 0$. This implies that

$$v(0) = \sup_{a \geq 0} \mathbb{E}\left[e^{-rT}v(\delta) - \int_0^T e^{-rt}C(a)dt\right],$$

where T is the time until when the next discovery is made.

In the case where $s_0(x) > \dfrac{1 + s_1(x)}{2}$ for some x, we let

$$x^* = \sup\left\{x : s_0(x) > \frac{1 + s_1(x)}{2}\right\}.$$

If no such x^* exists (that is $s_0(x) < (1 + s_1(x))/2$ for all $x \in \mathbb{R}_+$), we let $x^* \to 0$. In particular, the constraint

$$v(x^*) = \sup_{a \geq 0} \mathbb{E}\left[e^{-rT}v(x^* + \delta) - \int_0^T e^{-rt}C(a)dt\right]$$

must also hold. The above thus implies that

$$v(0) = \sup_{a \geq 0} \frac{\lambda a v(\delta) - C(a)}{\lambda a + r}. \tag{21}$$

4.1.2 Numerical Discretization

As explained in [9], we can reduce the above to an iterative ODE that can be solved with Runge-Kutta methods by defining $v^0(x) = v(x)$, which is the no-exploration case solved in Sect. 3.3. For all $n \geq 1$, we recursively define

$$rv^n = (q_0^n(x))^2 + \frac{1}{\gamma}\left(\max(\lambda(v^{n-1}(x+\delta) - v^n(x)) - \kappa, 0)\right)^\gamma$$

$$q_0^n = \begin{cases} \frac{1}{2}(1 - s_0(x) - v'(x)) & \text{if } x \geq x_b^1 \\ \frac{1}{3}(1 - 2s_0(x) - 2v'(x) + s_1(x)) & \text{if } x \in (x^*, x_b^1) \\ 0 & \text{if } x \in [0, x^*] \end{cases}$$

$$v^n(0) = \sup_{a \geq 0} \frac{\lambda a v^{n-1}(\delta) - C(a)}{\lambda a + r}.$$

$$v^n(x \in (0, x^*)) = \sup_{a \geq 0} \frac{\lambda a v^{n-1}(x + \delta) - C(a)}{\lambda a + r},$$

where $C(a)$ is defined in (19). The above can then be solved using standard Runge-Kutta methods. In particular, it is shown in [9] that for a monopoly with zero costs, the above iterative scheme does converge uniformly to the value function with exploration.

4.2 Numerical Solution for Two Players

A numerical solution for the case where $s_0(x) = .15e^{-.05x}$ and $s_1(x) = .15(1 - e^{-.1x}) + 0.5$ is evaluated graphically using the iterative approach above in Figs. 7 and 8. We note in particular that as $x \to \infty$, $\|v(x) - v^0(x)\| < \epsilon$ for $\epsilon \to 0$. This follows from $v'(x) \to 0$ as $x \to \infty$ for all iterations n; hence, we also have from the Mean Value Theorem that $\Delta v(x) \to 0$ as $x \to \infty$, and $\kappa > 0$ forces $a^*(x)$ to be realized at 0 as $x \to \infty$. Applying this limit to the Hamilton-Jacobi equation, we see that as $x \to \infty$, we recover the original equation for the non-exploration case, implying that $\|v(x) - v^0(x)\| \to 0$ as $x \to \infty$. Further, we can see that for all x, $v(x) \geq v^0(x)$. This follows immediately from a revealed preference argument, since when we allow exploration, the oil producer can never be strictly worse off, as he can always choose to never explore.

From Fig. 8, we have that for all $x < x_{sat}$, the oil producer (player 0), does indeed explore and has a value $a^*(x) > 0$. We also have analytically that $a^*(x)$ is strictly decreasing in x. This follows from Fig. 7 which indicates that $v(x)$ is concave everywhere. This is a result of our chosen cost function which has a low enough Lipschitz constant to obey the condition outlined in Proposition 8. Hence, $\Delta v(x)$ is decreasing in x so $a^*(x)$ must also be strictly decreasing. Since the process that governs evolution of reserves δdN_t has increased probability of identifying with δ as $x \to \infty$, we can see that there are more "jumps" or discoveries as x decreases.

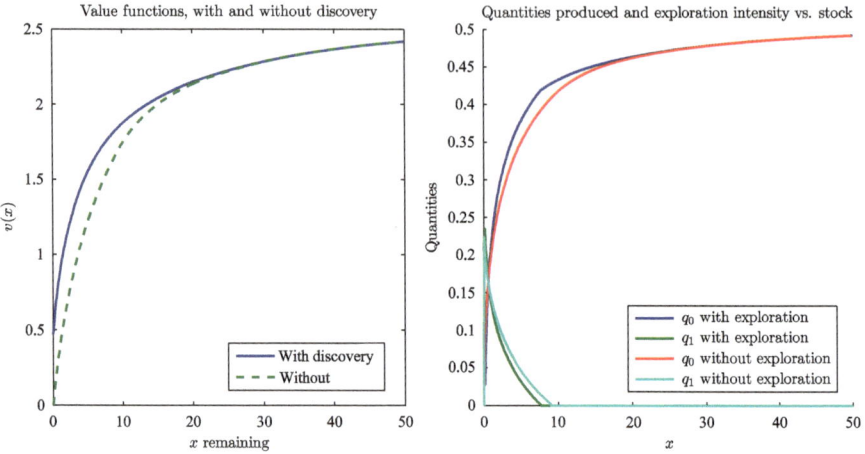

Fig. 7 Oil producer has cost function $s_0(x) = .15e^{-0.05x}$ and the renewable player has cost function $s_1(x) = .15(1 - e^{-.1x}) + 0.5$

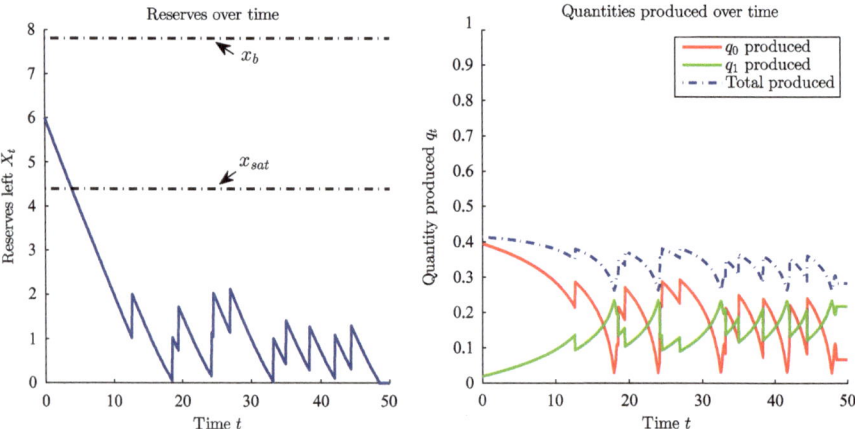

Fig. 8 Simulation vs. time of an exhaustible resource producer (player 0) and renewable producer (player 1) with costs $s_0 = .15e^{-.05x}$ and $s_1(x) = .15(1 - e^{-.1x}) + 0.5$

Intuitively, as oil runs out, the oil producer needs to amp up discovery in order to stay in the game and is thus willing to pay additional cost, since the opportunity cost of not exploring is to leave the game. In the right panel of Fig. 8, we can also see such jumps. In particular, in the beginning, the oil producer does fairly well, and for each discovery, the oil producer does produce additional quantities; even though the cost for the renewable player approaches a cost significantly higher than that of the oil producer (0.5 for player 1 vs. .15 for player 0), the renewable producer does indeed outproduce when the oil effectively runs out. Finally, we can note a trend of total quantity $q_0 + q_1$ decreasing in x, which corresponds to a higher market price over time.

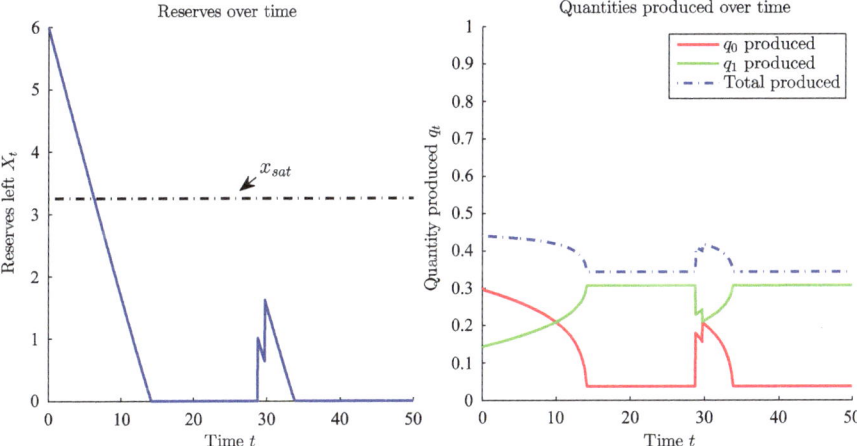

Fig. 9 Simulation vs. time with increased relative cost for player 0 (exhaustible producer). In particular, we now let $s_0(x) = .25e^{-.05x}$ and $s_1(x) = .15(1 - e^{-.1x}) + .35$

If we now increase the relative cost of $s_0(x)$ by both increasing $s_0(x)$ and by decreasing $s_1(x)$, we obtain the simulation presented in Fig. 9. We can note a few differences. First, $x_b \to \infty$, so the renewable producer is never blockaded. Further, the saturation point x_{sat} is now strictly lower. That is, the oil producer stops researching for new reserves at a lower threshold. This is a mathematical triviality from the expression for $a^*(x)$; as we increase the relative cost of $s_0(x)$, the value function $v(x)$ is decreased since profits are lower. This in turn causes $\Delta v(x)$ to go down since the marginal utility of an additional δ amount of oil at any given x is also lower. Hence, the threshold for which the cost becomes too prohibitive to research is also lower.

In particular, $a^*(x)$ is also strictly lower in this case due to an argument similar to that in the previous paragraph. This means the probability of successful discovery of reserves is also lower. Hence, there is a dual effect to increasing relative cost. Not only does the oil producer produce less and hence have less discounted profit overall, but he also spares less for researching, which harms him when $x \to 0$.

Finally, we can note that in the long term, the market price is quite similar since the total quantity produced, plotted in the right panel of Fig. 9 is about the same as before. However, now, the green producer almost always overtakes the red producer because the probability of discovery is now lower.

4.3 N-player Case

Finally, we consider a multi-player oligopoly differential game with exploration. Consider an N player game where player 0 is an exhaustible producer, whom we will term the "oil" producer, while players $1, \ldots, N-1$ are renewable players with fixed costs $s_1, \ldots, s_{N-1}$. For simplicity, we set each of these costs to be constant for all x. Further, let the "oil" producer have a decreasing cost of extraction $s_0(x)$ such that $s_0'(x) < 0$. Unlike previously, we now allow the oil producer to invest in R&D which may result in discovery of new oil sources. Borrowing notation from previously, we let

$$S^{(N-1)} = \sum_{i=1}^{N-1} s_i, \qquad \text{and} \qquad \rho_N = \frac{1 + S^{(N-1)}}{N}.$$

Letting $(q_0, \ldots, q_{N-1})$ be the Markov strategies of production given any remaining stock $x \in \mathbb{R}_+$, the reserves evolve according to

$$dX_t = -q_0(X_t) \mathbb{1}_{\{X_t > 0\}} dt + \delta dN_t.$$

Now, there are N players, but player 0 has a cost of extraction $s_0(x)$ which depends on the amount of stock left. Denoting the cost of exploration with intensity a as $\mathcal{C}(a)$, the appropriate value function for the oil producer is, given that we denote $(q_0^*, \ldots, q_{N-1}^*)$ as a Markov perfect equilibrium,

$$v(x) = \sup_{q_0, a} \mathbb{E} \left[\int_0^\infty e^{-rt} \left(q_0(X_t) \left(1 - q_0(X_t) - \sum_{i=1}^{N-1} q_i^*(X_t) - s_0(X_t) \right) - \mathcal{C}(a_t) \right) dt \right],$$

where $X_0 = x$.

As throughout the other games that we have considered, the value functions for the other players are not needed for analysis of blockading and saturation points. However, for completeness, we have that

$$w_i(x) = \sup_{q_i} \mathbb{E} \left[\int_0^\infty e^{-rt} q_i(X_t) \left(1 - q_0^*(X_t) - \sum_{j=1, j \neq i}^{N-1} q_j^*(X_t) - q_i(X_t) - s_i \right) \mathbb{1}_{\{X_t > 0\}} dt \right. $$
$$\left. + \int_0^\infty e^{-rt} \frac{1}{(N+1)^2} \left(1 - (N+1)s_i + \sum_{i=1}^{N-1} s_i \right)^2 \mathbb{1}_{\{X_t = 0\}} dt \Big| X_0 = x \right],$$

for $i \in \{1, \ldots, N-1\}$.

The corresponding Hamilton-Jacobi equation is

$$rv(x) = \sup_{q_0} \left\{ \left(1 - q_0 - \sum_{i=1}^{N-1} q_i^* \right) q_0 - q_0 v'(x) \right\} + \sup_a \left\{ -\mathcal{C}(a) + a\lambda \Delta v(x) \right\}.$$

It is easy to verify that assuming the same cost function for exploration, the expression for $a^*(x)$ is given by (20) as before, with the difference being internalized in the jump term $\Delta v(x)$. Further, we can note that the boundary condition (21) from before also holds. However, now that $s_0(x)$ is decreasing, we have that for all x such that $s_0(x) > \rho_N$, the oil producer does not play and the appropriate Hamilton-Jacobi equation in that instance is simply

$$rv(x) = \frac{1}{\gamma} [\max(\lambda \Delta v(x) - \kappa, 0)]^{\gamma} .$$

As we did in Sect. 4.1.2, we recast the problem into a set of iterative ODEs which uniformly converge to $v(x)$. In particular, letting $v^0(x)$ be the solution where $\lambda = 0$ (the no-exploration case), which was analyzed in Sect. 3.3, we define

$$rv^n = (q_0^n(x))^2 + \frac{1}{\gamma} \left(\max \left(\lambda (v^{n-1}(x + \delta) - v^n(x)) - \kappa, 0 \right) \right)^{\gamma}$$

$$q_0^n(x) = \begin{cases} \dfrac{1}{R+2} \left(1 - s_0(x) - v'(x) + \sum\limits_{i=1}^{R} s_i \right) & \text{if } s_0(x) < \rho_N \\ 0 & \text{if } s_0(x) \geq \rho_N \end{cases}$$

$$v^n(0) = \sup_{a \geq 0} \frac{\lambda a v^{n-1}(\delta) - \mathcal{C}(a)}{\lambda a + r}$$

$$v^n(x \in (0, x^*)) = \sup_{a \geq 0} \frac{\lambda a v^{n-1}(x + \delta) - \mathcal{C}(a)}{\lambda a + r}$$

Here, x^* is the value of x such that $x^* = \inf\{x \in \mathbb{R}_+ : s(x) < \rho_N\}$ and R is the number of renewable players who play at x. We can determine R by assuming that all players play, and then checking the quantity produced by the highest cost player. We denote

$$q_n = \frac{1}{n+2} \left(1 - (n+2)s_n + s_0(x) + v'(x) + S^{(n)} \right)$$

$$R = \min\{n : q_n < 0; n \in \{1, \dots, N-1\}\}.$$

4.4   N-player Simulations

We numerically evaluate the game using the expressions above. We choose $s_0(x) = .2e^{-x}$ and $s_1, \dots, s_9$ to be $0.51, 0.52, \dots, 0.59$ and depict the appropriate results in Figs. 10 and 11. In particular, we note that once again $v(x) > v^0(x)$ for all x, as expected. We note that as x increases, the market price decreases, since a low cost option (oil) is available. When comparing the blockading points (denoted by vertical bars) to the case where $\lambda = 0$ (no exploration), the blockading points are to the right. This follows intuitively, since when the oil producer can explore, he is better off and hence is capable of driving the other players out faster. Mathematically, the shadow

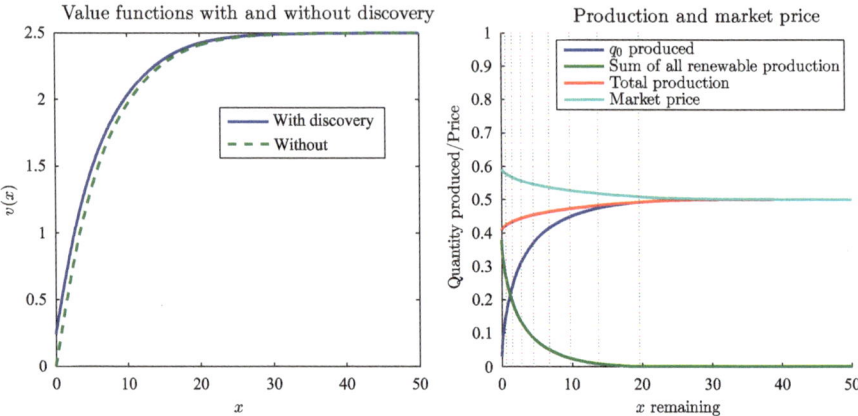

Fig. 10 We let $N = 10$ and initially start off with 9 renewable players and one exhaustible player. In particular, $s_0(x) = .2e^{-x}$ and $s_1, \ldots, s_9$ to be $0.51, \ldots, 0.59$ respectively. The *vertical dotted lines* in the right figure denote blockading points for each of the renewable players

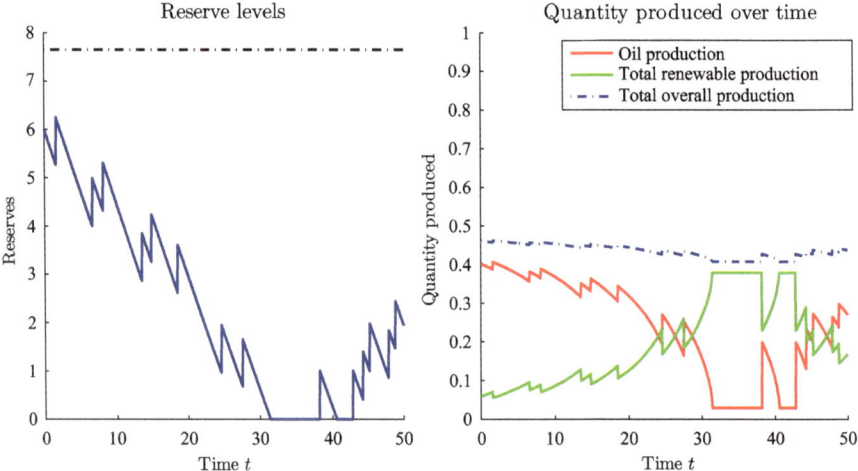

Fig. 11 We simulate the game over time $t \in [0, 50]$ for the same structure as described previously in Fig. 10

cost v' is lower in x with exploration, so at any given x, the oil producer can produce at a lower effective cost when exploration is an option, driving the blockading points to the right for the renewables.

In Fig. 11, we simulate the game over time. We note that in this case, x_{sat} is higher than the initial quantity of reserves, so the oil producer always puts in a finite amount of exploration effort, a^*. Overall production goes down over time, reflecting increased market price over time. Further, as the oil producer runs out, the renewables produce more, but each discovery marks a resurgence of oil into the market while the renewables temporarily cut back production.

Further, if we increase $s_0(x)$, then x_{sat} is decreased. That is, when $s_0(x)$ increases, the value function $v(x)$ is lower everywhere than before because profits are lower. In particular, $\Delta v(x)$ is also lower since over the interval $(x, x + \delta)$, the oil producer is not as well off with a higher cost of extraction, leading to lower discounted profits and in turn a lower value for $\Delta v(x)$. Since $a^*(x)$ is proportional to $\Delta v(x)$ when $a^*(x) > 0$ and since the threshold value x_{sat} is also dependent on the magnitude of $\Delta v(x)$, an increased $s_0(x)$ marks a lower x_{sat}.

Intuitively, increasing $s_0(x)$ means that there is a relative disincentive to explore, since even if exploration is successful, the marginal benefits are not as great because profit margins are lower. This suggests that in terms of policy implications, increasing $s_0(x)$ must be offset by a decrease in κ, which is reflective of the cost of exploration.

5   Application to Energy Policy

The above calculations demonstrate that the model that we have set forth is directly applicable to energy policy. We consider two such examples below.

5.1   Taxing Oil Production but Subsidizing Research

We now apply the above stochastic differential game models to energy policy. As alluded to throughout this paper, energy markets are similar to oligopoly models with varying costs and exploration. For simplicity, we will consider the two player model most recently explored whose costs both vary in time. We consolidate all finite stock based suppliers into one player (player 0) and all renewable players into a second (player 1).

In Fig. 12, we model a policy option of taxing oil production but subsidizing their exploration costs. That is, we lower $\mathcal{C}(a)$ but increase $s_0(x)$. Such a policy has the potential of being revenue neutral and we lower κ significantly to model this. We note as a result that the oil producer is active on average for a longer time, since $a^*(x)$ is greater for all x in this situation. This follows immediately from $\mathcal{C}(a)$ being lower. Further, we note that x_{sat} is much higher. Specifically, x_{sat} increases from 7.65 to $x_{sat} = 21.25$. In addition, since the oil producer has a lower cost of exploration, he stays in for a longer period of time, leading to a lower market cost.

However, such policy is limited at best since this model assumes that the probability of finding additional reserves is independent of the amount of discoveries already made, when in reality, such probability is not independent. As we find additional reserves, the marginal probability of finding another one is lower in the number of discoveries. However, it is of particular interest to note that lowering κ may have short term benefits in terms of price stability but has long term harms in that the total number of possible reserves is exhausted at a faster rate even if corrective action is taken in terms of increasing $s_0(x)$.

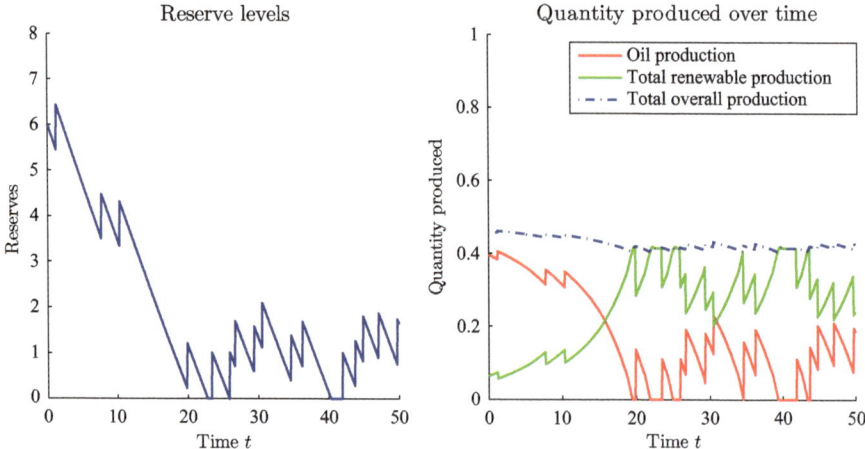

Fig. 12 We now increase $s_0(x)$ to $s_0(x) = .5e^{-x}$ but we lower $\kappa \to 0.01$ from $\kappa = 0.1$. This models taxing production but incentivizing R&D

5.2 Taxing Oil vs. Subsidizing Green Energy

We will now compare the policy options of taxing our oil player and subsidizing our green player. These are both policy options that are currently on the table and have been implemented to some extent. However, to truly compare welfare overall, we must also consider the third player, the consumer, who drives the inverse demand function that we have assumed throughout. Assuming that demand is positive for both goods (as reflected in our function), we can apply the Gorman Aggregation Theorem to note that there exists an aggregate utility function that consolidates the preferences of the representative household. In particular, the appropriate utility function for our inverse demand function can be verified to be

$$u(Q, m, x) = Q(x)\left(1 - \frac{Q(x)}{2}\right) + m(x),$$

where $Q(x) = q_0(x) + q_1(x)$, the total quantity produced when x stock is remaining, and $m(x)$ is the total money the consumer has at time x. In particular, the consumer will seek to maximize the time discounted integral of utility:

$$\sup_{Q(x(t))} \mathbb{E}\left[\int_0^\infty \left(e^{-\rho t} Q(x(t))\left(1 - \frac{Q(x(t))}{2}\right) + m(x(t))\right) dt\right],$$

where the evolution of stock is given by

$$dX_t = -q_0(X_t)\mathbb{1}_{\{X_t>0\}} dt + \delta\, dN_t,$$

and ρ is the discounting factor of our consumer.

We will consider three states of the world. Notationally, we denote the set of costs and utilities by the array $(s_0^i, s_1^i, u^i(Q_i, m_i, x))$. Here, s_0^i refers to the cost function for the oil producer in state i and s_1^i is the cost function for the renewable player in state i. Additionally, $u^i(Q_i, m_i, x)$ refers to the utility of the representative consumer in state i, with Q_i the total production and m_i the amount of disposable money. Finally, we will let q_0^i and q_1^i be the quantities produced by the oil and renewable players respectively in state i.

State 1 will be our control case. For our numerical simulations, we will let $s_0^1(x) = .3e^{-.05x}$ and $s_1^1(x) = .6(1 - e^{-.1x})$. The appropriate utility function in this case is simply $u^1(Q_1, m_1, x)$. State 2 will be the case where we impose a tax on the oil producer. This is not to be conflated with a Pigouvian tax, which is typically introduced to correct externalities. The utility function $u(\cdot)$ does not contain any direct disutility from consuming oil and the inverse demand function has no differentiation from energy derived from oil or renewable sources. Hence, this is merely a tax on finite resources, aimed at prolonging the time duration for which the oil producer will play and also inducing further smoothing over time of oil production. For numerical purposes, we will introduce a 33 % tax, so $s_0^2 = .4e^{-.05x}$, $s_1^2(x) = .6(1 - e^{-.1x})$ and $u^2(Q_2, m_2, x) = u^1(Q_2, m_2, x) + .1e^{-.1x}q_0^2$. We assume that the "government," or agent taxing the oil producer redistributes 100 % of the tax revenue to the consumer.

Finally, State 3 will be the case where subsidize green energy. We assume a 33 % subsidy for sake of consistency, so $s_0^3(x) = s_0^1(x) = .3e^{-.05x}$, $s_1^3(x) = s_1^1(x) - .2(1 - e^{-.1x}) = .4(1 - e^{-.1x})$, and $u^3(Q_3, m_3, x) = u^1(Q_3, m_3, x) - .2e^{-.1x}q_1^3$.

We define the aggregate welfare $W(x)$ to be $W(x) = u(Q, m, x) + \Pi_0(x) + \Pi_1(x)$, where $\Pi_i(x)$ is the profit of the ith firm at any point x in stock. We note that if at any given point x, if for two states i and j, $W_i(x) > W_j(x)$, then, state i is potentially Pareto improving over state j and hence preferable. In particular, since aggregate welfare is higher, it is possible to make every player better off up to some redistribution of profits.

In Fig. 13, the top left panel plots oil profits over remaining stock. The control is clearly the best case scenario for the oil producer, as either taxing the oil player or subsidizing green energy raises the relative cost of oil, lowering profits. In particular, for low oil reserves, taxing oil is worse for oil profits than subsidizing green, but the reverse is true for high reserve levels. The top right panel measures renewable profits; as expected, the renewable producer is best off when his cost is subsidized and does marginally better when oil is taxed, as doing so reduces the relative cost of green energy.

Of particular interest is the bottom left panel of Fig. 13, which measures consumer utility over remaining stock. The consumer is best off by subsidizing green energy at high reserve levels and at such high reserve levels, taxing oil worsens the consumer's situation. However, as oil begins to run out, the utility derived from subsidizing green energy begins to diminish. The bottom right panel accounts for exploration efforts. In particular, at high oil reserves, taxing oil incentivizes the oil

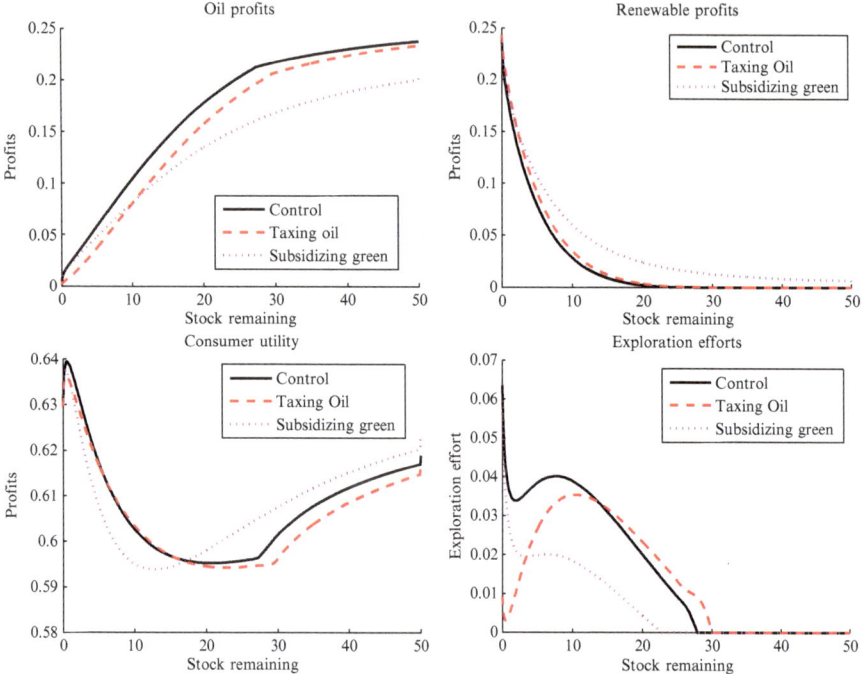

Fig. 13 *Top left*: plots oil (player 0's) profits over remaining stock. *Top right*: computes renewable energy (player 1's) profits for remaining stock X_t. *Bottom left*: plots consumer utility over X_t. *Bottom right*: plots exploration intensity a^* over X_t for all three policy options

player to conduct more research, but as oil begins to run out, the additional taxation reduces the discounted profits for the oil player should he discover, lowering discovery efforts.

Figure 14 consolidates the above welfare analysis, accounting for the welfare of both players and the representative consumer. For high oil reserves, the best policy seems to be to do nothing, but as oil begins to run out, subsidizing green energy is an effective policy, and as oil continues to deplete, taxation of oil might result in a marginal increase in aggregate utility.

Finally, since the utility function that corresponds to the inverse demand function does not factor beneficial macroeconomic effects such as price stability, we compare these in Fig. 15, which depicts evolution of production levels. The left panel suggests that taxing oil reduces oil production a bit, more so than subsidizing oil for low reserve levels, though subsidizing oil reduces oil production when reserves are large. However, subsidizing oil is far better at stimulating green energy for all energy reserves. The right panel demonstrates that subsidizing green energy also results in greatest price stability, while there is not as significant of a difference between taxing oil and the control.

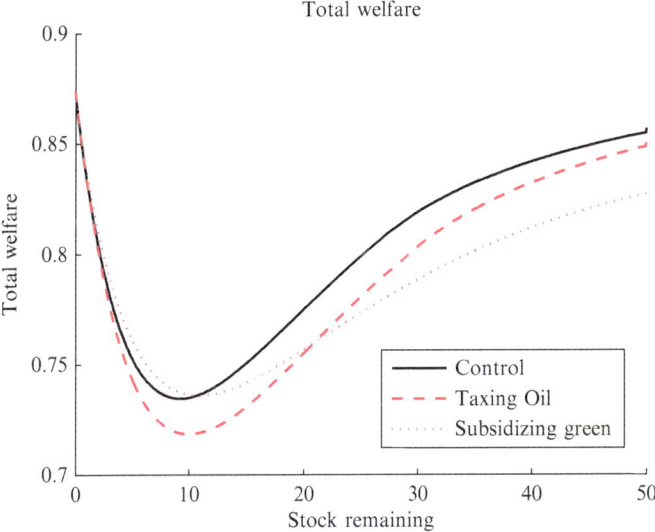

Fig. 14 Considers the aggregate welfare index W for three policy options: doing nothing (control), taxing oil, and subsidizing green energy

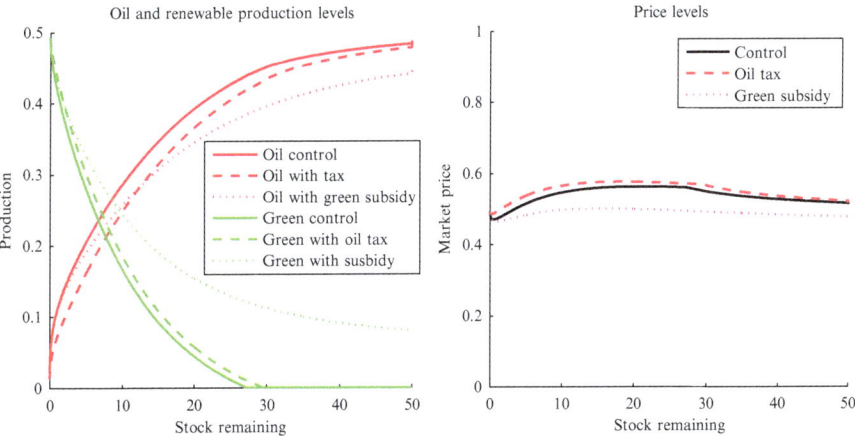

Fig. 15 Evolution of market price and quantities produced by both players as a function of stock

6 Conclusion

We developed a mathematical model for oligopoly markets with exhaustible resources, evolving extraction costs, and discovery. For the case where all costs are constant and where one player plays when his resource stock is non-negative and all others are renewable players, we presented a completely analytical solution, extending the work presented in [8]. We then presented partially analytical results with numerical solutions for the subcases where costs evolve over time but discovery is disabled, and finally when discovery of new reserves is permitted.

We next demonstrated how this model can be applied to energy policy. Of particular interest is our comparison of taxing oil and subsidizing green energy, two leading policy options that are currently being considered. We compared the two in terms of aggregate welfare, which measures whether a policy option is potentially Pareto improving. We found that for high reserve levels, taxation of exhaustible players should be done only to correct for environmental externalities, and that subsidizing green energy later in our game, when $x(t)$ decreases beyond a threshold value, is potentially Pareto improving.

Future work may be considered in two directions. First, additional work may be done to solve this problem in general for a constant prudence demand curve, as in Eq. (1), which may alter the results slightly. We feel, however, that this should not significantly alter the policy implications that are demonstrated in Sect. 5. Secondly, as evidenced by the two examples in Sect. 5, this model itself may be applied to derive further policy implications.

Acknowledgements The work of Anirudh Dasarathy is partially supported by NSF grant DMS-0739195. The work of Ronnie Sircar is partially supported by NSF grant DMS-1211906.

References

1. Bertrand, J.: Théorie mathématique de la richesse sociale. J. Sav. **67**, 499–508 (1883)
2. Cournot, A.: Recherches sur les Principes Mathématique de la Théorie des Richesses. Hachette, Paris (1838). (English translation by Bacon, N.T. published in Economic Classics, Macmillan 1897, and reprinted in 1960 by Augustus M. Kelley)
3. Dockner, E., Jørgensen, S., Long, N.V., Sorger, G.: Differential Games in Economics and Management Science. Cambridge University Press, Cambridge (2000)
4. Friedman. J.: Oligopoly Theory. Cambridge University Press, Cambridge (1983)
5. Harris, C., Howison, S., Sircar, R.: Games with exhaustible resources. SIAM J. Appl. Math. **70**(7), 2556–2581 (2010)
6. Hotelling, H.: The economics of exhaustible resources. J. Polit. Econ. **39** (2), 137–175 (1931)
7. Krugman, P.: Salvation gets cheap. New York Times, 17 April 2014
8. Ledvina, A., Sircar, R.: Oligopoly games under asymmetric costs and an application to energy production. Math. Financ. Econ. **6**(4), 261–293 (2012)
9. Ludkovski, M., Sircar, R.: Exploration and exhaustibility in dynamic cournot games. Eur. J. Appl. Math. **23**(2), 343–372 (2012)
10. Vives, X.: Oligopoly Pricing: Old Ideas and New Tools. MIT Press, Cambridge, MA (1999)